Canada Enters the Nuclear Age

Overleaf: Whiteshell Laboratories

Chalk River Laboratories

Canada Enters the Nuclear Age

A Technical History of Atomic Energy of Canada Limited

Published for Atomic Energy of Canada Limited
by McGill-Queen's University Press
MONTREAL & KINGSTON • LONDON • BUFFALO

ISBN 0-7735-1601-8

Legal deposit first quarter 1997
Bibliothèque nationale du Québec

Printed in Canada on acid-free paper

Canadian Cataloguing in Publication Program

Main entry under title:

Canada enters the nuclear age: a technical history of Atomic Energy of Canada Limited

Includes bibliographical references and index
ISBN 0-7735-1601-8

1. Atomic Energy of Canada Limited–History.
2. Nuclear energy–Research–Canada–History.

HD9698.C24A76 1997 621.4′8′072071 C97-900188-9

Contents

Foreword

As president of AECL in 1989, I was delighted when a number of retirees and elder statesmen came forward with a proposal for a technical history of the company. The time was opportune, they said, to produce a record of the research and development that had set the stage for Canada's notable contributions to the peaceful applications of nuclear energy. Not only was their enthusiasm inspiring, but as some of Canada's nuclear pioneers, their experience and knowledge made them the ideal authors of this chronicle. They have done an outstanding job in portraying the challenges of a unique and exciting quest to develop a complex and demanding technology.

The dawn of Canada's eminent achievements in nuclear development was almost a century ago, when New Zealand physicist Ernest Rutherford came to McGill University in 1898. He had just discovered that the radiation emitted by uranium and its compounds were of two types which he called alpha and beta rays. (A third type, gamma rays, was discovered by Paul Villard in Paris two years later.) These terms are still used. In 1901 Rutherford was joined at McGill by British chemist Frederick Soddy. Their brief partnership produced nine important publications which showed that radioactivity involves the transformation of one element into another.

Canada's more formal entry into the nuclear age came forty years later, in September 1942, when it was agreed that the heavy water and uranium oxide research that was being done at the Cavendish Laboratory in Cambridge, England, would be moved to Canada. And so began an era of intensive research and development that ultimately made a demanding technology into a safe, reliable and beneficial one.

The many facets of nuclear research and development are meticulously described in this book. Despite the modesty of the authors one can discern the impressive creativity, the tenacity, the intellectual capacity, and the dignity that underlay an achievement that would be marked with such accolades as a Nobel Prize and a place among Canada's top ten engineering accomplishments.

Many of the attainments of AECL's laboratories are significant milestones in the history of Canadian science. Not only that, they gave Canadian research global stature and provided world-leading products and services. It's a story that had to be written while the footprint of pioneering research and development is still conspicuous. It's the story of a venture that sprang from a vision, that persisted with a sustained sense of purpose, and that succeeded through dedication and talent. As a historical perspective it provides useful guidance for innovative scientific development in the future. The story deserved to be written and it deserves to be read.

Stanley R. Hatcher

Preface

The appearance of Robert Bothwell's book, *Nucleus: The History of Atomic Energy of Canada Limited*, which was largely a corporate and political history dealing with applied and industrial achievements, triggered interest in the preparation of a complementary technical history. In early 1989, Don Hurst, a former AECL employee and former president of the Atomic Energy Control Board (AECB), wrote a letter to AECL formally suggesting that such a technical history be prepared. The importance of timing was recognized since the main participants from the early days, whose recollections would be of great value, were disappearing from the scene.

AECL approved and sponsored the project, assigning the task of co-ordinating the preparation of the history to an employee, Gene Critoph (who retired in 1992 but maintained a coordinating function).

A Technical History of AECL Committee (THAC) was established to oversee the project. Over its lifetime THAC has had the following membership:

Members:

AECL retirees: A.S. Bain, J.H. Collins, E. Critoph (chair), M.F. Duret, T.A. Eastwood, C.E. Ells, R.E. Green, G.C. Hanna, R.G. Hart, A.M. Marko, J.C.D. Milton, D.K. Myers, H.K. Rae, J.A.L. Robertson, B. Ullyett.

AECB retirees: F.C. Boyd, D.G. Hurst.

Corresponding members:

G.L. Brooks (AECL retiree), D. Evans (Nordion), M. Tomlinson (AECL retiree), N.H. Sagert (AECL).

THAC held the first of more than a score of meetings in 1989 November.

THAC created targets for the end product of the project, including i) that the book be suitable for an audience of educated laypersons with special interests in science, and ii) that it give a balanced treatment of the various technical activities within its scope. THAC established an outline for the book that included a detailed list of suggested topics for each section. Each member of THAC then undertook responsibility for the preparation of part of the book, and provided valuable criticism to other authors on their contributions. THAC, as a whole, reviewed successive drafts of each submission. Over the period 1991-92 a consultant, D. Peirce of P.J. Spratt & Associates, was engaged to assist any author requesting help in improving comprehensibility, readability and interest for their contribution. THAC made numerous decisions regarding the style and structure of the book. Examples of the more controversial decisions were: i) to limit to very few the names that would appear in the text, in the interest of fairness to all, and ii) to keep formal references to an absolute minimum to improve readability.

In general this history extends from the late forties to 1985. The year 1985 seemed appropriate as a nominal closing date in view of the major restructuring that was implemented in 1986. Where significant events have occurred in the ensuing nine years, these are treated briefly.

The point of view adopted for this book is a research-and-development (R&D) one, with the main focus on the role of AECL's laboratories at Chalk River and Whiteshell in the Canadian nuclear program. Therefore the non-R&D aspects of AECL's role are not fully treated, and other parts of the nuclear industry are mentioned only inasmuch as they affected AECL.

The relative geographical isolation of the AECL research laboratory sites and the nature of the work being done required both a high degree of self-sufficiency for AECL Research and its operation as an integrated team. AECL's success rests on contributions from thousands of individuals from numerous disciplines – designers, fabricators, constructors, operators, researchers, technologists, regulators, development engineers and many others.

It was recognized that not all the R&D that was done could be recorded in a book of practical size. The authors of the individual chapters accept responsibility for content, but have, in general, sought the views of former colleagues. While it is hoped that readers will go through the entire book, many may wish to concentrate on particular sections. Nevertheless, everyone is urged to read chapter one, which provides an overview and a context for the more detailed chapters that follow, and also chapter twenty-one, the retrospective on learning from history.

Many people have contributed to the preparation and production of this volume. In addition to the specific contributions to the actual writing of the book that are acknowledged by the authors within their own sections, we collectively acknowledge the support freely given by the many AECL staff members who were approached for help. In particular we would like to acknowledge our debt to the staff in Technical Information Services, the Scientific Document Distribution Office, Central Records, Business Services and Corporate Relations. We would also like to specifically acknowledge the great contributions of Natalie Bellavance and Cathy Bennett in the difficult task of typing many drafts and the final manuscript, and McGill-Queen's University Press for bringing this book to fruition.

J.L. Gray

W.B. Lewis

Authors

ALASTAIR S. BAIN, B.A.Sc. (British Columbia)
Al Bain joined AECL in 1951 and spent the next twenty-eight years at the Chalk River Laboratories. His work included development of fuel rods for the NRX reactor, pyrometallurgical reprocessing of metallic fuel rods, and set-up of research hot cells. As head of the Fuel Development branch he was responsible for experimental work on CANDU fuel components and development of CANDU fuel bundles. He is author or co-author of about 200 documents on nuclear fuel. In 1979 Mr. Bain moved to Ottawa to work with a group newly formed in AECL to transfer technology under license to Canadian and foreign organizations. He retired in 1991.

FREDERICK C. BOYD, B.A.Sc. (Toronto)
After graduating in 1949 Fred Boyd worked on radioisotope applications with Eldorado Mining and Refining and then on the first CANDU plant with Canadian General Electric. In 1959 he joined the Atomic Energy Control Board. As head of Nuclear Facility Licensing he developed the Canadian Siting Guide, a foundation document in nuclear plant licensing. From 1972 to 1978 he was nuclear energy advisor in Energy, Mines and Resources. After a period as an International Atomic Energy Agency advisor to Korea's Ministry of Science and Technology, he rejoined the AECB as Orientation Centre director. He was admitted as a fellow of the Canadian Nuclear Society in 1994.

EUGENE CRITOPH, B.A.Sc., M.A.Sc. (British Columbia)
Eugene Critoph joined AECL in 1953 and worked in the field of reactor physics until 1967. Subsequently he held a number of increasingly responsible scientific and management portfolios within the laboratories. His senior positions included director of Fuels and Materials, director of Advanced Projects and Reactor Physics, and vice-president and general manager of the Chalk River Laboratories. In 1986 he was appointed vice-president of Strategic Technology Management for AECL Research. Also in 1986, in recognition of his significant contributions to nuclear research, he was awarded the Canadian Nuclear Association's W.B. Lewis Medal. It was under Mr. Critoph's leadership that this volume was developed.

MAURICE F. DURET, B.Sc., M.Sc. (Queen's), Ph.D. (Toronto)
Maurice Duret joined the Reactor Physics branch of AECL at Chalk River in 1952. When this branch was later split into two parts, he was appointed head of the part that became the Applied Mathematics branch. In that position he was responsible for developing analytical methods including theoretical aspects

of reactor physics and applied mathematics problems in areas such as fuel element modeling to predict cladding strains and stress analysis. His collaboration with international organizations included membership of the OECD Nuclear Energy Agency Committee on Reactor Physics and a major study for the World Energy Conference on the contribution of nuclear power to energy supplies.

T. ALEXANDER EASTWOOD, B.A., M.A., (Western Ontario), Ph.D (McGill), D.Phil. (Oxon)

Alex Eastwood joined the pioneers working on methods for the production of radioisotopes at Chalk River in 1947. This was his introduction to nuclear chemistry, the subject of his subsequent research career. His particular interests were the study of the decay properties of radionuclides and of unusual nuclear reactions in the reactors, and the determination of neutron capture cross-sections. He served as head of AECL's Research Chemistry branch from 1965 to 1967, assistant director of the Chemistry and Materials division from 1967 to 1970, and director of the division until his retirement in 1985. Dr. Eastwood is a Fellow of the Chemical Institute of Canada and an Emeritus Member of the American Chemical Society.

CHARLES E. ELLS, B.A.Sc., M.A. (Toronto), Ph.D. (Birmingham)

Charles Ells spent four years of World War 2 with the RCAF, obtaining the Burma Star from attachment to the RAF during the campaign in the Imphal region of Assam. Prior to obtaining his Ph.D. he worked on nuclear metallurgy at the Bureau of Mines in Ottawa and at Chalk River. After a period with Canadian Westinghouse he returned to Chalk River in 1957. Here his work was associated mainly with the development of CANDU fuel channel components. After retirement from full-time employment with AECL in 1990, he was appointed as Researcher Emeritus at Chalk River for a further four years.

RALPH E. GREEN, B.Sc., M.Sc. (Dalhousie), Ph.D. (McGill)

Ralph Green joined AECL at the Chalk River Laboratories in 1956, working first in reactor physics at Canada's first research reactor ZEEP, then in accelerator physics, and later as head of the Reactor Control branch for eight years. In 1979 he transferred to AECL's head office in Ottawa as a senior advisor to the executive vice-president of AECL Research Company. In 1981 Dr. Green was appointed acting general manager of the Whiteshell Laboratories. The following year he became vice-president and general manager of the Whiteshell site. In 1986 he was appointed vice-president of Reactor Development, responsible for all reactor-related R&D in AECL Research.

GEOFFREY C. HANNA, M.A. (Cantab.), F.R.S.C.
After graduation Geoff Hanna joined the British Ministry of Supply as a physicist working on radar development. He came to Canada in 1945 as part of the United Kingdom mission first in Montreal and later at Chalk River. His early research included the pioneering use of high-gain proportional counters to study low-energy radiations, investigations of the properties of transuranic nuclides, and measurements of nuclear data for reactors. He was head of the Nuclear Physics branch, director of the Physics division, and director of Research. He has an honorary D.Sc. from McGill University, and was president of the Canadian Association of Physicists in 1984–85.

ROBERT G. HART, B.Sc. (Toronto)
Bob Hart came to Chalk River Laboratories in 1948 and embarked on a notable career. After working on various projects including purification of heavy water in a reactor system, reprocessing of nuclear fuels and studying the physical properties of these fuels, he was transferred to Whiteshell Laboratories in 1965 as head of the Reactor Core Technology branch. He was appointed director of the Applied Science division in 1969, managing director of the Whiteshell site in 1973 and a vice-president of AECL in 1974. In 1978 he became executive vice-president in charge of AECL research. Mr. Hart was awarded the W.B. Lewis medal by the Canadian Nuclear Association in 1981.

DONALD G. HURST, B.Sc., M. Sc., Ph.D. (McGill), F.R.S.C.
Donald Hurst's career in nuclear energy began with post-doctoral work in the United States and in England. In 1939 he joined the National Research Council in Ottawa. He moved in 1945 to Chalk River, where he subsequently became director of the Reactor Research and Development division and director of Applied Research and Development. In 1970 he became president of the Atomic Energy Control Board. After retirement his activities included a period as chairman of the Senior Advisory Group of the International Atomic Energy Agency. Dr. Hurst received the Canadian Nuclear Association's W.B. Lewis Medal and Outstanding Contribution Award, and the American Nuclear Society's Tommy Thompson Award for outstanding contributions to nuclear safety.

ARTHUR M. MARKO, B.A. (Saskatchewan), M.D., Ph.D. (Toronto)
Following postgraduate research at the National Institute for Medical Research in London, England, Art Marko became a researcher and later professor of chemistry and paediatrics at the University of Saskatchewan. He came to Chalk River in 1961 as assistant director of AECL's Biology and Health Physics division, and became

division director in 1965. Dr. Marko has made outstanding contributions to molecular biology, radiation protection, and effects of radiation on living matter. He served on the executive council of the International Radiation Protection Association and has been a medical advisor to the Atomic Energy Control Board, as well as chairman of its Advisory Committee on Radiological Protection.

J.C. DOUGLAS MILTON, B.Sc. (Manitoba), M.A., Ph.D. (Princeton), F.R.S.C.
Doug Milton joined AECL in 1951 and began a notable career as a nuclear physicist, including a leading role in the Tandem Accelerator Superconducting Cyclotron project. He was appointed head of the Nuclear Physics branch in 1967, director of the Physics division in 1983, and vice-president of Physics and Health Sciences in 1986. Upon his retirement he was named AECL's first researcher emeritus. A fellow of the Royal Society of Canada and of the American Physical Society, Dr. Milton has served on many external committees and has been a visiting scientist at several institutions including the Lawrence Berkeley Laboratories and the University of Strasbourg. He was president of the Canadian Association of Physicists in 1992-93.

DAVID K. MYERS, B.A., M.A. (Toronto), Ph.D. (Amsterdam)
After post-doctoral studies in pharmacology and biochemistry at the University of Amsterdam, David Myers joined the Defence Research Board at Suffield, Alberta, in 1956. Two years later he joined AECL's Radiation Biology branch at Chalk River and was appointed head of that branch in 1976. In 1973-74 he was on attachment to the Institute of Medicine, Kernforschungsanlage (KFA), West Germany. Dr. Myers has been a prolific author in the fields of biological effects of ionizing radiation and the assessment of radiation hazards to humans. His professional affiliations have included the Radiation Research Society, the New York Academy of Sciences and the Canadian Radiation Protection Association.

HOWARD K. RAE, B. Eng. (McGill), Ph.D. (Princeton)
Howard Rae joined Chalk River Laboratories in 1950 and pursued chemical engineering research for twenty-two years, making major contributions in the areas of irradiated fuel processing, heavy water process development, and reactor chemistry. His management positions included head of the Chemical Engineering branch, director of the Fuels and Materials division, director of Applied Research and Development, and vice-president of Radiation Applications and Isotopes. He was chairman of the Advisory Committee to the Canadian Fusion Fuel Technology Progam. Dr. Rae was president of the Canadian Society for Chemical Engineering

in 1984-85 and received the society's major honour, the R.S. Jane Memorial Lecture Award, in 1978. He also received the Canadian Nuclear Association's Outstanding Contribution Award.

J.A.L. (ARCHIE) ROBERTSON, M.A. (Cantab.), F.R.S.C.
Archie Robertson served to captain in the Royal Engineers (1943-47). Following graduation he spent seven years on nuclear-fuel research at the United Kingdom Atomic Energy Authority's Harwell Laboratory. In 1957 he came to AECL, where he held positions that included director of the Fuel and Materials division, assistant to the Chalk River site head and director of Program Planning. Since retiring in 1985 he has been a consultant to the Ontario Nuclear Safety Review, the Atomic Energy Control Board, Ontario Hydro and the Auditor General of Canada, among others. He has been awarded the Canadian Nuclear Association's W.B. Lewis Medal and the Kroll Zirconium Medal.

BERNARD ULLYETT, B.A.Sc. (Toronto)
After a three-year period in Quality Control with Bell Telephone Bernard Ullyett joined Eldorado's radium operations at Port Hope. In 1952 he became supervisor of Radioisotope Operations at Chalk River and moved to the new Commercial Products Division (CPD) in Ottawa in 1955 as head of Radioisotope Operations. He subsequently established a product quality control program and became divisional manager for Operations Coordination. In 1975 he was posted to France as liaison engineer when CPD contracted with CGR.MeV (France) to produce medical linear accelerators. He was appointed project engineer for the new radioisotope processing operations at Kanata in 1978. Later he became a consulting engineer.

INTRODUCTION AND OVERVIEW

Chapter One

Overview of Nuclear Research and Development

D.G. HURST

INTRODUCTION

Canada is one of the very few nations that have successfully developed economical nuclear power plants, an achievement by no means to be expected, considering the small technological base of Canada's pre-war industry.

A fragile chain of events beginning in 1939 led to the establishment of a major nuclear laboratory in Canada. Under good leadership the laboratory acquired world recognition in science and initiated the CANDU®[1] heavy-water nuclear power system which by 1988 was generating 15 percent of the electrical power used in Canada. Canada currently produces a large fraction of the world's commercial trade in bulk radioisotopes.

This book records the scientific and technical advances that underlie Canada's nuclear achievement, discusses the associated planning and strategies, and considers some of the lessons learned. The story is being told by some of the scientists themselves, and the book covers the years 1939 up to 1985.

Background

Nuclear science was not entirely new to Canada. At the beginning of the century, while in Canada as a professor at McGill University in Montreal, Ernest Rutherford, the outstanding nuclear physicist of his time, made several of his famous discoveries. He showed, for example, that radioactivity is in many cases a manifestation of the long-sought transmutation of elements. After Rutherford returned to England, nuclear research was continued in Canada mainly at Queen's

1 CANada Deuterium Uranium; registered trademark.

University in Kingston and at Dalhousie University in Halifax, and former Canadian collaborators sent holders of graduate scholarships to work under him. As a result, several Canadian nuclear scientists were available for recruitment to the new laboratory. The discovery of uranium at Great Bear Lake in 1930 and the setting up of a plant to extract the radium and later the uranium gave Canada status as a supplier of radium (for use in cancer therapy) and uranium.

At the beginning of 1939 there was intense excitement in the scientific community when an explanation was given for radioactivities produced by exposing uranium to neutrons. For several years these radioactivities had been a major puzzle. They were explained as arising from a new phenomenon – fission.[2] In the fission process a uranium atom, on absorbing a neutron, splits into two parts of comparable size. These two fission products are generally radioactive. There is a release of a large amount of energy and an emission of neutrons. An early demonstration that neutrons are emitted during fission gave promise of fission chain reactions, which would fulfil a forty-year-old hope for the controlled production of nuclear energy. The growing threat of war heightened interest in another potential of such chain reactions, that of explosives more powerful by many orders of magnitude than conventional chemical explosives. (The energy released by one fission is about fifty million times the energy released by burning one atom of carbon.) As a result, international cooperation in open nuclear research was replaced by work under rigorously imposed secrecy. Many accounts have been written of the great political, diplomatic and technical efforts in several countries that resulted initially in military and later in civilian applications of fission energy (see the bibliography).

It would be inappropriate to produce another general account of the wartime activities. Instead, the sequence of events that led to the Canadian project will be outlined. The story begins in a laboratory at the Collège de France in Paris. H. Halban and L. Kowarski, two researchers there, announced in March 1939 the emission of more than one neutron when a fission occurs. The race towards a chain reaction was on. Soon, enough information was available to show that the fission neutrons must be slowed down for a sustained chain reaction to be produced in natural uranium. The slowing down can be accomplished by arranging for the neutrons to make billiard-ball-type collisions with atomic nuclei. The more nearly equal the nuclear mass is to the mass of the neutron, the more rapid is the slowing down. This points to protons, the nuclei of hydrogen atoms, as being most effective, because they have essentially the same mass as the neutron. However, protons readily capture neutrons. Carbon, with a nuclear mass approximately twelve times the neutron's mass, and with relatively small absorption, was the most promising element available in quantity, and indeed, carbon was the moderator (i.e., the slowing-down medium) in the first nuclear reactor, which operated in Chicago in December 1942.

In 1937, Halban had participated in a measurement which showed that deuterium, a rare, heavy form of hydrogen discovered in 1932, is not a strong neutron absorber; it might therefore be a good moderator, as its mass is only twice that of the neutron. Water in which the hydrogen was almost entirely deuterium was commercially available only in small quantities. An electrochemical company manufacturing fertilizer in Norway was the principal producer of this heavy water and held roughly 180 kilograms, almost the world's entire stock. After some discouraging experiments with ordinary water, the Paris group decided to obtain this heavy water to test it as a moderator. The French scientist, F. Joliot, received authority from the French government to purchase the heavy water, but the Norwegian company provided it as a loan. By this time, World War II had begun and there was an indication of German interest in the heavy water. In arranging to transport the heavy water to France by air, a French agent laid a false scent, which misled the German air force into intercepting the wrong flight and forcing it to land at Hamburg for a fruitless search, while the heavy water reached France via Scotland.

Before the crucial experiment could be done in Paris, the Germans invaded France and the heavy water was sent south.

[2] The index lists technical terms and the initials of reactors. Opposite these items the number of the page containing a brief explanation of the term is given in bold type.

A NUCLEAR PRIMER

All substances consist of *atoms*. Atoms are the smallest amounts of a substance that enter into chemical reactions.

A substance in which all atoms are practically indistinguishable by ordinary chemical means is called an *element*. Common examples are carbon, oxygen and iron.

An atom consists of a *nucleus*, which carries a positive electric charge, surrounded by *electrons*, each of which carries a negative electric charge.

Atoms have diameters in the order of 5/100,000,000 cm and nuclei about 1/50,000 times smaller. A trillion atoms can make only a microscopic speck of dust. Except for a fraction less than 1/1800 contributed by the electrons, the mass of an atom resides in its nucleus.

A nucleus consists of *protons* which are particles having a positive charge, and *neutrons* which are uncharged particles having about the same mass as a proton; since the nucleus of ordinary hydrogen is a proton, (the exceptional case of a nucleus without neutrons), all three – proton, neutron, and hydrogen atom – have very nearly the same mass.

The charges on electrons and protons are of equal magnitude but opposite signs. The magnitude is the fundamental unit of charge. An atom is electrically neutral, therefore the number of electrons is equal to the number of protons and this number is called the *atomic number*. The total number of protons and neutrons in an atom is called the *atomic mass number*.

All the atoms of an element have the same atomic number but may differ in atomic mass numbers because of differing numbers of neutrons. For example, the element lithium is atomic number 3 but has atoms of mass numbers 6 and 7. These two kinds of lithium are called *isotopes* of lithium and occupy the same place in the Periodic Table of the elements. An isotope is usually designated by its mass number and the abbreviation for the element. An isotope is nowadays called a *nuclide* when the relation to its sister isotopes is not involved.

Differences in the neutron content of isotopes have such slight effects on chemical properties that isotopes are almost indistinguishable in ordinary chemical reactions. However, small chemical and physical differences may be exploited by appropriate but expensive means to separate isotopes. Isotopes differ greatly in their nuclear properties.

The binding of the electrons to the nucleus is weak compared to the binding of the protons and neutrons in the nucleus; because of this, energies associated with nuclear changes are generally many times the energies associated with changes in the arrangement of the electrons. Consequently, our everyday experiences with materials result from interactions of electrons, not from direct nuclear effects.

Most of the nuclei that exist in nature are stable. With minor exceptions, the few unstable nuclei are either survivors from the primeval genesis of the solar system or are unstable descendants of such nuclei; examples of nuclides having unstable nuclei are uranium-238 and a descendant – radon. Nuclei with proton/neutron combinations that differ from the normal can be produced artificially, for example by bombardment with very energetic particles, but they are unstable.

Unstable nuclei transform to stable nuclei by what is called radioactive decay, giving off high energy radiations in the process – beta particles (fast moving electrons), alpha particles (fast moving helium nuclei), or gamma rays (high energy photons). Radioactive decay occurs randomly in time but in such a manner that the nuclide has a well-defined half-life. A free neutron is unstable and decays to a proton and a beta particle.

A different kind of nuclear instability occurs when an atom like uranium-235 captures a neutron. The resulting nucleus has a great excess of energy and may instantaneously break up into two parts of comparable size and some neutrons i.e., it *fissions*.

The first indication of the existence of radioactive decay was in 1896 when radiation from naturally occurring unstable elements was detected. Fission was first announced at the beginning of 1939.

While the Pétain government was taking control of France, the heavy water was removed to Cambridge, England, in the custody of Halban and Kowarski. There they made measurements which showed that a mixture of heavy water and uranium could sustain a chain reaction. Wartime conditions severely restricted the support that could be diverted to further the work at Cambridge. The American nuclear project was advancing rapidly and the main British nuclear-weapons team would soon be moved to the United States. However, proposals to transfer the heavy-water group to the United States withered in the climate of tight security and growing American exclusivity.

The dilemma was solved by Canada's agreement to supply a laboratory for a joint Canada/United Kingdom/United States project. Major factors contributing to the solution were: pioneer work at the National Research Council of Canada (NRC), in Ottawa, by G.C. Laurence, who was the first person in the world to study neutron multiplication in a large assemblage of carbon and uranium (see figure 1.1); Halban's interest in being near the United States; the ready acceptance and promotion of the project by C.J. Mackenzie, NRC president, who foresaw that Canada would thereby have a head start in a field with major peacetime applications, including a new source of energy; and, perhaps, Canada's position as a uranium supplier.

In 1942, the heavy water and many members of the Cambridge group were sent to Montreal, where they were joined by several Canadian scientists in a laboratory that was initially set up in a large house, but which soon moved to the University of Montreal. Halban was director. For more than a year the Montreal Laboratory's activities included measurements and theoretical analyses of nuclear properties of heavy water, of mixtures of heavy water and uranium, and of assemblies of graphite and uranium. Radiochemistry, reactor design and instrument development were also actively pursued. There was, however, no clearly defined mission. This lack of a mission, dissatisfaction with Halban as director, and uncertainties about the laboratory's future as British/American cooperation waxed and waned undermined morale.

Early in 1944, as a result of decisions made in August 1943 by Prime Minister Churchill and President Roosevelt at the

FIGURE 1.1
Cutaway drawing of Ottawa sub-critical assembly, 1941-42

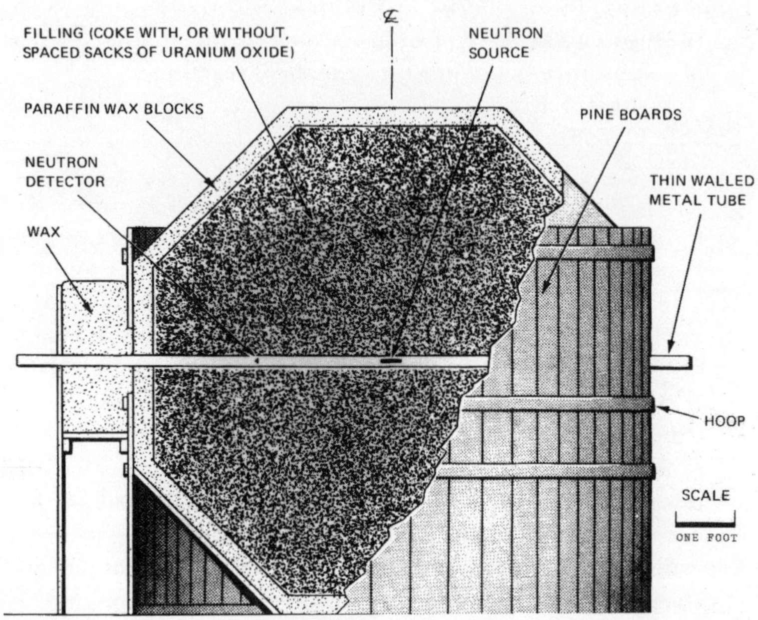

Quebec Conference, the Montreal Laboratory was assigned the task of developing the heavy-water reactor system beyond the stage expected to be reached by a 300 kW heavy-water reactor nearing completion at the Argonne Laboratory in the United States. (The Americans were arranging to produce heavy water; in particular, they had arranged for production of approximately half a tonne of heavy water a month from electrolytic hydrogen produced by the mining firm, Cominco, at Trail, British Columbia.) Specifically, the project was to design, build and operate a 10 MW heavy-water-moderated, natural-uranium-fueled reactor, to be named NRX. The reactor would provide neutrons to be used in research and produce plutonium-239, a fissile isotope of a new element plutonium, by capture of neutrons in the uranium-238 of the fuel. An auxiliary purpose was to produce, by neutron irradiation of thorium, the fissile isotope uranium-233, which does not occur in nature. These fissile products of reactors were initially sought as material for nuclear weapons.

The heavy water would be held in a vertical cylindrical tank 8.75 feet (2.67 m) in diameter and 10.5 feet (3.2 m) high with 198 tubes passing through it. A tank of this type is called a calandria (see figure 2.14). The uranium would be metal in rod form, sheathed in aluminum and cooled by ordinary water. Each sheathed rod would be inside a calandria tube. The calandria tubes would be vertical and arranged in a hexagonal array. In reactor jargon, a regular arrangement of fuel and moderator is called a lattice.

The Americans were successfully pursuing the graphite reactor route for plutonium production on a schedule that promised availability before the expected end of the war. The impossibility of large-scale production of heavy water within that time made heavy water a backup or a long-term alternative to graphite for plutonium production. The heavy-water system could usefully be developed in Canada without significantly affecting the main thrust of the American effort, and it gave promise of peacetime uses.

At the end of April 1944, Halban was replaced as director by J.D. Cockcroft, under whose able direction the work of the laboratory and of associated organizations involved in the major expansion of the program progressed well. By mid-August a site was chosen on the Ottawa River near the Village of Chalk River, Ontario, and construction began of what is today called the Chalk River Laboratories. A site was also chosen for the Town of Deep River to provide housing for the staff.[3] The work was centred on the NRX reactor, which had excellent built-in facilities for research and which would become extremely successful. Originally specified for operation at 10 MW with a design maximum power of 20 MW, NRX was operating at 40 MW by 1954 with a corresponding increase in neutron intensity.

Soon after his arrival, Cockcroft decided to build a low-power heavy-water reactor as a preliminary to NRX. A detailed proposal for this reactor, ZEEP (Zero Energy Experimental Pile), was put forward in August 1944 and Kowarski, who had recently arrived from England, was put in charge of its design and construction. The operating power chosen was one watt. ZEEP, which went critical on 5 September 1945, was the first reactor to operate outside the United States. It provided information on such matters as transient behaviour of heavy-water reactors in advance of NRX. Later, it served for many years as a testbed for measurements on nuclear properties of materials and on heavy-water/uranium lattices proposed for future reactors.

While NRX was still under construction Cockcroft was appointed head of the atomic energy establishment being built at Harwell, England. For nearly a year he shuttled back and forth between England and Canada, then in September 1946, W.B. Lewis was appointed director at Chalk River (formally the Atomic Energy Division of NRC). Later the tripartite partnership was dissolved, but cooperation was maintained with both the United Kingdom and the United States at levels that fluctuated considerably.

On 22 July 1947, the NRX reactor was brought into operation, completing a chain of events stretching from the Paris group's decision to obtain the Norwegian heavy water, through the hazardous travels of the material and the experiments with it, to the appointment of Lewis as director. Failure of any of these links would almost certainly have kept Canada from assuming a significant nuclear role, as there was no Canadian policy to enter into the nuclear-weapons race as there was in France, the United Kingdom, and the United States. Laurence's innovative work on a uranium-carbon assembly was not likely to have led to the Canadian government's setting up a major project, but it is to the government's credit that having had the heavy-water baby deposited on its doorstep, it took it in and nurtured it handsomely.

The NRX reactor was for several years the world's best research reactor. With NRX as its centrepiece, the Chalk River laboratory embarked on an expanding program of research and development, including the production of radioisotopes. These activities brought international recognition and gave Canada high status in nuclear matters; one outstanding success has been the development of the CANDU nuclear-power system.

---

3 There was no scientific basis for the local story that the Chalk River site was chosen because heavy water collects in the deep portions of the Ottawa River from which Deep River gets its name.

Organization of the Canadian Atomic Energy Project

The tripartite Montreal Laboratory was part of the NRC organization. Canadian members of the staff were NRC employees; other members of the staff were on the United Kingdom team, which included many nationalities.[4] The United States took little part in the work of the laboratory, but it made available special materials, for example heavy water, and it provided much information on a need-to-know basis; a liaison officer was stationed at the laboratory.

When the decision was made to build a reactor, Defence Industries Limited (DIL), the wartime branch of Canadian Industries Limited, was brought in to handle the detailed design, operation and services of the project and the planned residential village; construction was by the Fraser Brace Company. On 1 February 1947 DIL withdrew and NRC took over the whole project, except that DIL continued to oversee the current construction. Fortunately, continuity was maintained because most of the DIL employees transferred to NRC. The Atomic Energy Control Act (1946) established the Atomic Energy Control Board[5] (hereafter called "Control Board"), with responsibility for atomic-energy matters in Canada. Instead of itself becoming directly involved in the Chalk River project, the Control Board asked NRC to continue to operate the project on behalf of the Control Board.

Preparation for Harwell, the United Kingdom establishment corresponding to Chalk River, became an important interest for the United Kingdom staff. Some Chalk River facilities were made available, and there was collaboration on chemical and chemical engineering aspects of plutonium separation. The move of personnel to Harwell began in the latter part of 1945, and by the end of 1947 most of the United Kingdom staff had returned to the United Kingdom or joined the Canadian team. For many years a few United Kingdom staff members were stationed at Chalk River for specific tasks and there were occasional postings of Chalk River staff to Harwell.

By the early fifties, a semi-industrial nuclear power program was emerging that was considered to be inappropriate for the research-oriented NRC. Accordingly, a crown company, Atomic Energy of Canada Limited (AECL), was incorporated and on 1 April 1952 it replaced NRC as the administrative body. There were then two boards in the line to the minister – AECL's board of directors and the Control Board. The revision of the Atomic Energy Control Act in 1954 cancelled this cumbersome arrangement and put AECL directly under the designated minister.

The members of the AECL board of directors were chosen because their backgrounds were relevant to the mission of the new company. For example, some were utility representatives, Ontario Hydro's being its chief engineer. The replacement of the NRC as the administrative body for the project by a company board of directors weighted towards industry was an important factor in promoting the practical utilization of nuclear power in Canada. However, the change from NRC to AECL had no immediate impact at the scientific and technical levels.

Soon after its formation, AECL acquired, by transfer from Eldorado Mining and Refining, a division located in Ottawa that was engaged in the commercial preparation and distribution of radioisotopes, and that was developing radiation-therapy devices of its own design, for which there was a worldwide market. In 1958, responsibility for the design and engineering of nuclear power plants was moved from Chalk River to a division newly established in Toronto for that purpose. Construction of a research and development laboratory at Whiteshell, Manitoba, was authorized in 1960. AECL became responsible for the marketing of CANDU reactors abroad, and when problems arose with heavy-water production plants it took over the operation of two such plants in the early seventies. Chapter two discusses AECL's organization, management style and operations, with a particular focus on the research side.

In 1978, four semi-autonomous companies were formed from the original AECL: a research company, an engineering company, a chemical company, and a radiochemical company, all

[4] In August 1945, when the existence of the laboratory and its nuclear mission were made public, a Montreal newspaper reported that there were several "unclear" (sic) physicists on staff.

[5] In accordance with the proposed Nuclear Safety and Control Act, at the time of print Bill C23 (1996), the name will be changed to the Canadian Nuclear Safety Commission.

being branch companies under a corporate AECL management company. In 1979, an international company was added. In 1982, the engineering, chemical and international companies were consolidated into a new organization: CANDU Operations.

Over the more than four decades covered by this book, the principal organizations have had several changes of name. For simplicity the following informal names that have been in common use will generally be used.

- Montreal Laboratory: Established in late 1942 under NRC. Closed in 1946 with the transfer to Chalk River.
- Chalk River: At various times called Division of Atomic Energy of NRC, Atomic Energy of Canada Chalk River Project, Chalk River Nuclear Laboratories (CRNL) and, most recently, Chalk River Laboratories (CRL).
- Commercial Products Division (CPD) or Radiochemical Company: The Ottawa-based AECL organization responsible for the preparation, sale and distribution of radioisotopes and the design and manufacture of associated equipment – now separately incorporated firms Nordion and Theratronics International.
- Power Projects (later CANDU Operations, now AECL CANDU): The AECL organization based in the Toronto and Montreal areas, responsible for the implementation of AECL's nuclear power program, originally called the Nuclear Power Plant Division (NPPD).
- Whiteshell: the AECL Whiteshell Nuclear Research Establishment (WNRE) in Manitoba, now called Whiteshell Laboratories (WL).
- Head Office: In Ottawa. Now called the Ottawa office.

OVERVIEW

The tripartite decision to proceed with the development of the heavy-water reactor system as the principal mission of the Montreal Laboratory gave a great impetus to the laboratory's work early in 1944. The size of the staff was increased, effort was directed towards a specific reactor design, questions of reactor safety became realistic, and a site had to be chosen and developed.

Immediately after the plant site (Chalk River) was chosen, heat transfer and fouling tests with the river water were begun and were continued during the 1944-45 winter by engineers who were housed in summer cottages modified for winter habitation.

By the summer of 1945, construction at the plant site was well advanced and houses in the townsite (Deep River) were being occupied by staff members from Montreal and DIL. Their moves were made in step with plant and house construction, and by the middle of 1946 the Montreal Laboratory was closed.

Legacy of the Montreal Laboratory

Although the early work of the Montreal Laboratory had not been directed towards a specific mission, it provided experience and equipment that enabled the Montreal staff to begin work related to NRX without delay. Measurements of the nuclear properties of arrangements of uranium rods in heavy water were very soon underway to determine the parameters of the NRX lattice; the associated calculational methods had been the subject of much development in the earlier days. Measurements were done on heat transfer. The electronics group began work on electronic instrumentation for the new reactor and for radiation monitoring.[6] Ion chambers to be installed in the reactor as sensors of neutrons were developed; a related project was the production of quartz-fibre electroscopes suitable for personal radiation monitoring and other radiometric purposes. From this technology there came quartz-fibre microbalances for chemical studies of transuranic elements. A solvent-extraction process

[6] A hand-held monitor was called Pluto after the cartoon dog that went about sniffing. When General Groves, military head of the American project, learned of this, he demanded that the name be changed for security reasons - Pluto sounded too much like plutonium. A staff member proposed the name Lili. That name has been applied for many years to such instruments. The staff member had just been to the Gayety, a burlesque theatre, and seen Lili St. Cyr, a famous stripper. Use of the name Pluto for such instruments continued in the United States, presumably not having reached General Groves.

for plutonium separation was selected by the chemists after trials with hundreds of organic solvents.

Planning and design were undertaken for a charged particle accelerator of the Van de Graaff type, the first of a sequence of accelerators installed at Chalk River.

There was also research into fundamental properties of matter, some of which would have application in the design and use of nuclear reactors; the chemistry of uranium and plutonium (the latter limited by availability) was studied, as were the nuclear and fissioning properties of these elements. An important result was a measurement of the yields of several radionuclides produced by uranium fission; the resulting mass-yield curves for fission of uranium-235 were published in 1946. A new type of spectrometer for measuring beta-ray energies had been developed by a staff member in his graduate years. Spectrometers of this type were later used at Chalk River and in many Canadian university physics departments for much physics research. Radiolytic chemistry of water was important from the early days of the project. Chapter six describes AECL's commitment to basic research, and chapters seven and eight deal with chemistry and physics, respectively.

The medical use of radium and X-ray equipment in the first few decades of the century had resulted in severe exposures and therefore severe health effects to some users and patients. By the mid-forties, there was some quantitative knowledge about such matters. An important component of planning for NRX was preparation to limit exposures to radiation by providing adequate shielding, appropriate work procedures, and monitoring of the exposures.

The monitoring of radiation doses to workers and the public, the determination of dose rates in the environment and in the workplace, and the surveillance of waste management repositories are all essential aspects of radiation protection. Chapter three deals at greater length with this subject.

ZEEP[7] proved to be a very useful low-power reactor, enabling low-power measurements to be made for many purposes. These included the measurements of lattice parameters, transient effects and effects of various materials on reactivity (reactivity is a measure of the ability of the reactor to sustain a chain reaction); graphite intended for the first reactor at Harwell, England, was tested in ZEEP to ensure that it met specifications of low neutron absorption.

NRX Design and Safety

The design work by DIL was carried out in Montreal under the direction of the engineering group of the laboratory. Intense radioactivity of the reactor core during operation, and the residual radioactivity that persists after shutdown, introduced novel constraints on the engineering. To protect the workers the core was surrounded by thick shields of iron and concrete. Ensuring safe operation and making provision for handling accidents involving radioactive releases were priority aspects of the design; these added to the problems arising from inaccessibility of much of the reactor once it had operated at high power. Of special concern was the "boiling disease": loss of light water from the cooling channels by boiling would increase the reactivity and this would increase the rate of fissioning; the result would be more boiling, culminating in a power runaway if not stopped by rapid detection and counteraction of the increasing reactivity. Fast-acting shutoff rods were developed. These would be brought into action by evidence of reactor overpower, a rapid power increase, or reduction of cooling flow through a fuel channel. The design included so many shutoff activators that it was often stated that the reactor would never operate because there were hundreds of ways for it to shut itself off, but only one way to start it up.

Immediately surrounding the cylindrical wall of the calandria there is an annulus of graphite approximately three feet (90 cm) thick. Its purpose is to reflect back into the calandria neutrons

[7] The mechanical design of ZEEP was done at NRC in Ottawa under extreme secrecy, which forbade any public mention of the nuclear interest of the project. A security-minded visitor to the design office was appalled when a stranger looked in and asked, "Have you had your neutrons yet?" The visitor was somewhat mollified when he learned that there was a general staff NRC newsletter called "The Neutron," so named because, like neutrons, it was free of charge.

that escape through the side wall. Without this reflector the core of the reactor would have to be larger, to compensate for the loss of neutrons. The reflector and inner shield are air-cooled. The air is exhausted through a tall chimney, because some of the argon in the air is made radioactive by absorption of neutrons. In 1945, tests were run in which smoke discharged from the chimney under various wind and weather conditions was observed, to forecast the dispersion of the very weakly radioactive plume of effluent air.

NRX Research Facilities

Conceived in wartime but planned with an eye to peacetime uses, NRX incorporated many features intended to facilitate research. The advent of nuclear reactors increased the availability of neutrons by factors of many thousands over the most intense previous sources (i.e., accelerators). Work with those sources had, however, provided experience which, combined with lessons learned from one or two American reactors, enabled an excellent design for NRX to emerge. Horizontal holes through the graphite reflector and the shielding provide beams of neutrons to research equipment at the main-floor level. Access to the maximum neutron flux is available in a large vertical tube through the centre of the core (the so-called central thimble), or in the vertical tubes normally used for fuel. (The flux is sometimes said to be the number of neutrons crossing a square centimetre in one second; more precisely, it is the number of neutrons traversing a sphere of cross-sectional area 1 cm^2 in one second.) Two horizontal columns of graphite roughly two metres square extend from the graphite reflector through the iron and concrete shields to the outer perimeter, where they are closed with thick, shielding doors of lead. They are called thermal columns, because the neutrons at the outer ends are in thermal equilibrium with the graphite; that is, the energy distribution of the neutrons is similar to that of the molecules of a gas at the temperature of the graphite. The remaining area of the side wall of the reactor has special devices for irradiating many small samples, to produce radioisotopes. A ring of ninety vertical holes through the top shield enables tubes containing material for irradiation to be inserted into an annulus in the graphite reflector near the inner surface. These positions were originally intended for the irradiation of thorium to produce uranium-233, but some were soon being used to produce other isotopes; e.g., carbon-14, an isotope having many uses in biological research.

Chemical Extraction Plants

Uranium-233 and plutonium-239 were two products of NRX of major interest as fissile nuclides. Two heavily shielded chemical extraction plants were built near NRX, one to separate uranium-233 from the thorium carbonate that was irradiated in the annulus in the reflector, and the other to separate plutonium from the uranium and fission products in irradiated fuel rods. The chemical processes were new and the feed materials intensely radioactive; consequently, startup and operation were difficult. Nevertheless, both plants eventually delivered satisfactory output, but were later shut down as being not immediately relevant to the project's developing mission.

Waste Management

Operation of a nuclear reactor produces unwanted radioactive material, the safe management or disposal of which poses problems. The Chalk River site is in a sandy, rocky area not valuable for agriculture. Immediately adjacent to the plant site and within the restricted area is a sand-filled rock basin with a single water outlet, a creek flowing from a small lake called Perch Lake. From the beginning, Perch Lake basin has been used for radioactive waste management; the outlet water can be monitored for escape of the waste. Radioactive scrap from the repair of NRX in 1953, waste from the processing plants, and other radioactive wastes have been deposited in this area, usually in sealed containers. Research into the movement of radioactive material through the soil and other aspects of disposal of radioactive material have been pursued intensively. The movement through soil is very slow. Some underground tank storage was provided near the chemical-extraction plants for storage of radioactive liquids.

Early Status of the Chalk River Project

By the time of NRX startup (July 1947), the tripartite organization had been disbanded and Canada had assumed full responsibility for the project, but with good connections to the United Kingdom and the United States (within the limits imposed by their security regulations). Excluding construction workers, there were approximately 500 staff members (including 140 scientists) and 500 hourly rate workers; there were also forty research and technical United Kingdom staff. During 1947, the Canadian total increased by about 10 percent and the United Kingdom total decreased by about 20 percent. The new town of Deep River, built to house many of the staff, was well established on the shore of the Ottawa River, eleven kilometres upstream from Chalk River.

NRX

The permitted power level of NRX was raised in steps until it reached 30 MW by mid-1950. The reactor was operated to produce uranium-233 and plutonium, to provide radioisotopes, and to provide beams of neutrons for physics experiments. A less obvious reason was to develop experience in a new, promising technology; for example, the intense ionizing radiation posed novel operating problems. There was as yet no clearly defined mission, but the underlying hope for a source of energy was an incentive. In 1948, the Control Board stated that the general policy was to operate the reactor in the best interests of fundamental research and not for the specific purpose of producing plutonium.

Early Research based on NRX

Gamma rays

When an atomic nucleus captures a neutron, the result is a new nucleus in a state with considerable excess energy, a so-called excited state. In most cases, the new nucleus very quickly gets rid of excess energy by emitting gamma rays. Measurement of the gamma-ray energies provides information about the energy levels and mass of the newly formed nucleus. Research on these neutron-capture gamma rays was a major component of physics research at NRX. Two different types of equipment were installed: one measured high-energy gamma rays, the other lower-energy gamma rays. Many important results were produced; for example, within about a year, it was shown that the accepted mass of a neutron was in error and a much more accurate value was obtained.

Radioisotopes[8]

The nuclei resulting from the capture of neutrons are often radioactive and therefore not usually found in nature. Reactors are prime sources of these nuclei, and NRX has been a prominent supplier. Within a few months, there was a steadily increasing stream of requests from both project staff and outsiders for radioisotopes. The radioisotopes were themselves objects of research or were useful as tracers in other research. Over the years, medical uses for research or clinical applications predominated. In 1946 an isotope-production branch was established to develop, produce and ship radioisotopes for outside users.

Neutron diffraction

For many purposes, the neutron is treated as a material particle, but, like all other matter, it also has a wave character. Slow neutrons (i.e., those having energies like the energies of molecules in a gas) exhibit wavelengths similar to the spacings of atoms in solids and liquids. Just as X rays having similar wavelengths are diffracted by crystals, so are slow neutrons. By such diffraction, neutrons of specific energies may be selected or the structure of the crystal determined. A neutron spectrometer was installed at

[8] The modern term is radioactive nuclide or radionuclide, unless the intention is to relate it to nuclides belonging to the same element. However, universal usage dictates the older term, radioisotope, for general commercial use.

the experimental face of NRX, where a neutron beam could be extracted. This instrument provided mono-energetic beams of neutrons for research on structures of gases, liquids and solids, nuclear properties of deuterium, scattering of neutrons, and fission cross-sections as a function of energy.

Decay of the neutron

The mass of a neutron is greater than the sum of the masses of a proton and electron. From the well-known relation due to Einstein, $E=mc^2$ (E is energy, m is mass and c is the velocity of light), it follows that the decay of a neutron into a proton and an electron is energetically possible, and indeed it had been predicted but not observed. Equipment was set up at an NRX beam hole to detect and study the decay. This was first observed elsewhere, but the NRX experiment yielded a value of 12.8 ± 2.5 minutes for the half-life. The identification and measurement of the protons and electrons in the presence of neutrons and other radiations required experimental design and analysis of a high order.

Fission

From the beginning, there was wide scope for basic research into such matters as details of the fission process, the properties of the fission products, their relative yields, and differences in yields between the various fissile nuclides. Neutrons, initially from a thermal column but later from a beam hole, were used in detailed studies of the energies of the fission fragments and of the emission of light nuclei in the fission process; another experiment showed that the fission neutrons are emitted from the moving fission fragments.

Transuranic elements

For some years the high neutron flux of NRX gave it a superior position for synthesizing the newly discovered transuranic elements through the neutron irradiation of uranium and the sequentially produced elements. The chemical and physical properties of these new elements were investigated at Chalk River, and, at a later stage, in cooperative experiments with other laboratories.

Reactor characteristics

Many experiments on arrangements of uranium and moderators had been done in the United States, in the Montreal Laboratory and in ZEEP, but they were with fresh uranium at very low power and at room temperature. Calculations of temperature coefficients of reactivity and other properties of operating reactors were based on measured or estimated basic properties of the constituents, but supported by few direct measurements. It was therefore important to determine experimentally such properties of NRX.

The first major measurements were done during 1948, when the power was being raised to 20 MW. Temperature and power coefficients of reactivity and other parameters were measured, a special case being xenon poison. The radionuclide xenon-135 has an enormous cross-section for the capture of slow neutrons, of the order of a million times the average cross-section. (The cross-section is the target area the nucleus presents for interaction with the neutron flux. There are several kinds of cross-sections, for example, capture, scattering or fission cross-sections. Each of these has its own form of dependence on the neutron energy. The rate of interaction per nucleus is the product of the cross-section multiplied by the flux.) Xenon-135 is the result of radioactive decay of iodine-135, which is a product of fission. The interplay of the rate of destruction of the xenon-135 by neutron absorption and the decay periods of the two radionuclides leads to complex variations of reactivity in a high-flux reactor. Shutdown of a power reactor may be followed by a day or two of forced inaction, until the xenon poison dies out by radioactive decay.

Interest in power reactors made knowledge of the effect of irradiation on the reactivity of fuel of prime importance. Experiments to determine this effect were done in October 1951, March 1952 and December 1952, with NRX as the measuring instrument. The last set of experiments was terminated abruptly by the 1952 NRX accident (which will be discussed later). The reactor was restored by February 1954 and the first 1,000 megawatt-days' operation (roughly six weeks) was done under conditions specified for a major experiment on reactor characteristics. During this time, the reactor power was raised in steps to a new limit of 40 MW.

One of the long-irradiated rods measured in October 1951 was made the subject of a detailed examination that continued for many years, yielding results important for use in calculating the reactivity of power-reactor fuel. Eighteen thin discs were cut from the rod for metallurgical studies, chemical analysis, and measurement of radial distribution of various nuclides. Slugs were sent to an American laboratory for measurement of reactivity.

Future Systems Group

In December 1944, Cockcroft had formed a Future Systems Group, with a membership predominantly from the United Kingdom staff. Initially, the group considered reactors for the United Kingdom, with emphasis on reactors for plutonium production; production of uranium-233 and energy were secondary goals. Uranium was thought to be a rare element, with an ultimate availability of less than ten thousand tons. Only 0.7 percent of uranium, the uranium-235, is readily fissile. The remaining 99.3 percent, uranium-238, is practically inert insofar as fission is concerned; however, it can be converted to fissile plutonium by absorbing neutrons. Consequently, breeder reactors, which produce more fissile material than they consume, had high priority. Fast-neutron chain reactions with plutonium as fuel were known to be possible breeder systems, and there were indications that thermal-neutron reactors using uranium-233 as the fissile material and thorium to capture the excess neutrons and convert to uranium-233 might also breed. Fissioning of the precious uranium-235 would be necessary for the initial production of the plutonium or uranium-233 needed for the breeders.[9] (Breeding has recently been achieved in a power reactor but even yet it is not an accepted alternative to other power-reactor systems.)

The group's early discussions concentrated on fast-neutron plutonium-breeder reactors and gas-cooled thermal-neutron reactors moderated by graphite or beryllium oxide. Beryllium has low absorption for neutrons and, because it has lighter nuclei than carbon, seemed to offer superior moderation. The oxide behaves well at high temperatures, but otherwise little was known about its suitability. Gases considered for cooling included hydrogen, helium and carbon dioxide. Heavy water was not stressed as a possible moderator.

Information was sought on nuclear parameters of fuels and on such engineering properties as erosion of the moderators by hot gas, diffusion of fission products through them, and effects of intense irradiation. Action was taken to procure massive blocks of beryllium oxide.

In 1946, a "water boiler" was added to the list of possible reactors as a source of neutrons for experiments. The core would be a light-water solution of a plutonium compound contained in a stainless-steel vessel. Pulsing of the power would provide momentary neutron fluxes a hundred times the flux expected in NRX. The plutonium would have to be supplied by another reactor of a different type.

NRU, a Replacement for NRX

As a result of transfer of staff to the United Kingdom and the change in directorship from Cockcroft to Lewis, the Future Systems Group did not meet for nearly a year and a half. In November 1947, Lewis convened his first meeting of the group with a revised and enlarged membership to consider a replacement for NRX. The corrosion-limited life of NRX had been estimated as about five years and soon after NRX was in operation the Control Board had approved consideration of a second reactor to provide continuity in the event that NRX had to cease operation.

The reactor types initially chosen for study by the new group were very similar to those considered earlier. A reactor cooled and moderated by lithium-7 was added in spite of the formidable problem of separating lithium-6 from the lithium-7. In retrospect it is clear that none of the reactor types would have

9 A remark that illustrates the importance given to the supposed shortage of uranium is recorded in the minutes of a 1946 Future Systems Group meeting: "Large-scale production of power is not likely to be economically interesting until breeding has been made effective."

been a good choice. In April 1948 a suggestion was made that consideration be given to an improved high-power heavy-water reactor. At the next meeting six weeks later, a heavy-water reactor with a larger neutron flux than NRX, good experimental facilities, good economy of neutrons and possibly some use made of the power, was said to be the tentative choice of the group. Lewis christened this reactor NRU, with the "U" standing for universal. A subcommittee was established to prepare technical outlines and rough cost estimates for four versions of a heavy-water reactor. In the spring of 1949 the version corresponding to NRU was selected for further study. A Reactor Development Panel was appointed to supervise the next stage of conceptual design of NRU. For twelve months the design concept was developed while the question of approval was moved closer to government. The cost of the reactor was a stumbling block, but the plutonium that NRU would produce was thought to be a valuable commodity that could be extracted and sold to the British or Americans. There was considerable negotiation, which culminated in the sale of the used fuel to the United States for its plutonium content.[10] This sale of intact fuel removed the need for a costly new chemical extraction plant at Chalk River. The American involvement not only made the project financially acceptable, but also ensured the availability of fuel and heavy water.

The NRU reactor, as proposed by the Reactor Development Panel, was to be moderated and cooled by heavy water, fueled by natural uranium, have fuel-changing under power, provision for experiments with neutrons (beam holes, through tubes) and several positions for fuel tests. The operating power was to be 200 MW; the idea of generating useful energy had been dropped. Government approval to proceed was given on 13 December 1950. The final design was done by a consulting engineering firm, C.D. Howe Co., with guidance from Chalk River.

Nuclear Power

The prospect of generating useful energy from the heat produced by nuclear fission helped get the Canadian project on the road and to keep it moving. Inevitably, some dead ends were encountered along the way. The Future Systems Group in its early consideration of plutonium-production reactors for the United Kingdom proposed converting some of the heat to electric energy. This anticipated the first British power reactors, which had a dual purpose: to produce plutonium for military purposes as well as electric power for civilian use.

When the revised group sought a replacement for NRX, it had studies done on proposals for useful energy generation. This was not pursued, but a steam engine project was begun to demonstrate that coolant from a loop in NRX at a temperature of 200°C could generate power "at a not insignificant efficiency". A reciprocating engine to run on steam was ordered in mid-1952, but a serious accident in NRX delayed the project to a time when it would have little value, as a demonstration of this kind had been made in the United States; also, power reactors were being designed. The project was cancelled.

Apart from keeping interest alive in nuclear power, the Future Systems Group contributed little to the birth of the Canadian nuclear power program. This program was conceived outside the activities of the group and began to take shape in 1951. Nevertheless, in July 1952, Lewis suggested that the group maintain action on a liquid-metal-cooled thermal neutron reactor and breeder reactors. About this time the idea arose to produce fissile material by the absorption in uranium or thorium of neutrons resulting from the spallation (breaking into smaller pieces) of heavy elements bombarded with high-energy protons; this idea was to be developed further over the next fifteen years.

By 1947 the United States navy had a program under Admiral Rickover to develop nuclear-powered submarines. Light water[11] and liquid metal were considered as coolants. Eventually the admiral vigorously promoted, as the heat source, a light-water-cooled and -moderated reactor fueled with enriched uranium, all contained in a pressure vessel. To test fuel for these reactors,

---

[10] Bothwell, *Nucleus*, 139-43.

[11] The term light water is commonly used for ordinary water in contrast to heavy water, since 99.986 percent of the hydrogen in ordinary water is light (i.e., ordinary) hydrogen.

hydraulic circuits known as loops containing fuel elements and cooling water at temperatures and pressures suitable for power generation were installed in some American reactors. NRX offered superior conditions (higher neutron fluxes and larger in-reactor volumes), and in 1950 negotiations led to the use of positions in NRX for American loops. Many details of the fuel elements under test were American secrets, but what information was revealed was of great value to Canadian power planning, as was the experience with the technology of loops. Cooperation and competition between the United States and Canada on fuel technology increased over the years and proved extremely profitable for both parties.

The breakthrough in Canadian thinking on nuclear power came in 1951 when Lewis combined knowledge of the United States loop experiments with other facts. Proposing an elementary design of a heavy-water power reactor, he estimated the cost of production of electric power to be in the commercial range, thus for the first time indicating the economic viability of nuclear power.

In summary, the facts were:

1 NRX fuel had survived the production of 3,000 megawatt days per tonne and had retained much of its reactivity.

2 In loops, fuel was being cooled by water at temperatures and pressures like those in thermal power plants. Zirconium was a possible sheathing material in these conditions.

3 Fuel changing under power was planned for NRU.

The proposal fell on fertile ground and blossomed in the next few years. The engineering community had a general professional interest in nuclear power, but Ontario Hydro had a specific interest, because it was approaching the time when all suitable hydroelectric sites in Ontario would have been developed and new plants would be thermal; i.e., based on the burning of fuel. (Ontario has no indigenous practical sources of fossil fuel despite the fact that the first oil well on the continent was in Ontario.) At the beginning of 1952, there were discussions with Ontario Hydro about nuclear power and by the end of 1952, AECL and Ontario Hydro had moved towards cooperating on a nuclear power plant.

The NRX Accident, 12 December 1952

The program of the newly established AECL was progressing well in 1952. Research was producing important results, radioisotope production was increasing, NRU construction was underway, and nuclear-power planning was moving into an era of optimism. Then on 12 December, the NRX reactor was seriously damaged in an accident, a costly interruption to the work of the project.

Experiments had been in progress to measure the effect of irradiation of fuel on its reactivity. The reactor was being operated according to procedures that differed from those for normal operation. Somewhat similar procedures had been used successfully in a previous set of measurements. An operator error was the first of a sequence of events, a difference in any one of which might have prevented the serious consequences. There was a power runaway, enhanced by deliberately reduced cooling arrangements for certain rods. Steam explosions in several channels caused irreparable damage to the calandria. Central portions of several fuel rods disintegrated, releasing gaseous fission products to the atmosphere and becoming oxidized by contact with water and air.[12]

Removal of the heat generated by decay of fission products in the fuel required that a small flow of water be maintained through the core. The water, laden with highly active material, either reached storage tanks or leaked into the basement below the reactor building, where it accumulated to a depth just under a metre. All the active water was eventually piped to the waste-management area.

[12] On the basis of a preliminary telephone call from Chalk River, AECL's president in Ottawa told the press there had been a pinhole leak. When a group of Americans visited Chalk River to discuss the accident, one of them began his remarks by saying that Canadians must measure pinholes as they do gallons because this was certainly an imperial pinhole.

NRX Restoration

The damage to the reactor was so extensive that the feasibility of restoration was initially in doubt. However, after a brief study, a decision was made to proceed with restoration. Radioactive fission products released from damaged fuel rods were spread about in the calandria and the building basement, making disassembly difficult. Workers soon reached their radiation dose limits, which were then 0.3 rem per week (0.003 sievert per week). (A rem is a unit of radiation dosage. It is defined in terms of the energy deposited in tissue by ionizing radiation, with adjustment to take into account the biological effects of the specific radiation. A sievert is the Système International dose unit corresponding to a rem; one sievert = 100 rems.) The Canadian army and the American navy supplied extra workers. The latter had two incentives to participate: the NRX facilities were important in the development of fuel for nuclear submarines; and the United States navy had a group being trained to handle radioactive contamination resulting from an accident or a nuclear-bomb explosion. NRX gave them hands-on experience in a real situation.

A restored and improved NRX was brought to criticality on 16 February 1954, and in April of that year it was operating at 40 MW, a new upper limit made possible by a modification during the restoration.

A remark made soon after the accident, that it might be a blessing in disguise, was not fully justified, but it did have some validity: the restoration itself was an encouragement to the nuclear community; major benefits were (i) the renewal of strong concern for safety among staff members, many of whom became prominent in the nuclear power program, (ii) improvements in safety systems planned for NRU or being developed for power reactors as a result of reviews prompted by the accident, and (iii) the demonstration that a reactor could be repaired after an accident.

Space in the reactor for loops, and accommodation in shielded rooms for auxiliary equipment were in demand by Canadian, American and British researchers. An example of actions taken to satisfy the demand for space for auxiliary equipment was the move of the gasholder for the moderator cover gas, helium, to outside the building, to make available the pit in which the gasholder was originally located.

An increasing diversity of NRX usage required frequent changes in the operational program. In the early days, only a weak link was needed between the operational timetable and the plans of researchers (who, in the main, were using neutron beams), because changes could be made to the research equipment whether or not the reactor was operating. The advent of loops introduced a need for close coordination of reactor and loop timetables.

Chemistry and physics research using neutron beams and radionuclides from NRX had also expanded and was producing first-class results. The value of NRX to Canada was at its peak during this interval and it was to continue at a high level for at least a further twenty years.

NRU

Even with the NRX experience as background, the design and development of NRU presented major problems. The higher flux, $3x10^{14}$ neutrons/cm^2/s, at 200 MW compared with $6x10^{13}$ in NRX at 40 MW, meant a greater transfer of heat from the fuel and consequently a different design of fuel element. The flat bar shape selected for the uranium-metal fuel elements made the assurance of contact between sheath and uranium more difficult than with the round NRX fuel. An innovative design concept involving the changing of fuel at power was complicated by the need to avoid any loss of the heavy-water coolant.

Access to high-flux regions was provided in several ways: vertical tubes, horizontal through-tubes and re-entrant tubes called thimbles. The space under the reactor was arranged to facilitate a possible calandria replacement. Progress in design and construction was slower than originally scheduled, but by 3 November 1957 the reactor reached criticality. There were some teething problems with the new fuel. Water entered through cracks in the aluminum sheath, causing it to swell away from the uranium. Perhaps there were some rods in this condition when a power surge on 24 May 1958 resulted in disruption of several rods. During a difficult rod removal, a piece of uranium fell out of the removal flask into a pit at the top of the reactor and caught fire. The building was severely contaminated and three months passed before the reactor was started again at lower power. By the end of November, NRU was again up to power. This was the most serious incident in NRU and subsequently, with fuel in which the uranium was bonded to the aluminum by nickel, NRU operated in an exemplary manner. It has been, like NRX in its heyday, a pre-eminent research reactor and one of the world's outstanding reactors for engineering tests and radioisotope production.

Nuclear Electric Power, NPD

The ten or twelve years following 1952 were crucial years for the future of nuclear power. NRX and reactors elsewhere had demonstrated the feasibility of producing large amounts of controlled energy from uranium fission, but the energy was available only as heat at relatively low temperature and its production had been without serious cost constraints. Each branch of science and technology faced new restrictions in its application to reactors. For example, material used to construct the cores must have low absorption for neutrons, and coolant chemistry had to take into account the ionizing properties of radiation and the possible transport of radioactive material by the coolant. Such restrictions complicate the cost estimates required in nuclear-plant planning.

The United States arrived at civilian nuclear electric power in 1957 by way of power plants for submarines – small cores containing enriched uranium with ordinary water as moderator and enclosed in pressure vessels. For its first power plants (1956), the United Kingdom modified its designs of gas-cooled, graphite-moderated, natural-uranium-fueled, plutonium-producing reactors to provide dual-purpose plants that would produce plutonium for weapons and electric power for civilian use. In neither case was there strong economic motivation to avoid wasting neutrons.

The situation in Canada was different. A nuclear power program would be a significant item in the national economy, with no military nuclear program to dwarf the expenditures on civilian energy. Ontario Hydro was the only Canadian utility large enough to accommodate a nuclear power plant that was also in need of a new source of energy; but the cost of the power it supplied to its customers was low, setting a difficult target for the nuclear option.

Lewis, whose proposal of 1951 set the Canadian power program in motion, continued to expand and refine his considerations of cost, putting emphasis on the value of neutrons and insisting that they must not be wasted. Some neutrons produced in a reactor disappear without causing fission by leakage out of the core, by unproductive capture, or by capture in uranium-238 to yield plutonium-239, a fissile nuclide.[13] For the power to remain constant, an average of one neutron per fission must result in a further fission. Wastage of neutrons by excessive leakage or unproductive capture reduces the useful life of the fuel.

Because plutonium is fissile, its production prolongs the useful life of the fuel. An important aspect of costing is choosing the balance between the gain due to an increase in plutonium and the added costs to achieve it. Estimating the cost of nuclear power became a growth industry. The Canadian plan to use heavy water as moderator and natural uranium as fuel had some good data from NRX experiments on which to base predictions of fuel consumption. Americans regarded Canadian cost predictions as unduly optimistic, but by the mid-seventies the success of the Pickering Nuclear Generating Station validated the Canadian estimates. The enriched-uranium, light-water system has a lower capital cost but a higher operating cost than the CANDU system; the former is competitive where there are high capital charge rates, as for private utilities in the United States.

[13] The survival time of a neutron in the core is so short compared to its lifetime for natural decay to a proton and an electron, about 1 to 10^6, that the effect of natural neutron decay on reactivity may be neglected.

In 1954, a Nuclear Power Group consisting of engineers from Canadian industries and utilities was assembled at Chalk River to design a prototype nuclear power plant. They soon had a design concept for a 10 MWe plant, later 20 MWe, in which the reactor was moderated and cooled by heavy water at high temperature and high pressure, and was fueled with natural uranium metal contained in zirconium-alloy tubing. The reactor core was enclosed in a pressure vessel.

In spite of its moderately large neutron absorption, stainless steel was being considered by the United States for fuel sheathing in its power-reactor cores, but this was feasible only with enriched uranium. Very early, zirconium had been considered as undesirable because of an apparently significantly large absorption cross-section. However, in 1947, researchers in the United States discovered that much of this cross-section was due to an impurity, hafnium, making purified zirconium an attractive possibility as a fuel-sheathing alloy, especially for natural uranium.

A partnership was formed between AECL, Ontario Hydro and Canadian General Electric, each being given distinct responsibilities. Some engineers transferred from the nuclear power group to work on the prototype nuclear power plant, called the Nuclear Power Demonstration (NPD). A site was chosen for NPD near the hydroelectric generating station 25 km up the Ottawa River from the Chalk River laboratory. Excavation and construction were soon under way.

The high temperature of the cooling water made the proposed use of uranium metal fuel questionable. Uranium dioxide (UO_2) was shown, largely in NRX loops, to have physical and chemical properties more suitable for in-reactor use than uranium metal, but its lower density made it less desirable from a nuclear standpoint. There were strong opinions on both sides, but eventually a zirconium alloy developed in the United States and UO_2 were chosen.

The Nuclear Power Group began work on the design of a 200 MWe nuclear power plant based on the concept of zirconium-alloy pressure tubes instead of a pressure vessel, uranium dioxide instead of uranium metal, and an alloy of zirconium for sheathing. NPD, designed as a reactor contained in a pressure vessel, was now headed into a dead end. In March 1957, construction was halted for preparation of a new design based on pressure tubes and uranium dioxide fuel.

Problems had been envisaged with the pressure-vessel NPD. As a prototype, its success would not have assured scaling up to high-power versions, since the pressure vessels required would be very large, much larger than pressure vessels for corresponding light-water reactors. (Because its slowing-down distance is longer, heavy water is a less effective medium for slowing down neutrons than is light water, and so it requires a larger volume of moderator between fuel rods.) In addition, the safety of the pressure vessel was being questioned.

The change to pressure tubes and uranium dioxide was welcomed and put Canada firmly on the road to CANDU, a Canadian-developed class of reactors moderated by heavy water, with fuel in zirconium-alloy pressure tubes and fuel-changing at power. The current representatives have horizontal pressure tubes and are cooled by heavy water; the fuel is uranium dioxide in the form of short cylindrical pellets sheathed in zirconium-alloy tubes, which are assembled into bundles 50 cm long, with spaces between the tubes for flow of coolant (see the illustration in chapter thirteen). The fuel bundles are fed through the reactor in opposite directions in alternate pressure tubes, resulting in a symmetrical distribution of power longitudinally. Normally, the bundles passing through any one pressure tube emerge with a similar degree of irradiation, but the intensity of irradiation varies across the reactor from pressure tube to pressure tube. By appropriately varying the length of time the fuel spends in a tube, the distribution of power may be flattened across the core. Fueling is normally done with the reactor operating at power, a major advance that avoids long non-producing intervals while fuel is being changed. Fueling at power, longitudinal symmetry, and flattening of the power distribution all enhance the efficient use of fuel. The fuel has been natural uranium, but the design is also suitable for enriched uranium, fuels containing plutonium or a fuel cycle using thorium.

Within about six months the new design of NPD, modeled on the emerging CANDU concept, was sufficiently developed for construction to be resumed. The complex discussions and negotiations among the three corporate partners and the government have been described by Robert Bothwell in Nucleus. The cognate scientific and technological developments are described later in this book. Of special importance were the behaviour of uranium-oxide fuel and the response of zirconium alloys to their environment, especially the deterioration that could result from the absorption of hydrogen and its subsequent precipitation as zirconium hydride.

After the necessary commissioning, NPD was brought to criticality on 11 April 1962, and power was first fed into the electricity grid on 4 June 1962. With its refueling at power and low loss of heavy water, NPD successfully demonstrated the feasibility of the CANDU concept. Chapter ten gives a more detailed overview of the CANDU nuclear power system.

Douglas Point

Even before NPD went critical, construction of the 200 MWe plant had begun at Douglas Point on Lake Huron. At that time, 200 MWe was regarded as full-scale, but before the plant was completed it had been outclassed by the start on the Pickering station, where there would soon be four 500 MWe plants. Douglas Point is now usually referred to as a prototype.

In 1958, the Nuclear Power Group was disbanded and a new AECL division, the Nuclear Power Plant Division, was set up in Toronto, to be responsible for the design and development of the Douglas Point plant and to oversee the work on NPD.

Whiteshell Organic Cooled Reactor (WR-1)

Water, either ordinary or heavy, has disadvantages as a coolant in a power reactor. At the temperatures needed for efficient generation of power, the water exerts a high pressure, which increases rapidly with temperature. In pressure-tube reactors, a balance has to be struck between the coolant temperature and the thickness of the pressure tube. Increasing the temperature increases the pressure and thus a thicker pressure tube is needed. The higher temperature permits a higher thermodynamic efficiency, but the thicker tube captures more neutrons. The balance results in a lower efficiency than that obtained in coal- or oil-burning plants.

In the fifties, terphenyls, which are organic liquids used for heat transport, were studied as possible alternatives to water as reactor coolants. Canadian studies indicated that a heavy-water-moderated reactor cooled with an organic liquid might produce power more cheaply than a water-cooled CANDU. The organic coolant could be run at a higher temperature with thinner pressure tubes than the water.

A proposal to build a test reactor cooled by organic liquid as the centrepiece of a nuclear research establishment to be established in Manitoba was approved in 1960. The establishment was named the Whiteshell Nuclear Research Establishment. Construction of the reactor was begun in 1963 and criticality was achieved in 1965. The reactor was called WR-1.

Safety

Hazards associated with fission reactors – e.g., the production of dangerous radioactive material and the potential for rapid overheating (but not for a bomblike explosion) – have produced a special concern for safety that has been evident ever since the early days of NRX design. Safety aspects were routinely considered and, on occasion, were referred to the Pile (reactor) Operating Committee. In 1958 a ZED-2 Safety Review Committee was formed, very soon to be replaced (1959) by the Review Committee on Reactor Safety, with a mandate initially to review the safety of ZED-2 and, eventually, all AECL reactors. In 1967 it was replaced by the Nuclear Safety Advisory Committee (NSAC).

The presence in power reactors of large volumes of water at high pressure and high temperature (about 300°C) introduced major increases in hazards over those of research reactors such as ZEEP or NRX. In the event of a destructive power runaway, steam would greatly enhance the escape of fission products from the disrupted reactor and their dispersion in the environment. Even without a runaway, the pressure of the hot water may cause catastrophic rupture of the pressure-retaining components. Much applied research and development has been done by AECL to enhance the safety of power reactors, to analyze the course of potential accidents, and to ensure containment of radioactive material in the event of a failure. This work is described in chapter fifteen.

Because the moderator in a pressure-tube reactor is cool, there is much less water at high temperature than there would be in a pressure-vessel reactor. This in itself enhances safety, but in addition the cool moderator can act as an energy sink rather than promote dispersion. The change from metal to oxide fuel was another safety improvement, because if the hot metal came into contact with water it would have reacted chemically to release hydrogen, which forms an explosive mixture with air.

Before NPD, the Control Board had let AECL handle reactor safety internally. Unlike the existing reactors, which were owned and operated by a federal agency (AECL), the NPD reactor would be operated by Ontario Hydro and owned federally. For this and perhaps other reasons, the Control Board decided to require the licensing of reactors not fully owned and operated federally. A safety assessment was required for the licence, and in 1956 the Control Board established the Reactor Safety Advisory Committee (RSAC) to do the assessment. Under the guidance of its chairman, G.C. Laurence, who had earlier been involved with the subject, the RSAC developed a distinct Canadian approach to reactor safety. General principles are described in chapter four.

International Aspects

The Canadian project was formed with tripartite international participation. The tight security of the early days hindered but did not prevent broader contacts; once the nuclear character of the project had been revealed, in 1945, most of the basic research could be freely published and discussed; exceptions were, for example, properties of plutonium and certain parameters of the fission process.

Canada earned a high status in nuclear matters at the United Nations. It had membership on the United Nations Scientific Committee on the Effects of Atomic Radiation (UNSCEAR) and on the United Nations Scientific Advisory Committee (UNSAC). The first of four United Nations Conferences on Peaceful Uses of Atomic Energy was convened in Geneva in 1955, the organization of the conference being a responsibility of UNSAC. Canada helped organize the conference and presented thirteen technical papers. Before the conference, a vast amount of information was declassified by agreement between the United States, the United Kingdom, and Canada. This declassification broadened the scope for publication and left only a small fraction of the work at Chalk River secret. However, commercial secrecy appeared as the development of nuclear power plants produced proprietary designs and processes.

In addition to convening the 1955 conference, the United Nations promoted the establishment of the International Atomic Energy Agency (IAEA). The purposes of the IAEA were to provide for the control of fissile material and to assist its member states, particularly the developing countries, in peaceful uses of atomic energy. The problem of control was a difficult one, which has resulted in a system of safeguards involving the inspection of nuclear facilities to detect and hence deter military applications.

Canada was made a permanent member of the IAEA's board of governors and many Canadians have served the IAEA in a variety of positions. Particularly noteworthy was Lewis's twenty-year-long chairmanship of its Scientific Advisory Committee.

Values of nuclear properties such as cross-sections for neutrons are important for designing reactors and in research. The three countries – Canada, the United Kingdom, and the United States – shared information on nuclear data, with due regard to individual national restrictions. To facilitate the transfer, a Tripartite Nuclear Cross-Section Committee was established in 1956. The committee was very active, not only in the transfer of data, but also in such matters as standardization of notation, exchange of samples for measurements, and arranging for joint measurements; it was dissolved in 1963. Canada was a member of the European-American Nuclear Data Committee formed a few years earlier, which soon became the Nuclear Energy Agency Nuclear Data Committee. In 1966, an International Nuclear Data Committee was set up by the IAEA, Canadians having played a part in its formation.

Early in 1955 Canada and India began discussions which culminated in an agreement that Canada would supply an NRX-type reactor to India under the Colombo Plan; the Canadian-supplied reactor is now called CIRUS. This was AECL's first large international project and indeed also established a new magnitude for international nuclear aid. Cooperation on the technical aspects was good, but India was loath to accept Canadian inspection to ensure that plutonium was not diverted to non-peaceful uses. There were no international standards to guide the discussion. However, Canada received the right to inspect Canadian-supplied fuel and India gave assurance of peaceful uses only.

In December 1963 an agreement was signed between Canada and India for the construction of a Douglas Point-type CANDU reactor at Rajasthan, India, to be called RAPP. Three years later an agreement for a second reactor was signed. The first reactor, RAPP 1, was in operation in 1973. Then in 1974, India detonated a nuclear explosive device using plutonium from CIRUS, and claimed that it was a peaceful explosion. This ended Canada-India nuclear relations. RAPP 2 was completed by India alone in 1981.

The potential for a large international market for nuclear power plants was recognized early. By 1958, international marketing of nuclear power plants had begun; Canada was expounding the prospect of cheap power from Douglas Point but was not yet in the marketplace. By 1962, Canada was also looking for sales. Except for its impact on the technical work of the laboratories, the marketing story is not germane to the purpose of this book and has already been told (Bothwell, see Bibliography). Suffice it to say that Canada has maintained a sizeable foothold in a field where the competition includes intense sales efforts by the U.S.A., France and Germany.

A 132 MWe CANDU nuclear plant was built in Pakistan at Karachi by Canadian General Electric. Named KANUPP, it began operation in 1972. This was the only non-AECL Canadian sale of a CANDU.

The international marketing of radioactive isotopes is an unqualified Canadian success story. By the mid-fifties, an important share of the worldwide demand was supplied by AECL's commercial products division. Its cobalt teletherapy machines for the treatment of cancer won wide acceptance, and continue to today. Demands for the radioactive cobalt isotope used in these machines taxed the capacity of NRX until NRU was in operation; many other isotopes were also produced. Medical applications continue to predominate, but there are also many industrial uses. Chapter five describes the activities of the commercial products division in more detail.

Power Reactor Development Program Evaluation Committee

A few weeks after AECL's tenth anniversary (1 April 1962), NPD began to feed electric power into Ontario Hydro's grid. AECL had come a long way. It was an opportune time for AECL to reassess the paths it was taking, and to seek ideas for major new research facilities.

To assess and guide the power reactor program, the president of AECL set up the Power Reactor Development Program Evaluation Committee (PRDPEC), with Lewis as chairman. It held its first meeting in March 1962. Its initial task was to compare four nuclear power plants based on heavy-water-moderated, natural-uranium-fueled reactors. Each plant comprised a steam turbine, an electricity generator, the necessary auxiliary equipment, and a reactor. The reactors differed in their primary coolants:

1 Pressurized heavy water (PHW). PHW had the best technological base because NPD was in operation and Douglas Point was under construction.

2 Organic liquid (terphenyl). Considered promising as a reactor coolant by several countries, organic liquid contains only about three-quarters as much light hydrogen as water does; the outlet coolant in the comparison reactor was 130°C above that of the D_2O in PHW, and at one-sixth of the pressure, both advantageous possibilities. Organic coolant was to be used in the organic test reactor (WR-1) being planned for Whiteshell; in fact, the initial name for WR-1 was OTR.

3 Light-water fog. Fog cooling, in which the coolant enters the fuel channels as low-quality steam (i.e., steam loaded with water droplets) and leaves as higher-quality steam (i.e., with fewer droplets), was of interest, because it would reduce the amount of light water in the reactor core compared to pressurized or boiling light water. Fog cooling was, however, at an early stage of development and problems were foreseen with dryout of the fuel-sheath surface, and with the means to introduce fog into the channels; a direct cycle was considered, with steam going directly to the turbine.

4 Boiling light water (BLW). BLW offered several advantages over PHW, one being direct-cycle operation, another being the avoidance of losses of heavy water from the coolant system, but BLW would be unacceptable unless the amount of light water in the core could be kept low.

During the first year, the committee concentrated on analyzing the four types to arrive at comparative costs of power. It arranged for a report to be prepared on each of the plant types. For this to be done required company-wide involvement, including sufficient design effort to permit credible costing of the power produced. Procedures were established to ensure that the comparisons were equitable.

The choice of the electrical output of the plants was the subject of much discussion. Douglas Point would be 200 MWe, an output that was considered, at the time of commitment, to be full-scale, but as a result of increasing turbine sizes and forecasts of demand growth, the plant was already being looked upon as a prototype. Many sizes were considered, ranging up to 750 MWe; the size chosen for the final comparison was 450 MWe.

The committee reported on the comparison in March 1963. The spread in estimated costs of power was about 0.04 cents/kWh (roughly 8 percent of the cost). Organic had the lowest cost, and PHW and fog were close together at the high end. In spite of this spread the committee recommended, "In view of the competitive economic position and advanced stage of design and development, the D_2O-cooled reactor deserves priority for extension to higher power ratings and general improvement ... The economic attraction of the organic-cooled design justifies its continued development ... Development of the fog- and boiling-water-cooled channels may bring rapid rewards and should be pressed vigorously."[14]

PRDPEC remained active for many years, watching over the power program and giving advice on the full-scale plants planned by Ontario Hydro.

Power Reactors

Nuclear Power Demonstration (NPD)

The success of NPD was an important advance in the Canadian approach to nuclear power. It proved the feasibility of the pressure-tube system for heavy-water cooling, of fuel changing at power, of zircaloy-sheathed uranium dioxide as fuel, and of the CANDU concept in general. Modifications of components and procedures were tested; for example, various designs of fuel, and boiling in some channels. The NPD station became the training centre for Ontario Hydro's nuclear operating staff; it was kept in useful operation for twenty-three years in spite of its mere 20 MWe contribution to the Ontario grid.

Douglas Point

The Douglas Point reactor began operation in 1966. Early problems with heavy-water leakage and other matters prevented steady operation for some time. During operation, radionuclides were transported out of the core by the coolant and deposited on portions of the coolant system outside the shielding, resulting in high radiation fields in working areas. These problems were eventually overcome and Douglas Point provided much electricity to the grid. In later years it served as a source of steam for a heavy-water extraction plant. The lessons of Douglas Point enabled designers to provide a more successful station at Pickering, Ontario.

Pickering

Even before Douglas Point was completed, the growing demand for electricity hastened the move by Ontario Hydro to acquire

[14] AECL, 1730 *Power Reactor Development Evaluation*, 21-22.

nuclear stations of "commercial" size. In line with PRDPEC's conclusion that the PHW reactor was the advisable choice for higher power, the design of a 500 MWe reactor was undertaken during 1963, and the project developed into a four-reactor station at Pickering, just east of Toronto. The same site had been considered for the reactor that became the Douglas Point reactor, but, because of its proximity to Toronto, it had been rejected in favour of a site on Lake Huron. The development of vacuum containment (i.e., a large vacuum building to act as a sink to suck in steam, radioactive gases and aerosols in the event of a serious reactor accident) provided an economic solution for a four-unit station that was judged to make the Pickering site acceptable. The first of four 500 MWe plants was operational in 1971. By 1973 these four plants constituted the Pickering A station. Four more plants were added to the site by 1985 to form the Pickering B station. AECL was the designer of the nuclear parts of Pickering A. In its early years the Pickering station was very successful; recently, however, pressure-tube replacement has been necessary somewhat sooner than expected.

Gentilly BLW (Boiling Light Water)

Soon after the PRDPEC recommendations, Hydro-Quebec, the Quebec provincial electricity utility, expressed interest in nuclear power. This, combined with the recommendation of PRDPEC for a strong follow-up on light-water cooling, led in 1965 to an agreement for the construction of a heavy-water-moderated, BLW-cooled power reactor at Gentilly, Quebec. The quality (i.e., the weight percent of steam) of the outlet coolant was 16.5 percent. The channels were vertical, to permit going later to higher-exit quality. The plant reached its design output of 250 MWe in May 1972. The combination of BLW and natural uranium imposed heavy demands on the control system, adding to other operational problems. Although controllable at 250 MWe, the system did not really lend itself to larger sizes. As the 250 MWe plant was not economically competitive with the cost of power in Quebec, the plant was taken out of service in June 1977. Chapter sixteen describes the CANDU-BLW in more detail.

RAPP (India) and KANUPP (Pakistan)

(See International Aspects)

Bruce

By the late sixties Ontario Hydro foresaw the need for more power, and began to plan for a four-plant station in the same area as Douglas Point. The new station was an Ontario Hydro project, with much of the nuclear engineering and development being done by AECL. The plants were each of 750 MWe capacity, with extra heat output to supply steam to a heavy-water plant. The station, called Bruce A, came on line from 1976 to 1979. Another station, Bruce B, with four plants, each of which was of 860 MWe capacity by 1991, was completed from 1985 to 1987. Initially the operation of both stations was excellent, but problems have developed in Bruce A.

CANDU 6

By the early seventies AECL, having decided that there was a potential market for plants of moderate size, designed a 600 MWe plant that incorporated many of the lessons learned over the years. The result was CANDU 6 (originally CANDU 600), and plants have been installed at:

- Gentilly-2, Quebec — 1982 criticality
- Wolsong-1, Korea — 1982 criticality
- Point Lepreau, New Brunswick — 1983 criticality
- Embalse, Argentina — 1983 criticality

All of these units have operated at high capacity, Point Lepreau being a world leader.

The following figures, from a general proposal for CANDU 6, indicate the huge size of a nuclear power plant; values for specific plants may differ. The calandria is a horizontal stepped cylindrical tank made of stainless steel, 6 metres long and, in the main, 7.6 metres in diameter. It accommodates 380 pressure tubes, each containing 12 fuel bundles. The total uranium content is 84 tonnes (Mg). The calandria contains 240 tonnes of 99.75 percent heavy water as moderator; the moderator system contains a further 23 tonnes. The primary heat transport system contains 200 tonnes. The total fission power is 2,180 MW, from which 685 MWe are generated, 638 being sent out on the grid.

The CANDU nuclear power system is very flexible regarding its fuel. The stations presently in use are in essence fueled with natural uranium. A variety of fueling schemes has been considered, including the use of plutonium, enriched uranium and thorium. Chapter eighteen describes the relevant fuel cycles.

Research Reactors

ZEEP

Although ZEEP was too small to be useful for all measurements on fuel assemblies for power reactors, because of their large sizes and spacings, it continued in use for many years as a facility for measuring specific parameters. For example, in the early sixties a mockup of a power-reactor fuel channel was installed in ZEEP and measurements were made with it to determine the neutron energy spectrum (the energy distribution of neutrons) in various regions of the fuel assembly. In 1980 ZEEP was permanently decommissioned.

NRX

The NRX reactor continued to be a major facility for research and isotope production. With the confidence gained from the 1952-54 restoration, a spare calandria was kept on hand and modified from time to time to provide better facilities. It was installed in 1970. Natural-uranium metal, the only possible choice for the original NRX fuel, was also suitable for chemical extraction of plutonium. When Canadian interest in plutonium waned, the spent NRX rods were sold to the United States as part of the agreement for sale of NRU rods. In anticipation of the expiry of the agreement, studies were made of several less costly fueling schemes for both NRX and NRU. In these reactors the property of interest is the flux, not the power; consequently, enriched fuel proved to be preferable. In NRX an increase of about 50 percent in flux, with no change in power, was obtained with a fuel loading of natural uranium dioxide and rods of an alloy of enriched uranium and aluminum. Both types of fuel were sheathed in aluminum. The change was completed over an interval of six months in 1962, the flux being kept constant by reducing the power until the new fuel was in place; the power was then raised to the old value to increase the flux. This procedure eased the transition for experiments such as loop irradiations. Later the fuel was converted entirely to enriched alloy.

PTR (Pool Test Reactor)

The primary purpose of PTR (described later in this chapter) was to enable measurements of reactivity to be made with irradiated fuel, an intensely radioactive material. The information provided data for accurate predictions of the lifetime of fuel in power reactors - in other words, the amount of energy that could be extracted from the fuel before the destruction of fissile material and the accumulation of fission products lowered the fuel's contribution to an unacceptable level. Measurements continued over the years, some as parts of international programs.

NRU

Like NRX, the NRU reactor was initially fueled with natural-uranium metal. In 1963, with the expiry of the plutonium sales contract with the United States, the fueling was changed to rods of an alloy of highly-enriched uranium and aluminum. There was a reduction in power to maintain the same flux, which was of a magnitude appropriate for tests of power-reactor fuel. The needs of these tests outweighed the attractiveness of higher flux for other research, but in small regions of the reactor, provision could be made for special flux requirements. There was a capability for increased radioisotope production.

Corrosion and other deterioration of various components of the NRU calandria and its attachments resulted in a decision to change the calandria, which was done in 1973-74. The opportunity was taken to improve the in-core facilities for loops.

NRU is a major source of radioisotopes, including cobalt-60 of high specific activity and molybdenum-99, which decays to technetium-99m, a radioisotope in great demand for medical diagnoses.

ZED-2

ZED-2, a larger and more elaborate version of ZEEP, was built for measurements on power-reactor lattices. It has been an essential tool in the expansion and refinement of the power program, permitting diverse measurements on a variety of lattices.

Organic Test Reactor (OTR) WR-1

The PRDPEC recommendations on organic cooling meant that the reactor approved for Whiteshell was in the main stream of AECL development. This ensured its construction, which began in 1963; reactor startup was in 1965. Developmental tests of organic liquid discovered small changes that greatly improved its performance, such as the removal of chlorine, and the addition of a trace of water.

By 1970, experience with WR-1 led to chemical conditions of the coolant that gave excellent results; in particular, there was hardly any transport of radioactive materials by the coolant. However, by that time large commitments had been made to heavy-water coolant for the Ontario Hydro system and money for the development of an organic-cooled power plant was not forthcoming. With hindsight, it might have been better to have followed the organic route for the Gentilly-1 reactor, rather than the light-water route. The experience with organic coolant in WR-1 and the concurrent development of a CANDU-organic cooled reactor design concept are described in chapter sixteen.

Taiwan Research Reactor (TRR)

AECL sold an updated version of NRX to Taiwan late in the sixties. A diplomatic break with Taiwan occurred before the construction was complete, but it did not interfere with the project.

Small Reactors

During the very early days of nuclear energy there was uninformed talk about such things as nuclear-powered cars. What did appear was a number of very large fixed installations for plutonium production and, later, for electric power generation. There were also reactors such as ZEEP and NRX, and as time went on, research reactors with enriched fuel operating in "swimming pools". Small power reactors were built in the United States and the USSR, especially for military uses such as submarine propulsion.

In the mid-sixties, AECL began to consider reactors to produce up to 30 MW of heat for housing in remote areas, or for industrial use. The designers considered them safe enough for unattended operation. Although no commercial reactors were produced, the idea of safe, unattended operation was followed up, as explained in the next paragraph. A 2 MW demonstration reactor was built at Whiteshell in the mid-eighties, but in the absence of a promising market, the project was not developed further.

SLOWPOKE

A low-power research reactor capable of safe, unattended operation was developed and given the name SLOWPOKE (Safe Low Power Critical Experiment). A prototype SLOWPOKE, after tests at Chalk River, was installed at the University of Toronto, where for many years it or its replacement, a production model, provided an adequate neutron flux for trace-element analysis, tracer experiments and radioisotope research. SLOWPOKEs have also been installed in several other Canadian institutions, and one has been installed in Jamaica.

MAPLE

After extensive consideration of Chalk River needs, and an assessment of research reactors elsewhere, particularly in the United States, AECL designed a research reactor called MAPLE (Multipurpose Applied Physics Lattice Experiment), which would use enriched fuel. Chapter twenty describes AECL's studies on small reactors, and the realization of the SLOWPOKE and MAPLE reactor concepts.

Heavy Water

Until about 1970, the Canadian project was entirely dependent on the United States for heavy water. Because of time constraints, the Americans had not chosen heavy water as a moderator for their wartime production reactors, but they maintained an interest in it for many years, and had chosen it for production reactors committed in 1950. The heavy-water extraction plants built to supply those reactors provided the heavy water for

NRU, NPD, Douglas Point and the first unit at Pickering. By 1964 a need was felt for a Canadian supply. The political, financial and technical maneuverings were complicated (Bothwell pp. 320-331). In outline, the construction of a plant at Glace Bay, Nova Scotia, to produce 200 tonnes of heavy water per year using sea water for cooling, was a joint undertaking of the provincial government and a private company. This was a technical and financial disaster; by 1976, the plant was brought into production under AECL management following a major reconstruction and conversion to fresh water. A second plant, established by Canadian General Electric in Port Hawkesbury, Nova Scotia, was more successful and first produced heavy water by 1970. The failure of the Glace Bay plant put the startup of subsequent units at Pickering in jeopardy. AECL had to scrounge heavy water from all over the world as well as borrow the heavy water from ZEEP, NRU, Douglas Point and Gentilly-1.

Because Ontario Hydro felt the need for a heavy-water plant under closer control, AECL built a plant on the Bruce site, which began operation in 1972. By previous agreement the plant was sold to Ontario Hydro in 1973.

As AECL became involved in heavy-water production, laboratory staff undertook research that solved many technical problems arising in the plants. They developed an alternative process of separation to the pilot-plant stage for possible future use. Laboratory staff also developed several instruments for measuring the deuterium content of water. Chapter seventeen gives further details on heavy-water production.

Research and Development, 1952-1985

In the early years of the project, when there was considerable ignorance of the properties of new materials and devices, progress was made possible by large design allowances. Good fortune, combined with those allowances, usually resulted in success. A noteworthy example is the design of NRX for 10 MW operation and its subsequent operation at 40 MW. With time, and particularly with the move towards commercial power plants, came pressure to reduce some of the large allowances by emphasis on research to improve design data. Also, it was essential to understand the exact behaviour of materials, both customary and new, in the novel environments of reactors. Applied research in chemistry and materials science became vital, as did basic research into the underlying phenomena; engineering development was a major user of the results.

In all fields, technology was moving from an empirical base to a scientific base as the knowledge and application of chemistry, physics and materials science increased. In the nuclear field much of the technology was new, in many cases because special requirements brought in previously little known materials and novel environments; decisions putting at risk millions of dollars sometimes had to be based on rather immature technical information. Intense, directed research was required. Underlying research was still necessary to provide a background of knowledge for the applied research. Clarification of terminology used to classify research is given at the beginning of chapter six.

This section provides some examples of research done by AECL from 1952 to 1985; a few other examples appear in the preceding text, notably on heavy-water extraction, organic coolant and safety. Subsequent chapters discuss the examples in greater detail.

Fission

The program of fission studies that was started soon after NRX began operating continued for many years with increasing sophistication of techniques and instrumentation (see chapter eight). This was mainly basic research, because the needs of the reactor program for such data had been largely satisfied. Time-of-flight measurements gave more accurate values of masses and energies. Fission caused by accelerated deuterons provided further information about the fission process.

Radiation chemistry

Ionizing radiation in a reactor or from radionuclides consists mainly of gamma rays, fast electrons or neutrons; a result of the ionization is the production of chemical changes in irradiated material. An important example that required study very early in the project is decomposition of the heavy-water moderator into deuterium and oxygen gases. Studies of such chemical changes constitute radiation chemistry, a subject that has been a significant part of chemistry research at AECL (see chapter seven). When organic cooling was adopted, the effects of radiation on the coolant were carefully researched. The processes that result from the ionization in a material are complex and their interpretation has required development of considerable theory.

Radiotracers

Radioisotopes of an element provide a means of locating the element and following it through a series of movements or reactions. Many research problems have been solved by the use of such radiotracers. An example of practical importance was the unravelling of the mechanism of oxide growth on zirconium alloys. Important uses of SLOWPOKE reactors include radionuclide production and irradiation of specimens with neutrons. In the latter case, the resulting radioactivities serve to identify chemical elements in the specimens.

Channeling

During a determination of the ranges of accelerated ions in metals, an effect was observed that was later interpreted as relatively unimpeded travel of suitably directed ions along the spaces between rows of atoms in the crystals of the metal. This channeling, as it is called, is a phenomenon that has had many applications in basic research and in such applied research as development of semiconductor electronic devices.

Gamma rays of neutron capture

The neutron-capture gamma-ray measurements developed into a program of worldwide interest. The gamma-ray energies became a basis for international tabulations and provided important data for theories of nuclear structure, as they represent energy differences between energy levels of the nucleus that results from the capture. They are characteristic of that nucleus and can be used as an analytical tool for its identification (see chapter eight).

Physics with accelerators

AECL's major program of basic research has been the study of nuclear structure by the use of accelerated ions. After some interesting results were obtained using relatively low-voltage accelerators, an impetus was given to the program by the coming into operation, about 1952, of the three-million volt Van de Graaff accelerator, planned in the Montreal Laboratory. This was followed in 1958 by the first tandem accelerator, a great advance in accelerator design that gave not only higher energy but also a much wider choice of the atomic number of the accelerated ion. This tandem was replaced by an MP Tandem, which, after many years, had a superconducting cyclotron coupled to it. This series of machines has enabled Chalk River physicists to spearhead much important research on the structure and behaviour of nuclei (see chapter eight). Development of detectors and auxiliary equipment has been an important factor in the success of the program.

Life Sciences

Research into the life sciences has been directed towards furthering knowledge on the biological effects of radiation, the behaviour of radionuclides in the environment, the use of radionuclides for biological research, and developing instruments for radiation protection. Mathematical models have been developed to describe the metabolic behaviour of radionuclides and the resulting doses from them to organs and tissues in the human body. The effects of radiation have been studied not only on complete organisms but also at the cellular and molecular levels. In particular, important information has been obtained on the association between variations in radiosensitivity, DNA repair deficiency and cancer proneness in humans; an understanding of the various levels of radiosensitivity in people with cancers is necessary to establish a sound protocol for their treatment by radiation. Chapter three discusses this subject in greater detail.

Neutron diffraction and scattering

The original diffraction spectrometer and later models provided monoenergetic beams of neutrons, used mainly for measurements of the intensity of neutron scattering by substances as a function of angle. The successful design and operation of a spectrometer that incorporated a second crystal that could be rotated about an axis to measure the energies of the scattered neutrons greatly broadened the scope of the research. It enabled the characteristics of vibrations of crystals to be determined and it helped to resolve a long-standing problem of anomalous behaviour of liquid helium at low temperature. This triple axis spectrometer and achievements with it brought several honours, including a Nobel Prize, to the senior researcher (see chapter eight).

Reactor Physics

Studies of uranium/heavy-water lattices begun in the Montreal Laboratory were continued in ZEEP for some general cases, but specifically for NRU and NPD. To obtain the required precision with lattices containing multi-element fuel assemblies for power reactors, a larger version of ZEEP was installed at Chalk River and operating by 1960. It was called ZED-2, because Z-2 would have been Americanized in pronunciation to ZEE-2. ZEEP and ZED-2 were limited to studies of unirradiated fuel.

A major research program on reactor physics to find the effect of irradiation on reactivity was essential to enable estimates of fueling costs to be made. Early measurements with NRX have been mentioned, as has the extensive program of detailed measurements on samples from a long-irradiated NRX rod. To make more precise and varied reactivity measurements, a small, low-power reactor designed for such measurements was installed at Chalk River and was operating by late 1957. Called the Pool Test Reactor, PTR, it is fueled with uranium enriched in uranium-235 and is moderated by ordinary water contained in a deep "swimming pool". Irradiated samples of fuel are shuttled into and out of the reactor core on a fixed-time cycle. The resulting oscillations of reactor power, when compared with corresponding signals from unirradiated material, yield information about the effects of irradiation on reactivity. Chapter eleven describes the evolution of the physics of the CANDU reactor core.

The program on irradiated material involved much chemistry and mass spectrometry for determining the chemical and isotopic changes in the fuel. Theoretical formulas for the lattice properties were matched to the results of this extensive experimental program as they became available, the matching being done by adjustment of parameters in the formulas. With the advent of computers, the formulas formed the basis for elaborate computer programs that were used in the CANDU engineering design and cost estimates. The quantity of primary interest for estimating the fueling cost is the energy derivable from the fuel before the fuel must be removed because of loss of reactivity.[15]

Spallation; Electro-nuclear breeding

When fuel is irradiated in a reactor the accumulating fission products absorb neutrons, which are thus lost from the chain reaction; moreover, the relative production of neutrons decreases, because the loss of uranium-235 is not fully compensated by the production of plutonium-239. These effects necessitate the replacement of irradiated fuel by fresh fuel. Possible means of delaying replacement are to supply the reactor with plutonium produced elsewhere from uranium-238 by a source of neutrons, i.e., electro-nuclear breeding, or to feed neutrons into the reactor. It was for purposes of this kind that spallation was considered as a means of producing neutrons. In the fifties, little was known about the yield of neutrons in spallation by high-energy protons. Canada had no accelerator capable of accelerating protons to the desired energies (many hundreds of millions of volts), but there are protons of the requisite energies in cosmic rays. An experiment to measure neutron yields from spallation of heavy elements by cosmic-ray protons involved measurements at high-altitude sites; although difficult, it was successfully completed. More accurate measurements were later made at the Brookhaven Cosmotron in cooperation with a group from the Oak Ridge Laboratory (see chapter nine).

ING (Intense Neutron Generator)

The NRX and NRU reactors have been excellent research and engineering test reactors. In the early days their large neutron

---

[15] On a blackboard in his office Lewis kept a chart of this quantity, updating it as each new result from the program introduced a change. The chart was known as "Dr. Lewis's fever chart".

fluxes had given Chalk River physics research an edge over the rest of the world. By 1960 these fluxes were no longer exceptional, and in other countries facilities with larger fluxes were in operation. From an engineering viewpoint, the NRU fluxes were nearly ideal for testing power-reactor fuels; much higher fluxes would have been an inconvenience. But for some research higher fluxes would be advantageous or even essential.

For Chalk River to regain a position as one of the institutions with the highest neutron fluxes, a major new facility was needed. Careful study led to a proposal to build a high-current, high-energy proton accelerator to produce neutrons by spallation. This proposal for an intense neutron generator (ING) was an outgrowth of the idea behind the cosmic-ray experiment of the fifties. Such a facility, if located at Chalk River, would restore the laboratory's favourable position as regards neutron flux. It would also be a major step towards the development of electro-nuclear breeding. Initially, two accelerator types were considered: a linear accelerator and a separated orbit cyclotron. The former was chosen for development.

Characteristics proposed for ING were in several respects beyond the technical capabilities of the day. Development programs were started on ion sources to yield a proton current of a few tenths of an ampere, on efficient high-power radio-frequency oscillators, on accelerator structures, and on liquid-metal- cooled targets able to cope with the approximately 100 MW of power in the proton beam. As a result of opposition in the scientific and engineering communities and a failure to win government support, the ING project was terminated in 1968. Sufficient progress had been made to justify continued accelerator development, especially of medical and industrial irradiation devices. The ING project and the subsequent accelerators are described in chapter nine, as is AECL's involvement in a Canadian fusion program.

Chemical and metallurgical technology

In the early days, the extraction plants for plutonium and uranium-233 presented many problems in chemical engineering. Later, the analysis of irradiated CANDU fuel posed related problems.

Getting the chemistry of the reactor-cooling water to be compatible with the various metals in the primary heat-transport system was of great importance, as was a similar requirement for the water in the secondary system. Many considerations were involved in the choice of materials for these systems; steam generators are subject to costly failures if there is incompatibility. Canadian experience with steam generators has been among the best, although recently there have been failures. Chapter fourteen describes research that resolved many of the chemical and materials problems of the heat-transport system.

Thermalhydraulics

The performance of a power reactor depends critically on the transfer of heat from the fuel sheaths to the coolant. High rates of heat transfer mean a lower cost of power, but too high a rate can result in fuel damage. The difficult choice of the design rate for the multi-rod CANDU fuel bundle is based on empirical relationships derived from elaborate and costly experiments. These include tests on instrumented models of fuel bundles with electrically heated sheaths and heat removal by circulated hot water. The quality of the empirical data available has been greatly improved by continuous research in this field of thermalhydraulics during some forty years.

Another aspect of thermalhydraulics deals with the behaviour of the coolant under accident conditions. Chapter fifteen describes research on thermalhydraulics.

Loops

The in-reactor loops at Chalk River and Whiteshell, together with associated equipment such as shielded cells equipped for remote handling and examination, constitute equipment of great importance to research and development of fuel for nuclear power. Beginning in 1951, with the installation in NRX of an American loop, the number and sophistication of loops increased progressively. At times there were seven loops in NRX, four in NRU and four in WR-1. The situation was not static; in some cases, a loop would be installed, used for a series of tests, and dismantled to make room for a different loop. The staff acquired considerable expertise in choosing, planning and executing experiments. Loops and spaces for loops were rented to organizations from other countries, usually with the stipulation that the results be available to AECL. Chapter two describes loops in greater detail.

Fuel

The choice of uranium compounds suitable for CANDU fuel was limited and soon settled mainly on uranium dioxide, but uranium carbide was used successfully in the organic-cooled WR-1. Much research had to be done to achieve consistently good behaviour over long exposure to neutrons.

Fission products include the noble gases xenon and krypton; their release from the uranium dioxide, although only partial, requires the provision of free volume. Free volume is also needed to accommodate thermal expansion. Temperatures inside the fuel pellets are high enough to promote movement of material, particularly gases. Impurities, especially water, can have a profound effect on the interaction of the fuel with its zirconium-alloy sheath. The study of these and other effects was the purpose of many loop irradiations. Chapter thirteen describes the complex behaviour of uranium dioxide in a power reactor and the problems associated with sheathing it.

Pressure tubes and sheaths

For the CANDU reactors with their water-cooled, natural-uranium dioxide fuel, the only metals suitable for pressure tubes and fuel sheaths are zirconium alloys. Two classes of alloy have served the Canadian program: zircaloys, developed in the United States, and zirconium-niobium, pioneered by the USSR. As with the fuel material, much research was needed to find the optimum materials and conditions. Hydrogen dissolves in zirconium and moves down a temperature gradient, concentrating in the coolest region. With increasing concentration, precipitation of zirconium hydride causes localized swelling and can have deleterious effects on the mechanical behaviour. The sheaths and pressure tubes may pick up deuterium, an isotope of hydrogen, from the coolant; the sheaths may also get hydrogen from any water in the fuel. Much information is required on these phenomena, and on the effects of metallurgical condition on hydrogen pickup. Another matter for important studies is the effect of irradiation on pressure-tube deformation. Many of these studies were based on loop experiments. There were, of course, other studies; for example, pressure tubes were burst deliberately to check the likelihood that bursting of one pressure tube would propagate to others. Longevity of fuel channels is of supreme importance in the economics of CANDUs. Chapter twelve discusses the underlying technology of fuel channels.

Advanced Fuel Cycles

Although CANDU reactors have traditionally been fueled with natural uranium, the fueling system is very flexible and lends itself to other fuels: for example, the plutonium that is a by-product of power production could be chemically separated and fed back into the reactors as oxide mixed with the uranium oxide; alternatively, slightly-enriched uranium could be used; irradiation of thorium in reactors with enriched uranium or plutonium-containing fuel would provide uranium-233 for development of a thorium-based fuel cycle. Schemes like these have been considered during much of the Canadian power program.

The primary incentive was to compare the resulting costs of power with costs from the traditional CANDU fuel. More recently, there has been a marketing incentive because certain characteristics of these advanced fuel cycles, particularly the possibility of reduced uranium consumption, appeal to some potential foreign customers. As chapter eighteen makes clear, AECL has maintained an extensive program on advanced fuel cycles.

Waste Management

The Perch Lake basin at CRL provides a suitable area for the management of local wastes; it has also accommodated waste from various nuclear activities outside AECL. It has been the site of many researches into the behaviour and management of wastes. When the Whiteshell laboratory was established, it had similar needs for waste management, but further research was required because of different soil characteristics.

A waste-management problem of much greater magnitude has arisen because the nuclear power program is producing quantities of highly radioactive used fuel and waste of varying levels of radioactivity. The treatment must take into consideration radioactive lifetimes of many thousands of years. There is public concern that significant amounts of radioactive material might enter the environment between now and the next hundred-thousand years or so.

In the mid-seventies, AECL began a program to prove that the wastes could be managed without harm to the public over such long times. The program has continued and has grown to include much research into waste repositories and the movement of water in soil and massive rock structures. An underground research laboratory has been established near Whiteshell. Chapters three and nineteen describe the waste management research and development. Chapter nineteen includes some of the political aspects of the long-term storage of nuclear-power-plant wastes.

Conclusion

By 1985, the closing date for this book, AECL had accomplished much of its primary mission. Twenty CANDU reactors were in service in Canada and elsewhere – Argentina, India, Korea and Pakistan. Several of these had appeared repeatedly among the first ten of the world's power reactors listed in the order of capacity factor, a measure of production reliability. Several more CANDUs were under construction.

In spite of this success, the mission to develop nuclear power plants could not be considered complete. Continued development was essential to keep abreast of the competition and to maintain a capability to deal with problems. To relax would allow the technology to atrophy; applied research and development were maintained at a high level.

The radioisotope business was flourishing. A variety of radioisotopes was being marketed, produced by irradiation in reactors (NRX, NRU and several power reactors) and in cyclotrons in British Columbia.

Pure research was advancing. The advent of the superconducting cyclotron was anticipated with considerable interest, and work on advanced slow-neutron spectrometers was in hand.

From the early days, industrial application of the work of AECL was considered important. This applied not only to the two main items, nuclear power plants and radioisotopes, but also to many aspects of the technology. A specific program for finding commercially viable applications was established. Chapter twenty discusses the successes and failures.

The transformation of AECL from a research organization of moderate size to a large organization capable of designing and supplying nuclear power reactors, as well as having a varied capability in radioisotope production, applied research and development, and pure research, has provided many lessons on how, or sometimes on how not, to proceed. Chapter twenty-one discusses some of these lessons.

Epilogue

A decade has passed since the nominal closing date of the book. It seems appropriate to summarize important events that have since occurred in Canadian nuclear affairs.

All four units of the Darlington Nuclear Station that were under construction on the shore of Lake Ontario, about 65 km east of Toronto, are now in operation, with a net capacity of 3,544 MWe; the station is the largest Ontario Hydro station and adds about 30 percent to the nuclear electric capacity of Canada. Some early problems that caused vibration and damage to fuel have been overcome.

An economic recession and political concerns about nuclear energy have delayed the implementation of Ontario Hydro's plan to proceed with the construction of further nuclear stations. An order prohibiting Ontario Hydro from doing work towards future reactors caused important funding to be withdrawn from AECL's forward-looking program. On the other hand, AECL's expertise and special equipment have been in demand for investigations into difficulties that arise in operating reactors; this demand is one justification for continued federal support of AECL's technical capability, support that was strongly advocated by the Ontario Nuclear Safety Review, an Ontario commission on reactor safety that reported in 1988.

In spite of its commercial exploitation of its technology, AECL, as a crown company responsible for nuclear research and development, needs considerable core funding from the federal government. Current government retrenchment severely restricts AECL's potential.

Most of AECL's applied research and development is now done for outside organizations, a major change; in particular, the CANDU Owners Group (COG), an organization well described by its name, joins with AECL in choosing and directing, on a cost-shared basis, applied research and development of mutual interest.

The success of the CANDU 6 reactors, especially in Korea, brought an order from that country for three more. Construction of a five-unit CANDU 6 station in Romania resumed after delays caused by political events there.

To be able to respond to an expected demand for units of a modest size, AECL has designed a 450 MWe unit called CANDU 3. The design includes a high level of standardization but is nevertheless adaptable to individual requirements. This standardization is subsequently being applied to a larger 880 MWe unit called CANDU 9.

The radioisotope business flourished to such an extent that the Radiochemical Company was divided into two components, which have been separately incorporated.

NRX has been permanently shut down. Nordion has proposed the building of one or more reactors for radioisotope production when NRU is shut down. AECL has outlined proposals for an Irradiation Research Facility to replace NRU.

The Tandem Accelerator Superconducting Cyclotron combination provides a wide variety of heavy-ion beams. It attracts researchers in many disciplines from across Canada and around the world.

The neutron diffraction and scattering facilities at NRU are also of international status. A new system, comprising two spectrometers of advanced design, was installed in 1991. The facilities are used for basic and applied research on the solid state, and for commercial applications to non-destructive testing.

The goal of controlled fusion energy has been the aim of some very large programs by the major powers. In the late seventies a national fusion program was started in Canada; it has been managed by AECL for the federal government since 1987. The program had a budget of about twenty-five million dollars in 1991 and, with its focus on a few aspects that fit Canadian expertise, is consequently able to make useful contributions to the subject.

Looking to the future of fission energy, AECL sees the CANDU reactor as supplying a significant fraction of the world's needs for nuclear power plants. The only serious competition is from light-water reactors of various types. To support current plants and to prepare for future plants, AECL has a large research and development program, on, *inter alia*, advanced fuels, safety matters, waste management and heavy-water process development; the large investment in heavy water needed for a CANDU plant makes an important contribution to the cost of the power produced, a cost that is in the range of costs of power from light-water reactors. Lowering the cost of heavy water could be a determining factor in the competition.

BIBLIOGRAPHY

Political, corporate and general scientific and technical aspects of the Canadian program are well covered by

Bothwell, Robert. *Nucleus: The History of Atomic Energy of Canada Limited.*

Toronto: University of Toronto Press, 1988.

Eggleston, Wilfrid. *Canada's Nuclear Story.* Toronto: Clarke, Irwin & Company, 1965.

Relevant activities in other countries

France

Goldschmidt, Bertrand. *Atomic Rivals.* New Brunswick, N.J.: Rutgers University Press, 1990. (Translation of: *Pionniers de l'atome.* Paris: Stock, 1987.)

Weart, Spencer R. *Scientists in Power.* Cambridge: Harvard University Press, 1979.

United Kingdom

Gowing, Margaret H. *Britain and Atomic Energy, 1939-1945.* New York: St. Martin's Press, 1964.

———. *Independence and Deterrence: Britain and Atomic Energy, 1945-1952.* New York: St. Martin's Press, 1974.

U.S.A.

Groves, Leslie R. *Now It Can Be Told: The Story of the Manhattan Project.* New York: Harper & Row, 1962.

Hewlett, Richard G., Oscar E. Anderson, Jr., and Francis Duncan. *A History of the United States Atomic Energy Commission.* 2 vols. Springfield, Va.: National Technical Information Service, 1972. (Originally published by Pennsylvania State University Press, 1962-1969.)

Smyth, Henry DeWolf. *A General Account of the Development of Methods of Using Atomic Energy for Military Purposes Under the Auspices of the United States Government, 1940-1945.* Washington: U.S. Government Printing Office, 1945. (Also known *as The Smyth Report.*)

Early days of the Canadian project

Laurence, G.C. "Canada's Participation in Atomic Energy Development." *Bulletin of the Atomic Scientists* 3 (November 1947): 325-8.

Biographies

Carmichael, H. "George Craig Laurence, 1905-1987." *Transactions of the Royal Society of Canada,* ser. 5, 4 (1989): 381-6.

Fawcett, Ruth. *Nuclear Pursuits: The Scientific Biography of Wilfrid Bennett Lewis.* Montreal: McGill-Queen's University Press, 1994.

Hartcup, G., and T. Allibone. *Cockcroft and the Atom.* Bristol: A. Hilger, 1984.

Herzberg, G. "Chalmers Jack Mackenzie, 1888-1984." *Proceedings of the Royal Society of Canada,* ser. 4, 23 (1985): 137-141.

Legget, R.F. "Chalmers Jack Mackenzie, 10 July 1888-26 February 1984." *Biographical Memoirs of Fellows of the Royal Society* (London) 31 (1985): 411-34.

Lovell, B., and D.G. Hurst. "Wilfrid Bennett Lewis, 24 June 1908-10 January 1987." *Biographical Memoirs of Fellows of the Royal Society* (London) 34 (1988): 453-509.

Mooradian, A. "Wilfrid Bennett Lewis, 1908-1987." *Transactions of the Royal Society of Canada,* 5th ser., 3 (1988): 163-7.

Note: With few exceptions, any scientific and technical document produced by AECL, either for wide distribution as a report or for publication in a journal, is given a number in the AECL Report series. These AECL Reports are cited in this and subsequent chapter bibliographies with AECL Report numbers. Copies are available from the Scientific Documents Distribution Office, Chalk River, Ontario, Canada K0J 1J0, and are indexed by major abstracting services. The International Nuclear Information Service (INIS) makes complete sets available in microfiche. Longstanding recipients of AECL Reports include major national nuclear establishments and principal national depository libraries.

INFRASTRUCTURE AND LABORATORIES

Chapter Two

Organization, Management and Operations

E. CRITOPH

While AECL was not established until 1952, there has been continuity in the Canadian, federally sponsored, R&D effort in nuclear energy since 1942. This began with the formation in 1942 of the Montreal Laboratory and continued at Chalk River, first under the National Research Council (NRC), and then, since 1952, under AECL (see chapter one). Over this whole period, various names have been given to this effort, the most recent being "AECL Research".

This chapter describes the milieu, as seen from its laboratories, within which the R&D programs of AECL Research were carried out. The organization, management style and operations of AECL Research (in its widest sense) were vital and ever-present factors in the integrated AECL technical performance.

BACKGROUND

The initial organization, management style and operations of AECL Research arose from the historical circumstances under which the Canadian effort in nuclear energy was initiated, the nature of the site on which AECL was originally established, and the nature of the AECL mission, which was to ensure that atomic energy ("nuclear energy" in today's terms) was exploited to the maximum collective benefit of Canadians.

A 1946 NRC organization chart (see figure 2.1) identifies, in boxes, the elements that formed the NRC part of the Chalk River effort. Some support services were supplied by other parts of the NRC organization, and construction and operating support was the responsibility of Defence Industries Limited (DIL), which contracted with the Fraser Brace Company for construction.

The total NRC-directed effort at Chalk River in 1946 consisted of 289 employees, divided among the following branches:

FIGURE 2.1
NRC Organization – 1946

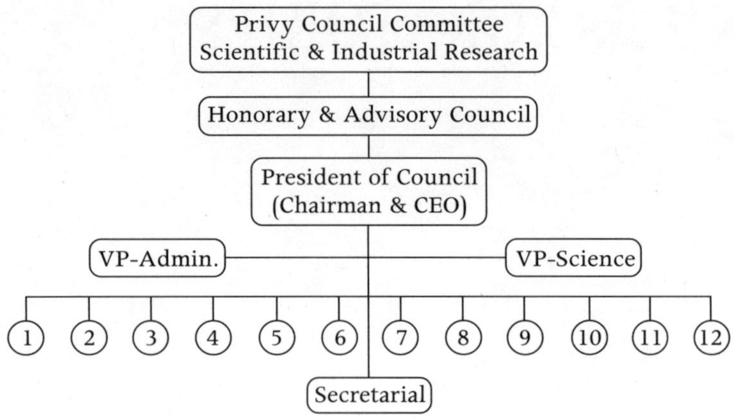

1. Extramural Activities
2. Medical Division
   — Chalk River Branch
3. Biology Labs
4. Prairie Region Labs
5. Chemistry Labs
   — Industrial & Chemical Engineering Branch
6. Mechanical Engineering Labs
7. Electrical Engineering & Radio Labs
8. Physics Labs
9. Atomic Energy Division
10. Patent Corporation
11. Codes & Specifications
12. New & Special Projects

- Engineering (atomic energy division)
- Technical physics (atomic energy division)
- Nuclear physics (atomic energy division)
- Theoretical physics (atomic energy division)
- Chemistry (atomic energy division)
- Research services (atomic energy division)
- Chalk River branch (medical division)
- Industrial & chemical engineering (chemistry labs)

In addition, there were four people in the United States Liaison Office, about 650 workers associated with DIL, and fluctuating numbers of construction workers.

The composition of the NRC-directed staff employed was unusual. For one thing, there was a large contingent of United Kingdom staff, arising from the agreements that led to the formation of the Montreal Laboratory, and there was a high proportion of professional scientists and engineers in a wide range of disciplines. This, in itself, brought new challenges in management and operation. The NRC-directed staff (see figure 2.2) comprised, from Canada and the United Kingdom, R&D professional, R&D technical, and service staff.

Initially, the responsibility for operating the laboratory site fell to DIL. They made an important contribution to the overall program by establishing the good operational practices that provided the basis for the ongoing successful operation of the site. Operational responsibility was transferred to NRC in February 1947. A large part of the DIL effort and personnel was absorbed as part of NRC, at CRL.

One of the significant characteristics of the organization in the forties and fifties was the total involvement of the company in the lives of the staff. Deep River was a company town and hence the company (DIL initially, and then NRC) was landlord to all inhabitants. But the relationship went far beyond this. The company controlled who could live in (and, in the very early days, even who could enter) Deep River and the type of accommodation they could rent. They also took great pains to make life in Deep River as attractive as possible in order to attract and retain top-level staff in spite of the relative isolation. This included the provision of a recreational director and staff, involvement in community affairs through provision of hospital

FIGURE 2.2
CRL – 1946
Distribution of NRC-directed effort

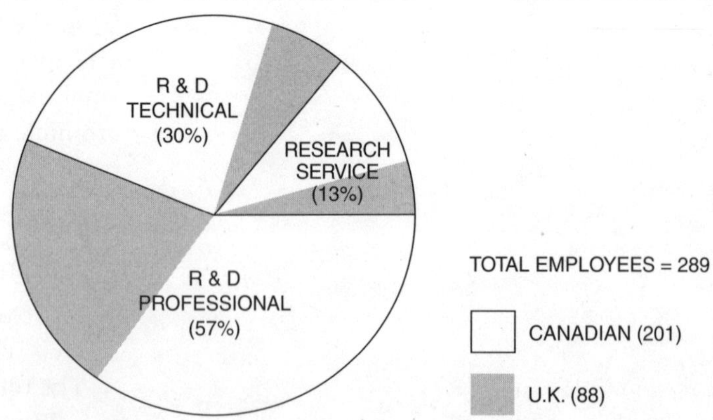

facilities, schools and services, help in planning and constructing recreational facilities, and even a free, weekend bus expedition to Pembroke for shopping. In spite of the rather primitive nature of many of the amenities of life, Deep River was regarded as a desirable place to live by most residents.

ORGANIZATION

By 1952, when AECL was established as a crown corporation, there were some 1,200 employees at the plant (CRL), about sixty buildings of a variety of sizes, and major scientific facilities comprising NRX, ZEEP and a Van de Graaff accelerator. The proposal to build NRU had been accepted and the reactor was under construction. (By the end of the period covered by this book (1985), the number of buildings had increased to about 150 and there were more major scientific facilities.) Initially, this incarnation of AECL Research was AECL in total. Of course, for most of its history, AECL managed a broader effort than just AECL Research; employees were distributed among various units of AECL over the period from 1952 to 1990 (see figure 2.3).

Within a few months of the establishment of AECL, the commercial products division of the Eldorado Mining & Refinery Company was transferred to it; this division formed the basis of the radiochemical unit, which was centred in Ottawa (see figure 2.3). The unit was dedicated to the packaging, marketing and sale of radioisotopes, and the development, manufacture, marketing and sale of equipment for the application of those radioisotopes (most notably medical therapy units and industrial irradiators). The radiochemical unit depended heavily on AECL Research for the production of the radioisotopes at CRL. Initially this was an exclusive dependence, but starting in the seventies other sources were tapped. In 1988, the unit was separated from AECL as two incorporated companies, Nordion International and Theratronics International (see chapter five).

AECL's main thrust immediately after its establishment was the development and exploitation of a Canadian nuclear power-reactor concept. This effort fell under AECL Research until the

FIGURE 2.3
AECL Employees

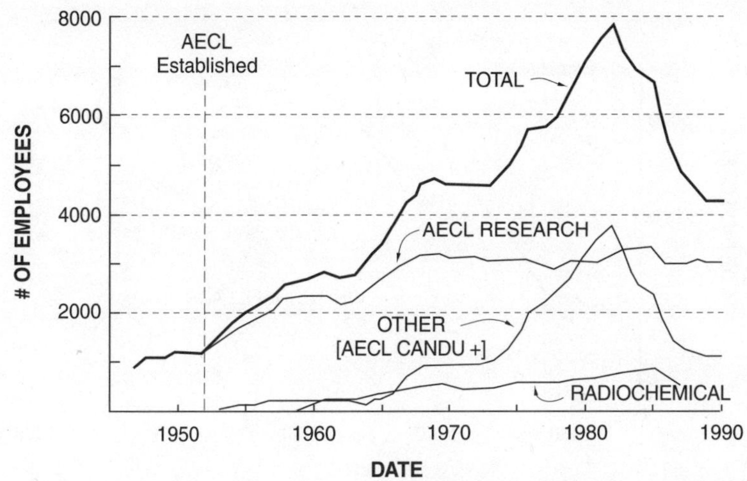

formation, in 1958, of the nuclear power plant division (NPPD) in Toronto. This marked the start of the effort labelled "Other [AECL CANDU +]" in figure 2.3. This is the effort that is within the AECL CANDU unit of AECL at the time of writing, plus the Corporate (Ottawa) staff. At various times in the past, it has included units that were named differently; e.g., NPPD, Heavy Water Projects, International Marketing. This effort grew rapidly in the sixties and seventies, reaching a peak in 1981-82 with work on the AECL 600 MWe power reactors for Quebec, New Brunswick, Argentina and Korea (see chapter ten), in addition to work for Ontario Hydro on its reactors.

The AECL Research effort that in 1952 was all at CRL expanded over time to include the Whiteshell Laboratories (WL) near Lac du Bonnet, Manitoba, and head office in Ottawa. Figure 2.4 shows the distribution of AECL Research over the period 1946-90. As might be expected, the total organization grew rapidly from 1952 to about 1971, when the first commercial CANDU came into operation, and has been relatively constant since. The decision to establish WL was made in 1959 on the grounds that CRL was at an ideal size for an applied laboratory of its type and further expansion would be counter-productive. A company townsite was

FIGURE 2.4
AECL Research Employees

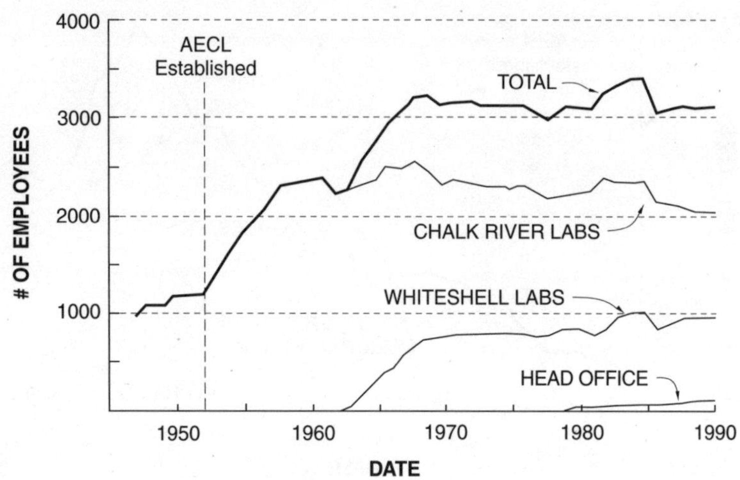

FIGURE 2.5
AECL Research Expenditures ($s of the year)

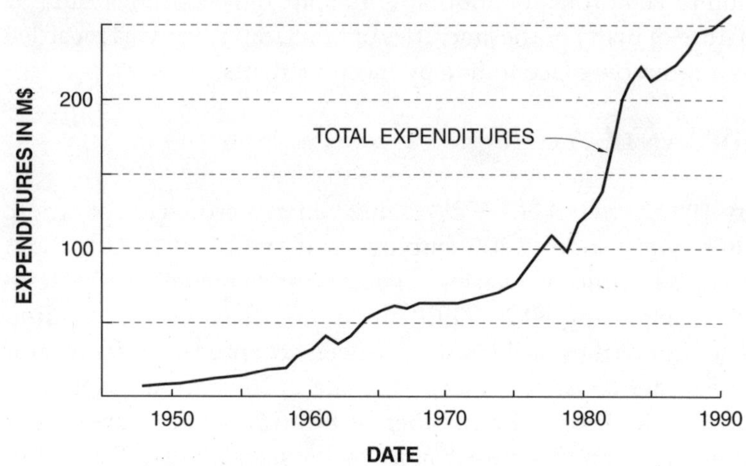

established at Pinawa, analogous to Deep River for CRL and about the same distance from its plant. The staff started to locate at WL in 1964. Staff shown in figure 2.4 as belonging to WL in 1962 and 1963 were actually located at CRL. WL eventually grew to be about half the size of CRL (in terms of employees and budget).

The major programs in which WL has taken the lead role over the years have been in the development and promotion of the organic-cooled CANDU power-reactor concept (CANDU-OCR) (see chapter sixteen), reprocessing for advanced-CANDU fuel cycles (chapter eighteen), CANDU reactor safety R&D (chapter fifteen), and waste management (chapter nineteen). At the same time, it has contributed to the majority of the other AECL R&D programs.

Figures 2.5 and 2.6 plot AECL Research annual expenditures over the years. AECL has a fiscal year running from April 1 to March 31 and the convention used has been to plot, for example, the expenditures for 1980-81 at 1981. Figure 2.5 shows expenditures in dollars of the year and figure 2.6 shows the same expenditures in constant 1985 dollars; the conversion is made by multiplying the expenditure in dollars of the year by the ratio of consumer price index (CPI) for 1985 to that for the year in question. Figure 2.6 is more meaningful in terms of the real value of the money spent, while figure 2.5 gives the actual amounts spent.

The ratios shown in figure 2.7 aid in an understanding of the implications of Figures 2.5 and 2.6. Figure 2.7 displays the following ratios: Capital/Total Expenditures; Parliamentary Appropriations/Total Expenditures.

As might be expected, the ratio of capital/total expenditures was much higher in the early years, when there were large initial investments in needed facilities and equipment. However, it is now some thirty years since those initial investments were made and higher relative capital expenditures will be required in the future if the viability of AECL Research is to be maintained.

Parliamentary appropriations made up some 90 percent of total AECL Research expenditures over the period up to 1980. (Note that in figure 2.7 no values are given for the period from 1967 to 1978, since it is difficult to determine the exact value of the AECL Research share of the parliamentary appropriations to AECL over that period. In every year of this period except

FIGURE 2.6
AECL Research Expenditures (constant 1985 $s)

FIGURE 2.7
AECL Research: Informative Ratios

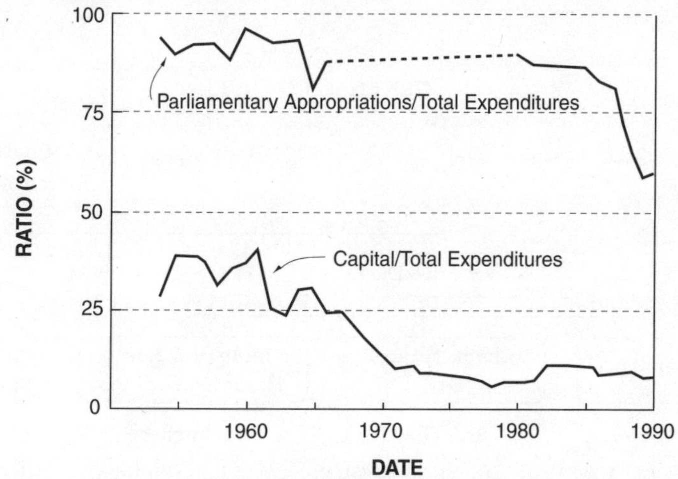

1976-77, when appropriations were 90.5 percent of total expenditures, AECL Research total expenditures were less than the parliamentary appropriation to AECL.) A considerable drop in this ratio occurred in the eighties, particularly after 1985, coincident with a government emphasis on increasing the leverage of government funds by a policy of having those who benefit most contribute to the cost. AECL adopted a more commercial approach to providing know-how and technology. The process was accelerated in 1985 when the federal government, in its budget, projected a 50 percent decrease over five years in its appropriations to AECL. While this decrease was never fully implemented, it initiated important changes in the operation of AECL Research. The biggest of these came from the negotiation of greatly increased support from provincial utilities, most notably Ontario Hydro, and the change in program control that resulted.

Figure 2.8 shows the organization chart just after AECL was established. The basis of the R&D part of the organization was scientific discipline rather than project involvement. Control of budgets was with these organizational units and this system worked well at the time.

The growth of CRL through the fifties and early sixties led to changes in organization, but these were essentially an expansion of the same basic type of organization. By 1959, while the number of employees had doubled from 1952, the number of divisions outside the administration, finance and medical services part of AECL Research had risen from four to seven and the number of associated branches from twenty to thirty. The industrial operations division had been divided into two, creating an operations division and an engineering services division, and two other new divisions had been created: the reactor R&D division and the research engineering division.

The establishment of WL and its growth required that changes in organization be made. In 1967 changes were also made in the internal organization at CRL. The result was an AECL Research organization of the type shown in figure 2.9, which existed in essentially the same form until 1986. Under this organization, CRL and WL were treated as parallel units under the AECL president. CRL, being the larger, had an extra level of management, with the divisions being grouped into research, applied R&D and plant administration and operations, each under a group head. Some other units or functions,

FIGURE 2.8
AECL Organization at CRL in 1952

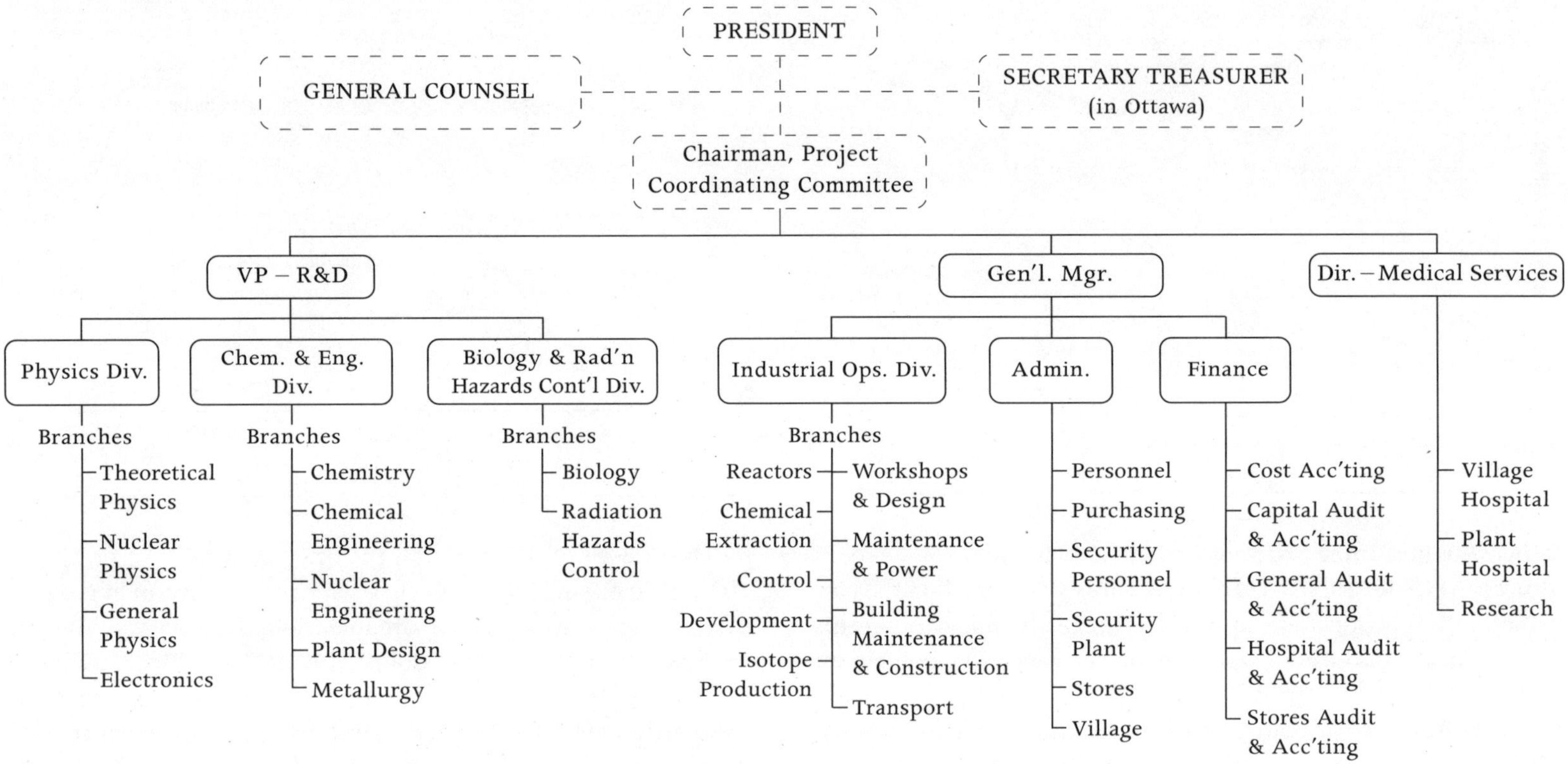

such as finance, quality assurance, and environmental authority, reported directly to the site head. By 1979 the number of employees had increased by fifty percent from 1959 and there were about twenty divisions and eighty branches within the AECL Research organization.

Starting in the seventies, there was an increasing emphasis on accountability and on programs and projects which had their own budgets and management; there was a more careful assignment of support costs to programs and projects. In 1976 an organization consisting of six steering committees – most controlling several working parties – under an R&D program committee was established, to plan and monitor all AECL Research R&D programs. This supplemented the line organization of figure 2.9, particularly in the areas of assigning priorities, monitoring progress and gathering budget information.

Just prior to 1979 the position of executive vice-president, AECL Research, was created, and the site heads reported to this position rather than directly to the AECL president. This was part of a general shift by AECL to a more decentralized structure and made AECL Research somewhat more autonomous in its day-by-day operations, while still being responsible to AECL Corporate for general direction and performance. At the same time, an AECL Research head office was established in Ottawa directly under the executive vice-president, whose role included coordinating functions in finance, planning and business development. Soon after, commercial offices were set up at

FIGURE 2.9
AECL Research Type of Organization 1967 – 1979

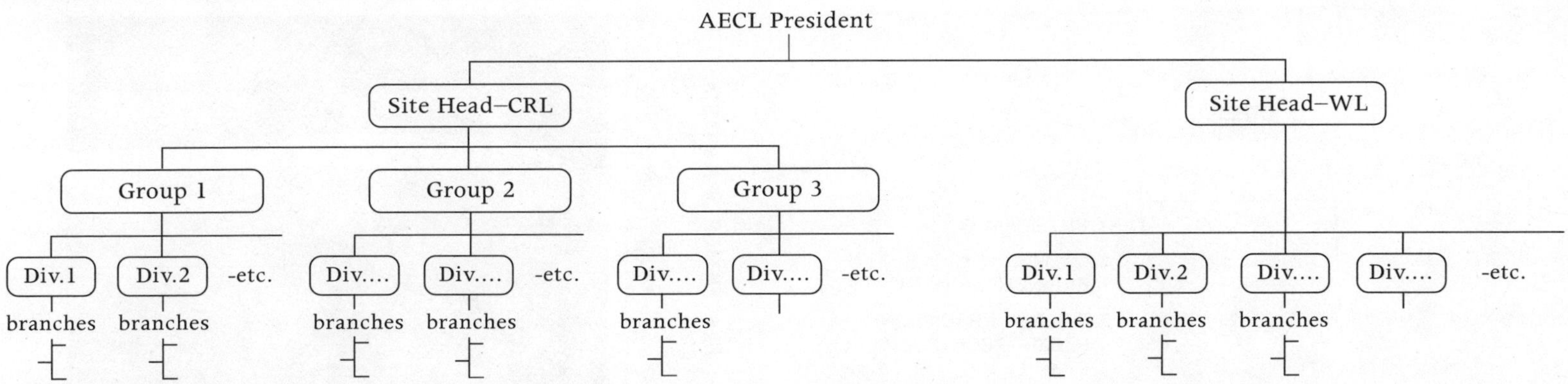

the sites (at CRL the head of this office had the status of a group head.) Otherwise, the organization remained essentially unchanged until 1985.

In 1985, faced with large cuts in federal appropriations over a five-year period, AECL Research took the first large step towards consolidating its site operations. The new organization was defined in 1985 and instituted in early 1986 (see figure 2.10).

Under this organization, although the sites were operated separately, the AECL Research R&D programs were managed across sites. (Site operations were consolidated in July 1990.) In these later years, particularly post-1985, there was an increasing tendency for major projects to control their own budgets, while using staff assigned from a variety of units.

Figure 2.11 shows the relative effort, in terms of personnel, associated with the main units of the organization in 1952; the direct effort associated with R&D was some 25 percent of the total, while that dealing with industrial operations was more than 50 percent. These ratios were influenced by the isolated

FIGURE 2.10
AECL Research Organization 1986

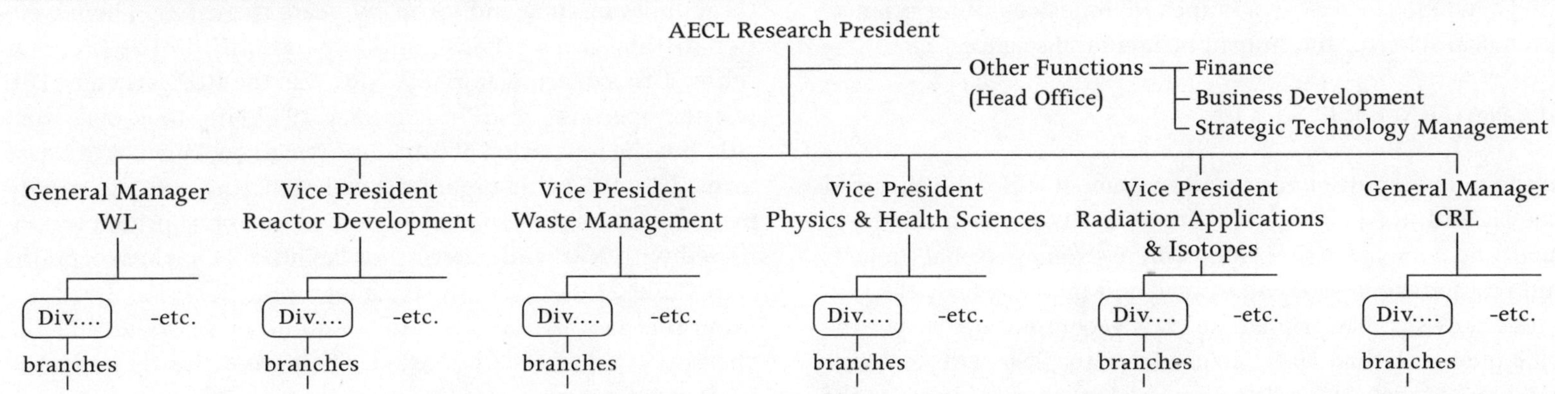

FIGURE 2.11
Distribution of AECL effort at CRL – 1952

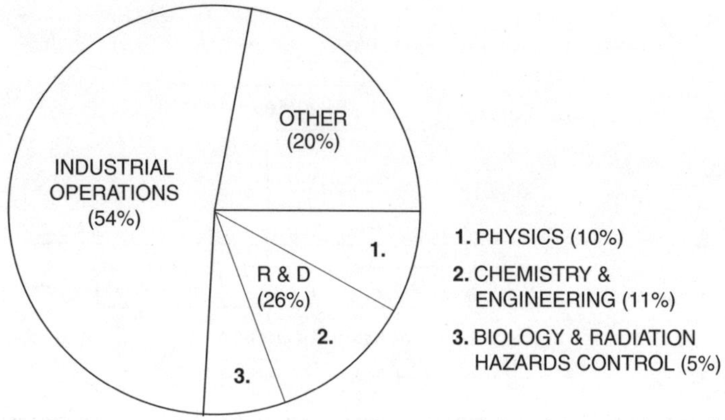

FIGURE 2.12
R&D Staff Disciplines

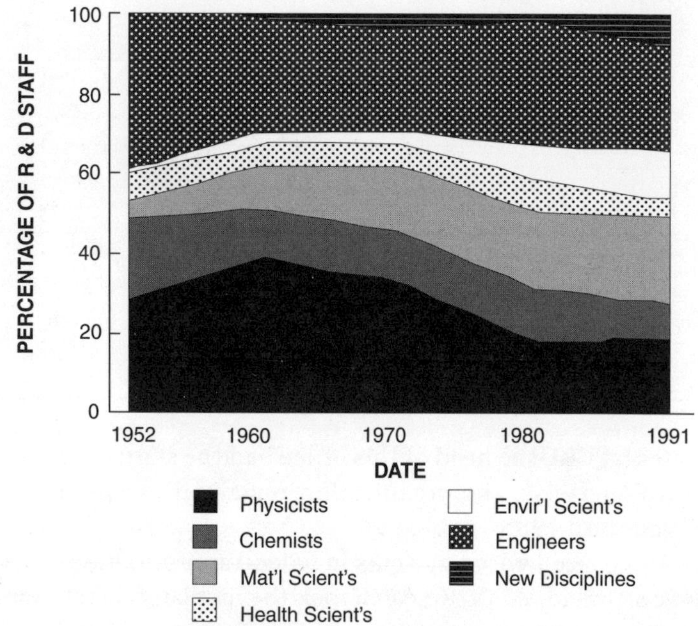

nature of the site and the need to provide services that in less isolated locations could be obtained externally. Nor was the boundary always clear cut between activities classed as R&D and those classed as Operations. The ratio of direct R&D effort to total effort, in terms of personnel, increased as AECL grew, as would be expected. It reached about 35 percent by the late sixties and fluctuated by a few percent around that value until 1985.

The application of this R&D effort in terms of programs and disciplines varied with time as well. In particular, there was a marked shift from basic to applied R&D (see figure 2.12). The category labelled new disciplines includes computer science, geological science, and human behavioural science.

MANAGEMENT STYLE

In the following discussion of management style, the effects of size should not be forgotten. In 1952 AECL was a relatively small and compact organization where strong personal contacts and relationships were widespread and important. By the seventies staff size had tripled and was geographically dispersed. This inevitably had an influence on management style. What had worked well in the fifties was no longer appropriate to the seventies and beyond for this reason alone, and there were several other factors encouraging change.

In establishing the first major nuclear site in Canada at Chalk River, an approach was taken that resulted in an effective and efficient style of management that served the program well through its infancy and for many years thereafter. There were two distinct arms: "R&D" and "operations" in the broadest sense of those terms. Initially, NRC was the R&D arm and DIL was the operating arm. In February 1947 NRC took over from DIL, but the separation of functions remained. When AECL was formed in 1952 this organizational structure was essentially maintained. As the program expanded, the same principle continued with NRU and other major facilities. Notable exceptions were ZEEP, ZED-2, PTR, the metallographic hot cells, the plutonium fuel fabrication laboratory and accelerators within the chemistry and materials division. These were directly under the R&D management.

This division of responsibilities between R&D and operations was a large factor in the success of Chalk River as an effective nuclear laboratory. Each group was responsible for what it was best qualified to do, enabling each to concentrate on its own strengths. This also minimized the risk of conflict of interest in major facilities between the best interests of the experiment and those of the facility, to the long-term benefit of both. There was, of course, always a higher authority to resolve any serious conflicts.

Extensive cooperation and collaboration existed between operations staff and R&D staff, initially on a largely informal basis. There was a great deal of mutual respect between the two groups. Operations staff knew that they could count on R&D staff to help in resolving any complex technical problems that arose, while R&D staff relied on operations staff for aid in meeting unusual research-facility requirements. Projects often were carried out with participants from both the R&D and operations arms and from many subgroups within each. The participants were accountable to the management structure within their side of the organization and had to keep them well informed. However, a great deal of latitude was given to permit a close relationship at all levels so that much of the work could be carried out, and many decisions could be made, without going up one side of the organization and down the other. This system worked well in the early days (perhaps because budgets were not as tight as they later became) and relied greatly on management trust and responsible staff behaviour. Such a system led to difficulties in predicting and assessing the cost and manpower requirements of some projects. More formal procedures began to be adopted in the seventies as government emphasis on accountability increased.

As might be expected, there were differences, as well as similarities, in management style within each of the two arms. On the R&D side the style was akin to that in universities. There was a great deal of flexibility in management of the effort; e.g., work and leave schedules, job assignments, travel, communication. Managers were largely selected from internal candidates on the basis of their technical strengths and achievements, rather than their specific management training or experience. At the lower levels, there were, initially, few routine management chores involved, with administrative assistants being provided for many of these. This led to technical leadership as the main responsibility, and a collegial approach to program content and work assignments. On the operations side the style was more similar to that adopted in industry. While positions were largely filled internally, based on previous experience and performance, there was less flexibility, presumably due to the more clearly defined and shorter-term nature of their duties and objectives. Also, there was more emphasis on the actual management aspects of the managers' jobs. The differences in management style led, on occasion, to some friction, although this never became serious. On both sides of the organization, more emphasis was placed on management, financial and administrative skills and responsibilities as time went on, leaving managers less time for purely technical leadership.

The decisions on whether to use in-house or outside resources for the services required by AECL Research must have seemed fairly straightforward in the early days, given the isolation of the site and the sensitive nature of the work being done. Thus, with the exception of construction services, the decisions were made in favour of "in-house" for the majority of required services. While AECL Research profited from the dedication and commitment of the staff providing the services, there is little doubt that if the same decisions were being made today, a number of them would be reversed on the basis of better communications and transportation, growth in external availability of services, and changed attitudes.

While relatively isolated geographically, the laboratories of AECL Research have never been isolated technically. With management support this has been achieved through:

- close relationships with universities, including: scientific collaboration, shared use of both AECL and university facilities, contracts with universities for work in fields where they have relevant expertise, employment of summer students and visiting professors, authorization of AECL staff as adjunct professors and participation in projects of mutual interest,
- extensive participation in international committees and activities,

- bilateral technical exchange agreements with many countries,
- two-way technology transfer with Canadian industry,
- training programs and courses for external groups,
- commercial contract work for industry, both Canadian and international,
- attached staff from foreign laboratories,
- attached staff from Canadian industries and utilities, and
- close collaboration with utilities and, through them and AECL CANDU, with manufacturers.

As a crown corporation, AECL is an instrument of federal-government policies and is committed to following and promoting those policies.

OPERATIONS

General

In general, the industrial operations group at CRL and later the engineering services division at WL were responsible for:

- reliable maintenance and operation of the major facilities needed by the R&D group to achieve their mission,
- provision and maintenance of the buildings on the site, using either in-house or off-site resources,
- provision of some of the conventional services needed by the R&D group to achieve their mission, using either in-house or off-site resources, and
- major production activities, such as chemical extraction and isotope production.

There was also an additional implicit responsibility: that of maintaining a competent staff in those areas required to fulfil the above responsibilities. Since some of the areas were new, and external training unavailable, internal training programs were necessary. An apprentice program was established for the trades that not only served AECL well but also provided significant spin-off to industry in general and the nuclear power program in particular. Technical staff training programs were also set up and, in some cases, expanded to serve the needs of Canadian industry and utilities, as well as international customers and associates (e.g., India and Taiwan). For a few years in the early sixties, a reactor school for international mature students was operated at CRL.

Many of the activities involved in fulfilling these responsibilities were highly technical. However, other important, but less technical, support activities were provided by the operations group as well as other divisions in the laboratory organizations (see table 2.1). Some of these have been shaped by the nuclear mission of AECL Research:

- Of extreme importance to any R&D organization is easy access to a library that is comprehensive in areas relevant to the organization. Libraries established at CRL and WL by AECL Research have become the Canadian repository for nuclear-energy-related material and provide international access to this data. As well as standard library services, special

TABLE 2.1
Support activities – not highly technical

SHAPED BY NUCLEAR MISSION
Library & Technical Information Services
Public Affairs
Medical (Hospital)
Protective Services
Decontamination & Laundry
SPECIAL COMMENT
Workshops
Quality Assurance
Townsite Administration
Transportation
OTHER
Reproduction & Photography
Human Resources Development
Employee Relations
Personnel Services
Compensation
Office Services
Industrial Safety
Finance
Purchasing
Stores

technical information services are provided to aid researchers in identifying and locating references.

- Public affairs specialists at the laboratories ensure that issues unique to the nuclear industry are handled properly and that staff and the public are kept well informed.
- AECL maintains medical and hospital services at the plants for employees.[1] The services ensure immediately available special expertise and treatment in the event of nuclear-related accidents.
- To satisfy national and international obligations, AECL must provide tight security against theft or dispersal of the potentially dangerous nuclear materials in its possession. Thus, AECL has its own protective services in terms of guards and firefighters.
- In many areas radioactive contamination of equipment and protective clothing is inevitable and once-only use would be prohibitively expensive. Decontamination and laundry services deal effectively and efficiently with the clean-up of equipment and clothing.

Comment is warranted on some of the other support services listed in table 2.1:

- There have been two types of workshops at the sites. The general workshops are the largest and include a planning and estimating function. They are under the operations arm and consist mainly of machine shops handling a wide range of jobs, from making experimental devices for researchers to manufacturing special parts for construction and maintenance; they include a small paint shop for site signage and some carpentry. In addition there have been some small machine shops within the R&D arm doing either more highly specialized work on a dedicated basis for the R&D staff, or providing a very fast service on the basis of rough sketches and personal discussions.
- Quality Assurance[2] was established formally as a separate unit within AECL Research in 1980, with the initial aim of introducing formal quality assurance programs to the manufacturing function. The concept was later broadened to one of total quality management; i.e., making customer satisfaction the central focus of the company and all its component parts. This concept was embraced by AECL Research in the late eighties with its Continuous Quality Improvement (CQI) initiative.
- Town administration was important in the early days of both sites, when Deep River and Pinawa were company towns. This importance diminished with the towns' evolution towards independence.
- Chalk River Laboratories operates a unique bus service to and from work for employees who live in Deep River, Chalk River, Petawawa and Pembroke. This service at one time involved a fleet of fifty buses. In the early days, the service was heavily subsidized by AECL. The subsidy was gradually removed and, by the late eighties, the service was fully supported by its users. A service involving transport of materials around the sites and to and from external points is also provided.

The other support services listed in table 2.1, while essential, had responsibilities common to those of similar units in many other organizations. Of course, the nuclear focus of the work did create specific differences as well as problems to be resolved.

For most of the period covered by this book, the operation and use of the large research reactors (NRX, NRU and WR-1) provided the main focus for the work of the laboratories and was central to their mission, strength, *modus operandi* and morale. Figure 2.13 indicates the scope of the effort involved: the central box represents the groups responsible for the actual operation of the reactors; the boxes in the lower half of the figure represent the groups whose support was required for the operation, and the boxes in the top half of the figure represent the groups using the reactor.

---

[1] To ensure that medical standards were not compromised for operational efficiency, these services were the responsibility of a senior authority, who was independent of "operations."

[2] Quality Assurance denotes a system of processes designed to ensure that the requirements for a product are well understood by the producer, and that the product meets these requirements. The driving force behind the development of quality assurance as a discipline in Canada came from the aerospace and nuclear power industries, but the discipline has very wide application.

FIGURE 2.13
Operation & Uses of Research Reactors

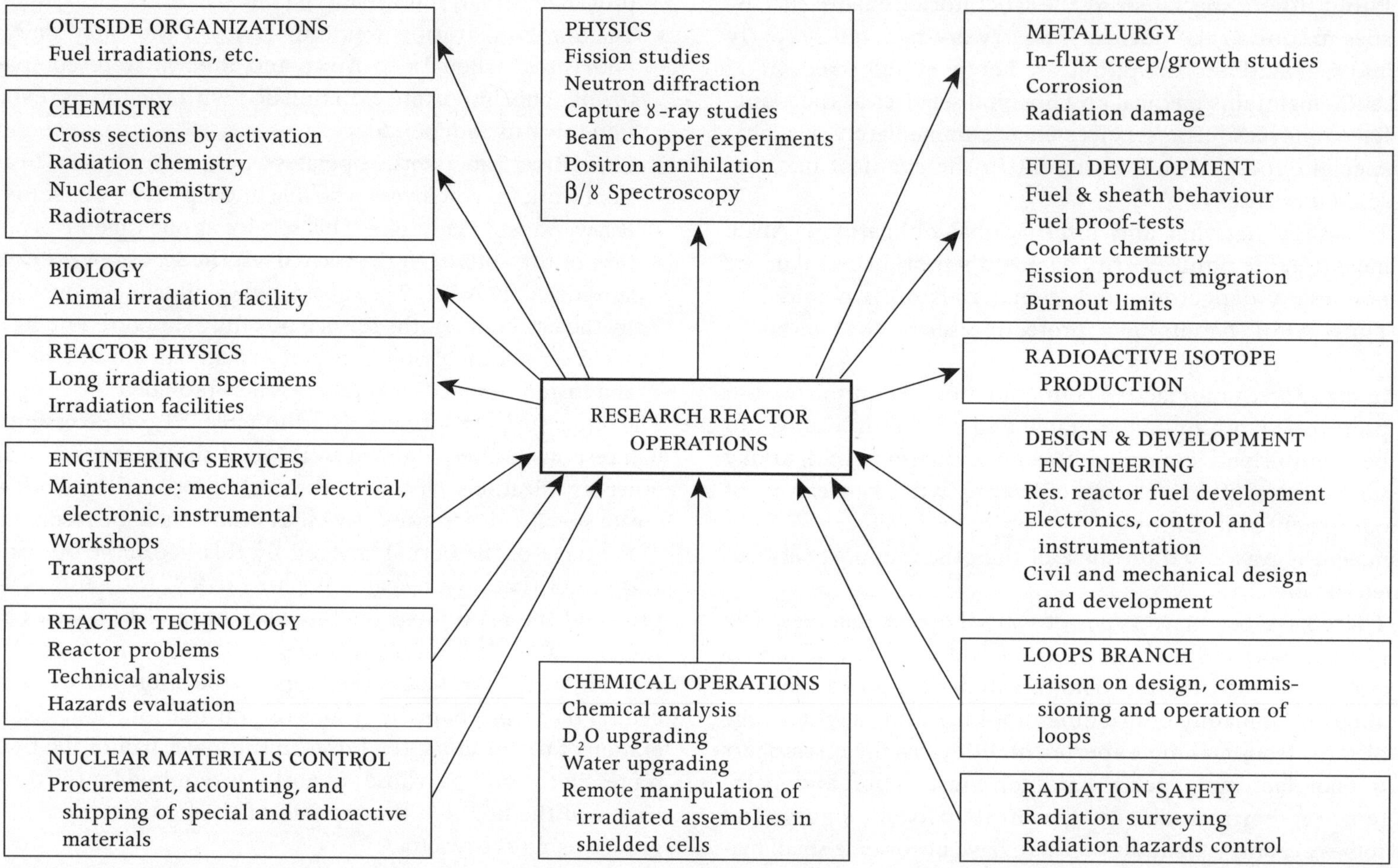

RESEARCH REACTOR OPERATION

Pre-1952 Canadian Research Reactor Experience

The decision to build NRX, for the dual purpose of providing a source of neutrons for research and for production of plutonium, was made in 1944. Flexibility was one of the key characteristics of NRX, and this paid dividends throughout its life. The flexibility allowed adaptation to changing requirements and allowed applications that were unknown at the time of the design.

Figure 2.14 shows NRX – a heavy-water-moderated, light-water-cooled research reactor. The main features of interest to production[3] and R&D activities are:

[3] The main production activity has been isotope production, but initially there was also plutonium and uranium-233 production.

- fifteen experimental holes, through which beams of neutrons (or other forms of radiation) could be extracted,
- fifteen self-serve mechanisms, through which small samples for irradiation could be easily introduced and removed,
- 198 vertical channels mainly used for fuel rods, but which could be adapted to house control and shut-off rods or various types of irradiation facilities, including loops (which are discussed in a later section),
- the central thimble, which could house a wider variety of types of irradiation facilities than the fuel-rod sites, and these in the highest neutron-flux location,
- two graphite thermal columns, in which there were relatively very few high- energy neutrons, most neutrons being in thermal equilibrium with the graphite, and
- the air-cooled J-rod annulus, with access via a ring of ninety vertical holes through the top shield. The annulus was originally used to produce uranium-233 by irradiation of thorium, but later used to produce other isotopes.

For many years after the startup of NRX, the main characteristics distinguishing it from other research reactors were its availability, for production and R&D, of high neutron-flux levels and relatively large volumes of more or less uniform neutron flux. These characteristics were exploited to the fullest extent. Even after the flux levels were matched and surpassed by NRU and other reactors outside Canada, the combination of flux and available volume kept NRX competitive with foreign reactors in some fields (e.g., fuel element and assembly testing, isotope production) for many more years. In addition, NRX served even longer as a very useful supplement and back-up to NRU.

Some of the key staff who were to be responsible for the operation of the reactor received technical training at the Montreal Laboratory. The "Montreal Lectures" were the basis for this training and the written versions proved to be of great use to new, inexperienced staff for many years afterwards.

Construction of NRX was completed in the late spring of 1947 and everything seemed to be ready for startup. For a startup with fresh fuel in the reactor, specially sensitive neutron detectors were required and had been installed. Because the isotopic purity of the heavy water was to be maintained in the region of 99.8 percent, drying of the system was attempted to remove the last traces of ordinary water. There was evidence that water was entering the system. An intense search traced the in-leakage to porosity of water-cooled castings in a blower intended to circulate the helium cover gas. After the faulty equipment was replaced, water vapour persisted in the system despite the drying. Further searching located a U-shaped pipe that had escaped earlier notice. When that was drained of seven kilograms of water the reactor system was successfully dried. Heavy water was then put into the system and the reactor was brought to criticality at 6:13 a.m. on July 22, 1947.

The power level of NRX was gradually increased from the initial value of a few watts to 250 kW in September, 2 MW by March 1948, 12 MW by May 1948 and 20 MW by September 1948.[4] The power was limited to 20 MW by the capacity of the moderator cooling system. As an interim measure, the river-water cooling for the heavy-water heat exchanger was fed from an available higher-pressure system, enabling operation at power levels as high as 27.5 MW when the river was cool enough.

The operation of a system exposed to intense ionizing radiation presented novel challenges (in addition to more standard ones[5]). It was known that the radiation decomposed water into its constituents, oxygen and hydrogen, and a system had been

[4] During the initial approach to 100 kW, a jitter of a percent or so appeared on the power meter (which was sensing neutron flux) at about 80 kW. Lowering the power slightly caused the jitter to vanish, but it reappeared at 80 kW. This went on for two or three days; the control room became a meeting place for senior scientific staff who hypothesized about the cause of the "frantic mouse," as the phenomenon was called. Theories ranged from a loose isotope-production capsule bouncing in the cooling air flow to a restructuring of heavy water at a specific temperature. The effect disappeared after a couple of days.

[5] Employee's recollection: "In the winter of 1948 I was given the job of finding a way of preventing algae formation in the inlet end of the NRX fuel rod annuli. I tested a few things in the lab and the problem disappeared as if by magic. It turned out to be due to cold water algae in the river water that disappeared as soon as the water warmed up. I never had an easier problem to solve in my career."

FIGURE 2.14
NRX Research Reactor

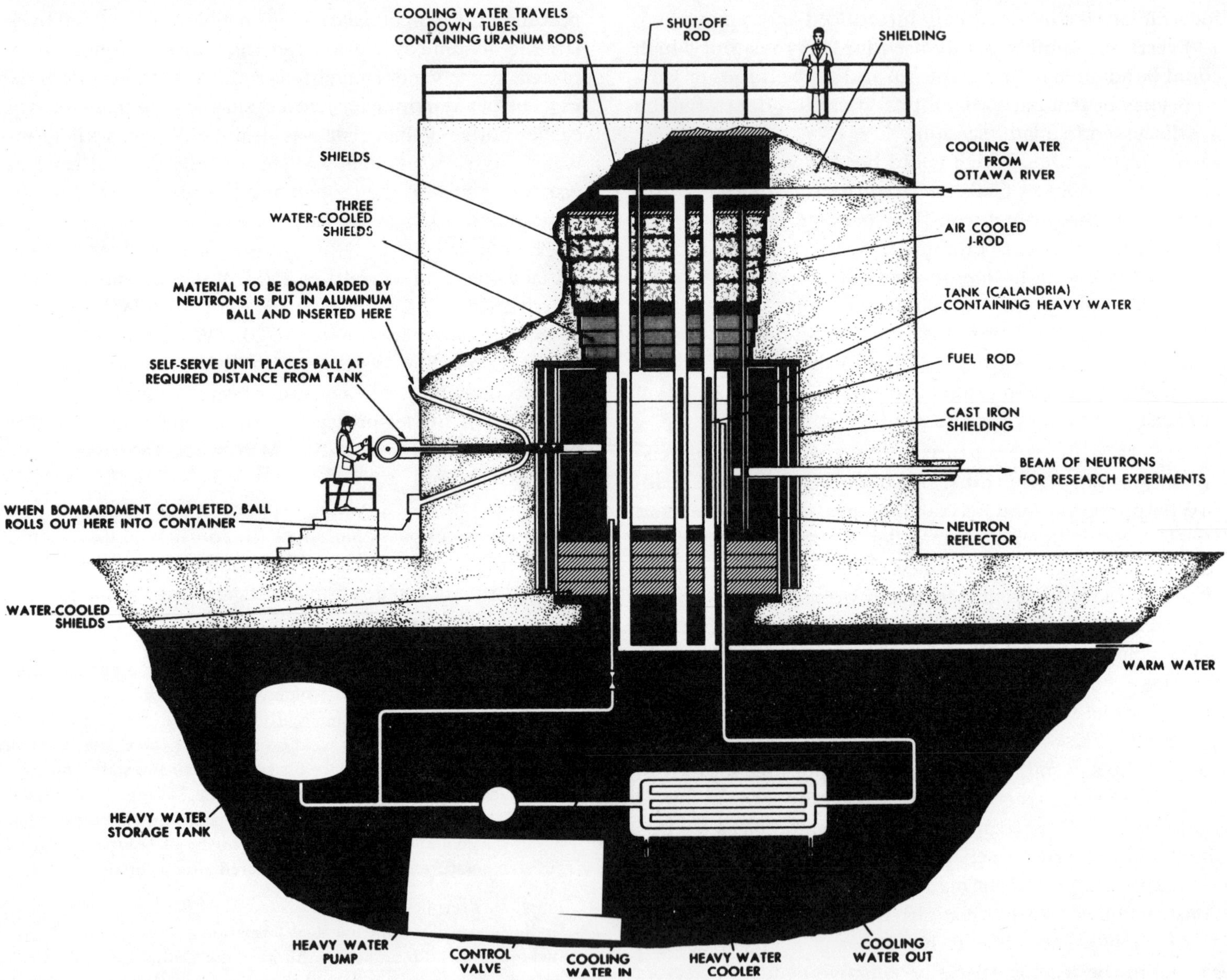

provided to circulate the moderator cover gas (helium) through catalysts to recombine the oxygen and heavy hydrogen. The only relevant experience was with a 300 kW heavy-water reactor in the United States, and that indicated high decomposition rates. There were other more subtle effects, which led to acidification of the moderator. Instead of following the American practice of adding deuterated ammonia to neutralize the harmful acids, the NRX operators distilled the heavy water and used newly available ion-exchange resins to maintain purity.[6] Improvements were also made to equipment for removing nitrogen and other impurities from the helium. The result of these measures was such a low rate of decomposition that concern shifted to fear that the low rate was deceptive, and that oxygen and deuterium were accumulating in the heavy water and might be released in a burst of an explosive mixture of the two gases. The fear was found to be groundless, and the moderator system behaved well, an encouraging situation for future power reactors.

Heavy-water leakage control was a major ongoing effort in NRX. To ensure that heavy-water leaks at flanges, valves and other fittings could be detected quickly, heavy-water detectors, called Beetles, were installed wherever the leaks might occur. Any water dripping into the tray of the Beetle collected in a small depression at the bottom and made contact with a probe. An alarm rang in the control room, indicating a water leakage, which was immediately investigated by the operating personnel. (The NRX average net heavy-water operating loss from 1955 to 1960 was 12 kg per month.)

In the early days of NRX operation a great deal of effort went into understanding its behaviour. While considerable experimental information on reactor characteristics had been gathered from American sources, and in the Montreal Laboratory and ZEEP, most of this was at very low power and flux levels, and at room temperature. Characteristics of NRX had been mainly estimated from the basic properties, not all of them well known, of the constituent materials. It was important that these estimates be verified by direct measurement.

The most important characteristics, from the viewpoint of effective operation and safety, were those associated with the effects on reactivity of changes in the composition of the reactor core (through insertions or removals) or of the materials within the core, changes in the temperatures of the core constituents, changes in the relative positions of core constituents, and changes in the neutron flux. To make quantitative measurements of these effects, a nominal geometric reactivity scale was established for NRX, whereby changes in moderator level could be interpreted quantitatively as changes in reactivity (see Box A).

The general method for making the measurements was to start with a critical reactor, make the physical change to be measured, and then restore criticality by changing the moderator level. The change in reactivity due to the physical change was then the negative of the change in nominal geometric reactivity. As mentioned in chapter one, many measurements were made in the period 1947-52. Measurements of particular value to the reactor operation were temperature and power coefficients of reactivity: the ratios of change in reactivity to change in component (moderator, rod assembly, etc.) temperature and to reactor power, respectively; the change in reactivity associated with long irradiation of fuel rods; the differences in reactivity associated with differences in fuel rod design; the reactivity effect of inserting the Hot Loop in the central thimble; the reactivity effect of removing light-water cooling from the in-reactor portion of fuel rods, and xenon

[6] Employee's recollection: "During the first year of operation the NRX moderator became progressively more acidic. Aluminum dissolved from the calandria, then precipitated in the heavy water giving it almost the appearance of skimmed milk. The moderator had to be completely distilled but in six months the problem was as bad as it was before distillation. Fortunately, at about that time, ion-exchange resins first came on the market. I was given the job of seeing if this new product could be used to clean the NRX moderator. The test work was done on the main floor of NRX. I clearly recall the superintendent and assistant superintendent anxiously standing by to watch the first tests, both armed with cotton batting to mop up any drops of heavy water that were spilled. The tests worked, columns were put on a side stream of the moderator circuit, and in seven days the moderator was as clean as when it went in the reactor. Without the timely commercialization of ion-exchange resins, the operation of NRX and its dependent programs would have been much more difficult."

BOX A GEOMETRIC REACTIVITY SCALE FOR NRX

As discussed in chapter eleven, the reactivity of a reactor, k is defined as the ratio of neutron populations in successive generations (with the value of k being =1 for a critical reactor, <1 for a subcritical reactor and >1 for a supercritical reactor). Changes in reactivity are usually expressed in millik (mk) with 1 mk being a change of 0.001 in k. A change in moderator level changes the probability that neutrons will leak from the reactor, and hence changes k. The relationship between moderator level and reactivity can be determined theoretically and, from this geometric reactivity scale, the reactivity change associated with any moderator level change can be determined. The scale can also be determined by measuring the time response of the neutron flux to a change in moderator level from the critical level. Measurements were made of the time for the flux to double after a moderator level change from critical (done at low reactor power to minimize changes in reactivity during the measurements due to temperature changes), allowing enough time to elapse after the addition of the excess reactivity to give an accurate approximation to the asymptotic doubling time. The results from both methods agreed within the uncertainties associated with them (see figure 2.15), giving confidence in the geometric reactivity scale adopted for NRX. To give some feeling for the relationship, at a moderator height of 230 cm an increase in level of 10 cm gives a reactivity increase of 5 mk.

FIGURE 2.15
Asymptotic Reactor Doubling Times vs Excess Reactivity

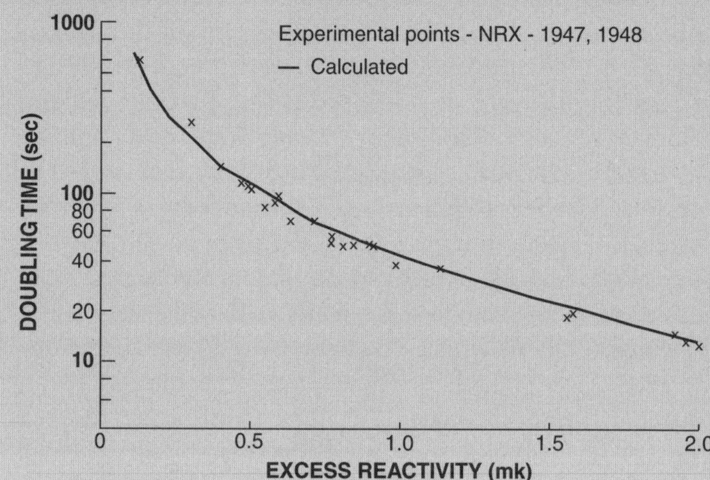

poison. It was particularly important to the efficient operation of the reactor that xenon poison be understood, since, for example, it was responsible for the relatively short window of opportunity for quickly restarting the reactor after a shutdown from high power. The length of this window depended on the recent reactor operational history, and was typically an hour or less. The usual result of missing the window was a delay of a day or more in getting back to power (see chapter eleven).

About the time NRX began operation, information was received from the United States that, in their experience, when metallic-uranium fuel rods had produced an energy of 300 megawatt days per ton of uranium (MWd/tU),[7] mechanical deterioration such as pimpling and distortion would require the removal of the rods from the reactor. The irradiations of the middle portions of the NRX rods, where the rate of energy production was highest, were kept under review. Occasionally,

7 There was confusion over what was meant by a ton. Canadian engineers used 2,000 lbs. which is the Canadian ton. The United Kingdom used 2,240 lbs. which is the British ton, and some scientists used a metric ton, which is 1,000 kilograms, (2,205 lbs). Finally, it was decreed that in this matter a ton meant a metric ton (then to be called a tonne, now a megagram (Mg)), and some discrepancies of 10 percent or so in people's numbers vanished. The preferred energy output unit is now megawatt hours per kilogram of uranium (MWh/kgU). {300 MWd/MgU=7.2 MWh/kgU}

rods were removed and examined for pimpling. No serious trouble developed at 7.2 MWh/kgU and irradiations continued, with the middle portions of some rods eventually reaching 72 MWh/kgU. The attainment of such long irradiations was an important factor in the first quantitative Canadian proposal for a nuclear power plant.

Some rods did fail because the uranium expanded and split its aluminum sheath. These failures were detected by instruments that monitored the effluent cooling water for radioactivity. The position of a failed rod was approximately indicated by installed equipment, after which the specific rod was located by a manual radiation survey. The rod was then removed.

Even intact rods could not be left in the reactor indefinitely, because irradiation burns up the fissile material and so reduces a rod's contribution to reactivity. Fresh uranium was needed to keep the reactor operating. In October 1949, a planned refueling at one rod site could not be completed, because the fuel rod was stuck in place. A quick survey showed that nine rods were either immovable or movable only by a few feet. Shortening of the uranium under irradiation had caused wrinkles, like those around a loose sock, to develop in the outer aluminum tube. Some of the wrinkles caught on the surrounding calandria tubes. After considerable ingenuity and effort, all the rods were successfully removed.[8] The stuck rods were from the second batch of rods, in which the rolling conditions had produced a metallurgical texture conducive to shrinkage. There was no satisfactory record of the rolling conditions that produced the original batch, and much effort was required by metallurgists to discover a procedure that would not lead to undue shrinkage. There is no doubt that the project would have been in jeopardy if the original rods had jammed themselves into the core. Several incidents in the intervening two years had given the operating and design staff experience and confidence in dealing with unusual events in a complex radioactive environment.[9]

By late 1948, a simple electrical circuit had been added to keep the reactor power at any chosen level automatically. This was the beginning of a Canadian tradition of automatic control, with the result that Canadian power-reactor designers have pioneered the use of digital computers for reactor control. For many years, operators of research reactors in the United States were required to have a human manipulating a control knob to maintain the power level. Canadian and American views differed over the question of whether it is preferable to give the operator freedom to move about, checking instruments in the control room, or to have him sit for hours doing the boring task of keeping a spot on a galvanometer scale at a fixed point.

In 1949 a pneumatic carrier tube was constructed to carry irradiation samples from the research chemistry building directly to a central fuel-rod position in NRX (some 900 feet).[10] Samples could be installed from the top of the reactor or from the chemistry building into either a central core position or into a rod in the J-rod annulus. The irradiated samples could then be received either at the top of the reactor or in one of four

---

8 The removal of these rods provides a good example of the highly imaginative procedures and equipment, coupled with great skill, that were frequently required to successfully deal with problems that arose. The fuel rod assembly was about 9.4 m long, with the mid 3 m containing the uranium. The solid uranium-metal fuel was contained in an aluminum sheath about 3.5 cm in diameter that had 3 fins to centre it in the 4.3 cm diameter flow tube. The normal air gap between the calandria tube and flow tube was less than 2.5 mm. The decision was made to saw through the wrinkle. A tubular saw was built to close tolerances. It had to slide down through the 2.5 mm annulus. Before this could be used, the upper stainless-steel shielding portion of the fuel rod assembly had to be removed because it was greater in diameter than the fuel section. Using extensive mock-up tests that were an exact physical duplication of the situation, including the wrinkles, a method was developed to disjoint the upper portion of the rod assembly and remove the wrinkles.

9 Employee's recollection: "This was one of the earliest incidents that established the overall procedural pattern that has been diligently and successfully followed up to the present. You establish a team and restrict the time they spend on the job in spite of their enthusiasm – in this case no more than seventy hours per week. You do extensive and precise mock-up testing of all operations. This incident also pointed out the importance to the operation of what we used to call the gadgeteer."

10 This facility was nicknamed the rabbit, in recognition of its speed and rapid starts and stops.

receiving stations in different laboratories in the chemistry building. Compressed air was used to drive the capsule (at up to eighty kilometres per hour) and also to cushion the capsule as it entered the rod position and stopped at the elevation of maximum flux. The capsules were about 1.5 inches (38 millimetres) long and 9/16 inches (14 millimetres) in diameter. Although the maximum sample size allowable was about one gram, owing to heat-removal limitations, this was more than adequate in most experiments, thanks to the high neutron flux in NRX.

As mentioned previously, the power of NRX was limited by the capacity of the moderator cooling system. An additional heavy-water heat exchanger, two additional heavy-water circulating pumps and an additional cooling water pump were added to the system. By mid-1950 the reactor power limit was 30 MW.

The first two loops (see later section) were installed in the central thimble of NRX in 1951. These were re-entrant loops (i.e., the coolant both entered and left from the top) and were referred to as the Cold Loop and the Hot Loop. They were used by the Americans in support of Admiral Rickover's nuclear submarine program. This marked the beginning of a long relationship (lasting until 1981) in which the Americans made extensive use of loops in NRX for various tests.[11] The use of loops in research reactors became an important element in the Canadian nuclear power program.

By 1952, fuel rods with a cooling-water annulus 0.071 inches (1.8 millimetres) thick came into service, replacing rods with 0.100 inch (2.54 millimetres) annuli. This reduced the amount of ordinary water in the core, thereby increasing the reactivity.

RESEARCH REACTOR OPERATIONS HIGHLIGHTS

NRX Highlights

The NRX accident in December 1952 was a severe but temporary setback to research reactor operation at CRL. Prompt and effective action in the period immediately following the accident and for the next few days (i.e., reactor shutdown, personnel evacuation, temporary storage and subsequent disposal of highly contaminated coolant[12]) limited its effect on personnel and the environment. As reported in chapter three, a study almost forty years later revealed no excess cancers or leukemias – above those expected for the general population – among personnel involved in the cleanup. It has been estimated that the maximum dose a member of the public could have received due to the accident was less than 1 millirem (0.01 mSv).

The core of NRX was rendered useless by the accident, and there was no experience at that time from which to derive firm confidence that it could be replaced. Nevertheless, a decision to repair NRX was quickly made and it was back in operation in fourteen months.

The restoration of NRX proceeded in parallel with actions to contain and remove the radioactive materials deposited in the building by the accident. Most of the fuel rods could be removed in the normal manner, but removal of the damaged rods was more difficult. Some fuel residues were lying on the bottom of the calandria. Removal of these residues and the decontamination of the basement involved most of the staff at CRL and resulted in some workers receiving radiation doses around the regulatory limit then in force. Extra workers were supplied by the Canadian army and the American navy.

[11] Employee's recollection: "Early on the USAEC recognized the value of NRX. In the fall of 1948 a special irradiation of uranium oxide in conjunction with neptunium separation studies was done for Argonne. The shipments were made using a special United States Army aircraft. The irradiation was done in NRX because there was no reactor elsewhere with sufficient capability to do the investigation. This was the first formal cooperative experiment where Canada supplied the reactor facilities in exchange for the results of the experiment – something that Dr. Lewis was always very keen on."

[12] About 4,500 cubic metres of cooling water, containing 10 kilocuries of fission products, flooded the NRX basement. An 1,830-metre, 10-centimetre diameter insulated pipeline was laid within nine working days and the cooling water was pumped to the waste management area.

Removal of the calandria on May 22, 1953 was a difficult and important preliminary step towards reconstruction.[13] During the reconstruction, there were improvements over the original NRX. Changes were made to many of the control and shutdown systems to reduce the probability of another incident.[14] New inner shields above and below the core removed the concern that at higher powers differential heating through the original shields would cause undue buckling, and this permitted the power limit to be raised to 40 MW.

The reactor was back in operation on February 16, 1954, and by early April it was at a new operating power of 40 MW.

Following the restoration of NRX, increased user demands for more tray rods, fast-neutron rods, loops and other irradiation facilities put space at a premium, both within the reactor core and within the NRX building. It also meant that more reactivity was needed than could be provided by the standard natural-uranium fuel. Special rods called booster rods were designed, made and used to provide increased reactivity. These booster rods were fueled, in general, with enriched uranium (in the form of a uranium-aluminum alloy), but a few were enriched with plutonium (in the form of a plutonium-aluminum alloy) that had been separated from spent fuel and made into booster-rod fuel at Chalk River.

The fast-neutron rods were used to irradiate samples in a higher fast-neutron flux than was normally available. Over the years the design varied, but it consisted basically of a sample irradiation space within a fuel annulus situated in a normal fuel-rod site. By the early sixties, six fast-neutron rods were installed in NRX.

Tray rods were used for isotope production. These were air-cooled assemblies placed in normal fuel-rod sites. They comprised an open-sided tray assembly with clips to hold thirty sample capsules at various elevations in a rod. The tray rods could be removed from the reactor and the samples changed without shutting down the reactor. By the early sixties there were fifteen tray rods in the reactor core and one in the J-rod annulus. In addition, many miscellaneous experimental rods were installed at various times.

In 1955 one of the plutonium-enriched booster rods in NRX failed, puncturing the calandria tube in which it was located. An estimated 100 grams of alloy containing 20 grams of plutonium, plus some of the light-water coolant, was lost to the heavy-water moderator. It was determined that failure had been caused by the aluminum sheath that surrounds the fuel alloy bulging away from it, causing the fuel temperature to rise enough to cause the failure. All fuel of this type was removed from the reactor core. The NRX heavy water was downgraded to about 97 percent deuterium oxide, but all the heavy water was recovered and reconcentrated. The calandria vessel was cleaned[15] and the damaged calandria tube replaced *in situ* in seven weeks.

A facility called the Hydraulic Rabbit (to distinguish it from the Pneumatic Rabbit already described) was installed in NRX

---

[13] About a year after the event, *Popular Science* published an article about the removal of the calandria under the title "An Atomic Dragon's Eerie Funeral". The following is excerpted from this article. "This is the story of a funeral – the strangest funeral in the history of man. The coffin was a huge canvas bag, the grave a great hole and the mourners atomic scientists. The corpse, was the [NRX] calandria – the largest and most dangerous radioactive object that has ever been handled. A long whistle blast warned that the funeral procession was underway. In buildings along the way pens of radiation monitoring instruments went off the charts. The [burial] ground will not go unvisited – but so formidable is this atomic dragon, even in death, that it will be perilous to approach for some time to come."

[14] In the years following the start-up in 1954, the control and shutdown systems were more or less completely changed - the location of the control console was changed, automatic power control was introduced, the weir box for heavy-water-level control was removed, a new heavy-water-level control was introduced, and a new dump system was installed for shutdown.

[15] Employee's recollection: "After the plutonium-aluminum fuel-rod failure there were a lot of flakes of plutonium-aluminum lying on the bottom of the calandria that could not be removed by flushing. If they remained they would have caused hot spots on the inside of the calandria with resultant damage. Bill Stevens had the idea of putting soda water in the calandria to float the flakes so they could be flushed out. It worked like a charm and possibly avoided the need for a second calandria replacement. As it turned out only the calandria tube had to be replaced. I think they got the soda water from a local bottling works."

in 1958, to carry out a large number of short-term irradiations of experimental fuel. Each sample was loaded in the reactor hall basement, and propelled by cooling water through a 7/8 inches (22 millimetres) stainless-steel tube into the highest flux position in a fuel site, where it was held by a retractable stop. When the sample had been irradiated for the required time, the stop was opened and the sample was pushed by the cooling water to the top thermal shields. After a period of decay it was removed to a shielded flask.

Studies of alternative fueling schemes for NRX had been initiated in 1957 as a result of indications from the United States that their agreement to purchase spent fuel (with its plutonium content) from NRX and NRU would be terminated within a few years. Without a market for the spent fuel, the economics for operation of the research reactors would change significantly. Net neutron production from the fuel, rather than power, was the primary characteristic of value for R&D applications. In 1959 the decision was made to convert NRX to use a combination of two types of fuel: i) natural uranium-dioxide and ii) highly-enriched uranium-aluminum-alloy rods. The actual conversion started in the spring of 1962 and the main part was over in about six months, with no break in the normal operation of NRX, and with no structural changes.

The rationale for the choice was fourfold:

- the reactor would have the capability of accepting higher production and experimental loadings,
- a higher flux, by approximately 50 percent, could be achieved in production and experimental sites without changing the reactor power,
- the life of uranium-dioxide fuel was expected to be considerably longer than that of the uranium-metal rods, and
- the experience to be gained in the production and irradiation of uranium dioxide on a fairly large scale would be useful in the power-reactor development program.

A reference loading was chosen that consisted of three roughly defined fuel zones: an outer zone of solid uranium-dioxide rods, an intermediate zone of annular uranium-dioxide rods plus enriched uranium-aluminum-alloy rods, and a central zone of the enriched uranium-aluminum-alloy rods. The experimental and isotope-production facilities were in the intermediate and central zones.

The procedure adopted was to shift over to the reference loading, as occasion warranted, keeping the flux constant by lowering the reactor operating power. On a date announced six months in advance, the reactor was returned to normal power, with a resulting abrupt 27 percent increase in neutron flux. This procedure allowed adequate preparation of test fuel and other irradiation facilities to accept the flux increase. Further smaller increases were made as the reference loading was completed, with the integrated increase of about 50 percent. Some years later, when it had served its purpose, the uranium-dioxide fuel was gradually eliminated and fueling was entirely with enriched uranium-aluminum alloy. This conversion had the effect of multiplying the capabilities of NRX several-fold, thereby extending considerably its productive life.

In the summer of 1963 the first corrosion failure of a calandria tube, under normal usage, occurred in NRX. Sealing devices, consisting of removable plugs with expanding metal washers, were used to seal the ends of the tube. This permitted reactor operation without leakage, but the lattice site was not available for use. Tubes continued to fail for the same reason over a number of years. For some time these were plugged in the same manner, leading to more unusable sites. A tube-replacement program was then initiated, which required a large amount of ingenuity and technical expertise in remote work, but retained the usability of the site where a tube was replaced.

A spare NRX calandria was on hand on the main floor of NRX, which included modifications to allow for three large loops and four large fast-neutron irradiation facilities, as well as other improvements. It was finally decided to replace the NRX calandria. Taking advantage of the experience gained in 1953, this was done in a relatively short time period in 1970.

In the eighties, calandria tubes again began to fail in NRX. The continued operation of the reactor was possible only because the efficient and effective work of the operators, coupled with the expertise of the designers, made replacement of calandria tubes routine.

Taken as a whole, NRX operated extremely well and reliably, and enjoyed a long and productive life.[16] In the eighties its role was almost exclusively one of backup to NRU for isotope production. On March 13, 1986, NRX was designated as a nuclear historic landmark by the American Nuclear Society. The plaque that was presented hangs in NRX and reads: "NRX is the world's longest-operating research reactor, beginning operation July 22, 1947 – an important facility for isotope production, fuels and materials testing, and basic research."

NRU Highlights

The success of NRX, along with uncertainty regarding its useful life, led to consideration of a replacement as early as the end of 1947. It was also expected that the sale of plutonium to the United States would greatly defray the costs. By the end of 1950, the government had approved proceeding with an improved high-power, heavy-water research reactor, to be known as NRU. Design and site preparation were well advanced by 1952, and construction was completed in 1957. Startup was in late 1957, with criticality attained on November 3.[17]

While NRU was similar to NRX, in that it was a heavy-water-moderated reactor designed to be fueled with natural uranium, there were fundamental differences:

- Power – at 200 MWt the power rating was five times that of NRX,
- Flux – the maximum thermal neutron flux level was five times that of NRX,
- Size – the effective core size was 40 percent larger in volume than NRX,
- Coolant – the reactor coolant was heavy water, not light water (which NRX used), one result being that there were no calandria tubes (there are seals between the cylindrical reactor vessel and the top and bottom headers),
- Fueling – fuel could be replaced on-power, whereas NRX required a shutdown, and
- Reflector – light water was used as the reflector, not graphite (which NRX used).

The main features of interest for production and R&D activities included: 227 lattice sites, of which 221 were available for use; three lattice sites with access from top and bottom for engineering loop experiments; horizontal holes of various sizes (max. 30 cm diameter), of which two pass through the lattice; a thermal column; and forty-six available positions in the J-rod annulus. Special tray rods, capsule rods and fast-neutron rods were developed for irradiations. NRU also had a pneumatic carrier serving the chemistry laboratories and the universal cells.

Close control of heavy-water spills and leakage is important not only because heavy water is very expensive, but also because when it is irradiated for long periods, as research-reactor moderator is, the concentration of tritium builds up and presents an internal radiation hazard. Immediately following NRU startup, heavy-water leakage was detected from the lower seal between the vessel and the bottom header. This problem was solved by tightening all vessel bolts, imposing a limit on the temperature difference between the heavy-water moderator and the light water in the reflector, and maintaining a helium pressure on the bottom interseal space.

Heavy-water leakage continued to be a problem, a result of the structural complexity arising from the provisions made for neutron experiments and irradiations. In 1959 a heavy-water leak from one of the through tubes into the J-rod annulus was found. It was repaired by rolling in a long, closed tube (referred to as a thimble), and a similar leak that developed later in the other through-tube was repaired in a similar manner. In 1962 a major leak was traced to cracks in stainless-steel pipes, located in a position with limited access, that carry heavy-water coolant

---

[16] So successful was NRX that two copies were built, with Canadian help, in foreign countries: the Canada/India Reactor, CIR (later renamed CIRUS), in the early sixties in India, and the Taiwan Research Reactor, TRR, in the early seventies in Taiwan.

[17] Soon after startup, it was found that there was 33 percent excess absorption in the fuel-coolant tubes, caused by excess boron in the aluminum of the tubes. The result was a loss of seven mk in reactivity. The solution was more stringent specifications for the aluminum.

from the underside of the top header. A seal, either permanent or readily removable, had to be designed to eliminate the leakage. An epoxy resin that could be remotely installed was used. It was subject to radiation damage but could be easily removed.[18] The resin was used for periodic replacement of the seal over a period of about ten years before a suitable mechanical sealing device was designed and installed.

In the early operation of NRU, fuel failures were a major problem. In May 1958, during the removal of a failed fuel rod, a piece of uranium fell out of the removal flask into a pit at the top of the reactor, and caught fire. The building was severely contaminated and it was three months before the reactor could be started again. The cleanup operation involved work in areas of high radioactive fields and required the use of many staff (some of them inexperienced) to keep individual doses within limits. It marked the first time television was used at Chalk River to effectively control and monitor the work of inexperienced personnel.

In the commissioning of NRU, problems were encountered with the heavy-water pumps caused by stainless steel running against stainless steel – the result of a communication failure between designers and manufacturers.[19] The problem was solved by changing the bearing material. The most serious operating problem with the heavy-water pumps was with the shaft seals. The manufacturers had always made water pumps with designed leakage through the seals to ensure lubrication, with a recovery system. With heavy water, a design that allowed for the full recovery of such leakage was needed. Shaft seal development became a continuing program for the heavy-water power reactors.

Early in the operation of NRU, the formation of gibbsite[20] (see chapter fourteen) in the heavy water caused blockages of sample lines. An evaporator was installed in the heavy-water system and the sludge that accumulated was periodically removed and sent for further evaporation treatment to recover the heavy water, with the final sludge being processed as waste.

Late in 1958 the absorber of a control rod became disconnected from the drive mechanism and remained leaning at an angle in the reactor core on startup. This led to the introduction of shroud tubes around control and adjuster rods. Subsequently, a shroud tube buckled, jamming the absorber, which led to a limit being placed on the number of shroud tubes permitted in the reactor around control and adjuster rods.

About 1960 it became obvious that the efficiency of the heavy-water heat exchangers was deteriorating. The river water used for cooling came directly to the shell side of the heat exchangers and the cause of the problem was found to be a very persistent form of algae that seemed to thrive in the high radiation fields of the heat-exchanger rooms. After some experimentation, it was found that the most effective way to remove the algae was with a laundry detergent.

In the fall of 1961 a broken vessel bolt was found in the NRU upper service space. Metallurgical examination indicated that all 144 bolts would eventually fail in the same way, caused by the type of lubricant initially used in installation. Inconel-X-stressed sections and a suitable lubricant were developed. All bolts were replaced, some at each shutdown, over an extended time period.

It was decided in early 1962 to convert NRU to operation with highly-enriched uranium fuel. The trigger for this decision,

---

[18] A technique, similar to sand blasting but using ground walnut shells in place of sand, was used. This technique removed the epoxy and cleaned the metal surfaces without damaging them.

[19] Employee's recollection: "According to an engineer from the manufacturer who took part in the design and manufacture of the pumps and helped with the modification, this was really a security problem. Everything they received was stamped secret and they were not encouraged to ask too many questions. They were well aware that stainless steel could not run satisfactorily on stainless steel under normal conditions, but they were asked to design stainless-steel pumps for a service for which they were not apprised. They just assumed that there was something different about this service in a nuclear facility so they produced the pumps – complete."

[20] Gibbsite is a hydrated aluminum oxide. A more serious problem occurs with gibbsite when there are extensive fuel failures. In this case the gibbsite adsorbs and transports large amounts of fission products, resulting in very high fields around all heavy-water equipment, making maintenance difficult.

the impending loss of the plutonium market in the United States, was the same as that for the conversion of NRX to enriched fuel. However, for NRU, a different ground rule was adopted in arriving at a preferred fueling scheme. This was to maintain the thermal-neutron-flux level while minimizing the fueling cost. In general, with highly-enriched uranium fueling (compared to natural-uranium fueling), the thermal-neutron flux level could be maintained at a lower reactor power and with the use of fewer fueled lattice sites (freeing more sites for experimental and production use). On the other hand, the fast-neutron flux levels were, in general, considerably lower. The scheme chosen used uranium-aluminum-alloy fuel rods (initially with a uranium-235 content of 0.5 g/cm over the nine-foot (2,743 millimetre) active length), with the uranium enriched to 93 percent uranium-235 in uranium. A fuel-management scheme was adopted in which fresh fuel rods were loaded into the outer regions of the core and moved progressively (normally twice) towards the centre of the core before being removed. Fast-flux levels were maintained where they were required by experimentalists, using natural-uranium, fast-flux-peaking rods. This scheme satisfied the ground rule, and provided flexibility for increased reactor loads (with associated increases in power and fueling costs) and higher thermal flux, if desired.

The conversion was carried out from November 1963 to March 1964, with NRU shut down (contrary to the procedure with NRX). Immediately after conversion, NRU was operated at 60 MW, compared to its pre-conversion operation at 220 MW (both figures being exclusive of loop powers), with the same usage. It soon became apparent that user demand justified greater capability. The power and uranium content of the enriched fuel were both raised over time to about 130 MW and 1.5 g(uranium-235)/cm, respectively (with roughly the same neutron flux level).

The conversion was extremely successful. The expanded capability and relative cost-effectiveness greatly extended the useful life of NRU.

Around 1970, a hole developed in one of the expansion bellows of the NRU reactor-vessel side-wall. A mechanical plug and washer device was designed and installed to successfully seal the hole. Over the next several years, more holes developed, with gradually increasing frequency. The sealing of each hole required several weeks of reactor shutdown, hastening the decision to replace the reactor vessel. A new vessel had been ordered in 1958 and was delivered to the site in 1966. In the years before its installation, a great deal of work was done on the vessel at Chalk River. Over this same period, a project team had been making preparations that would facilitate the replacement when it was required. Of particular significance was their development of a very complex, large, remotely operated tool for inspecting, cleaning and repairing the stainless-steel surfaces in the reactor, to which the aluminum side wall must seal to contain the heavy water. NRU was shut down for vessel replacement in June 1972 and restarted in August 1974. Operation then continued with a very high usage and production for many years.

WR-1 Highlights

AECL interest in the concept of an organic-cooled heavy-water-moderated power reactor (CANDU-OCR) was initiated by a suggestion from engineers at Canadian General Electric, who were following the American program on organic reactors. AECL commenced a development program for this new concept, to provide a backup alternative to its main candidate – the pressurized, heavy-water-cooled, heavy-water-moderated power reactor. Building a small reactor to test the concept was deemed the best way to focus the development effort; such a reactor could also serve a valuable role as a materials-and-fuel-test reactor. In 1960, a proposal was approved to build such a reactor (to be known as WR-1) as the main initial facility at a new research establishment in Manitoba: the Whiteshell Laboratories. Construction of WR-1 began in 1963 and criticality was attained on 1 November 1965.

WR-1 differed in several fundamental ways from NRX and NRU, besides having organic as the coolant. The reactor coolant could operate at high temperature (up to 425°C at the outlet), although still at relatively low pressure. No neutron beam tubes were provided, so there was no external access to the neutrons produced, and hence no basic-physics program. Moderator

FIGURE 2.16
WR-1 Reactor Cutaway

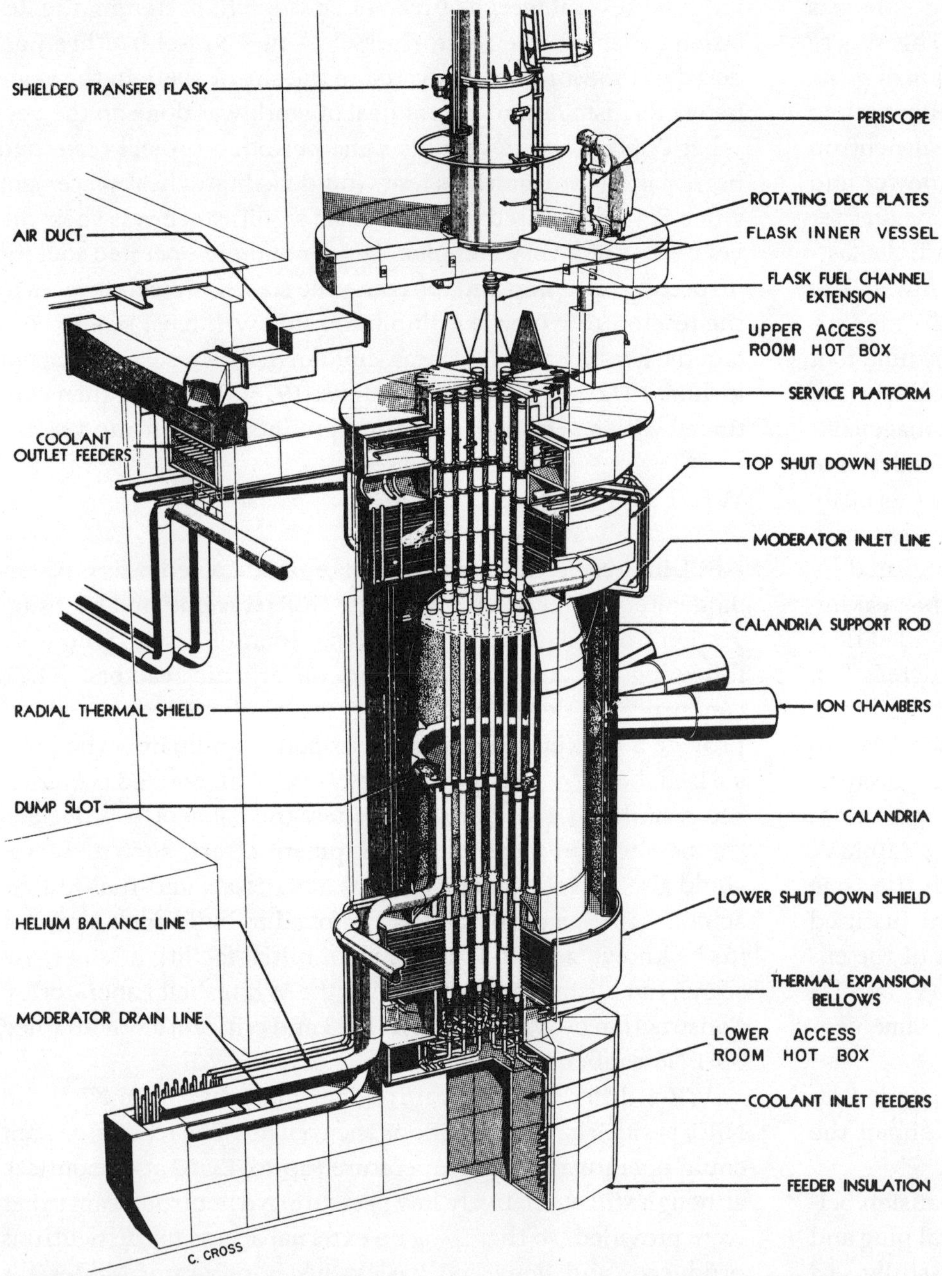

level, moderator dump and addition of soluble neutron absorber (boron) to the moderator were relied on for all power control and shutdown functions; there were no control or shut-off rods. Provision was made in the design for four loops: three organic-cooled and one water-cooled.

Figure 2.16 shows a cutaway view of WR-1. The reactor is of a vertical pressure-tube design, with provision for fifty-five fuel sites divided among three coolant circuits, and having a nominal thermal power rating of 60 MW. However, the initial configuration did not use this full capability.

In the original commissioning there were thirty-seven fuel-rod sites, each with an aluminum calandria tube, and the nominal power rating was 40 MW (thermal). Fuel channels, the in-core portion of which were pressure tubes (I.D.= 8.29 cm), were installed in each of these fuel sites, to support fuel assemblies and to contain and direct the flow of coolant (inlet pressure = 2.17 MPa) over the fuel. Stainless-steel pressure tubes were installed in thirty-one sites, and zirconium-alloy pressure tubes in the other six. The primary coolant was a partially hydrogenated mixture of terphenyls (Monsanto's tradename: HB-40) that is fluid at room temperature; the coolant is discussed further in chapter sixteen.[21] Two independent primary cooling circuits served nineteen and eighteen fuel-rod sites, respectively. The fuel was 2.4 wt. percent (uranium-235 in uranium) uranium dioxide clad in zirconium-2.5 percent niobium alloy (Zr-2.5Nb).[22] Each fuel string comprised five eighteen-pencil bundles on a central shaft that formed an eight-foot-long fuel assembly.

WR-1 commissioning proceeded smoothly, with no serious difficulties and only one major delay.[23] As might be expected with a new reactor design, some modifications were required in the pre-critical phase to make the various systems function as intended. For example:

- to prevent gas-locking of the moderator pumps during a moderator dump, it was necessary to extend the discharge pipe across the dump tank and install baffles to keep the entrained helium from the pump inlet, and
- it was necessary to re-route the ion-chamber cables to eliminate false signals due to "cross-talk" from the crane control circuitry.

The first approach to critical was carried out with fuel in the inner nineteen sites only, to obtain physics measurements with the moderator near its equilibrium level. The startup was smooth and uneventful; the low-power commissioning was carried out throughout November 1965 with WR-1 operating almost continuously at 0.01 percent of full power. This allowed for measurements of nuclear coefficients (change in reactivity with changes in moderator temperature, power, coolant temperature, and coolant void; change in reactivity due to insertion of driver fuel in the central site; reactivity change in the first second of a moderator dump), neutron fluxes, coolant dosimetry, and regulating-system performance. After these measurements and tests were completed for the nineteen-site core, six additional sites were fueled for the high-power core. The major physics measurements and tests were repeated for the twenty-five-site core. Satisfactory agreement was obtained between design values and measurements, except that the output from the ion chambers was found to be too high relative to the system design, due to a significantly lower flux gradient in the ion chamber cavities than predicted. The ion chambers were withdrawn sufficiently to permit full-power operation, some six weeks after first critical.

The main problems encountered during high-power commissioning of WR-1 were difficulties in obtaining satisfactory response from the automatic temperature control system, and in obtaining satisfactory performance from the thermal power control system. The use of standard modules in a relatively complex control circuit proved to be unsatisfactory, and many were replaced by higher quality versions.

Early in the operation of WR-1, a small defect occurred in a fuel element. The failed-fuel-detection system performed as designed, and the fuel element was located and removed without difficulty.

Over the next four years WR-1 operated well, with an average availability factor of 85 percent. Much of that operation was at a coolant outlet temperature of about 370°C, and long, uninterrupted periods of operation were obtained. The operators

---

[21] Employee's recollection: "HB-40 was originally intended to be used only as a startup coolant in WR-1, to be replaced later by a product composed of unsaturated terphenyls. HB-40 had the advantage of being liquid at room temperature, but it was thought that its rate of decomposition would be too high to allow it to be used as the permanent coolant. However, as WR-1 operated it was found that the rate of decomposition in-reactor was much lower than predicted and the coolant type was never changed. The HB-40 did indeed decompose, but many of the decomposition products recombined to form other compounds that behaved equally well as reactor coolant."

[22] Employee's recollection: "CGE's reference fuel cladding material for WR-1 was sintered aluminum product (SAP - Al + 10 percent Al_2O_3). However, every test of SAP-clad fuel in the X-7 loop in NRX failed because of the low ductility of SAP. Within about twelve months of the start of WR-1 there was no satisfactory cladding and CGE were about to revert to stainless-steel cladding (which, if used in an organic-cooled power reactor, would make the use of natural-uranium fuel impossible). Some zirconium coupons had been tested in organic coolant and the results were promising. Trefoils of zirconium-alloy-clad fuel were made, to the Douglas Point design, for testing in the X-7 loop. After a three-month test it was found that the fuel cladding remained structurally sound (all the hydrogen that had entered the cladding had migrated to the cooler wire wrap that was part of the design). On the basis of this one test, Zr-2.5Nb fuel cladding was chosen, and there was no cause to regret this decision over the life of the reactor."

[23] This delay was associated with difficulties in aligning the main coolant pumps and motors. It was found that the base and springs on which these were mounted were designed for an 8,000 lb. motor, whereas the weight of the motor had been revised to 12,000 lb. New springs had to be manufactured and the commissioning of the heat-transport system was delayed for twenty-eight days.

concluded that the commissioning and operation of an organic-cooled reactor was more trouble-free and straightforward than that of a pressurized-water-cooled system. They attributed this to two characteristics of the organic system: high coolant-outlet temperature is attainable with relatively low operating pressure in the primary system, and the radiation fields near the primary piping, feeders and headers are very low, minimizing the problems of access control for normal operation and maintenance.[24]

One potential problem with the organic coolant is the fire hazard it presents. Organic coolants are flammable, and appropriate precautions must be taken in handling them. Experience with WR-1 showed that use of organic coolants is feasible. In this four-year period there were a few small fires, confined to glowing insulation, and two significant spills of hot organic coolant (at 350°C), with no attendant fire. The keys to prevention of serious fires in organic coolant reactors are:

- Good housekeeping – the thermal insulation on pipes must be carefully installed so that air access is minimized; if any insulation does become soaked with organic coolant, it must be replaced immediately.
- Elimination of normal ignition sources by use of explosion-proof fittings and thorough grounding of all piping.
- Provision of smoke detectors and protection of areas containing organic circuits with fog and water sprays.

Over this period it was also demonstrated that fouling of heat-transfer surfaces and hydrogen uptake by zirconium alloys in the primary system can be controlled by coolant chemistry. For the former, the coolant must be free of chlorine and low in metal oxides, and this is aided by maintaining a small amount of water in the coolant and by side-stream circulation through clay columns. For the latter, maintenance of an oxide film on the zirconium provides an excellent barrier to hydrogen migration into the material, and is achieved by maintaining a small amount of water in the coolant and low chloride contamination.[25]

With WR-1 operating satisfactorily, there was considerable incentive to make improvements that would demonstrate the feasibility of a CANDU-OCR, provide better conditions for experimenters and lower the fueling costs. Towards this goal a number of changes were made, so that by 1975:

- the stainless-steel pressure tubes (relatively high absorbers of neutrons) were all replaced by zirconium-alloy tubes,
- a third organic primary coolant circuit was added to service seventeen of the remaining sites, which were then commissioned as fuel sites, and
- the fuel material was changed from uranium dioxide to uranium carbide at a suitable enrichment.

At the same time, a great deal of effort went into development of inserts for the loops and the fuel sites, to best exploit the R&D capabilities of the reactor.

WR-1 was shut down and decommissioned in the mid eighties due to budget pressures.

Research Reactor Fueling

The requirements for research-reactor fuel are generally quite different than those for nuclear-power-reactor fuel. In power reactors, the primary objective is the generation of thermal energy at a high temperature (to get efficient conversion to electrical energy), which means, for water-cooled power reactors, that the fuel must operate in a high-temperature, high-pressure coolant. On the other hand, research reactors, such as NRX and NRU, are built to provide neutrons and a neutron-rich environment,

[24] Employee's recollection: "One of the very pleasant surprises about the operation of WR-1 was the absence of any activity transport. In all of the water reactors, and even in water loops, corrosion products are transported by the coolant through the reactor where they become radioactive. They then deposit in the out-reactor circuits, making maintenance difficult. The operation of WR-1 showed that the organic did not corrode the circuit and, even after twenty years of operation, one could do maintenance work on any part of the out-reactor circuit in very small radiation fields."

[25] Employee's recollection: "When it was decided to use zirconium alloy as fuel cladding, it was also decided to add a trace (100 ppm) of water to the coolant, to maintain an oxide layer on the cladding to reduce hydrogen ingress. It was found that the addition of water also substantially reduced coolant-decomposition-related fouling on the fuel and other hot surfaces." [Fouling was a problem that plagued the first American efforts to use organic liquids as reactor coolants.]

and operate most efficiently with low-temperature and -pressure coolant. The situation is somewhat different for WR-1, which was built partly to test first-hand the concept of an organic-cooled CANDU. Since, at temperatures suitable for electricity production, the pressure of organic coolant is very much less than that of water, it proved to be feasible to operate WR-1 fuel under conditions very similar to those suitable for a power reactor. Thus the experience gained from the WR-1 driver fuel is much more directly relevant to possible power-reactor fuels for a CANDU-OCR than experience with NRX and NRU driver fuel is to CANDU-PHW fuel.

In all cases, research-reactor fuel performance is important in providing reliable operation of an important R&D facility. The provision of research-reactor fuel can also be an important factor in building up the infrastructure for a nuclear fuel industry. However, there is a relatively small market for research reactor fuel, and so most of the development discussed here was done in-house.

NRX Fuel

For the first fifteen years of NRX operation, natural-uranium-metal rods were the standard fuel. These consisted of a solid uranium-metal core 1.36 inches (34.54 millimetres) in diameter and 10 feet (3,048 millimetres) long. The uranium was clad with a drawn aluminum sheath, which was either 0.08 or 0.04 inches (2 or 1 millimetres) thick. The clad uranium core was inserted into an outer aluminum coolant tube to provide (post-1952) a water-cooling annulus of 0.071 inches (1.8 millimetres).

Difficulties were experienced with the uranium-metal rods, mainly due to dimensional instability during irradiation: changes in length and diameter, transverse fractures, and longitudinal splits. These conditions caused restrictions in coolant flow and failures in the aluminum cladding, which required removal of rods before their scheduled burnup (approximately 30 MWh/kgU).

The uranium-metal cores were formed by hot rolling, starting with large uranium-metal billets and using multiple passes to achieve the final desired diameter. Work to determine the effects of fabrication procedures on uranium stability under irradiation continued over the years. Factors investigated included type of billet (vacuum-cast or forged), rolling temperature, diameter reduction per pass, heat treatment and finishing following rolling, and inspection methods. Although, during their whole period of use, approximately 23 percent of the rods had to be removed from the reactor before reaching their scheduled burnup, the later batches of rods had a much-improved performance due to the increased knowledge and experience acquired.

By 1961 it was decided to introduce new fuel designs into NRX, and 10-foot (3,048 millimetres) long, solid-pellet uranium-oxide rods[26] were designed and fabricated for irradiation. The sheaths were made from an aluminum alloy, to provide the greater strength required by the buildup of pressure during irradiation from the fission product gases released by the uranium dioxide. These rods were confined to the lower-neutron-flux regions of NRX, to limit the temperatures in the centre of the uranium-dioxide pellets. The irradiation performance of the solid uranium-dioxide rods in NRX was very satisfactory. No significant changes in rod length or diameter were experienced.

Annular pellet uranium-dioxide rods that were cooled internally and externally were developed for the hotter (higher-neutron-flux) positions in NRX. The pellet had an outside diameter of 35.8 millimetres (1.410 inches) and a bore of 15.24 millimetres (0.60 inches), and was approximately 22.2 millimetres (0.875 inches) long. Several rods were examined after irradiation and no significant changes in rod length or diameter were observed.

A seven-element booster assembly was developed for NRX to maintain the reactivity with increased loads. Enriched uranium (93 wt. percent uranium-235) metal pieces were dissolved in molten aluminum at 1,050°C in a graphite or ceramic crucible,

[26] Uranium-oxide pellets had been developed and commercially produced for the NPD demonstration power reactor. The uranium-oxide pellets for the NRX rods used similar fabrication procedures and specifications, but were larger. They were 1.41 inches (35.8 millimetres) diameter and 0.8 inches (20.3 millimetres) long.

to yield a uranium-aluminum alloy. The melt was cast into water-cooled moulds to provide billets 38.1 millimetres (1.5 inches) diameter and 89.0 millimetres (3.5 inches) long. The billets were extruded at 540°C to 7.0 millimetres (0.275 inches) diameter. The extruded cores were drawn cold to 6.35 millimetres (0.250 inches) diameter in one pass of 17.5 percent reduction. The core was then scanned with two scintillation crystals and photo-multiplier tubes, to determine the dispersion of uranium-235 in the core. Aluminum end plugs were attached to the uranium-aluminum core and the core assembly was clad with 1.14 millimetres (0.045 inches) of aluminum by an extrusion process. An eddy current test was developed to detect voids and inclusions in the clad element. The aluminum-coolant tube that contained the seven elements had a bore of 35.46 millimetres (1.396 inches) and a wall of 2.29 millimetres (0.090 inches). The seven elements were located in the coolant tube by six spacers.

The standard fuel in NRX was subsequently converted to enriched uranium-aluminum fuel. The NRX booster assembly was modified, and although the dimensions of the cladding and fins were changed, the fabrication process for the fuel elements was essentially the same as the booster assembly. The coolant tube was modified so that no internal spacers were required. These enriched fuel assemblies performed very well in NRX.

NRU Fuel

When NRU started up in 1957, the initial fuel assemblies consisted of natural-uranium-metal flat bars clad in aluminum sheaths, with no bonding between the bar and the sheath. The uranium flat bars were fabricated from vacuum-melted ingots. The ingots were hot rolled to oval billets and the billets were then hot rolled to provide flat bars of finished thickness. The flat bars were roll straightened, and the width of each bar machined to the finished dimensions. The length of all bars was 10 feet (3,048 millimetres). The 0.04 inch (1.0 millimetres) thick aluminum sheaths were extruded close to finished size. These extruded sheaths contained spacing fins, which formed the coolant channels. The uranium flat bars with attached aluminum end plugs were inserted by hand into the sheaths.

FIGURE 2.17
Cross-section of Flat Type NRU Natural Uranium Rod

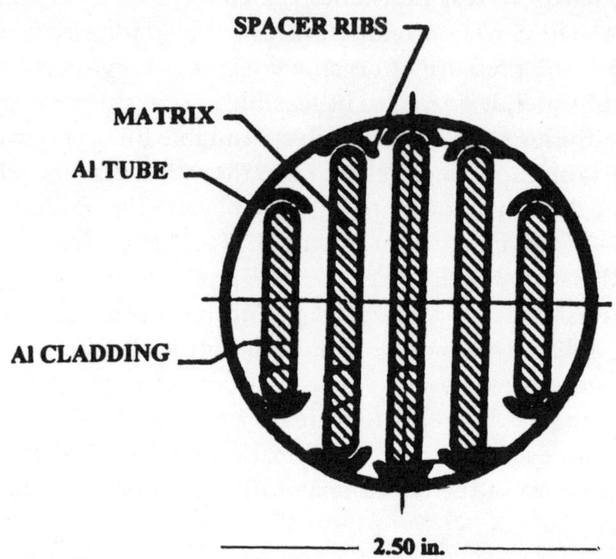

The clad flats were then drawn through a draw die, to sink the aluminum cladding onto the uranium flat bar. The ends of the sheath were welded to the aluminum end plugs, and the flat fuel elements were assembled into a round coolant tube.

The irradiation history of the non-bonded fuel assemblies in NRU was not good. Failures occurred at an unacceptable rate. In many cases the uranium metal shortened longitudinally under irradiation, causing transverse wrinkles in the aluminum sheath. These wrinkles significantly impeded the coolant flow. Other failures were traced to the poor heat transfer between the uranium fuel and the unbonded aluminum sheath. It was obvious that a fuel assembly of improved design was required.

Fortunately, the development of a bonded fuel assembly was initiated before the startup of NRU (see figure 2.17). The most successful development was the cladding of the flat uranium bars with aluminum by extrusion. The machined uranium bars were degreased, nickel plated, and vacuum dried. The extrusion cladding operation consisted of extruding a rectangular ribbon of solid aluminum at 535°C, and then introducing the uranium bar, pre-heated to 350°C, into the flowing aluminum as a floating, self-propelled mandrel. The extrusion operation provided a cladding thickness of 2.3 millimetres (0.090 inches), which was subsequently machined down to 0.63 millimetres (0.025 inches). It took about ten minutes to machine one clad bar, and an eddy-current instrument was used to inspect the final cladding thickness. The flat fuel elements were assembled by hand into an extruded aluminum coolant tube, which contained spacing fins along its length to position them.

Over 12,000 nickel-plated and extrusion-clad uranium bars were irradiated in NRU, and no fuel failures were detected. However, the burnup achievable with this fuel was limited to about 43 MWh/kgU, and work started on development of an alternative design of natural-uranium-metal fuel (an assembly with nineteen round elements). This work was terminated when it was decided to fuel NRU with enriched fuel assemblies.

In 1962 it was decided to change NRU from operating with natural-uranium fuel to operating with enriched-uranium fuel. The objective was to have a complete loading of enriched-uranium-aluminum fuel assemblies by 1964. Two approaches were considered. The first was to purchase an existing co-extruded tubular design manufactured in the United States. The second was to develop a design that could be made in Canada without a major development program. The experience gained from the fabrication and irradiation of the enriched seven-element assembly in NRX indicated that a multi-element assembly for NRU could probably be developed without a major effort. Various design configurations were investigated, including the number of fuel elements, coolant flows, coolant temperatures, sheath surface temperatures, and allowable surface heat flux. The investigation showed that a twelve-element fuel assembly, nine feet (2,743 millimetres) long, of small diameter elements, would meet the design requirements.

The enriched-uranium-aluminum-alloy billets were fabricated by essentially the same process used for the NRX fuel. The alloy was 9.8 wt. percent uranium and the uranium was 93 wt. percent uranium-235. Similarly, the uranium-aluminum fuel cores were clad with 0.03 inch (0.76 millimetres) thick aluminum by the extrusion cladding process. The extruded aluminum-alloy coolant tube was circular in shape. Six spacers were required to locate the fuel elements along the length of the assembly.

Post-irradiation examination of several twelve-element assemblies showed no significant corrosion, erosion, or fretting wear at the design coolant flows and reactor residence times. The twelve-element assembly became the production design for the enriched-uranium-aluminum fuel for NRU.

WR-1 Fuel

Fuel development work for both WR-1 and CANDU-OCR concepts was carried out in the X-7 and U-3 loops at CRL. However the operation of both standard and experimental fuel in WR-1 itself contributed greatly to the inventory of knowledge regarding fuel performance with organic coolants.

As previously mentioned, it was possible to operate WR-1 with coolant conditions similar to those suitable for a power reactor. Of the three separate organic coolant circuits in WR-1, the 'A' and 'B' circuits were generally operated at a coolant inlet temperature of 290°C. However, the 'A' circuit was operated at an inlet temperature of 375°C (outlet temperature about 400°C) from 1971 to 1977. The 'C' circuit inlet temperature was 320°C.

The initial driver fuel used in WR-1 was of an eighteen-element uranium-dioxide design similar to the Douglas Point nineteen-element bundle, except that the centre element was omitted so that the bundle would fit over a central support shaft. The uranium was enriched in the range 1.2 to 2.4 wt. percent uranium-235 in total uranium. Two fuel types of this general design were made: one used Zr-2.5Nb-sheathed fuel elements assembled into 19.5 inch (495 millimetres) long bundles, with each reactor fuel string having five bundles suspended on the central hanger tube; the other used sintered aluminum product (SAP), sheathed fuel elements assembled into 32.1 inch (813 millimetres) long bundles, with each reactor-fuel string having three bundles suspended on the central hanger tube. About half of a full reactor charge of the SAP-clad fuel had been made by the time WR-1 started up, and this was eventually irradiated, but no more was made.

WR-1 operated with the uranium-dioxide driver fuel from startup in 1965 until 1973, when uranium dioxide was gradually phased out in favour of uranium- carbide fuel. Experience with the uranium-dioxide fuel in WR-1 was not good compared to the performance of uranium dioxide in water-cooled reactors[27] and to the later experience in WR-1 with uranium-carbide fuel performance. The overall average string burnup for all the uranium-dioxide fuel irradiated in WR-1 (about 1,100 bundles) was 128 MWh/kgU, which satisfied the original target of 120 MWh/kgU, but little more. The number of fuel strings removed due to failure or sticking exceeded those retired without incident.

Experimental irradiations of uranium-carbide fuel started in 1966, and in 1973 the irradiation of uranium carbide as the reactor-driver fuel, in quantities sufficient to obtain statistical information, commenced. Conversion of WR-1 from uranium-dioxide to uranium-carbide driver fuel was essentially completed by the end of 1977.

The driver-fuel bundles for WR-1 consisted of one ring of 14 Zr-2.5Nb-clad, uranium-carbide fuel elements, and five of these were assembled on a hollow central hanger to form a fuel string. The uranium was enriched in the range 1.3 to 2.25 wt. percent uranium-235 in total uranium.

The performance of the uranium-carbide fuel was excellent both from the viewpoint of burnup achieved and failure rate. Average burnup of the first 125 bundles retired was 253 MWh/kgU, whereupon the target burnup was raised from 240 MWh/kgU to 360 MWh/kgU. Not only was the failure frequency low, but the consequences of failure were not serious. It was found that while activity releases were high enough to detect failure, they remained low for a long enough period (six weeks or more) to continue operation of the failed fuel until the next scheduled shutdown before removing it. A significant problem was encountered in the first few strings of uranium-carbide fuel operated in the 'A' circuit during the period of its high-temperature operation. Hydrogen migration to the bundle end plates caused their embrittlement, and six bundles had to be retired due to end-plate breakage during shuffling operations. The problem was solved in the short term by changing the fuel management scheme. A longer-term solution, involving larger hydrogen-sink volumes in the end-plate region, was developed.

Research Reactor Loops

An in-reactor loop is a closed-circuit hydraulic system that provides a coolant and radiation environment similar to that in a nuclear power reactor. It consists of pumps to circulate the coolant, heaters to heat the coolant, coolers to remove the excess heat, pressurizer to pressure the coolant, test sections for the test assembly, piping to interconnect the components, and the electrical and process control system to power and control the process. Figure 2.18 shows an example of a loop circuit.

The concept of high-pressure, high-temperature, light-water loops with an in-reactor test section for fuel development, water chemistry, materials testing, etc., was introduced to CRL in 1950 by the United States Atomic Energy Commission and the Westinghouse Atomic Power Division (WAPD). The first high-pressure, high-temperature water loop started operating in the central thimble of NRX in 1951 and was removed in 1952. This was followed by the installation of the WAPD CR-IV and WAPD CR-V loops starting in 1954.

Many loops were installed in AECL research reactors over the years, and these were often modified over the course of their use. They were an extremely important resource for the Canadian power-reactor development program. Table 2.2 lists

---

[27] Two factors make the behaviour of organic-cooled uranium-dioxide fuel elements significantly different from that of water-cooled elements: the relatively low system pressure and high-sheath surface temperature in the organic-cooled case. These conditions allow the fuel to swell more readily and, at the same time, the higher fuel temperature causes a greater fission-gas release from the fuel, which tends to reduce the fuel-to-sheath heat-transfer coefficient. The situation is unstable and eventually causes failure at a hot spot where diametral growth, bowing or surface deposits have restricted coolant flow.

FIGURE 2.18
U-2 Loop Simplified Flowsheet

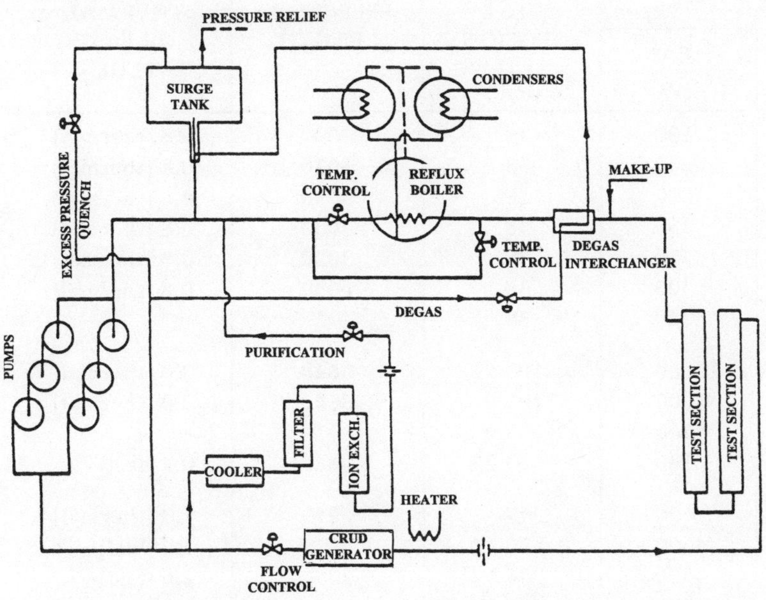

key characteristics of the main AECL loops, circa 1980, to indicate the scope of the loop facilities that were available for use in power-reactor development. Table 2.3 lists the uses made of the loops.

Fuel testing in NRU loops could be done with full-scale power reactor fuel assemblies, whereas tests in NRX were limited to single- or few-element fuel. The following points will supplement the basic data listed in table 2.2:

- X-1 was one of the WAPD CR-IV loops installed between 1954 and 1957. A major modification was made in 1964, when both the main flow and the heat rejection capacity were increased. Boiling was limited to low outlet steam qualities.
- X-2, although originally a WAPD CR-IV loop, was completely redesigned and rebuilt by 1977 as an extended-fuel-defect-and-safety-related studies loop. The modifications included:
  - filters at the outlet to trap fuel particles released from failed fuels, special valves for loss-of-coolant tests,
  - extensive on-line gamma-spectrometry equipment to monitor sections of coolant piping for radioactivity, and
  - an improved ventilation system to minimize radioactive releases to the environment.
- X-3, another original WAPD CR-IV loop, had its flow and heat-rejection capacity uprated in 1964, and later modifications were made to permit operation with two parallel in-reactor test sections. The loop was largely dedicated to the Westinghouse fuel program.
- X-4 was originally designed and built by the English Electric Company Ltd. for the United Kingdom Atomic Energy Authority as a pressurized water loop, and installed in 1957. Modifications were made in the period 1962 to 1966, which extended the operating range and permitted its use as a multi-mode water loop (as table 2.2 indicates).
- X-5, the largest loop in the NRX reactor, was designed and fabricated for WAPD as the WAPD CR-V loop, and installed in 1955. Until about 1979, it incorporated a test section of a 62-millimetres internal diameter. It was found that the smaller-diameter test section listed in table 2.2 was better suited to later programs.
- X-6, originally similar to the other WAPD CR-IV loops, was modified and uprated as a boiling-water loop in 1963.
- X-7 was operated in NRX until the organic-cooled-reactor program was terminated in 1972, at which time it was decommissioned. It was left as intact as possible to facilitate recommissioning if required. X-7 was used for tests associated with WR-1 development.
- U-1, as originally installed in 1964, operated in four coolant modes: pressurized water, boiling water, steam-water inlet, and steam inlet. In 1974 the loop was uprated to operate with two in-reactor test sections in series, and this modification limited operation to the pressurized and boiling-water modes.
- U-2 was installed in 1959 as a pressurized water loop, but subsequent tests showed that limited boiling to about 10 percent outlet steam quality was possible. In 1974 the loop was uprated to operate with two in-reactor test sections in series.

TABLE 2.2
General data for AECL research reactor loops – circa 1980

REACTOR	LOOP	COOLANT & MODE *	DESIGN			TEST SECTION I.D.	EFFECTIVE LENGTH	THERMAL OR FAST FLUX
			PRESS (MPa)	TEMP. (°C)	HEAT (kW)	(mm)	(mm)	(10^{14} n,cm^{-2}.s^{-1})
NRX	X-1	PW,B	17.3	330	300	38	3030	0.8 (thermal)
	X-2	PW	13.9	336	200	38	3030	0.8 (thermal)
	X-3	PW,B	17.3	330	600	38	3030	0.8 (thermal)
	X-4	PW,B,TPI,S	15.3	335	250	38	3030	0.8 (thermal)
	X-5	PW	17.3	330	550	38	3030	0.8 (thermal)
	X-6	PW,B	18.6	359	300	38	3030	0.8 (thermal)
	X-7	organic	2.9	430	220	38	– **	–
NRU	U-1	PW,B	13.9	354	12000	103.9	3640	3.0 (thermal)
	U-2	PW,B	13.9	354	8000	103.9	3640	3.0 (thermal)
	U-3	organic	4.2	430	4500	103.9	– **	–
	U-5	PW	17.3	354	140	63.25	3640	0.2 (fast)
WR-1	1L2	PW	9.0	330	400	47	2250	1.2 (thermal)
	1L4	organic	3.8	427	4500	76 ***	2250	1.0 (fast)
	1L5	organic	3.8	427	4500	76 ***	2250	1.0 (fast)
	1L6	organic	8.6	427	4500	76 ***	2250	1.0 (fast)

* PW pressurized water
B boiling
TPI two-phase at inlet
S steam

** decommissioned

*** This is the diameter of the support tube for the standard fast neutron fuel assembly insert normally used. Without the insert the diameter available is 104 mm.

- U-3 was operated in NRU until being decommissioned in 1972. It was left as intact as possible to facilitate recommissioning if required. U-3 was used for tests associated with organic-cooled reactor development.
- U-5 was designed for materials testing, not fuel testing. Light water at the required chemistry, pressure and temperature is supplied to as many as six in-reactor test sections operating in parallel, and normally installed inside hollow fast-neutron rods.
- 1L2 was normally operated in the pressurized mode, although boiling was allowed in the fuel channel with outlet steam qualities up to 15 percent.
- 1L4,5,6 were essentially identical, except for a difference in some of the piping in 1L6, which allowed it to operate at a higher pressure than the other two. A fast-neutron fuel assembly was normally inserted. The loops were used primarily for materials-irradiation experiments at high temperature in a high fast-neutron flux.

Loop experiments impose demanding data-gathering requirements that need flexible and highly reliable systems. Computer systems were installed at NRX and NRU for data acquisition in the early sixties, but by the seventies they were becoming increasingly unsatisfactory. In the mid-seventies there was a growing interest in the development of computer networks for data processing and control (i.e., distributed systems). It was felt that CANDUs might benefit from such systems, as replacements for the dual central computers, through reduced cabling costs. To evaluate this new technology, it was decided to develop a distributed data acquisition system, to be known as REDNET, for the NRX and NRU loops, to replace the old system.

REDNET development proceeded slowly. The complexity of distributed system concepts had been underestimated, and the project had started at a time of dwindling funding for AECL Research, and of changing priorities (see chapter twenty). By the time it was completed, about 1985, it represented the largest software project undertaken at CRL (sixty-five person-years) and it was required only in NRU. A local network with five processors gathers, processes and stores the data from more than one-thousand data points, which are scanned twenty times per second. Terminals in NRU and other buildings allow users to monitor data, to request displays, and to modify data-processing rates or display modes. Archival data is transferred to the computing centre for permanent storage. High reliability is achieved through sharing of tasks amongst processors, and automatic adjustment of this sharing under failure conditions. The system is flexible and capable of considerable expansion; it should be readily adaptable to new hardware and software.

Isotope Production

Chapter five describes the production and marketing history of radioactive isotopes in Canada. The following will give a brief account of the CRL contribution to that history.

TABLE 2.3
Uses made of loops

USES	LOOP														
	NRX							NRU				WR-1			
	X-1	X-2	X-3	X-4	X-5	X-6	X-7	U-1	U-2	U-3	U-5	1L2	1L4	1L5	1L6
Fuel development	✓				✓	✓		✓	✓		✓				
Pressure tube development	✓			✓		✓		✓	✓						
Activity transport	✓			✓	✓	✓		✓							
Water chemistry	✓			✓	✓	✓		✓	✓						
Irradiation of metallurgical samples	✓			✓	✓										
Materials testing											✓		✓	✓	✓
Corrosion - in reactor	✓			✓	✓	✓		✓	✓						
Corrosion - out reactor				✓	✓	✓		✓	✓						
Fuel defect studies		✓													
Heat transfer						✓									
Critical heat flux				✓				✓	✓						
Fuel moving				✓											
Fuel power ramping								✓	✓						
Fission product release		✓				✓		✓							
Fuel power cycling								✓	✓						
Fuel irradiation in: pressurized water				✓				✓							
Fuel irradiation in: boiling water				✓				✓							
Fuel irradiation in: steam/water				✓				✓							
Fuel irradiation in: steam				✓				✓							
Minor fuel sheath defect tests	✓				✓										
Instrument development				✓											
Organic-cooled reactor development							✓			✓					
Westinghouse fuel program			✓												

Driven by demand, production of radioisotopes started in NRX as soon as the reactor was operational. An isotope production branch (IPB) established within operations initially used existing laboratory space. At the same time, planning started for construction of new facilities for the production and separation of isotopes and their distribution to universities and others.

By September 1947, millicurie quantities of phosphorus-32, iodine-131 and sulfur-35 had been extracted from seventy-five samples irradiated in the central thimble of NRX, and two external requests were in the process of being supplied. The first actual radioactive isotope shipment (of cerium-144) was made to the University of Saskatchewan on 31 October 1947 (three months after NRX startup).

By 1949, routine shipments were being made of iodine-131 to pharmaceutical companies for medical treatments, and of phosphorus-32 and sulfur-35 to university laboratories across Canada. Many service irradiations were done by request. Separation processes were being developed[28] to extract tritium from irradiated lithium fluoride and carbon-14 from irradiated potassium nitrate. Cobalt was being irradiated to produce cobalt-60.

During this period, isotope production in NRX was facilitated by changes in the reactor and by new techniques. Originally, isotope production in NRX was restricted to the central thimble (high flux, but serviced only during reactor shutdown) and the horizontal self-serve units (low flux, but serviced during reactor operation). Additional space for high-flux, in-core irradiation became available when it was found that three of the four control rods were not required. This led to the development of tray rods to position isotope targets in reactor fuel sites, which, after modification of the fuel-rod flask, permitted their transfer in and out of the reactor during a short shutdown.

In 1950 the IPB, with the same small professional staff but more process operators, moved into a new, large building devoted solely to isotope production, and the isotope program was immediately expanded. A thermal diffusion column was built for the tritium production program, with the final product being absorbed by uranium turnings in special containers. Carbon-14 was separated. Mixed fission products were transferred from the plutonium separation plant and were selectively separated using ion exchange methods (e.g., strontium-90, cerium-144 and cesium-137).

The value of cobalt-60 for both radiography and radiotherapy was recognized very early. In the fall of 1949, one-curie sources for radiography were produced for Dominion Bridge Company and Toronto Iron Works. The first irradiation of cobalt to produce cobalt-60 for beam therapy units (which was to become the dominant use of this isotope) started in the fall of 1949, and shipments of two 1,000-curie sources were made in 1951.

Around 1950 it was agreed that the commercial products division (CPD) of Eldorado would take on responsibility for the sale of NRX-produced cobalt-60, and in 1951 this agreement was extended to include most NRX-produced isotopes. When CPD was transferred to AECL in the fall of 1952, CPD took on responsibility for marketing all AECL-produced radioisotopes. In the period 1952-55 there was a transfer of some IPB staff to CPD in Ottawa, and some CPD staff to CRL, to work on processes. The IPB at CRL was disbanded about 1955, when the operations for isotope separation were transferred to Tunney's Pasture in Ottawa under CPD.

The reactors at Chalk River remained the major source of bulk radioisotopes for CPD,[29] henceforth referred to as RCC (radiochemical company), throughout the period to 1985 and beyond. The CRL operations group continued to play an important role in the success of RCC, not only in the supply of

---

[28] The aim was to develop methods for the production of "carrier-free" isotopes; i.e., free of isotopes of the natural element. That is why potassium nitrate, containing no carbon, was used to make carbon-14 initially. For phosphorus-32 the target was a compound of sulphur. Any amount of phosphorus-32 could be produced by the (n, gamma) reaction with natural phosphorus, but it would be much diluted with natural phosphorus and hence less valuable as a tracer in agriculture or medicine.

[29] Reactor control rods (those used for xenon-poison override, whose function required neutron absorptions over long periods at power) for the production of high-specific-activity cobalt-60 were put to efficient use. Dr. Lewis was fond of pointing out to visitors how this function of the control system was performed by the production of an isotope subsequently used for medical purposes.

radioisotopes, but in ensuring the reliability of this supply and in adapting to requirements for expansion of production, both in terms of volume and type of product. The advent of NRU with its capacity for large-scale isotope production was an important element in the ability to satisfy these demands.

The two radioisotopes constituting the bulk of RCC sales were cobalt-60 and molybdenum-99. In the eighties responsibility for separating molybdenum-99 and xenon-133 was moved to CRL under its operations group.

CHEMICAL OPERATIONS AND OTHER TECHNICAL SUPPORT

Reprocessing

In line with the original mission of the laboratories, a plutonium separation plant and a uranium-233 separation plant soon followed the NRX reactor as CRL facilities.

Initially, the plutonium-separation plant used a process developed earlier at the Montreal Laboratory. This involved dissolving the aluminum sheathing from the spent uranium-metal NRX fuel rods using caustic soda, dissolving the uranium in nitric acid, conditioning the dissolver solution with ammonium nitrate, extracting the plutonium with "trigly" (triethylene glycol dichloride), scrubbing the trigly extract with a solution of ammonium nitrate and nitric acid, and precipitating the plutonium as a phosphate. The precipitation stage of the process never worked properly and, shortly after starting up the plant, it was replaced by a process involving the use of hexone (methyl isobutyl ketone), and a final purification step using thenyltrifluoracetone (TTA) and concentration by evaporation. While the modified process operated well, with an overall recovery of 92 percent of the plutonium in a 95 percent pure product, the continued use of ammonium nitrate was undesirable, since it constituted a serious explosion hazard.[30] The process was further modified, using a scheme that replaced the ammonium nitrate with uranyl nitrate and the hexone with tributyl phosphate (TBP). This process operated reasonably well, but was shut down in 1954 when development work on an alternative process showed more promise.[31]

In this alternative process the plutonium was separated from the uranium and fission products by an anion-exchange resin. A pilot plant was built that provided two passes of the dissolver solution through ion-exchange resin, with intermediate conditioning between passes to reform the anionic complex. After some minor early problems, largely associated with the intermediate conditioning step, the process worked well. Although the decontamination factors approached 10^5, the final product still contained about 1.5 percent uranium, and enough radioactive niobium to make further extensive handling hazardous without shielding. Thus the TTA purification process was retained for final purification.

The uranium-233 separation plant also used a process based on work at the Montreal Laboratory. The feed for the plant came from the thorium rods that were irradiated in the J-rod annulus of NRX. The process developed at the Montreal Laboratory for uranium-233 separation involved dissolving the aluminum sheathing from the thorium-oxide rod with caustic soda, dissolving the thorium in nitric acid with some hydrofluoric acid added, extracting the uranium with diethyl dithiocarbonate in hexone and backwashing with 10 percent ammonium carbonate. This reduced the thorium/uranium ratio from about 9,000 to 30-40. Time had not permitted the chemists at the Montreal Laboratory to take the process further, so a method for further processing had to be developed at Chalk River. This involved a controlled thermal decomposition of ammonium carbonate in the backwash solution, followed by anion exchange. The overall uranium-233 yield in the process was about 95 percent.

---

[30] An explosion of ammonium nitrate in an evaporator in December 1950 resulted in the death of a worker.

[31] In about 1950 a Chalk River scientist, while looking for a better method of analyzing solutions from the plutonium separation plant for plutonium, discovered that plutonium could be effectively separated from most other materials by absorbing it from a strong nitric-acid solution on anion-exchange resins, which were just then becoming commercially available. He reasoned that this could provide the basis of an effective separation process. This led to further laboratory investigation and the development work mentioned.

By 1956, arrangements had been made to sell the used fuel from NRX and NRU to the USAEC, and all reprocessing facilities at Chalk River were shut down. While they operated, they produced a total of about 17 kg of plutonium and about 500 g or less of uranium-233. The first 3 kg of plutonium was sold to the United Kingdom for use in their military program before they had product from their own facilities, and the remainder was used in the development and fabrication of booster fuel rods for NRX.

Hot Cells

Shielded facilities for the remote handling of highly radioactive materials, using manipulators, are commonly known as "hot cells". The first AECL Research hot cell was built in the metallurgy building at CRL, and was operated by the R&D arm of AECL Research. This hot cell, known as the "cave", was used for all early examinations of highly radioactive materials at CRL (apart from cursory examinations that could be carried out in the NRX fuel bays, where spent fuel was stored underwater). In the mid-fifties a second cave was built in the same location, to provide increased capacity and the ability to handle more highly radioactive materials.[32] Over time, metallurgical facilities were added adjacent to the metallurgy caves, to provide expanded examination and measurement capabilities. The whole facility was used primarily for R&D, and has always been operated by the R&D arm of AECL Research.

In the mid-fifties it was decided to build another hot-cell facility at CRL specifically for the handling of irradiated cobalt, the examination of irradiated fuel, the selection of metallurgical samples, and the repair and modification of irradiated or contaminated equipment. This facility was called the universal cells and was the responsibility of the operations arm of AECL Research.

The first section of the building that houses the cells was constructed in 1956 and two cells were installed, over the next year or two, in this section. Figure 2.19 shows a front view of the two cells. As well as the cells themselves, the building includes isolation rooms, a shipping and receiving area, change rooms and an office. These facilities soon became inadequate to meet the requirements of the expanding cobalt-60 production,

FIGURE 2.19
Front View of Cells 1 and 2

[32] These metallurgy cells were "dirty", in that the floors, tables, walls and ceilings were covered with radioactive debris arising from cutting open radioactive specimens. The facilities would operate in this condition for stretches of about five years, with installation of new equipment and repair and maintenance having to be done remotely. The cleanup at the end of the five years was a major undertaking, requiring involvement of personnel from throughout the plant.

and of the fuel-development program. An extension to the building was built in 1962 to accommodate a third cell, used chiefly for the examination of full-length fuel assemblies.

Equipment was installed in the universal cells for various purposes, including dissolving the aluminum sandwiches that contained nickel-plated cobalt pellets irradiated in NRU control (adjuster) rods before the pellets were shipped to RCC in Ottawa. Later, uranium-235 targets irradiated for molybdenum-99 production were dissolved in the cells for xenon-133 production before being shipped to Ottawa for processing. The cells also served as a backup to RCC facilities in Ottawa for molybdenum-99 processing.

The facilities were available for use by all CRL in-reactor loop users, including CRL experimenters and those from external laboratories. They allowed fuel tests from the in-reactor loops to be examined between irradiation periods, as well as after completion of their irradiation. Relevant equipment included various optical devices and photographic, machining, cutting and welding capabilities. In many cases preliminary inspection of tests was done in the universal cells, where smaller samples were selected and prepared for shipment to the metallurgical cells for more detailed examination.

At WL the functions of the universal cells (apart from the production functions, for which there was no need) and the metallurgical cells were combined in the hot-cell facility (HCF). The HCF was built to provide post-irradiation nondestructive and destructive examination for the various WL programs, fuel management and component handling for WR-1, and remote handling services and facilities, as required. It started operation in 1965 and consisted initially of eleven hot cells, to which a twelfth cell was added that began operation in 1976. Of special note was the dedicated use of four of the cells for an extended period for the thorium-fuel reprocessing experiment (see chapter eighteen).

Heavy-water upgrading

In the early fifties, a heavy-water upgrading plant was built adjacent to NRX, using electrolytic cells obtained from the plant run by Cominco at Trail, British Columbia. The operation was expanded in 1952, first using more cells obtained from Trail, as well as newer and better cells developed at CRL. A new plant was built and, for some time, ran at full capacity (4 MW of electrolysis or 1,000 Mg deuterium oxide/a at a feed concentration of 65 percent deuterium oxide), with a large backlog, upgrading heavy water primarily from NRX, NPD and Douglas Point. In the eighties, upgrading was done on a commercial basis for the utilities and the heavy water plants,[33] but the demand was not sufficient for full-capacity operation.

Site Waste Management

Chapter nineteen describes the history of AECL Research site waste management. The actual implementation of an effective waste-management strategy for the sites involved considerable skill and dedication from the operations groups. The activities for which they were responsible included:

- management of the waste-management sites, including construction and preparation of the relevant facilities at those sites for storing the wastes, collecting the site wastes and transporting them to the waste-management sites,
- emplacement of the wastes at the waste-management sites,
- security of the sites,
- maintaining appropriate records,
- operation of the waste treatment centre (WTC) at CRL, and
- operation of the underground research laboratory (URL) at WL.

Additional responsibilities

The CRL operations group was also responsible for operating a number of other facilities. For example, operation of the larger accelerators used for nuclear-physics research, mentioned in

[33] The production rate at Glace Bay and Port Hawkesbury was limited by the capacity of the water distillation final-enrichment step, and was increased by producing 94 percent heavy water, which was raised to reactor grade (99.8 percent) at CRL.

chapter eight, was the responsibility of the operations group until relatively recently, when the responsibility was transferred to the R&D group.

There were also a number of activities managed by the R&D arm of the organization, involving operation of facilities or provision of services, that were similar in some ways to activities under operations. Some examples are:

- the operation of the lattice test reactors, ZEEP and ZED-2, and the pool-test reactor, PTR,
- the operation of the Computing Centres (until 1986),
- analytical chemistry,
- electronics and instrument development,
- nondestructive testing, and
- glass blowing.

These were under the R&D arm, in general, for at least one of two basic reasons:

- the facility or service was used almost exclusively by one specialized group (examples being the operation of ZEEP, ZED-2, and PTR, and glassblowing) and/or
- the operation was of a highly specialized and technical nature, often with a significant R&D component (examples being the computing centre, analytical chemistry, electronics and instrument development, and non-destructive testing).

ACKNOWLEDGMENTS

Significant contributions to the content of this section from J.H. Collins, J. Inglis and J.S. Nelles are hereby gratefully acknowledged.

BIBLIOGRAPHY

Campbell, W.M. "Canadian Fuel Reprocessing." *Nucleonics* 14 (September 1956): 92-7. (Also AECL-357.)

Hurst, D.G., R.E. Manson, E. Critoph, and H.E. Petch. "Utilization of Canadian Research Reactors." In *Proceedings of the Third International Conference on the Peaceful Uses of Atomic Energy, Geneva, 31 August-9 September 1964,* vol. 7, 10-18. Paper A/CONF.28/P/2. Geneva: United Nations, 1965. (Also AECL-2002, -2002F.)

Jones, R.W., and J.L. Crosthwaite. *Uranium Carbide Fuel for Organic Cooled Reactors.* AECL-4443. 1973.

Logie, J.W. "Three Vessel Replacements at Chalk River." In *Decontamination and Decommissioning of Nuclear Facilities,* edited by Marilyn M. Osterhout, 531-45. New York: Plenum Press, 1980. (Also AECL-6899).

Manson, R.E., and H.E. Smyth. *The NRX Reactor: A General Description.* AECL-2692. 1967.

———. *The Universal Cells at the Chalk River Nuclear Laboratories.* AECL-4256. 1972.

Mooradian, A.J., E.C.W. Perryman, and T.J. Kennett. "Irradiation Facilities: Their Importance in Developing Canadian Nuclear Competence." In *Peaceful Uses of Atomic Energy: Proceedings of the Fourth International Conference on the Peaceful Uses of Atomic Energy, Geneva, 6-16 September 1971,* vol. 6, 277-90. Paper A/CONF-49/P/152. New York: United Nations; Vienna: International Atomic Energy Agency, 1972. (Also AECL-3976, -3976F.)

Nishimura, D.T. *Summary of Loops in the Chalk River NRX and NRU Reactors.* AECL-6980. 1980.

"The NRX Accident." *Canadian Nuclear Society Bulletin* 13 (winter 1992): 2-5.

Ryz, Marvin A. "Operating Experience with the Whiteshell Nuclear Research Establishment Hot Cell Facility." In *Proceedings of 27th Conference on Remote Systems Technology,* 42-7. Ed. by Diane R. Wallin. Chicago: American Nuclear Society, 1980. (Also AECL-6648.)

Summach, A.J. "The Commissioning of WR-1." *International Conference and Exhibition of the Canadian Nuclear Association, Winnipeg, 1966,* Paper 66-CNA-310. Toronto: Canadian Nuclear Society, 1966. (Also AECL-2552.)

Tegart, D.R. "Operation of the WR-1 Organic Cooled Research Reactor." AECL-3523. 1969. (Presented at Conference on Reactor Operation Experience, San Juan, Puerto Rico, 1969, CONF-691009-10.)

RADIATION PROTECTION AND NUCLEAR SAFETY

Chapter Three

Radiation Protection

A.M. MARKO
D.K. MYERS

GENERAL PRINCIPLES

Ionizing radiation in the form of X rays was first discovered in 1895, and natural radioactivity in 1896. Within the first decade of the discovery of X rays it had been established that X radiation was useful in medical diagnosis and in the treatment of disease; however, the hazards of skin damage, of bone marrow damage and of malignancy following high doses had also been noted. The need for protection against exposure to high doses of ionizing radiation was thus apparent.

Radiation protection was an ad hoc affair for many years and was directed primarily towards the prevention of reddening of the skin, since there was no acceptable method to measure small radiation doses. An ionization chamber developed in 1925 made it possible to measure small exposures accurately in terms of the amount of ionization they produced. An international committee on radiation units and measurement was founded in 1925 and this committee subsequently adopted units of radiation dose based on these physical measurements. A second international committee on radiological protection was founded in 1928. The latter was the forerunner of the International Commission on Radiological Protection (ICRP). Both committees operated under the auspices of the International Congress of Radiology (concerned originally with medical applications of X rays and radium) and both committees, although reorganized and renamed in 1950, remain to this day the most important sources of international recommendations concerned with radiation protection. Canadian scientists from a variety of organizations, but particularly from AECL, have participated for many years in the work of these committees.

In 1977 the earlier units of dose were replaced by those of the International System of Units. The unit of "dose equivalent" of ionizing radiation is the sievert

(Sv), which is the absorbed dose, in joules per kilogram, multiplied by a weighting factor that expresses the relative biological effectiveness of the radiation. The weighting factor is unity for X rays, beta rays and gamma rays, and is currently estimated to be 20 for alpha particles and 5-20 for neutrons, depending on their energy. A convenient unit for use in radiation protection is the millisievert, mSv, equal to one thousandth of a sievert. Formerly, the dose equivalent unit was the rem and 1 sievert equals 100 rem.

A useful reference point is the natural background dose received from cosmic rays and naturally occurring radioactive materials. It varies from about 1 mSv per year to about 10 mSv per year at various locations in Canada. The average is about 2 mSv per year.

Radiation doses to people can be external, internal or both (see figure 3.1). External radiation doses in people result from exposure to sources[1] outside the body. Internal doses result from radionuclides that have entered the body through ingestion, inhalation or absorption through broken or intact skin, and are distributed, metabolized or deposited in organs or tissues of the body (see figure 3.2). Elimination of radionuclides from the body occurs primarily via urine and feces. Physical decay of the radionuclides in the body must also be taken into account – especially for the short-lived nuclides. Mathematical models are used to assist in estimating the resulting doses. Radiation dosimetry uses a variety of methods to measure or estimate external and internal doses resulting from radiation exposure. Environmental research is concerned with the definition of the major pathways of distribution of radioactive materials from source to man, and to the formulation of mathematical models to represent their behaviour in the environment.

FIGURE 3.1
Risk of Health from Radiation

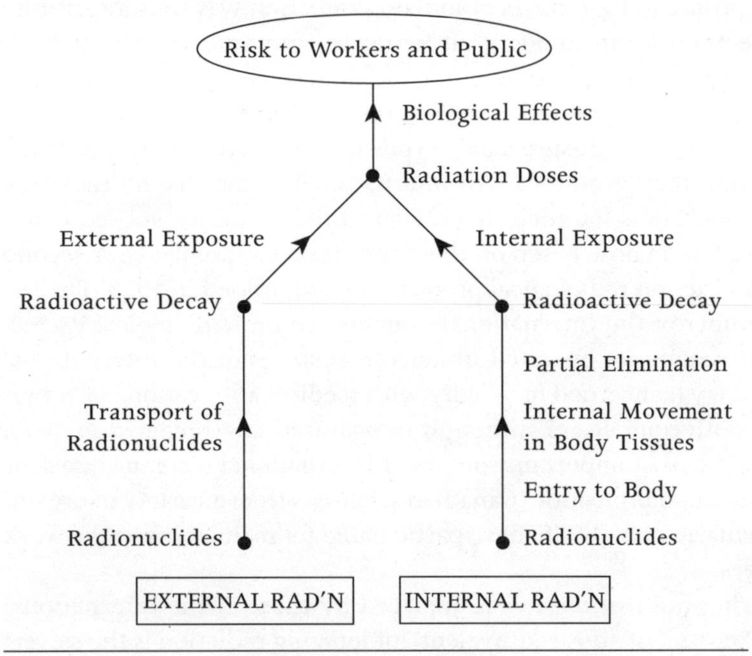

Depending on the magnitude of the dose, a series of effects may be produced (see figure 3.3). High levels of radiation (e.g., > 1,000 mSv in a short period of time) may lead to early clinical acute effects, such as death, skin burns and a variety of acute radiation syndromes. The acute effects of high doses are not seen clinically at doses of radiation below a measurable threshold (500 mSv is expected to depress the blood formation by bone marrow). Radiation, even at low doses, may cause delayed effects, such as cancers in the irradiated individual (somatic effects), and hereditary (genetic) effects in the progeny of the irradiated person. These delayed effects are believed to be caused by damage to DNA that is not properly repaired. They are sporadic and occur with a probability that increases with the dose; they are therefore said to be stochastic. Much research in radiation biology is devoted to the study of the biological effects of radiation, and incidentally, of other related toxic agents whose effects resemble those of radiation.

A specific health-and-safety objective in a nuclear establishment is to protect the worker and the public from unnecessary exposure to radiation. Meeting this objective calls for a wide

[1] This term is used by the ICRP to indicate the source of an exposure, not necessarily a physical source of radiation. When radioactive materials are released to the environment as waste, the installation as a whole might be regarded as the source.

FIGURE 3.2
Principal Metabolic Pathways of Radionuclides in the body

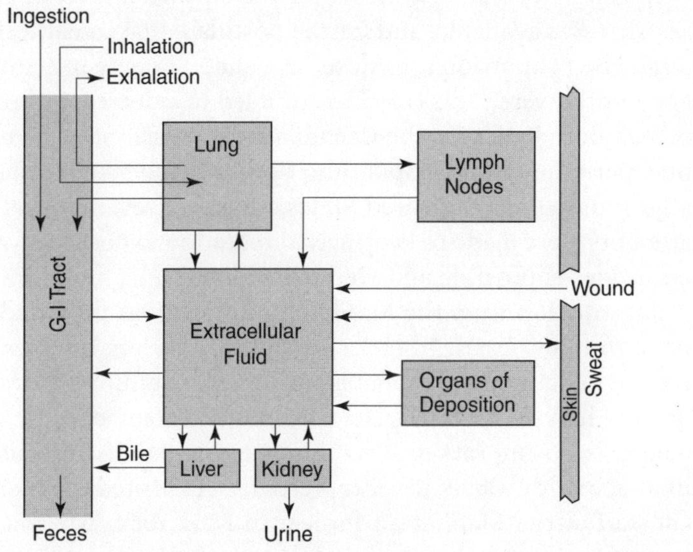

FIGURE 3.3
Biological Effects of Radiation on Living Systems

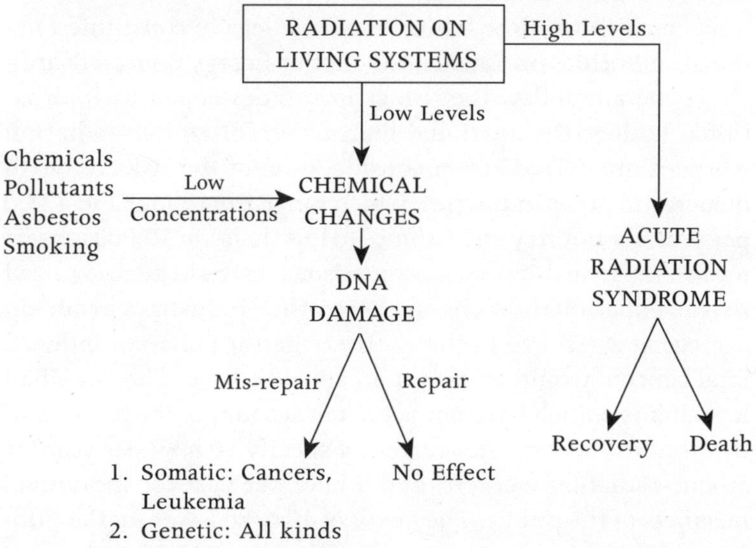

range of activities in epidemiology, biology, environmental research, dosimetry and operational health and safety. They are discussed in later sections of this chapter, but it is useful at this point to outline how the effects of ionizing radiation on humans have been determined, and how regulatory bodies arrive at recommended dose limits.

The effects (cancers and genetic changes) produced by natural background radiation or by radiation from Canadian nuclear reactors are too rare to be distinguished from the natural incidence of these effects. Risk estimates for somatic effects are obtained by following up groups of people who have been exposed to high levels of radiation, and making the prudent assumption that the rate of occurrence of these effects is proportional to the dose received, i.e., that there is no threshold below which exposure to radiation is without risk. The best information is provided by some 40,000 Hiroshima and Nagasaki survivors, who have been followed up since 1950. Estimates of the doses they received have recently (1986) been revised downwards, with a corresponding increase in the estimated risk per unit dose. The revised estimate is that the lifetime risk of fatal cancer from an accumulated dose of 1,000 mSv is 4 percent for workers (5 percent for the whole population), which is added to the normal risk, for Canadians, of about 26 percent as given by vital statistics. The risk coefficient is given for workers from age eighteen to sixty-five years, while for whole populations it is stated for exposure over a complete lifetime.

There is no evidence of any significant increase in hereditary defects in the children of the atomic-bomb survivors, or of any other group of irradiated humans that has been studied, but there is no biological reason to expect that radiation will not produce genetic effects in humans, just as it does in all other organisms that have been studied under carefully controlled laboratory conditions. Estimates of human genetic risk indicate that an accumulated dose of 1,000 mSv to parents prior to reproduction will produce a 0.4 percent chance of **serious** genetic effects in their children or a 2.5 percent chance of serious genetic effects in any of their progeny summed over all generations. A serious genetic effect is one that requires clinical attention at any time in the life of the afflicted individual. These risks may be compared with a normal incidence of about 25 percent in each generation in humans.

Recommended dose limits are set by legally constituted national authorities (in Canada, the Atomic Energy Control Board), who generally follow the risk estimates developed by international bodies. Occupational limits, i.e., for atomic-radiation workers, are derived from consideration of the risks of death incurred in other industries, which range from about 1 in 1,000 per year, for forestry and fishing, to less than 1 in 10,000 for safe industries. In making such comparisons it should be recognized that occupational deaths in these other industries occur on average at age thirty to thirty-five, whereas radiation-induced fatal cancers would appear at an advanced age. The Canadian legal limits, which have not yet taken account of the revision of the bomb-survivor data, currently specify 50 mSv per year for atomic-radiation workers, and 5 mSv per year for individual members of the public. The factor of 10 or so lower for the public is believed to provide an adequate margin of safety.

As well as being subject to these dose limits, atomic-energy activities in Canada operate under the ALARA principle, enunciated by the ICRP, that "all exposures shall be kept As Low As Reasonably Achievable, economic and social factors being taken into account." Thus, at Ontario Hydro nuclear-power stations occupational doses average about 2 mSv per year (that is, equal to the average background exposure from natural sources), and the maximum dose to individual workers does not exceed 20 mSv per year under normal operating conditions. The maximum exposure of any member of the public due to Canadian nuclear-power stations is about 0.05 mSv per year, and the average to all members of the public living within 50 km of a power station is roughly 100 times less than this maximum.

EARLY HISTORY: INTERNATIONAL ACTIVITIES

Radiation protection was a major concern from the beginning of the Canadian atomic energy project. In 1944 the United States Manhattan Project provided "information necessary for guarding the health of the operators at the Montreal Plant," and in 1945 a research laboratory was set up. A 1945 document from the Montreal Laboratory is remarkable for the advanced level of expertise that was available, and for the postulate that somatic effects might be proportional to dose, as genetic effects in fruit flies apparently were. This consideration led to a decrease in the accepted dose limits for the Canadian and British projects to 150 mSv per year, the corresponding figures being 500 internationally, and 300 in the United States. However, even at that time attempts were made to keep actual radiation exposure of workers as low as possible and well below limits.

Information from the Manhattan Project was provided on a need-to-know basis. In order to obtain additional information on the toxicity and metabolism of ingested radionuclides, one of the Chalk River biologists (G.C. Butler) commenced experiments, exposing rats to uranyl nitrate in 1946. After this preliminary study, Chalk River representatives visited the biomedical staff at the Manhattan Project in New York. After several hours of scientific discussion, the biomedical group seemed satisfied with the competence and credentials of the Canadians. Files on experimental animal studies at Berkeley were made available. With this preliminary information on uranium, plutonium and other radionuclides, Butler was able to compile one of the earliest reports on the topic in 1947, entitled "The Biological Basis of Maximum Permissible Exposure for Workers in the Field of Atomic Energy."

The purpose of Butler's report was to calculate the intakes of uranium, plutonium, strontium-90, iodine-131 and radioactive isotopes of krypton and argon that would be equivalent to one of the two "tolerance doses" (a maximum of about 150 mSv per year of radiation from external sources or a maximum of 0.1 microgram of radium-226 fixed in the body, primarily in the bone) accepted for workers at that time. This calculation required a fairly detailed understanding of the probable concentrations of each radionuclide in different tissues of the body and of the length of time that the various radionuclides were likely to remain in those tissues; much of the required information on these topics was obtained from United States' files.

In 1949 a tripartite conference on radiation-protection principles was held at Chalk River with participants from the United States, the United Kingdom and Canada. The notes and minutes of this conference provided a multitude of concepts, ideas, questions and resolutions to problems that set the scene for decades to come in international and national recommendations and regulations. There were eight Canadian representatives, including a Chalk River scientist, G.C. Laurence, who was to become a member of the International Commission on Radiation Units and Measurements in 1950, and later became president of the Control Board.

The conclusions of the tripartite conference had great impact on the recommendations of the ICRP. At least one person from Chalk River staff has participated in the deliberations of the ICRP or its committees since it was reorganized and renamed in 1950. C.G. Stewart served for two terms (1969-77) as chairman of the ICRP and was awarded a gold medal for his services.

The United Nations Scientific Committee on the Effects of Atomic Radiation (UNSCEAR) was established in 1955. Its purpose was to collect and evaluate reports on the levels of ionizing radiation and radioactive substances in the environment, and on the biological effects of ionizing radiations on humans. The ICRP relies heavily on the work of UNSCEAR for risk estimates and thus for deriving its own recommendations on corresponding dose limits. Many Canadians, including AECL staff, have made contributions to the work of UNSCEAR. Chalk River staff have contributed to UNSCEAR since its inception.

Because of the close liaison AECL scientists had with international organizations such as ICRP and UNSCEAR, they were familiar with developments and thinking at the forefront of radiation protection. Consequently, AECL management had confidence in these scientists when they advised on principles of radiation protection, risk estimates and dose limits. AECL adopted ICRP recommendations before they were incorporated into Canadian regulations, and sometimes even earlier than their formal publication by ICRP.

The rest of this chapter will outline AECL's contributions to the subject of radiation protection. Many of these contributions were directed to specific Canadian needs. However, from the early days it was realized that Canada could not alone do all the necessary research. To maximize access to the international pool of knowledge in this area, AECL published work of importance to others. Information exchange in the life sciences concerning radiation was rather free and open on the international scene.

EPIDEMIOLOGY

Significant effects have been observed in the epidemiological studies of miners who were in the past exposed to very high concentrations of radon and its decay products. Four of the eight major studies around the world relate to mining sites in Canada. Chalk River personnel have been involved with the radiation protection of uranium miners since the fifties, and have periodically reviewed the results of these epidemiological studies. An appreciable excess of lung cancers is observed in these early miners. Human population studies by Health and Welfare Canada failed to reveal any measurable effects of much lower concentrations of radon-decay products from natural sources on lung-cancer incidence in various Canadian cities, where the mean concentrations in the air in homes varied by tenfold. The effects of low concentrations of radon are too small to be measured in human populations.

A Chalk River scientist instigated an important epidemiological project on excess breast cancers in Canadian women exposed to multiple fluoroscopic examinations during treatment for tuberculosis in the thirties and forties. Accumulated average doses to the breasts were about eight times greater in the Nova Scotia portion of this study than in the Japanese bomb survivors. The study on excess breast cancers was carried out by the National Cancer Institute of Canada and has been widely cited by international scientific committees concerned with risks of radiation-induced cancers. The doses to the Nova Scotia patients were greater because they faced the fluoroscopic machine, whereas in Ontario they did not.

A variety of studies has been carried out in various countries, including Canada, on workers who were occupationally exposed to whole-body radiation, and on populations who live in areas where natural background radiation is higher than average. Those studies have been inconclusive; if there is any effect, it is too small to be detected against the normal high background of cancers attributable to other causes.

Epidemiological records were studied of Chalk River employees who had died between 1950 and 1989. Separate follow-up studies of 412 Chalk River workers who received an average of about 20 mSv each year for twenty-two years, and of 1,090 Chalk River personnel involved in the cleanups of the 1952 NRX and the 1958 NRU accidents, were also carried out. No excess cancers were observed in these studies and there was no excess of leukemias, over and above those expected in the general population. For example, there were 153 cancer deaths observed among all Chalk River male employees (including retirees) in 1966-89, compared to an expected value of 172 for the general population of Ontario; total deaths from all causes in the same group were 561 compared to 676 expected. Both Whiteshell and Chalk River staff were involved in follow-up studies of all 14,000 AECL employees who were employed up to July 1980. The data, as analyzed by the National Cancer Institute of Canada, are yielding similar results. Personnel from the University of Ottawa undertook follow-up studies of the armed-forces personnel who were involved in the cleanup of the NRX and NRU accidents; they did not reveal any excess cancers. These results are reassuring; no effects of radiation exposure were detected. The primary value of this type of study has been to allay concerns about the health effects of low-level radiation and to demonstrate that these effects are not being grossly underestimated by the ICRP risk estimates.

In order to estimate the health impact of radiation-induced mutations in humans, we need to know a) the normal incidence of genetic disorders in human populations, b) the normal rates of mutation (or genetic change) that are responsible for this normal incidence, and c) the increased rates of mutation that are induced by radiation exposure in experimental studies on mice and other organisms. The results of these experimental studies on mice are extrapolated to humans. To obtain quantitative information on radiation-induced mutation rates, in the fifties and sixties the Oak Ridge National Laboratory set up the multi-million dollar mouse experiment in which the frequency of coat-colour mutations in mice was scored after partial irradiation. This enormous and expensive study was funded by the United States' Atomic Energy Commission. AECL biologists were able to make important contributions to items a) and b) within their limited budgets.

As a result of his involvement in UNSCEAR discussions on the genetic burden in human populations from fallout from atmospheric testing of nuclear weapons, and the need for data on the normal frequency of genetic disorders in human populations, one of the Chalk River geneticists turned his attention to British Columbia, which had a good vital-statistics registry of births, deaths and marriages, plus a first-rate registry of handicapped children. These registries were separate and needed to be merged together to obtain the required information. Up to the late fifties, such consolidations were carried out manually, but he decided to develop a computer-assisted record linkage for this purpose. He received great help from computer experts at Chalk River in his early endeavours. The first problem was to devise means by which computers could recognize records pertaining to the same individual, even when the birth dates, spelling of names, and such did not agree precisely, as was frequently the case. By adding other bits of information, for example, mother's maiden name, ages, addresses, and so on, it became possible to identify specific individuals. Once the programming of methods to compare names and other data was established, it was possible to carry out record linkages between the registries of vital statistics and the registry of handicapped children. In this way quantitative information on the frequency of genetic disorders in the British Columbia population, their heritability and the longevity of afflicted persons, was obtained. The resulting data were extensively quoted in the 1977 UNSCEAR report on genetic effects of radiation. The principal investigator pointed out that the same methods of records linkage could be used to study the health of workers exposed to radiation and other carcinogenic agents, and worked together with collaborators at Statistics Canada to carry out studies of this kind. The technique of record linkage is now widely used by universities and other government agencies, such as Statistics Canada and the Royal Canadian Mounted Police.

Another Chalk River geneticist carried out a careful re-examination of the published literature on spontaneous mutation rates for various genetic disorders in humans. His review and corrections of accepted mutation rates was published in 1982 and cited extensively in the 1990 report on genetic hazards by the United States Committee on the Biological Effects of Ionizing Radiation (BEIR).

LINEARITY

The assumption, recommended by the ICRP, that the probability of radiation-induced cancers and genetic changes increases in direct proportion to any increase in radiation exposure is generally regarded as reasonable and prudent for purposes of radiation protection, but is very difficult to prove or disprove. The data on radiation-induced cancers from the bomb survivors and the early uranium miners are compatible with this assumption, but do not have sufficient precision to prove that it is true at the low radiation doses that are relevant to the application of nuclear energy. Other data, notably epidemiologic data on bone cancer in humans exposed to low doses of radium, and animal studies on lung cancer in rats exposed to low concentrations of radon, can be interpreted in terms of a threshold for prolonged exposure at low dose rate.

The best evidence for linearity comes from studies of genetic change in lower organisms. In 1955, a French scientist showed that the frequency of a genetic change involving the liberation of a bacterial virus was proportional to radiation dose down to 3 mSv. A 1980 Chalk River study of the frequency of genetic change in yeast cells showed linearity over a very wide range of doses - from several tens of sieverts down to 10 mSv, below which level the effects were too small to be measurable.

It is of interest to study more complex organisms than bacteria and yeast. One of the Chalk River biologists used wasps (Dahlbominus), which readily provide enough offspring for adequate statistical accuracy, to study the frequency of eye-colour mutations following parental irradiation. A total of half a million wasps was examined, and a linear dose effect was established down to 150 mSv. Reaching 75 mSv would have required another two million wasps, and this was not attempted. To test for linearity experimentally in complex organisms, high doses of radiation need to be used to obtain significant statistical results on a limited number of organisms.

A further attempt to probe linearity was carried out using a fish (trout), which is a vertebrate and thus more closely related to humans. To execute this experiment, a senior technologist travelled to a fish hatchery in Jasper, Alberta. He took along a portable cobalt-60 irradiator and thousands of phials to hold the fish embryos. On his return to Chalk River in 1967, the technologist scored the observed malformations in the fish embryos. A linear dose-effect relationship for these genetic endpoints was established down to 250 mSv in fish embryos from irradiated sperm. It took 40,000 fish embryos to establish the 250 mSv point. The natural mortality and infertility were **reduced** by 250-500 mSv of radiation, even though genetic malformations were increased. Not all of the biological effects of low doses of radiation are harmful. Hormesis is a term applied to the beneficial effects of radiation and there are reports of such effects in the scientific literature, but it remains a controversial topic. In 1994 UNSCEAR reviewed this topic as "Adaptive Responses to Radiation." The term is broader and is not restricted to beneficial results only.

Another study on radiation-induced genetic changes examined the intelligence of rats, as measured by their ability to learn and to remember how to find their way through complex mazes. However, despite repeated high doses to the male gonads in each of several generations, no significant effects on the average learning ability or on the distribution of learning abilities could be detected. The first results from this study were published in 1961. Although innate learning ability is pre-determined by the action of many genes, it was, reassuringly perhaps, found not to be affected by repeated doses of radiation to the gonads in successive generations. Lower doses were not tested, since no effect was found at high doses.

MOLECULAR BIOLOGY

Early files (1944-47) of the Canadian Atomic Energy Project reveal that comprehensive plans for biomedical research programs had been made. These plans involved studies on the mechanism of action of various types of radiations on cells, with special emphasis on nucleic acid and nucleoprotein components of cells. These were considered important because of the concept that DNA (one of the nucleic acids) was in fact the genetic material in the nucleus of the cell. In the first half of the 20th century it was thought that DNA was a simple organic molecule composed of four nucleotides containing the bases: adenine, thymine, guanine and cytosine. With the introduction of mild and gentle techniques for isolation of DNA in the forties, it was recognized that DNA was not a simple molecule but a large polymerized, complex molecule which could serve as the cellular genetic material.

In the fifties a growing body of evidence indicated that genetic effects of ionizing radiation and ultra-violet (UV) light were the consequence of damage to DNA in organisms. One of the Chalk River researchers in the fifties started to chemically characterize the radiolytic products of the bases in DNA. In the early seventies, many publications on the identification of radiolytic products began to appear, but this pioneer work at Chalk River predated those studies by a number of years.

In the late sixties, scientists at Chalk River became concerned that empirical quantitative studies of the biological effects of high doses of radiation, epidemiological follow-ups, and costly, large-scale animal experiments were not going to elucidate the effects of low doses and low dose rates of radiation. It was felt that more detailed information on the mechanism of radiation-induced damage in DNA and its subsequent enzymatic repair would provide insight into the effects of low doses and low-dose rates of radiation in living systems. AECL recruited a team of molecular biologists to work on DNA damage and repair. The objective was to gain more detailed information on mechanisms, which would complement the data on observed biological effects of high radiation doses and facilitate extrapolation down to low doses at low-dose rates, where the biological effects cannot be measured directly. These workers produced some notable firsts in biological sciences, which were recognized by awards from Canadian and American scientific societies.

The AECL researchers worked with radiosensitive and radioresistant strains of bacteria, yeasts and cultured human cells. The initial amount of damage (such as strand breaks and base damage) that could be measured in the DNA immediately after exposure of living cells to ionizing radiation was found to be independent of the type of cell used. The differences between radiosensitive and radioresistant strains of the same type of cell could be attributed to natural mutations that had inactivated one or more of the repair systems that normally remove most of the radiation-induced damage in living cells. The same principles that were found to govern the relative radiosensitivity of bacteria and yeast cells were found to apply to cells derived from humans with two specific hereditary diseases. One of these diseases (ataxia telangiectasia) rendered the persons, and cells derived from them, more sensitive than normal to the toxic effects of the high doses of radiation used in cancer radiotherapy. The second (xeroderma pigmentosum) rendered the persons suffering from this particular disease, and cells derived from them, more sensitive to the effects of UV light. In both cases, the disease was shown to be associated with inherited deficiencies in one of the DNA repair systems. This work, which was supported in part by a large contract from the United States National Cancer Institute, contributed to an important understanding of the relation between DNA damage and cancer development (chapter twenty).

A micro-organism isolated from food that had received very high levels of radiation proved to be very radioresistant. It was known as *micrococcus radiodurans*. It was found that the radioresistance was not inherent, but due to the extremely rapid enzymatic repair of the lesions induced in the DNA by radiation.

A Chalk River scientist in collaboration with a researcher from the University of Toronto found that a wide variety of organisms, simple as well as complex, contained long stretches of polypyrimidine tracts in their DNA. The polypyrimidine tracts consisted of a series of cytosine and thymine bases joined together without any purine bases being involved. Although the discovery of polypyrimidine tracts contributed to the knowledge of the structure of DNA, the significance of the structure is largely unknown.

A scientist at Chalk River was the first to demonstrate that phorbol esters known for tumour promotion caused damage of cellular DNA. He was able to demonstrate this damage because he had developed a fluorimetric technique to measure single and double strand breaks in DNA of cells. Attempts to market this assay were unsuccessful.

In risk analysis of energy production from a variety of sources it is commonly assumed that the average biological risk to people is constant. It was demonstrated by a Chalk River researcher that the risk of tumour causation in mouse skin by carcinogens was not constant, but could be altered by varying the regimen of initiators and promoters, such as their concentrations and the time between applications of the initiators and promoters to the skin.

Attempts were subsequently made at Chalk River to develop a test on white blood cells to identify cancer-prone and radiosensitive persons in the general population. Both ionizing radiation and chemical carcinogens were used in these studies. Further effort to develop a reliable test is ongoing. A simple and reliable test for the differences in radiosensitivity of individual humans would be particularly useful in cancer radiotherapy, and many studies on the radiotherapeutic aspects are being pursued in other laboratories around the world. The lack of reproducibility of these sensitivity tests in the normal work force has resulted in the postponement of these studies at AECL. Furthermore, the utilities who have funded part of this work are not interested in continuing support of this type of research.

The medical biophysics group at the Whiteshell Laboratories complemented work on the molecular biology of DNA by research efforts on early events occurring in fractions of a second following the deposition of energy into the system. These early physical events include, for example, the production of free radicals (see chapter seven), which cause chemical damage in DNA and in other biological constituents of cells. The chemical mechanism of the interaction between radiation and chemical carcinogens was investigated. The role of naturally occurring enzymes, such as superoxide dismutase and peroxidase, was studied to learn how these enzymes protect biological structures from being damaged by the highly reactive free radicals. It was found that these enzymes degraded the potent free radicals to less toxic molecules. This work has greatly improved our understanding of how radiation produces biological changes in living organisms.

The mechanism of action of chemicals, known as radiosensitizers, which increase the sensitivity of cells to the lethal effects of radiation, was also investigated. The ultimate aim of these studies was to improve the radiotherapy of cancers. The lack of cures for, and the recurrences of, cancers are believed to be due to the fact that cells in the centre of tumours are not killed because they are anoxic or lack oxygen, and for this reason are radiation resistant. Although radiosensitizers are effective in counteracting the radiation resistance of anoxic cells, cancer treatment centres have found that radiosensitizers have considerable toxic side effects. The sensitivity of mammalian cultured cells to heat and radiation was also studied. Results of these studies have implications for the use of hyperthermia in the treatment of cancers. Preliminary promising therapeutic results were obtained on clinically terminal cases.

Another focus of research at Whiteshell Laboratories was work on membranes. Membranes are important in cell functions, because they control the exit and entry of important food stuffs, waste products, minerals and water, while retaining macromolecules such as DNA and proteins. For this reason it was suggested that damaged membranes could lead to delayed effects (cancers and genetic defects).

One of the scientists at Whiteshell used model membranes prepared from biological materials to study the integrity of the membranes after exposure to low doses and low-dose rates of radiation, as well as to compare the effects of X rays with those of tritiated water. From his studies, he claimed that as the dose rate was decreased, the effects on the model membrane *per unit dose* increased. This observation, which received some publicity in the popular press, is at times called the "inverse dose rate effect," where the effectiveness of low-dose rates is greater than higher dose rates. The interpretation of these data generated considerable controversy within and outside AECL. Most of the AECL scientists did not accept the results or the interpretation of the experiments. The applicability of results on simple membrane models to complex living systems remains questionable. In 1980 the United States National Research Council's BEIR committee reviewed this work and concluded on page 464 of its report that "although it is well recognized that membrane integrity is essential for normal cell function, there is inadequate basic understanding of membrane structure and function on which to base a detailed theory of radiation-induced damage mechanisms." The work on model membranes was not reviewed by UNSCEAR, because of the work's absence of evidence that it was relevant to carcinogenesis, and because there was more direct evidence from animals on the induction of cancer at different dose rates.

TRITIUM

Tritium is a radioactive isotope of hydrogen that is produced naturally in small amounts by cosmic radiation. Large amounts were also released into the atmosphere during the testing of nuclear weapons in the fifties and early sixties, and spread around the world in the form of tritiated water in the atmosphere. This increased the natural concentration of tritium in the Ottawa River and in lakes by a factor of more than 100. Due to its radioactive decay (radioactive half-life 12.3 years) and gradual transfer into the ocean depths, the concentration of tritium in rivers and lakes has been decreasing steadily since it reached its peak in 1963-64, although the concentrations are still much higher than the natural levels.

Large amounts of tritium are formed in heavy-water reactors. Although every reasonable effort is made to contain this tritiated water, inevitably small amounts leak out, resulting in some exposures of workers, the public and the environment. Tritium is thus of special concern both to AECL and to the utilities that operate CANDU power reactors.

Exposure of workers at Chalk River to tritium has been systematically measured by urinalysis since 1956. In general, tritium exposures have been very small at Chalk River, but radiation doses from tritium at the Pickering plant increased to a point where they represented a third or more of the total occupational dose. This was a major reason for the construction of tritium-extraction plants to remove tritium from the heavy-water moderator (see chapter fourteen).

To test the various metabolic parameters applicable to tritiated water (HTO) in humans that were used by the ICRP in the sixties, volunteers (primarily research scientists in health sciences) were exposed to small amounts of HTO vapour in a special exposure chamber. A critical finding was that the rate of uptake of HTO by lungs was equal to the rate of uptake by skin. The effective half-life in the body was found to be ten days. This value was subsequently adopted by the ICRP. It was also observed that the ten-day half-life could be further reduced by a factor of three or so by drinking copious amounts of fluids. The effectiveness of protective clothing on the uptake of HTO by workers was also studied. For employees who need to work in an atmosphere of low levels of HTO, the most practical protective clothing is a ventilated plastic suit with a fresh air-line mask.

Improved stationary and portable monitors, compensated for gamma radiation, have been built by AECL to measure tritiated water vapour in air. An automated tritium urinalyzer was also built so that workers could monitor themselves at nuclear power stations. Ontario Hydro personnel claimed the use and maintenance of these urinalyzers was too fussy for routine use and they were therefore discontinued.

Studies on human volunteers and animals at Chalk River confirmed that, while HTO is actively metabolized and thus retained in the body, inhaled tritium gas (T_2) is not retained in the body but is quickly breathed out again. Very small amounts of inhaled tritium gas are oxidized in humans to HTO, and only by micro-organisms in the gastro-intestinal tract. The net result is that tritium gas is 10,000 times less radiotoxic than HTO. With the building of the tritium extraction plant, which removes tritium from tritiated heavy water and processes it in the form of a gas, it became necessary to be able to measure tritium gas and tritiated water separately. Monitors were developed using Nafion (a plastic film that is permeable to HTO but not to T_2), which provided the capability to measure these two forms of tritium separately. AECL research on the measurement and relative toxicity of tritium gas is also applicable to monitoring in the fusion research facilities that use tritium.

It was commonly believed by most scientists that tritium gas (T_2) is oxidized very slowly in the atmosphere, but some doubts had been expressed. One of the Chalk River scientists arranged an international collaborative experiment to measure the oxidation of T_2 released into the air. Scientists from Germany, France, United Kingdom, United States, and Canada cooperated in this experiment at Chalk River. It was confirmed that the rate of oxidation of T_2 in air proceeded very slowly. However, when the T_2 was deposited on the ground, it was oxidized rapidly by micro-organisms present in the soil.

Studies on the relative biological effectiveness of tritium beta-rays were carried out in Chalk River at the request of the health and safety division of Ontario Hydro. Various reports in the literature suggested that the relative biological effectiveness of tritium beta-rays range from one to four, depending on the biological endpoint and whether X rays or cobalt-60 gamma rays are used as the reference radiation. This creates considerable problems in the conversion of absorbed tritium doses to dose equivalents in units of mSv. It was decided to investigate this matter using an endpoint that is directly relevant to radiation protection, namely induction of cancer. The reference radiation used for comparison was X rays, which account for more than 90 percent of all man-made exposures to the general population. At Chalk River, two long-term studies were carried out over a period of about ten years by staff members from dosimetric research and radiation biology. The first study explored acceleration of the appearance of breast cancers in female rats. The second, funded in part by the Control Board and the utilities, explored the induction of myeloid leukemia in male mice from a special strain obtained from the Harwell laboratories in the United Kingdom. Both studies showed that the relative biological effectiveness of tritium beta-rays was about 1.2 compared to X rays. Recent recommendations by the ICRP are that tritium beta-rays and X rays should be considered to be equally effective for practical purposes in radiation protection.

AECL staff doing research in dosimetry, environmental sciences and radiation biology have all been involved from time to time in tritium research, and considerable expertise in this area has been acquired. Chalk River staff were thus invited by the Control Board Advisory Committee on Radiological Protection to prepare a report on the toxicity, metabolism and dosimetry of tritium; this report was approved for publication by the Control Board in 1987.

ENVIRONMENTAL RESEARCH

AECL's environmental research program has always been concerned with how radioactive material moves through the environment (pathway analysis), and its uptake by plants and animals. Mathematical models are designed to explain the dispersal of radioactive materials in the environment, and the validity of these models is tested by the analysis of radioactivity in environmental samples. Routine environmental surveillance was carried out on samples of water, air, plants and soil to ensure that releases of radioactive materials from facilities at Chalk River were below specified limits. The geographical locations of Chalk River and Whiteshell provide unusual opportunities for research studies of this kind, in that short- and long-term experiments can be carried out in field stations located in natural terrestrial and aquatic ecosystems. Both research sites are located in remote fenced-in areas, and are protected from external disturbance. The fact that the field facilities are in close proximity to research laboratories is also unusual, which attracts national and international scientists from government agencies and universities to undertake collaborative studies with AECL.

Perch Lake Basin is the main location at Chalk River for research on environmental radioactivity. This area contains two experimental disposal sites, where low-level liquid radioactive wastes from Chalk River operations have been deposited since the late forties. Since that time, small amounts of radioactivity have migrated through the surface and ground-water systems to other areas of the basin, including Perch Lake, and have provided an unusual opportunity to study the behaviour of radioactivity in natural ecosystems.

For example, researchers undertook studies of the migration of cesium-137 and strontium-90. These two radionuclides are of biological significance because they are readily taken up by living organisms; their migration rates are influenced by the geochemistry of the system through which the water travels. It was found that the migration rate of strontium-90 is 3 percent of that of ground water in Chalk River soils, while for cesium-137 it is 0.3 percent. This type of information is critical for the waste-management program (see chapter nineteen).

Before the migration rates of strontium-90 and cesium-137 were determined, it was thought that large amounts of radionuclides might enter Perch Lake and escape into the Ottawa River via the Perch Creek outlet. Preliminary plans were made to build a treatment plant on the Perch Creek outlet to remove the radioactive substances from the water by absorption onto a special clay. However, measurements indicated that the rate of movement of strontium was so slow that it would take about 300 years for large amounts of strontium to reach Perch Lake from the disposal sites, and by this time the radioactive strontium-90 (its half-life is twenty-nine years) would have decayed by a factor of about a thousand. Therefore, a treatment plant was not necessary.

In another research project associated with waste management, high-level fission products were incorporated into nepheline syenite glass blocks and buried in the water table in the Perch Lake basin in 1959. In a continuing study since that time, the leaching of radioactivity from these blocks has been measured. Although ground water has flowed past the blocks for more than thirty years, the amounts of radionuclides released are small, being less than 0.003 percent of the total amount originally buried. A small amount of the released radioactive material became incorporated into micro-organisms and transported biologically in this form for short distances. The minute amounts of radioactive materials leached from the glass blocks point to the feasibility of disposing of high-level wastes by incorporating them into glass blocks and emplacing them in deep, underground repositories. Chapter nineteen discusses this topic in more detail.

Perch Lake water contains very small, but measurable, amounts of radioactive cobalt-60 and strontium-90, introduced by migration via ground water from the experimental disposal sites. The lake supports a rich population of plant and animal life. In 1955, a research program was begun to study the processes that distribute radioactivity throughout a lake ecosystem. Much of the cobalt-60 and strontium-90 entering the lake is absorbed by organic particles and deposited on the lake bottom, but some remains in solution and is taken up by vegetation and animals living in the lake water. Bio-discrimination, the process by which living organisms reject one chemical element in favour of another, plays an important role in the exclusion of radioactive substances from organisms. It was found at Perch Lake that there was a progressive *decline* in the concentration of cobalt-60 and strontium-90 as they are transferred to higher levels of food chains. In other words, predators, such as carnivorous fish, have lower concentrations of cobalt-60 and strontium-90 per unit body weight than do forage fish, insects and plants. It is now generally agreed that the same principle is valid for most other radionuclides, with a few exceptions, such as tritium and soluble cesium-137 (an analogue of potassium), which are incorporated without discrimination into all living organisms. Most radionuclides do not bio-accumulate as one progresses up the food chain, thus differing from DDT and other organic toxins. This has great importance in the assessment of the impact of radioactive releases into the environment.

To study the evaporation of water from lakes, the presence of traces of tritiated water in Perch Lake was utilized. To do this experimentally it was necessary to carry out a total tritiated water budget; that is, the amount entering the lake by various routes and the amount leaving the lake in the outlet stream plus the amount being evaporated to the atmosphere. The channeling of the inflow of ground water into Perch Lake via the main inlet was accomplished by building a dyke (which also served as a road) on the northwest shores of Perch Lake. The amount of tritiated water in evaporation was determined by analyzing water collected with molecular desiccant that had been installed on meteorological towers at Perch Lake. The amount of tritiated water entering and leaving the lake in flowing waters allowed long-term parameters (days) to be used in the model, while the evaporated tritiated water was used to account for short-term components (hours) in the model. This study enabled the researchers to develop a model for the evaporation of water from a lake, and this model was successfully applied to other lakes in Ontario.

In assessing the impact of radioactive releases to the atmosphere from nuclear establishments and nuclear power plants, cautious assumptions are made concerning the dispersion in air. It is assumed that wind speeds are negligible and that there is a temperature inversion, which does not allow the plume to rise and disperse in the upper atmosphere. If releases under such restrictive conditions are less than regulatory maximum permissible releases, then releases under other atmospheric conditions more favourable for dispersion would also be acceptable.

Instead of relying on hypothetical weather conditions, it was decided to measure argon-41 concentrations at ground level and downwind from the main stack, where airborne emissions from the reactors are released. Argon-41 is produced by neutron activation of argon in the cooling air as it passes through the NRX and NRU reactors. Staff in dosimetric research were able to build a monitor to measure argon-41 specifically. These monitors were located in semicircles on the ground at varying distances from the stack. Ground concentrations of argon-41 were recorded continuously over short time intervals. In this way one could analyze the argon-41 concentration and obtain statistics about the frequency of minimum dilutions of argon-41 emitted from the stack. Correlation of these concentrations with observed weather conditions gave a more realistic model of dispersion under various weather conditions.

As part of the overall environmental program, it was important to obtain experimental data on the dispersion of radioactive material released into bodies of water. Since some of these bodies of water were public areas, fluorescent dyes rather than radioactive materials were used as tracers. Chalk River staff released concentrated solutions of fluorescent dyes and, at different times after the release, measured the concentrations of dye downstream in a river or in surrounding waters in a lake, using a fluorimeter mounted in a boat. Such dispersion studies were made in the Ottawa River at Chalk River, in the Winnipeg River at Whiteshell, and in Lake Huron at Douglas Point. The data was used to establish dispersion models at these various sites. The concentrations of tritium measured downstream in several places in the Ottawa River after an accidental release of tritiated water from Chalk River in 1988 was in good agreement with the values predicted by the model.

Environmentalists considered it important to undertake preoperational surveys in areas surrounding nuclear facilities before they were commissioned. A minimum acceptable survey would determine natural-background radioactivity, type and abundance of plant and animal life, and the dilution capacity of waters accepting discharges from the facilities. It was felt that there would be more credibility to the study if it were carried out by an independent outside body, rather than by AECL. Accordingly, a contract was negotiated with the Great Lakes Institute of the University of Toronto to carry out a preoperational survey at Douglas Point. It was known that as a result of normal thermal shocks in Lake Huron, fish die by the thousands and are thrown up overnight on the beaches of the lake. If we did not have this phenomenon well documented, critics might blame the fish deaths on radioactive releases at Douglas Point. The preoperational survey was carried out and reported by the Great Lakes Institute.

Water used to cool power stations is discharged into rivers or lakes, causing an increase in temperature of the receiving waters. This increase is often referred to as thermal pollution. However, to explore the beneficial uses of this waste heat in cold Canadian waters, an experimental facility was installed at Lake Maskinonge in the Chalk River enclosed area. Optimal conditions were determined for the growth of living organisms. Heated water is now being used for commercial greenhouses and for commercial fish-rearing ponds at the Bruce and Pickering Nuclear Generating Stations, respectively.

Using the Tandem Accelerator Superconducting Cyclotron, very small amounts of carbon can be analyzed for carbon-14 content. Using this technique it was possible to estimate the age of ground waters containing carbonate, from a variety of places in the Canadian Shield, in connection with the assessment of waste-disposal concepts (chapter nineteen).

At Whiteshell, two studies of the effects of ionizing radiation on plants and animals in the environment were carried out. The FIG (field irradiator gamma) and ZEUS (zoological environment under stress) projects were designed for that purpose. Field-irradiation studies had been carried out previously in the United States. The aim of the FIG project was to detect any ecological changes in plant communities that are continuously irradiated under the more severe conditions of growth on the Canadian Shield in Manitoba. Part of a natural forest, one kilometre in diameter, was enclosed with a fence and a large radiocesium source was placed in the centre of this area on a tower. The annual doses in the FIG area ranged from approximately 460,000 times background near the source to thirty-five times near the fence. Since 1973, observations of radiation effects have been continuously recorded. The most radiosensitive of the trees are pines and these died off in the high-radiation areas. Annual exposure rates required to kill the pines were about 100,000 times the background exposures from natural sources. No effects were observed at lower radiation doses, where dose rates were still much higher than those due to natural background. The FIG project was terminated in 1985. The ZEUS project, concerned with irradiation of natural populations of meadow voles, did not yield statistically conclusive results, but no catastrophic effects were observed.

Many other studies have been carried out by the environmental research group at Whiteshell, mostly concerned with the nuclear-fuel-waste-management program.

In the philosophy of radiation protection, the ICRP has long believed that if individual humans are protected, then the survival of species of other organisms in the environment will also be ensured. In nature, the species, as distinct from the individual, is of paramount importance. A Chalk River scientist's detailed review of the literature – including highly relevant data from the FIG project at Whiteshell – confirmed that the effects of radiation on environmental organisms other than humans need not be a critical consideration for the Canadian nuclear-fuel-waste-management concept. A recent review by the International Atomic Energy Agency has supported this conclusion, although this review noted a need for caution in the case of endangered species.

RADIATION DOSIMETRY

The instrumentation for measurement of radiation levels, radiation contamination, etc., was fairly elementary at the beginning of the project and has been under continuous improvement ever since. Radiation produces detectable physical and chemical effects in gases, liquids and solids that may be used to obtain estimates of radiation doses. The physical effects in particular allow the detection of very small doses.

Until 1971, the exposure of workers at AECL to external radiation was routinely monitored using photographic films. Each worker wore a film badge for two weeks and then the film was processed and developed in a standard manner. The optical density of the processed film, which is increased by exposure to radiation, was determined and converted to dose by a calibration curve, where optical density was plotted against known exposures to different radiation doses. The use of films for routine monitoring of over 3,000 AECL employees plus special extra films for particular types of work (finger and extremity film) was very labour-intensive and did not lend itself to automation. Fortunately, the phenomenon of thermoluminescence could be exploited for this purpose. When radiation is absorbed in certain crystalline substances, such as lithium fluoride, energy is stored in the crystal-lattice structure. If the crystal is heated later to an appropriate temperature, it releases the stored energy as light, whose intensity is proportional to the radiation dose and can be measured using a photomultiplier. One of the researchers at AECL worked on thermoluminescence dosimeters (TLDs), and developed an automated TLD reader. This made possible routine monitoring of workers; TLDs replaced film badges at AECL in 1971. TLDs are incorporated into the AECL personal dosimeters that all radiation workers are required to wear. AECL, Health and Welfare, Ontario Hydro, and Hydro-Quebec have used the automated TLD reader developed at Chalk River. Only the Point Lepreau personnel in New Brunswick use a different commercial TLD reader. Owing to recent developments in the miniaturization of electronics, it is now possible to construct reliable direct reading dosimeters. When these dosimeters become sufficiently rugged and commercially available, they are likely to replace the TLDs.

Neutrons having a wide range of energies are produced by accelerators and reactors. Measurement of doses from neutron sources is particularly difficult since no single detector has been found to have a response that matches the biological effects of neutrons of different energies. Some detectors are highly sensitive to low-energy neutrons and have negligible response to high-energy neutrons; others have the opposite characteristic. Chalk River personnel studied and developed the use of CR-39 (a polycarbonate plastic film) as a dosimeter for monitoring neutrons, and it is used throughout Canada for this purpose. CR-39 is damaged by exposure to fast neutrons and this damage is enhanced to form visible holes in the plastic by electrochemical etching. The density of holes can be measured by a reader and converted to neutron dose. Slow neutrons are detected by placing the CR-39 film adjacent to boron or lithium; these materials produce alpha particles when exposed to slow neutrons, and the alpha particles in turn produce detectable damage in CR-39.

Another staff member developed a novel and sensitive method for the detection of neutrons. Small droplets of Freon liquid are incorporated in a transparent polyacrylamide gel. On exposure to neutrons, Freon gas is released to form bubbles, which grow in size to become visible to the naked eye. The number of bubbles formed is proportional to the neutron dose. This device became known as the "bubble detector". The physicist involved had no experience in the making of polyacrylamide gels, but a radiation biologist working in the same building made them routinely to separate DNA fragments in the gel by electrophoresis. The biologist quickly trained the physicist and his staff to make these gels. This transfer of technology illustrates the advantage of working in a laboratory, where scientists of different disciplines are able to cooperate closely to solve technical problems. The bubble detector attracted a fair amount of attention inside and outside AECL and it was finally organized as a separate business venture, known as Bubble Technology Industries (BTI) (see chapter twenty).

It is also necessary to measure the intensity of radiation doses in working areas, so that work can be planned and monitored to avoid undue exposure. Semiconductors and scintillators, as well as geiger counters and ionization chambers, have been applied to measure radiation fields from beta and gamma sources. Staff in dosimetric research developed a series of different kinds of instruments. These included hand and foot monitors, which quickly verified whether or not a worker was contaminated with radioactive materials. These radiation protection instruments are excellent and have served their purposes within AECL very well.

The maximum permissible uptake of plutonium by workers is too small to be measured directly by bioassay or *in vivo* monitoring. Protection of workers is achieved by area-monitoring instruments. In any process dealing with plutonium, it is necessary to measure very small concentrations of plutonium in air containing naturally occurring radon. The problem has been solved using semiconductor detectors that can distinguish the differences in energy between alpha particles emitted by plutonium and those from radon and its progeny. This type of instrument has enabled the operators to demonstrate compliance with regulatory requirements on annual limits on intake of plutonium by workers.

Internal contamination by radioactive substances is a potential risk for some AECL employees, because of their work. It is the function of internal dosimetry to estimate doses to organs and tissues of individuals following intakes, or suspected intakes, of radioactive materials, and to ensure that these doses are kept to a minimum. It is also necessary to carry out research and development in internal dosimetry, monitoring, and modelling to improve internal dose estimation procedures for radionuclides associated with present and proposed nuclear programs in AECL (see figure 3.2).

One of the most sensitive and useful methods of screening employees for an intake of radioactive materials by inhalation, ingestion or through the skin is radiochemical analysis of excreta (bioassay), particularly of urine. Urine analysis is commonly used for the detection of tritiated water, strontium-90, radiocesiums and other radioactive contaminants, as needed.

When workers were internally contaminated by inhalation of insoluble particles of relatively pure carbon-14 during the retubing at Pickering, the residue of ashes from fecal samples from the workers was analyzed for carbon-14 content. Studying the excretion patterns in workers of the inhaled insoluble carbon-14 material, researchers at Chalk River, together with Ontario Hydro, quickly established a metabolic model of the behaviour of this hitherto unstudied material. The model allowed for the estimation of doses to workers who had been exposed to this source. All doses were within permissible limits.

In vivo monitoring by a whole body counter is required to complement bioassay monitoring and it is a very useful part of the program for the radiation protection of workers. It provides a measurement of radionuclides in organs and tissues of individuals by determining externally the number and energies of photons (gamma and X rays) that escape from the body; this requires efficient photon detectors and good energy resolution. The background counting rate should be as low as possible and when a special facility was relocated and rebuilt on the Chalk River site, but remote from radioactive materials released at the Chalk River laboratories, materials for construction were selected on the basis of their low background levels. Several models of artificial humans called phantoms are used for calibration purposes. Workers were monitored in the whole-body counters to detect the presence or absence of contamination from internally deposited radionuclides.

Metabolic modelling, as it relates to internal dosimetry, is the establishment of a mathematical model that describes the uptake of a radionuclide by organs and tissues, and its subsequent retention and excretion. Once the amount of radioactivity in the organs and tissues of an individual is known from bioassay or *in vivo* monitoring, the resulting radiation dose can be calculated from these metabolic models. Chalk River researchers have made major contributions to international development of a lung model and developed a computerized program known as GENMOD for use in the calculation of doses from internally deposited radionuclides.

OPERATIONAL RADIATION PROTECTION

Operational radiation protection, at times called operational health physics, is a very important topic in nuclear establishments and is devoted to ensuring the safety of workers. Radiation fields and contamination levels in working areas must be determined and records of radiation exposures accumulated in previous work must be available. Protective clothing ranges from simple gloves (to limit hand contamination) to full protective suits, possibly provided with supplies of clean air. Radiation-protection personnel provide valuable advice to operating staff who bear responsibility for safety. Consultations between radiation-protection personnel and operating staff are needed in the planning of work, so that exposures can be minimized (in other words, kept below individual dose constraints for the site). Most tasks will be governed by attention to the time spent on a task, the distance from the worker to the source, and the use of shielding to limit direct radiation exposure. Careful judgments have to be made to optimize procedures so that the total dose will be minimized during the task, and to ensure that individual doses will be as low as possible. In order to comply with dose limits or to reduce individual doses, time of exposure may be reduced by work-sharing. However, work-sharing does not decrease the collective or total occupational dose; in fact, the total collective dose may increase if work-sharing involves less experienced people who take longer to perform a given task. Radiation-protection personnel also have an extremely important role in bringing operational problems to the attention of designers of instruments and facilities. Important contributions are made to the design of the plant, in the layout of major components, and to facilitate maintenance work so that the plant can be operated safely with the least amount of radiation exposure.

Operational radiation-protection personnel are key players in the implementation of the ALARA process. In many areas Chalk River staff practiced the principle of ALARA long before it was formally recommended by the ICRP.

Operational radiation protection in AECL Research has for many years been combined with the more conventional aspects of industrial safety. The objectives in considering all occupational risks (rather than radiation risks alone) are to reduce the total health risk from all causes and to use funds spent for health protection as effectively as possible.

An examination of annual radiation exposures over the years provides an indication of the good control in AECL, as a result of sound operational radiation protection practices. The average annual dose of Chalk River staff was about 7.5 mSv in 1958, which gradually decreased to about 1.4 mSv in 1989 (the legal limit was 50 mSv per year throughout this time).

With careful planning, practice with mock-ups and rehearsals emphasizing effective use of time, distance and shielding, it was possible to remove and replace components from reactors or hot cells even in very high radiation fields. Thus the collective dose for the replacement of the NRX calandria (1970) was 2.7 person-Sv, while that for the NRU replacement (1972) was 3.6 person-Sv. The replacement of a cracked leaded window with a new one, in the hot cells, took only 0.01 person-Sv.

The medical surveillance of radiation workers in AECL has always been carried out by physicians and nurses. Medically qualified experts in radiation protection may be designated as "medical advisors to the Atomic Energy Control Board" (see the Atomic Energy Control Regulations). Few specific diseases or medical conditions render a worker unfit to be an atomic radiation worker. Determination of fitness is made on the basis of purely occupational or medical issues; i.e., fitness to perform the duties demanded under circumstances that require the use of respiratory protection or use of plastic protective suits in high heat and humidity. Clinical assessment of potential and real overexposures of workers to radiation is carefully carried out. The low frequency of acute radiation effects in AECL workers is another indication of good safety practices. Since 1944, there has occurred only one serious burn of the fingers, when a worker inadvertently picked up a heavily irradiated capsule that had dropped on the floor from the reactor core. Two other workers were found to have reddening of a small area of their skin because of irradiation from tiny particles of radioactive materials. Skin burns of the fingers of two other workers resulted from overexposure of the skin to soft X rays. Two workers developed malignancies, possibly induced by radiation; one was a case of malignant lymphoma and the other was a case of chronic myelogenous leukemia. Through Labour Canada, AECL reported these two cases for review by the Workers' Compensation Board of Ontario. The estates of these workers are being compensated for the presumed injuries to the workers.

Since the forties, when Chalk River produced radionuclides such as phosphorus-32, carbon-14, strontium-90 and cesium-137 for clinical use and for exploitation as tracers in biological experiments, staff at Chalk River have been involved in the training of university, government and military personnel in the safe handling of these radionuclides. AECL staff are still engaged in this important educational role, because of their expertise in radiation protection. Health-sciences personnel host an annual one-week radiation-protection course at Chalk River. The course reviews basic radiation units and dosimetric concepts, biological and medical effects of radiation, and the latest risk coefficients for fatalities and serious hereditary effects. Environmental pathways for the movement of radionuclides are discussed and the relationship to waste management is noted. Selected topics in operational radiation protection are presented, with passing reference to conventional safety practices. The lectures are interspersed with tours of facilities and demonstrations of important health physics procedures. In addition to AECL lecturers, outside speakers (Control Board, Environment Canada, Health and Welfare) are featured. About twenty to forty participants from all parts of Canada, representing government, military, universities and utilities, attend. Obviously, a one-week course does not make instant experts of the participants, but it does permit them to hear and meet outstanding Canadian experts in this field, and to learn important concepts and topics in radiation protection. Feedback from the participants over the years has indicated that the course is highly successful.

BIBLIOGRAPHY

Atomic Energy Control Board of Canada. Advisory Committee on Radiological Protection. *Basic Principles of Radiation Protection in Canada*. Report INFO-0340, ACRP-11. Ottawa: AECB, 1990.

———. *A Review of Radiation Risk Estimates*. Report INFO-0389, ACRP-13. Ottawa: AECB, 1991.

International Commission on Radiological Protection. "1990 Recommendations of the International Commission on Radiological Protection.". ICRP publication, 60. *Annals of the ICRP* 21 (1991) No. 1-3.

Marko, A.M. "Reminiscences: Some Hits and Some Misses." In *Workshop/Symposium on Radiation Protection: Past and Future, 1989 March 20-22*. Edited by D.J. TerMarsch and D.K. Myers. AECL-9959. 1989.

———, ed. *AECL Research Programs in Life Sciences*. AECL-7070. 1981.

———, ed. *Biological Effects of Ionizing Radiation*. AECL-7164. Revised 1982.

Marko, A.M., G.C. Butler, and D.K. Myers. "Biochemistry at Chalk River." *Bulletin of the Canadian Biochemical Society* 23 (1986): 12-19.

Marko, A.M., D.K. Myers, I.L. Ophel, G. Cowper, and H.B. Newcombe. *Research in Radiation Biology, in the Environment and in Radiation Protection at CRNL*. AECL-5911. 1978.

Myers, D.K., and J.R. Johnson. *Toxicity and Dosimetry of Tritium: A Review. A Report Prepared for the Advisory Committee on Radiological Protection*. Report INFO-0377, ACRP-10. Ottawa: Atomic Energy Control Board of Canada, 1991.

National Research Council (U.S.). Committee on the Biological Effects of Ionizing Radiations. *The Effects on Populations of Exposure to Low Levels of Ionizing Radiation: 1980*. Washington, D.C.: National Academy Press, 1980. (Also known *as BEIR III*.)

———. *Health Effects of Exposure to Low Levels of Ionizing Radiation: BEIR V*. Washington, D.C.: National Academy Press, 1990.

United Nations. Scientific Committee on the Effects of Atomic Radiation. *Source, Effects and Risks of Ionizing Radiation*. New York: United Nations, 1988.

Chapter Four

Nuclear-Reactor Safety Principles

F.C. BOYD

INTRODUCTION

Despite the many safety devices incorporated in the NRX research reactor, which started operation in 1947, NRX suffered a serious accident in December 1952, as chapter one described. Although there were no known injuries, that early accident served as a stimulus for reviews, analyses and discussions about the safety of nuclear reactors. The result was a distinctive Canadian approach to nuclear-reactor safety, the evolution of which is the subject of this chapter. Through the application of this approach, Canadian nuclear-power plants and research reactors have an excellent safety record – no member of the public has been exposed to radiation resulting from a nuclear reactor in excess of regulatory limits, and no worker has been killed or seriously injured due to a reactor accident.

BACKGROUND

All forms of energy present hazards. For nuclear energy, the particular hazard comes from ionizing radiation. As a consequence, the science and practice of radiation protection has been highly developed.

In nuclear reactors, radiation is produced by the fission process and by the decay of radioactive fission and activation products that result from reactor operation. The fission products that result from the splitting or fissioning of the uranium are almost all radioactive, but largely remain inside the uranium fuel. They constitute the vast majority of radioactive material in a nuclear reactor. Activation products are essentially small bits of corrosion in the coolant system that become radioactive as they pass through the reactor core.

A major safety challenge in the design and operation of reactors is to safely control the chain reaction. Although nuclear reactors cannot explode like an atomic bomb, they can increase their power extremely quickly and can experience overheating of their fuel. The result could be the release of a large quantity of radioactive fission products that could be very hazardous.

Because the potential hazard was recognized at the beginning of the nuclear program, there has always been an emphasis on prior analysis of possible failures. The term "nuclear safety", or, more explicitly, "reactor safety", is generally applied to the work aimed at preventing accidents in nuclear reactors, or to mitigating their consequences.

While reactor safety was not identified as a distinct topic in the work of the Montreal Laboratory when it was set up in 1942, the goal of a safe and controlled reactor was central to its work. This was reflected in the design of the reactor, which became known as NRX. Many studies and analyses were conducted on safety-related topics, such as shielding against the radiation that would be produced in the fission process, the dynamic behaviour of the reactor, heat transfer from the uranium fuel, and atmospheric dispersion of radioactive material.

The choice of a site for the reactor and laboratory also evoked a number of safety issues. Security (in the sense of secrecy) was a factor in choosing a remote location, but the safety advantage of having little or no population for kilometres around was certainly considered by those involved in the choice. As G. Laurence, senior Canadian staff member of the Montreal Laboratory, later wrote: "Isolation was desirable to avoid supposed hazards to population from possible explosion or release of radioactive dust into the atmosphere, and to simplify control of secrecy."[1]

Shortly after the Chalk River site was chosen, in August 1944, smoke tests were conducted to investigate the potential dispersion of radioactive material that might be released; these were the first of many experiments conducted specifically to provide information necessary to analyze the safety of nuclear activities.

SAFETY OF NUCLEAR REACTORS

The primary hazard presented by a nuclear reactor is the very large amount of radioactive products created by the fission process. These fission products are normally contained within the uranium fuel. Although radioactive material is produced by neutron activation, such as radioactive corrosion products in the coolant, and plutonium is produced in the uranium fuel, the fission products constitute about 98 percent of the total radioactivity in a typical reactor.

If the fuel becomes overheated and its sheathing breaks, the fission products can escape from the fuel assembly. Safety analyses of nuclear reactors, therefore, look at all possibilities of fuel damage. In particular, they examine all potential situations that could lead to the fuel becoming overheated, either by too much energy (too many fissions) being created in the fuel, or inadequate cooling of the fuel. The two major potential initiating conditions are a loss of coolant and a loss of control, leading to an uncontrolled increase in the reactor power.

Fission products generate heat even after the fission process (reaction) is stopped. Therefore, the fuel must continue to be cooled after the reactor is shut down. This is most critical when the power density of the fuel is high; i.e., when the fuel generates a large amount of energy per unit volume, as is typical in reactors used to produce electricity.

The original NRX reactor design included two ways of controlling the fission process – changing the height of the heavy-water moderator in the reactor tank or calandria (changing the size of the reactor), and inserting or removing a neutron-absorbing rod. To be able to stop the reaction at any time, shut-off rods were provided, initially eighteen, later twelve. These boron-carbide-filled steel rods were designed to drop into the reactor, accelerated by air pressure to increase their speed over that from gravity alone. The dropping of the shut-off rods could be activated automatically by signals from several measuring devices as well as manually. As a slow backup, the heavy-water moderator could be drained from the calandria.

That safety was high in the minds of the designers can be inferred from a comment by W.B. Lewis (the director of the Chalk River laboratory) that "the reactor has been characterized as having 900 devices for shutting it down but only one for starting it up."

The NRX accident of 12 December 1952 (described in chapter one) precipitated several reviews and, for Laurence, was the beginning of a preoccupation with reactor safety that lasted the rest of his life. Others were also prompted to examine their approach to reactor-safety design, with particular concern for the systems that control and shut down the reactor. Those who were working on the design of the control and shutdown systems for the 200 MW NRU reactor (construction approved in December 1950) decided to invoke the concepts of redundancy and coincidence through the use of multiple-channel systems.

If there is only one channel (or string of instruments and devices) in a system, any significant failure will disable it. With three (or four) parallel channels arranged such that two out of

---

[1] Eggleston, *Canada's Nuclear Story*, 125

the three (or four) are required to respond to effect a control or shutdown action, the reliability is significantly increased and there is the added bonus that individual channels can be tested while the reactor is in operation.

A further conclusion was that it was desirable to separate the protective (shutdown) system from the control system, so that failures in the more complex control system would not impair the ability to shut down. Because the NRU reactor was too far along in design and construction, this concept of separation was not incorporated in that reactor, but it has become a basic requirement for all subsequent reactors in Canada.

All shutdown systems of Canadian nuclear power reactors have triplicated channels and are physically and functionally separate from the control systems.

Another concept that emerged was that of a safety bank. The NRX reactor accident led to the recognition that a shutdown reactor was not necessarily safe. A safety bank involves resetting some of the protective devices (e.g., shut-off rods) after a shutdown, so that if any maintenance activity or other work increases the reactivity of the reactor, the rods will fall and prevent a possible accident.

With NRU well under way, attention turned to the power reactors that were being designed or proposed. An important difference between the research reactors (NRX and NRU) and the power reactors is that the latter have high-temperature, high-pressure water-coolant systems to transfer the heat energy produced in the fuel to the electricity-generating systems. From a safety viewpoint this introduces a major concern about a loss of cooling or loss of coolant. If a coolant pipe breaks, the loss of the coolant will deprive the fuel of cooling and the escaping hot, pressurized, water will flash into steam, pressurize the enclosing building and provide a means to carry radioactive material to the surrounding environment.

In 1957 E. Siddall, one of the NRU designers, wrote a seminal paper examining the objective of reactor safety, noting that there was a tendency to demand more and more safety measures without any clear evaluation of their benefit. A study issued by the United States Atomic Energy Commission that year (WASH 740) had caused considerable reaction with its prediction of disastrous consequences if a large fraction of the fission products in the fuel of a large (200 MWe) nuclear-power reactor were released to the atmosphere. This led to the demand for more and more safety measures.

Siddall proposed that a basic criterion for judging the safety of an activity should be the number of deaths per unit time arising from that activity. He observed that society implicitly accepted the risk of many activities, even though the accidental death rate from them is high relative to the general probability – private motoring being the prime example.

From studies of other industrial activities producing output of comparable value, in particular coal-fired electricity-generating plants, Siddall suggested that a death rate for a large nuclear power plant (200 MWe) of 0.17 deaths per year, or about one death per six years, should be acceptable. Although these numbers were never adopted, the concept of an acceptable risk, expressed numerically, has been a fundamental part of Canadian nuclear safety philosophy.

At the same time (1957), Laurence, who was then a senior director at the Chalk River laboratories, presented the first of what would become a series of papers on reactor safety. This was not long after he had been appointed the first chairman of the Reactor Safety Advisory Committee (RSAC) of the Atomic Energy Control Board. At that time the Control Board had only one scientific staff member, and chose the advisory committee approach to deal with the technical safety aspects of reactors then being proposed for operation outside of the Chalk River laboratory: a research reactor at McMaster University (Hamilton, Ontario) and the 20 MWe NPD reactor being designed by Canadian General Electric for AECL and Ontario Hydro. Over half of the original members of the RSAC were senior members of AECL's Chalk River laboratories.

Aware of Siddall's concepts, the designers of NPD proposed, in the preliminary "Hazards Report" for NPD submitted to the Control Board's RSAC in 1958, a basic objective of less than one death per 100 years for the plant, which the committee essentially accepted. This was the first formal "hazard", or, as they would be called later, "safety" report prepared in Canada for a nuclear plant.

The protective system of NPD was separated from the control system and employed the two-out-of-three concept. The design included continuous monitoring and inter-comparison of the three protective system channels and provided for testing, from input signal to shutdown device, during operation.

About two years later, Laurence proposed, as a reference value, an upper limit for the likelihood of a "major accident" of one per 100,000 reactor-years, on the premise that it was unlikely any accident would lead to more than 1,000 deaths. He also endorsed the multiple-channel approach for safety systems, with emphasis on testing to demonstrate failure rates and, thereby, availability. Such a target could be achieved, he argued, with very practical reliability criteria if there were adequate separation between, and independence of, the operating systems and the various safety systems.

Laurence and the RSAC conceived of the reactor design having three divisions: the "process" systems (the essential operating part of the plant); "protective devices" (e.g., to shut down the reactor if there are problems with the process systems); and "containment provisions" (to prevent any radioactive material that might escape from the process systems from being released to the environment). If each of these divisions were properly designed for their function, it would take a failure of all three before there could be a serious release of radioactive material to the environment. Further, if these divisions were sufficiently separate and independent, the frequency of failure of the process systems and the unreliability (or unavailability) of the protective devices and containment provisions could be combined to give the likelihood of a significant release affecting the public. This meant that the reliability requirements for individual systems could be practicable while still achieving a very low risk to the public.

On this basis, Laurence proposed criteria for the maximum frequency of serious failures of the process systems and unavailability of the safety systems. He argued that these standards should not only be easy to achieve but that they could be demonstrated by testing. In particular, the protective shutdown systems, if designed in the two-out-of-three or two-out-of-four arrangements, could be tested with the reactor operating.

Some of these concepts were expressed in the NPD design even though that design was completed before the concepts were formally presented in technical papers. Similarly, in the full-scale (200 MWe) Douglas Point NGS, then under way, the protective shutdown system was separate from and independent of the regulating control system and in addition the reactor was housed in a building specifically designed for containment.

During the late fifties and sixties, the designs and safety philosophy evolved together in a somewhat synergistic fashion. Designers met frequently with the RSAC and its growing technical staff to propose, debate, and eventually agree on various safety concepts. The meetings were often spirited; the designers and the regulators greatly respected each other.

With the confidence arising from the initial operation of the NPD station and the completion of the Douglas Point design, Ontario Hydro proposed building a large multi-unit station (four units of 500 MWe) near Toronto (Pickering NGS). The RSAC was not convinced that the Douglas Point containment design would be adequate for this plant. In response, the designers developed a unique arrangement in which the containment buildings surrounding each unit were attached to a large vacuum building. The concept of this vacuum building arrangement was that the steam and pressure resulting from a loss-of-coolant-accident (LOCA) would be sucked out into the vacuum building and the entire system would remain at atmospheric or sub-atmospheric pressure. Consequently, there would be no pressure to push out any radioactive material that might be released from the reactor systems due to the LOCA. After much debate, this concept was approved and has been used on all of Ontario Hydro's multi-unit nuclear power stations as a cost-effective containment arrangement.

For a LOCA to occur, the assumption is that a major failure (break) of a large pipe in the main reactor cooling system (sometimes called the heat transfer system) would allow the hot, pressurized water to escape. The LOCA had come to be regarded (and still is) as the most serious type of reactor accident, since it has the potential to cause failure of the fuel (due to overheating), which could release large quantities of radioactive fission products; at the same time, the steam produced by the escaping hot water of the cooling system would pressurize the containment building.

Canada's two single-unit nuclear power plants (Point Lepreau and Gentilly-2) each have a containment building designed to hold the highest pressure that would result from a LOCA. This design has also been approved by the Control Board as being fully adequate.

Basic reactor-safety concepts continued to evolve. By 1972 (after Laurence's retirement as president of the Control Board, to which he was appointed in 1962), the RSAC had adopted an approach in which the reactor plant was conceived as having process systems and several special safety systems – each of which had a specific safety function, such as shutting down the reactor, providing emergency cooling of the uranium fuel or preventing the dispersion of radioactive material. The special safety systems were to be separate and independent of each other and of the process systems.

As performance requirements for the designers, maximum values of radiation dose that might be received by a member of the public (using conservative assumptions and calculation methods) were set for single failures of a process system (such as a large pipe failure) and for dual accidents where such a process-system failure was associated with complete failure of one of the special safety systems.

A maximum frequency of serious failures of the process systems was set at one per three years, and a maximum unavailability of each special safety system was set at one per 1,000 years. (A serious failure of a process system was defined as one that could lead to the release of radioactive material in the absence of safety-system operation.) Assuming adequate independence, the likelihood of a dual accident should, therefore, be less than one in 3,000 per year. Following this approach, a major release could only occur if there were a serious process failure and complete failure of two independent special safety systems, which, if the requirements are met, would have a likelihood of less than one in 3,000,000 per year.

The reference dose value for an individual for a dual accident was chosen as that proposed by the Medical Research Council of the United Kingdom as tolerable to be received once in a lifetime, even though the postulated likelihood of a dual accident was one in 3,000 reactor years. These single-failure/dual-failure criteria are still used by the Control Board in judging designs for nuclear power plants, although additional requirements are now applied.

At the end of the sixties, the RSAC was faced with a challenge by the designers of the then-proposed Bruce "A" nuclear generating station. The proponents stated that they could not design and analyze a practical containment to meet the dual-failure criteria if they had to assume a runaway accident; i.e., an uncontrolled increase in power that could result from a failure of the control system accompanied by failure of the shutdown system. They proposed, therefore, to provide a second shutdown system specifically designed to guard against a power runaway.

After considerable debate over the requirements, the RSAC finally adopted the concept of a second shutdown system, but only if it was as capable as the first – separate and independent of the first shutdown system and of the control system – and used a different physical process. If all of these conditions were met, each shutdown system would be regarded as a separate special safety system and the single-failure/dual-failure criteria would apply; i.e., the reference dose limits must be shown to be met for any combination of a process system failure together with failure of any special safety system. All Canadian nuclear power plants since Bruce "A" have two separate, independent and different shutdown systems. This Canadian approach preceded by over a decade a long, protracted argument in the United States over what was called "anticipated transient without scram" – a power runaway without shutdown.

A major feature of the single-failure/dual-failure concept and the associated reference dose limits was that these were performance criteria, not design requirements, thus leaving the designers considerable flexibility. The major design requirement was a very general one: the need for separation and independence of the safety systems. This emphasis on objective was in sharp contrast to the detailed design rules issued by the United States Nuclear Regulatory Commission (USNRC), which succeeded the USAEC as the nuclear regulatory body in that country.

Members of AECL's scientific staff played a major role in the RSAC and in the development of Canadian nuclear safety philosophy in general, and most of the scientific knowledge and technical data that provides the basis for safety decisions has been generated by AECL's research groups (see chapter fifteen).

INTERNATIONAL

Canadian nuclear scientists and engineers have played an active role in the international field of nuclear safety.

The risk-based concepts outlined above were presented by Laurence and others to various international conferences and meetings as early as 1958. While there was widespread support among professionals involved with reactor safety, most national nuclear regulatory agencies followed the style of the USNRC in specifying prescriptive, detailed design requirements. However, by the early nineties even the USNRC was seriously examining risk-based regulations.

In the seventies, the international nuclear community began speaking of reactor safety in terms of defence-in-depth. This refers primarily to the many physical barriers between the radioactive fission products in the fuel and the environment, such as the fuel sheathing, coolant-system piping, and containment buildings. Other aspects, such as the exclusion zone around nuclear power plants, and the extensive analyses and quality-assurance programs, are also sometimes invoked as part of this approach. The defence-in-depth concept is often used when describing the safety of nuclear reactors to the public. The approach followed by Canada, especially that of separate safety systems, provided very effective defence-in-depth long before the concept became internationally popular.

SAFETY EXPERIMENTS AND RESEARCH

Most of the scientific and technical information needed for the safety design and evaluation of a nuclear reactor is the same as, or similar to, that used in the design of the basic plant; e.g., fuel behaviour, effects of radiation on materials, physics of the nuclear reaction. In some cases, the safety analysis raises questions beyond the available knowledge, or the safety design involves devices or techniques for which there is little or no precedent. For such situations special safety experiments have been conducted, particularly for the early CANDU nuclear power plants.

As the nuclear power program matured, most of the obvious questions had been addressed. Attention turned to the more esoteric safety questions involving conditions having a very low probability. To address them, a broad safety-research program evolved in the early seventies.

Chapter fifteen describes both the early, specific experiments and the later, comprehensive safety-research program in more detail.

OTHER FACILITIES

In recent years there has been renewed interest in small reactors, which has led to studies, experiments and proposed criteria specific to that class of reactor. Canadian scientists from AECL and staff of the Control Board have been in the forefront of international developments in this area.

SUMMARY

The safety of nuclear activities has been a high priority since the beginning of Canada's nuclear program. Special attention has been paid to the safety of nuclear reactors, especially the large CANDU reactors used in Canadian nuclear power plants.

A set of conservative criteria was developed over the first decade or so of the CANDU program, with the objective of minimizing the risk of a serious accident in a nuclear power plant. These criteria evolved out of a critical review of the NRX accident in 1952, and therefore reflect a unique Canadian approach. They emphasize separate and independent safety systems that are testable to achieve a very low likelihood of a serious accident. The concepts involved in this approach are now being adopted in many other countries with nuclear-power programs.

BIBLIOGRAPHY

Atomic Energy of Canada Limited. The Safety of Ontario's Nuclear Power Reactors: A Scientific and Technical Review, A Submission to the Ontario Nuclear Safety Review. Toronto: Ministry of Energy, 1988.

Bayly, J.G. "An Experimental Study of a Divergent Nuclear Reactor." *Canadian Journal of Physics* 31 (1953): 182-8. (Also AECL-25).

Boyd, F.C. "Containment and Siting Requirements in Canada." In *Containment and Siting of Nuclear Power Plants: Proceedings of a Symposium on the Containment and Siting of Nuclear Power Plants*, 67-75. Paper SM-89/22. Vienna: International Atomic Energy Agency, 1967.

Canadian General Electric Company. *NPD-2: Preliminary Hazards Report*. 1958.

Cipriani, A.J. "Health and Safety Activities in Reactor Operations and Chemical Processing Plants." In *Proceedings of the International Conference on the Peaceful Uses of Atomic Energy, Geneva, 8-20 August 1955*, Vol. 13, 263-5. Paper A/CONF.8/P/8. New York: United Nations, 1956. (Also AECL-207).

Eggleston, Wilfrid. *Canada's Nuclear Story*. Toronto: Clarke, Irwin & Company, 1965.

Hill, P.G., E.G. Hauptmann, and V. Lee. *Consequences of Pressure Tube Rupture on In-Core Components: Phase I, Technology Review.* Report INFO-0115. Ottawa: Atomic Energy Control Board, 1982.

Hurst, D.G. *The Accident to the NRX Reactor. Part II.* AECL-233. 1953.

Hurst, D.G., and F.C. Boyd. "Reactor Licensing and Safety Requirements." In *Canadian Nuclear Association. Proceedings of the 1972 Annual Conference*, Paper 72-CNA-102. Chalk River: Atomic Energy of Canada Limited, 1972. (Also issued as report AECB-1059. Ottawa: Atomic Energy Control Board, 1972.)

Laurence, G.C. "Operating Nuclear Reactors Safely." In *Reactor Safety and Hazards Evaluation Techniques: Proceedings of the Symposium on Reactor Safety and Hazards Evaluation Techniques*, vol. 1, 135-46. Vienna: International Atomic Energy Agency, 1962.

———. *Probability Considerations in Reactor Hazards Appraisals.* Report AECB-1059. Ottawa: Atomic Energy Control Board, 1966.

———. *Reactor Safety*. AECL-538. 1957.

———. "Reactor Safety in Canada." *Nucleonics* 18 (October 1960): 73-7.

———. "Reactor Siting Criteria and Practice in Canada." In *Nuclear Power Siting: A National Topical Meeting, 16-18 February 1965, Los Angeles*, 201-8. R.L. Ashley, ed. Report CONF-650201. Los Angeles, Calif.: American Nuclear Society, 1965.

———. *Required Safety in Nuclear Reactors.* AECL-1923. 1961.

Laurence, G.C., and F.C. Boyd. *Trends in Reactor Safety: A Canadian View*. Report AECB-1045. Ottawa: Atomic Energy Control Board, 1969. (Presented at International Nuclear Industries Fair, 6-11 October 1969, Basel, Switzerland (NUCLEX 69), Technical Meeting no. 5/1.)

Lewis, W.B. *The Accident to the NRX Reactor on 12 December 1952.* AECL-232. 1953.

Pearson, A., E. Siddall, and P.R. Tunnicliffe. "The Control of Canadian Nuclear Reactors." In *Proceedings of the Second United Nations International Conference on the Peaceful Uses of Atomic Energy, Geneva, 1-13 September 1958.* Vol. 11, 372-9. Paper A/CONF.15/P/213. New York: United Nations, 1958. (Also AECL-623.)

Ross-Ross, P.A. *Experiments on the Consequences of Bursting Pressure Tubes in a Simulated NPD Reactor Arrangement*. AECL-1736. 1963.

Siddall, E. *Reactor Safety Standards and Their Attainment*. AECL-498. 1957.

———. "Reliable Reactor Protection." *Nucleonics* 15 (June 1957): 124-9. (Also AECL-430.)

———. *A Study of Serviceability and Safety in the Control System of the NRU Reactor.* AECL-399. 1954, reprinted 1957.

RADIOISOTOPES

Chapter Five

Radioisotopes

B. ULLYETT

BEGINNINGS

Radionuclides that do not occur naturally in nature and are produced intentionally through nuclear reactions are called artificial radioisotopes. The first fruits of the Canadian nuclear program were realized in the production of artificial radioisotopes in the NRX reactor, and in their subsequent worldwide use in the medical and industrial fields.

These developments came just half a century after the discovery of natural radioactivity, when Henri Becquerel, in 1896, motivated by Wilhelm Röentgen's discovery of X rays in 1895, found that uranium emitted a penetrating radiation, referred to at the time as Becquerel rays. In 1898 Marie and Pierre Curie discovered two new elements in uranium ore, pitchblende, which they named polonium and radium. Radium soon came into great demand for cancer treatment in humans. It commanded a very high price; e.g., $75,000 per gram (1930 dollars).

Artificial radioactivity was discovered in 1934 by Irène, the daughter of the Curies, and her husband, Frédéric Joliot. They were studying the positrons produced by bombarding aluminum with particles from polonium. They found that the positron emission did not stop when the polonium source was removed, but decayed with a 2.5-minute half-life. This discovery prompted Enrico Fermi to examine whether neutrons could produce artificial radioactivity. He found, also in 1934, that slowed-down neutrons were particularly effective. By the end of 1935, about one hundred artificial radioisotopes had been identified.

With the advent of nuclear reactors, enormous quantities of radioisotopes could be produced, and a new industry was born. As just one indicator of the scale of this evolution, it was possible to replace the radioactivity of one gram of radium, i.e., 3.7×10^{10} atomic disintegrations per second, at $75,000 (1930 dollars) with sufficient cobalt-60 to yield the same amount of disintegrations per second at $1 (1977 dollars).

ARTIFICIAL RADIOISOTOPES FROM NRX

By 1944, subsequent to the 1943 World War II Quebec conference of the heads of Allied governments, the decision was made that the Montreal Laboratory should design, build and operate a 10 MW heavy-water, natural-uranium-fueled nuclear reactor to produce plutonium and to provide neutrons for research. The reactor was to be named National Research Experimental (NRX). The importance of radioisotopes to medicine had already been recognized (in particular cobalt-60, which had been identified by 1945 by the scientific community as a possible source of gamma radiation for the treatment of cancer tumours without invasion of the body), as well as their importance to agriculture and industry. NRX was immediately seen as an excellent means for radioisotope production and the reactor was designed with this in mind. Self-serve entry/exit tubes were designed into the reactor shielding walls to permit on-power sample entry for neutron irradiation and subsequent removal.

Before the NRX startup in 1947, there was a joint effort by NRC research chemistry and the isotope production branch (IPB) at Chalk River to develop methods to make and purify those radioisotopes that were expected to come into demand. Within two months of NRX startup, megabecquerel (MBq) amounts of phosphorus-32, sulphur-35 and iodine-131, and kilobecquerel (kBq) amounts of carbon-14, had been produced. Within three months of startup (31 October 1947), the first radioisotope shipment was made to the University of Saskatchewan, and within five months shipments had been made to Canadian university researchers from coast to coast.

With the chemical processing of many radioisotopes in rapidly increasing amounts, the IPB staff had to learn quickly to work safely with radioactive materials. Special processing fume hoods were designed, some with raised floor platforms in front, to increase the distance from hood working levels to the laboratory floor. These platforms were later lowered to floor level, with additional in-hood shielding provided by the design of interlocking modules of lead bricks to permit various shielding configurations.

After two years of NRX operation, it was obvious that Chalk River could produce higher radioactive specific activities (radioactivity of a specific radioisotope per unit mass of isotope target) than could Oak Ridge (United States), or Harwell (United Kingdom), due to the factor-of-ten higher neutron flux available with NRX. With the increasing amounts of radioactivity being produced as reactor irradiation experience grew, the IPB laboratory layout required the marking of floor areas with colour bands and circles, to alert technical staff traversing from a lower-activity area to a higher one. By 1950, the quantities of hydrogen-3, carbon-14, sulfur-35, phosphorus-32, iodine-131 and others were made available to researchers at Chalk River, to Canadian university research centres and to pharmaceutical firms studying the uses of these new products.

By 1949, it became increasingly apparent that cobalt-60 was coming into strong demand as a radioactive source[1] for use in the treatment of deep-seated cancer tumours. Chalk River arranged to sell this radioisotope through the commercial products division of Eldorado Mining & Refining Ltd. (Port Hope, Ontario), as a possible substitute for radium in tumour treatments and in radiographic work.

COMMERCIAL PRODUCTS DIVISION OF ELDORADO

At the time of its agreement (1949) to sell Chalk River's cobalt-60, Eldorado's commercial products division (CPD) was facing slumping profits from its radium operations, because of competition from other radium suppliers (United Kingdom, United States and France). There was also the gradual recognition that nuclear-reactor produced radioisotopes would eventually displace radium in the medical field.

By 1950, NRX had produced sufficient high-activity cobalt-60 in wafers of metallic cobalt to permit fabrication of cobalt-60

[1] Radioactive material encased in a sturdy capsule, without leakage, is defined, for this chapter, as a radioactive source.

sources. With this availability, CPD saw a marketing opportunity for the sale to cancer clinics of medical treatment units containing a cobalt-60 source.[2] Necessary funding was obtained for a CPD design, and a contract awarded to Canadian Vickers, of Montreal, to build the first unit.

This first CPD teletherapy unit, called the Eldorado A, comprised an encased, lead-filled shielding head mounted on a fixed vertical column that could be raised, lowered or angled in several directions. This unit was shipped to the London Cancer Clinic in London, Ontario. Another unit of different design was being constructed, at about the same time, by the Saskatoon Cancer Clinic for similar use. Within a short time of each other,[3] the two clinics began using their respective units for cancer treatment.

These were the first two cobalt-60 teletherapy cancer treatment units to operate in the world. Concerned about product quality, CPD decided to undertake their own manufacture of additional Eldorado A's. By the time that it was superseded by improved models, about fifty had been built and installed around the world by CPD staff. These units sold in a price ranging from $25-$30,000 ($US).

The head of a renowned cancer clinic in Paris who purchased one of the early Eldorado A units has kept the now-retired unit in his clinic as a museum piece. He states that, even today, the Eldorado A remains among the best and simplest of cancer treatment units.

By late 1952, the first rotational cobalt-60 teletherapy unit had been designed by CPD based upon specifications provided by a physicist with the Francis Delafield Hospital of New York. It was called the Theratron Model B. A design patent was awarded and royalty payments were made to the hospital physicist for each unit sold by CPD until 1973. In this type of unit, the cobalt-60 source (fixed within the shielded head) was rotated over an arc of a circle with the beam collimator aimed at the centre. The patient couch, an integral part of the unit, could be adjusted to permit patient location so that the tumour was at the centre of rotation and would alone receive the full radiation dosage. The source to centre of rotation was 75 centimetres. Junior Model B's having a smaller cobalt-60 capacity and a 55-centimetre treatment distance were also designed, and became popular at small clinics.

The Model B Theratron sold at a price that was double that of the Eldorado A units. The first 'B' unit was installed at Francis Delafield Hospital in 1953. The Model B was well accepted and complemented the Eldorado A in many clinics. By the late eighties, vertical units were no longer marketed. CPD later trademarked the name of the cobalt-60 teletherapy units to theratron units, which it now uses exclusively. (From this point forward in this chapter, theratron is used and specifically denotes a cobalt-60 teletherapy unit.) From 1946 until 1952, with the new radioisotope products to sell along with radium products, and the development of a first-class machine shop, CPD staff grew to a total of eighty-four, most being located in Ottawa.

COMMERCIAL PRODUCTS DIVISION OF AECL

CPD was transferred in August 1952 from Eldorado to Atomic Energy of Canada Ltd. The transfer included operations for the sales and production of radium, the new radioisotope products

---

[2] These units were called teletherapy units, to indicate radiation-beam treatment of deep-seated body tumours from a distance, and to identify the cobalt-60 source as a teletherapy source.

[3] For many years, rivalry existed concerning who had the first cobalt-60 unit, and who gave the first patient treatment. The following information was obtained from Miss Sylvia Fedoruk, who assisted at Saskatoon at the time (she later became Lieutenant-Governor of Saskatchewan), and Miss Elaine Marshall in London, as reported in *The Cobalt-60 Story* (see bibliography):

- Official opening of cobalt unit in Saskatoon by Dean Cronkite was 23 October 1951.
- First patient treatment on cobalt-60 unit in Saskatoon, by Dr. T.A. Watson, was 8 November 1951.
- Official opening of cobalt unit in London was 11 November 1951.
- First patient treatment on cobalt-60 unit in London, by Dr. I. Smith, was 27 October 1951. (Note: Opening date in London was after first patient treatment – Ed.)

and the manufacturing machine shop. Thus, the history of radioisotopes in Canada and their worldwide use, is, in effect, the history of CPD, one of the oldest, and for a long time the least known of AECL's divisions. CPD became simply AECL Commercial Products (CP) by 1968, and AECL Radiochemical Company (RCC) by 1978. For consistency, the name CPD is used followed by (RCC) when appropriate.

Although this new AECL division was to be profit-oriented, immediate profitability was not certain, and in fact there was no assurance of the operation ever making profits. However, the record indicates that in spite of some bad years, there were a few years of modest profitability.

CPD grew rapidly after becoming part of AECL. It was more able to consult with the nuclear physicists at Chalk River and to obtain services, such as glassblowing, radioisotope measurements, design concepts for laboratory equipment, and dosimetry from several of the AECL divisions. Radioisotope markets between nations gradually opened up. CPD established its own R&D branch to explore uses for radioisotopes. The design branch and machine shop at Ottawa expanded rapidly to manufacture theratrons, which were in demand following the London and Saskatoon installations. The small CPD group at Chalk River required increased quantities of radioisotope products from the IPB to meet orders, as well as a variety of newer radioisotopes such as calcium-45, chromium-52, gold-198, iron-59, mercury-203, phosphorus-32, yttrium-90, zinc-65, and the fission products strontium-90 and cesium-137. Several long-term contracts were obtained by CPD for some of these, which resulted in the scheduling of routine processes by the IPB.

One Chalk River official observed at the time that the accomplishments of such a small radioisotope production group in crowded facilities were amazing when compared with the new giants of the day. Oak Ridge National Laboratory (ORNL), in the United States, employed dozens of professional staff and built much larger radioisotope-separation laboratories that required isolation in individual buildings. They were required to process as much as ten times more radioactive target weights, because their reactor fluxes were a factor of ten lower than that of NRX. The Radiochemical Centre Ltd., United Kingdom, and Commissariat à l'Énergie Atomique, France, were each in the same position, as their reactor fluxes were about the same as those of ORNL. Nevertheless, the market for radioisotopes developed rapidly.

CPD staff reached 150 by 1954. Business activities were squeezed into rented basement quarters of an Ottawa store. Research and development operations and the machine shop were crammed into other leased buildings throughout the city. To handle radioactive cobalt-60 for the manufacture of ^{60}Co sources processing facilities were needed. The radium-source production laboratories at Port Hope needed expensive upgrading. The expanding CPD group at Chalk River was crammed amid IPB operations. More space was needed to accommodate the rapidly expanding business. CPD had the understanding that, once becoming an AECL division, new facilities were to be considered to consolidate all operations under one roof.

The CPD manufacturing shop could not wait for these. In late 1953, it moved temporarily into a warehouse on Laperrière Avenue in Ottawa. In 1954, new facilities, designed to accommodate the CPD Port Hope radium-source operations, the CPD radioisotope operations at CRL, the manufacturing shop and the scattered Ottawa groups, were begun in Tunney's Pasture in Ottawa, a federal building site situated close to the city core. The Ottawa choice for CPD was more suitable to commercial needs – such as Customs, transportation, and business contacts – than the more distant Chalk River. The new facilities, costing $1 million, were ready for occupancy by early 1955.

It was obvious that the allocated manufacturing shop space in the new facilities was only a fraction of what was required to meet the demand for fifty theratron units per year, and that number was ever-increasing.

CPD R&D had by then established that monitored, mega-rad dosages of cobalt-60 radiation to certain foods destroyed the enzyme action that was a cause of food spoilage, and left the food safe and palatable. Food-irradiation plants to use large quantities of cobalt-60 were being designed. These would require manufacture, complete assembly in the shop area, and testing before shipping. As a result, the manufacturing shop elected to stay at the nearby Laperrière location, scheduling three daily shifts that operated two continuous production lines to produce the required theratrons.

When CPD began operations in the new Ottawa facilities, AECL decided that all costs associated with the supply of commercial radioisotopes would be fully accountable by CPD. However, irradiation services from the NRX and NRU reactors would be provided on a marginal-cost basis (see chapter twenty). In this way the viability of the business could be evaluated. During the period 1955-57, the IPB turned over all of its radioisotope production technology to CPD, in some cases providing new glassware and remote-handling facilities. The IPB subsequently ceased its radioisotopes operations at Chalk River.

In the new CPD laboratories, each radioisotope was processed in an isolated laboratory to prevent cross-contamination of products, as opposed to a common area as at Chalk River. A transfer cell was constructed to enable the remote delivery of sealed irradiated targets from shipping containers to specific laboratories. By 1957, CPD began to prepare all of its reactor targets and to chemically process them after irradiation. These included formerly IPB-produced carbon-14, calcium-45, iodine-131, phosphorus-32, and sulphur-35. The list expanded in time to provide approximately fifty such products, many as custom orders. This necessitated division of laboratory space and the locating of some radiochemical processes in the basement.

The radium laboratories constructed in the Ottawa laboratory were built in an isolated section of the building, with facilities designed to avoid contamination of work areas. This had been a constant problem at Eldorado. All radium was now handled within sealed, lead-shielded, cubic metric boxes of lucite, fitted with "ball-joint" manipulators mounted through lead-shielded walls. In-box equipment was articulated with flexible joints and designed to be disposable. Interior air from the boxes was exhausted first through in-box shielded, activated charcoal filters, followed by absolute particulate filters, prior to reaching the final absolute filters in the upper exhaust room. All in-box waste was deposited in shielded waste containers (pre-approved for shipment to CRL), via inter-box transfer units isolated from the in-box operations. A remotely operated drawer vault for holding radium inventory was built.

These facilities operated as clean areas. Exposure of personnel to contamination with possible high-radiation dosages from radium never became a serious problem again. Face masks, double-lined gloves, double shoe covers and full whites were no longer necessary for routine work, but were available from nearby emergency cupboards.

The first product CPD (Eldorado) offered for sale, in 1946, was radium salts enclosed in steel needles and plaques, for interstitial treatment of cancer sites. These products were scheduled for phasing out by 1959 as the use of cobalt-60 encased in needles began to grow.

Radium products that represented significant sales for CPD until the late fifties were neutron sources produced by the homogeneous mixture of radium with beryllium. These became useful as starter sources to provide an initial supply of neutrons for the startup of small nuclear reactors. The nuclear reactors that powered the first two American nuclear submarines, the NAUTILUS and the SKATE, used these starters. Another was used by the British navy. (These could not have been prepared prior to the development of the remotely automated tungsten-inert gas (t.i.g.) welder, described in the next section.) Similar sources, with less radium content, had been used earlier by oil-well explorers (loggers), in the exposure and examination of deeply located substrata to identify oil-bearing sites.

In the late fifties, oil-well logging interests (Wells Surveys Inc.) asked CPD to produce higher-quantity neutron-emitting sources from an actinium:beryllium mixture for oil-well exploration. The exercise to irradiate radium, separate the actinium-227, and fabricate the sources became far too expensive. After the preparation of a number of these sources, the program, being uneconomical, was discontinued.

With the declining interest in radium, all products were phased out throughout the sixties. Borrowed radium was returned. The program was terminated in 1970, and all remaining CPD stocks were given to China as a gift.

Summary of CPD Revenues, Profits (Losses) – 1947 to 1978

PERIOD	REVENUES ($ million)	PROFIT/(LOSS) ($ million)
1947-52	2.66	1.39
1953-59	11.25	(0.286)
1960-69	61.50	3.50
1970-74	53.55	(1.63)
1975-78	81.42	(2.77)

Data in the above table ceases at 1978, the year in which CPD was restructured to become RCC. From 1980 to 1990, annual revenues increased, exceeding $100 million by 1990.

PROCESSING HOT CELLS AT CHALK RIVER AND CPD

The CPD need for larger quantities of cobalt-60, both as high specific activity (HSA: $>7.4\times10^3$ GBq/g (200 Ci/g)) for theratron sources and low specific activity (LSA: $<1.8\times10^3$ GBq/g (50 Ci/g)) for large-scale food and medical irradiation plants, raised problems. The handling of such vast amounts of radioactivity following reactor removal was a concern. Also, Chalk River needed better shielded facilities to handle irradiated reactor fuel elements that were under test for the power-reactor program. To meet these needs, in 1957, Chalk River built a complex of high-radiation-capacity processing units equipped with remote operating devices. These were called 'Universal Cells' (see chapter two).

The increased requirement of HSA cobalt-60 was met by the Chalk River development of a more efficient target. The 1×1 mm pellets of nickel-plated cobalt-59, normally loaded in an annular ring of 3 mm within irradiation capsules holding 30 grams, was revised to the sealing of that weight in one layer within 10 cm square pure aluminum sandwiches. These were curved and tack welded together to form tubes 2.5 metres in length and entered into high-flux reactor locations. LSA cobalt-60 was produced in 2.5-cm-long cylinders of cobalt-59 sheathed in aluminum, stacked vertically on trays that were assembled one above the other, and irradiated within the lower-flux regions of each reactor. Following irradiation, the sandwiches and cylinders required chemical processing in the Universal Cells for removal of the aluminum and recovery of the cobalt-60. About 7.4×10^7 GBq (2×10^6 Ci) of cobalt-60 could be produced annually in the Chalk River reactors.

At the Tunney's Pasture building in Ottawa, CPD had the same need for such hot cells. Construction of two general-purpose cells at the Tunney's building were completed in late 1957. All cobalt-60 source fabrication was subsequently transferred from Chalk River to the CPD cells. The handling of exposed cobalt-60 in the cells at CPD would result in serious in-cell contamination. Rigorous decontamination of sources would be required. Source seals had been made by Chalk River with screw-down metal lids using lead gaskets or force-fit metal plugs, followed by decontamination before shipment to CPD.

To eliminate source contamination and any possibility of source leakage, an improved method of a higher integrity source seal was needed. Following a year of trial units, a remotely automated tungsten-inert gas welder was developed at CPD. It proved to be reliable and at once heralded the dawn of source sealing within a welded inner stainless-steel capsule, followed by ultrasound decontamination and subsequent sealing within an outer stainless-steel welded capsule. The concern regarding the leakage of radioactivity was eliminated. The first all-welded stainless-steel theratron source was No. S-246 in 1960, which encased 2.96×10^5 GBq (8,000 Ci) of cobalt-60, one of the largest therapy sources ever made by CPD. It was sent to a Puerto Rico clinic. This welding technique was ultimately adapted for the sealing of all metal-encased radioisotopic sources.

CPD BEGINS RELOCATION AT KANATA

By 1964, nearly one hundred cobalt-60 theratron units were being produced annually, exclusive of other products. The manufacturing plant on Laperrière Avenue in Ottawa became totally inadequate for such a large volume. Crowded conditions existed at the Tunney's Pasture building and the situation was only partially resolved when the sales department vacated to rented quarters. Relocation to larger facilities had to be considered again.

CPD purchased approximately eighty acres of crown land in Kanata, just west of Ottawa. Phase I consisted of the construction of the largest and most comprehensive machine shop and assembly area in eastern Ontario. It was completed by 1966, along with a five-storey administration building. (These facilities were expanded three times over the next fifteen years.) Sufficient land was reserved for the eventual relocation of the balance of CPD Tunney's Pasture operations.

As business grew, the storage of cobalt-60 inventory at the Tunney's Pasture laboratories caused problems. Bulk cobalt-60 was stored in three-tonne lead- shielding containers that were used for shipments between Chalk River and CPD. High temperatures on container surfaces (60°C), due to the heat of decay of the cobalt-60, raised concern that possible melting or shifting of internal lead could cause gaps in the shielding, bringing about a serious radiation hazard. The quantities of cobalt-60 stored in these containers was accordingly reduced. It was realized, however, that if the demand for cobalt-60 continued to grow, as seemed highly likely, the situation would become immeasurably worse. Subsequent examinations showed that the containers were, in fact, sound. Nevertheless, new cobalt-60 facilities were considered to be vitally necessary.

In the second phase of the relocation to Kanata, a new cobalt building was completed by 1971. It consisted of four high-radiation-capacity hot cells, expandable to eight if required. Cooled deep-water storage pools were built for cobalt-60 storage of inventory and completed sources. The latest models of high-density shielding windows and master-slave manipulators were installed. All heavy fixtures – the eight-tonne hot-cell enclosure doors, 25-tonne bridge cranes for overhead handling of shipping containers, remote in-cell cranes and pool retrieval devices – were motor operated. The facility became a worldwide showcase for isotope-handling technology.

The CPD personnel who remained in the Tunney's Pasture building renovated the vacated cobalt facilities to provide space for the development of new radiochemical processing technology, particularly the separation of molybdenum-99 from irradiated uranium, and for the installation of the Chalk River-developed SLOWPOKE-2 reactor (see chapter twenty), for research.

COBALT-60 FROM POWER REACTORS

By 1965, CPD's food-preservation research program, based on gamma irradiation using cobalt-60, indicated that an expanding market was available. If the demand for the irradiation plants developed as forecasts indicated, then CPD would require much more LSA cobalt-60 than the Chalk River reactors could produce. Most of the annual 7.4×10^7 GBq (2×10^6 Ci) of cobalt-60 would be required to support the need for HSA cobalt-60 for theratron sources if the theratron business continued to expand, leaving less than 50 percent for LSA production. Cobalt-60 irradiators for food and medical-supplies sterilization were expected to be loaded with as much as 1.85×10^8 GBq (5×10^6 Ci) of cobalt-60 each.

An arrangement was made whereby cobalt target rods were irradiated in the reactors at Pickering A and B, Bruce B and Gentilly-2 nuclear power stations. The power reactors, in effect, would be producers of a saleable by-product.

Due to the long production period for cobalt-60 in power reactors (one to two years), and since the primary objective of the CANDU reactor being electricity production, CPD accepted responsibility to receive and pay for all cobalt-60 discharged at each reactor shutdown period (annually, on average), according to a long-term agreement. Any reduction in future production would require advanced negotiations. The potential annual production of LSA cobalt-60 from these flux adjuster rod irradiations was:

Pickering A and B	4.7×10^8 GBq	(1.3×10^7 Ci)
Bruce B	4.4×10^8 GBq	(1.2×10^7 Ci)
Gentilly-2	7.4×10^7 GBq	(2×10^6 Ci)

CPD was ready with their new high-capacity hot cells at Kanata when the first reactor discharge was made in 1972.

CPD/RCC — GROWTH, STAGNATION, PROFIT DECLINE

Theratron Units

CPD made excellent business progress from 1950 to 1970. Sales of equipment requiring cobalt-60 had been remarkable. By 1960, CPD had established basic standards for theratron units.

In the following decade, major modifications to theratrons were made as physicists gained experience and requested changes. This resulted in the first of a new series of theratrons for the seventies: the vertical units, Eldorados 6, 8, 76, and 78, followed by modified Model B theratron units, the C, C′, F, 60, 80, 760, and 780 (see figure 5.1). Many options became available, with some units being fully computerized for treatment setup. In 1968-69, CPD sales of theratrons reached a peak of 126 units. By 1970, a total of 600 units had been built. (Some models were contracted out for manufacture by CPD-licensed agents in Europe.) All were loaded with AECL-produced cobalt-60 sources[4] and installed by CPD's installation group or their licensed agents. More than 90 percent of the units were exported, yielding about 60 percent of all CPD revenues as foreign exchange.

Annual revenues by 1969 and 1970 exceeded $10 million, but residual profits were not high, dropping to less than $400,000. Orders for new theratrons then dropped alarmingly when the number of suppliers increased, with each offering enhanced technology and performance. Field maintenance problems arose with CPD's new line of theratrons for the seventies (malfunctions in solid-state boards, patient anti-collision devices on couches behaving erratically, and, most seriously, the detection of many cracked rotational arms on theratrons in current operation). About 150 arms required urgent replacement, which put severe strain upon CPD's capabilities. The replacements in the field were costly and caused serious delays on new unit deliveries.

The introduction of medical linear accelerators (LINACs) was a factor in the reduction of the theratron market. Varian Associates, of Palo Alto, in the United States, normally provided research accelerators, but in 1967 began to market its Clinac 4 MeV medical LINAC. This LINAC was only marginally more effective than cobalt-60 for deep-seated treatment of cancer. In spite of their higher cost and dubious reliability, accelerator installations increased rapidly. The French CGR Group began to market a 6 MeV medical LINAC (the Neptune) and a 32 MeV Sagittaire unit. All of this seriously affected the cobalt-60 sales picture.

FIGURE 5.1
Early Theratron 80

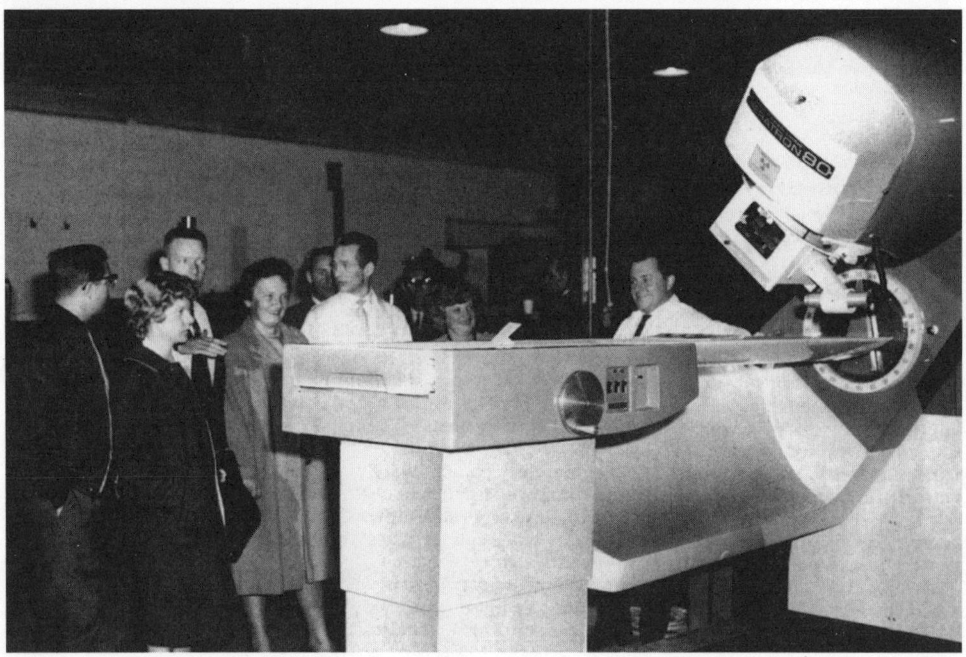

In 1970, CPD was able to market only fifty theratrons and the large-capacity shop began to look for non-CPD work, such as the manufacture of CANDU reactor components.

Medical LINACs

The impact of LINAC competition was sufficient for CPD to consider entering the LINAC field. CPD had no expertise with medical accelerators and would need to enter a cooperative

[4] The amount of cobalt-60 loaded into each source was approximately 7.4×10^4 GBq (2,000 Ci). The total cobalt-60 encapsulated into 600 theratron sources by 1970 was 3.7×10^7 GBq, (1×10^6 Ci), all produced at Chalk River, and most fabricated by CPD.

venture with an experienced company. The decision was difficult, for the venture would be the largest ever for CPD, and success was by no means certain. In 1972, a five-year agreement was signed with the French company, CGR.MeV, a subsidiary of the Thomson-CGR Group. This group had been CPD agents for many years, selling CPD's theratrons in Europe. To complement their Neptune and the large Sagittaire lines, a LINAC more suitable to the size of clinics – the Saturne, a 20 MeV unit – was being developed and was nearing the prototype stage. CPD realized that selling these accelerators in the United States would complement its theratron market and, at the same time, allow CPD to compete seriously for American accelerator sales. Before the United States would accept the accelerators, however, major redesign would be required to meet American radiation standards.

At the same time, Chalk River was developing a double-pass, high-energy LINAC that would provide 5 and 25 MeV electron beams, and was anxious for CPD to enter the LINAC field with its product. However, AECL's board decided that Chalk River was not commercially oriented. Chalk River accepted AECL management's recommendation regarding the CGR.MeV contract. The decision, while seemingly correct at the time, was to cause CPD considerable trouble and expense later.

The redesign of the French units, the necessary shop expansion for assembly and testing, plus the extensive technical training for installation, made the program extremely expensive. There were licensing delays, and demanding service requirements. A total of 36 Therac 6s (Neptune), and 20 Therac 20s (Saturne) were installed in North America. They all performed well, but the necessary competitive selling prices resulted in a severe loss situation for this product. In 1978, CPD decided that the expense to market Theracs with continuous losses would decimate all other programs that were emerging. Thus, a withdrawal from the French venture was planned in 1979. The AECL board urged CPD to continue the program with the Chalk River double-pass Therac 25. CPD agreed and installed twelve units, built to Chalk River specifications by 1985. These units initially failed to operate reliably. Six patient overdose accidents occurred, due to software deficiencies of the computer-control system. The product showed no sign of becoming profitable and the manufacture of additional units was abandoned by CPD in 1985.

Industrial

Since the mid-fifties, strong emphasis had been placed by CPD's R&D group on developing products that would use cobalt-60. This led to a concentration on radiation devices to sterilize medical supplies, assist the study of gamma-radiation effects on materials through laboratory experiments, and extend the shelf life and storage time of food. All avenues were explored concurrently.

Food preservation has always been important in developing countries. Much of their grain produce is destroyed by insect infestation, and many foods are destroyed by their own enzyme reactions due to limited refrigeration. Studies by CPD R&D proved that the storage life of many foods could be extended by cobalt-60 gamma irradiation, while leaving the food safe and palatable.

Results from these studies prompted CPD to build the Mobile Cobalt-60 Demonstration Food Irradiator (MDI) in 1961 (see figure 5.2). It contained a cobalt-60 source of 5.92×10^5 GBq (16,000 Ci) that required 18,000 kg of poured lead shielding. This was mounted on a heavy-duty road trailer and pallets of food on a conveyor were passed around the exposed source in ferris-wheel fashion within the shield, to provide an approved 10 kilorad dosage. The unit was hauled to eastern Canada, the Maritimes, through Michigan and to California from 1961 to 1964, demonstrating the storage-life extension of irradiated potatoes, fish, berries, onions and grapes.

The American military purchased a CPD-designed cobalt-60 food-irradiation facility to test on its mobile forces. A private enterprise at St. Hilaire, Quebec, purchased one for use as a potato irradiator (it failed for commercial, not technical, reasons). Inroads into the United States were prevented by lack of approval from the United States' Food and Drug Administration (FDA) despite the military testing, the publication of the experimental data, and acceptance in many other parts of the world.

Cobalt-60 irradiation was a proven method of destroying bacterial contamination of medical supplies. However, a large proportion of the medical-supplies industry was locked-in and satisfied with the existing method of exposure of medical goods to ethylene-oxide gas for sterilization, making cobalt-60 irradiation to sterilize prepackaged goods hard to sell. An important breakthrough came in 1962 when CPD was asked by Ethicon Inc. (a Johnson & Johnson subsidiary), in the United States, to bid on the design and supply of a cobalt-60 medical-sterilization plant. Obtaining this contract over a prominent world supplier (Marsh, United Kingdom) was a coup for CPD, as it established them as a world leader in the field. The plant was installed in 1964 in Sommerville, New Jersey. Whereas other parts of the American medical industry were initially slow to adopt this technology, installations were made by other Johnson & Johnson subsidiaries in Australia, Sweden, Italy, Brazil, Mexico, and Peterborough, Canada.

FIGURE 5.2
Mobile Cobalt-60 Demonstration Irradiator Designed and Built 1960/61

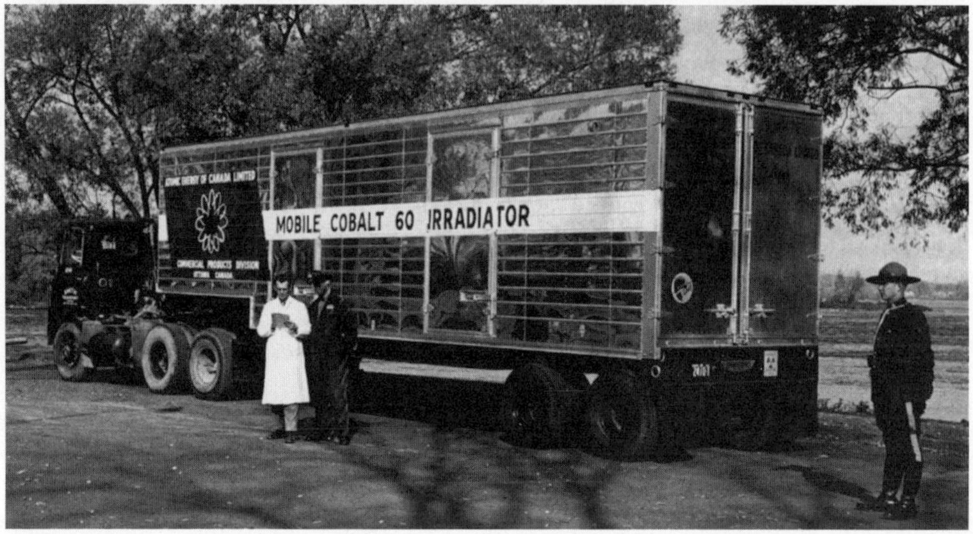

The apparent need for small in-laboratory devices to provide high gamma-radiation doses within a shielded enclosure of a 1-to-2 litre volume led CPD to design a variety of units, called Gammacells, that could be loaded with 2 to 8×10^5 GBq, (6-24,000 Ci) of cobalt-60. The first unit was exhibited at the New York Atom Fair in October 1957. These have become popular in research centres throughout the world for studying irradiation effects on biological and non-biological specimens.

A specific variation of this device, configured for the irradiation of human blood to destroy lymphocytes, used for transfusion into immuno-suppressed patients (newborn, leukaemia patients), has a growing market. Another is the "Panoramic Irradiator", capable of irradiating large specimens in a shielded room.

A technical process developed by CPD that failed to gain market acceptance was for hardening of certain woods, such as parquet floors, for use in heavy- traffic areas and bowling alleys. This hardening was achieved by the vacuum induction of liquid methyl-acrylic plastic into wood pores, which were then polymerized by cobalt-60 irradiation. Installations were made in the CPD (RCC) Engineering lobby, and another in the social area of the Nepean Sportsplex in Nepean, Ontario. An installation was likewise made by an American supplier at the Kansas City airport. All of these installations have been in constant use for some twenty years with minimal signs of degradation. The cost of shipping the wood to CPD, the vacuum process, the irradiation and the subsequent return of the product became expensive, and a good market failed to develop.

MAPLE was a device designed to utilize the heat of decay from a large cobalt-60 source to activate a thermo-electric power generator. The first unit, housed in a four-tonne shielded container, was used on the St. Lawrence River to operate a flashing beacon light. A second unit was located at Resolute Bay, to service weather-recording instruments. However, ordinary maintenance, such as replacement of light-bulbs, or replacement of the recording pen and paper at frequent intervals, negated the long-term advantages of cobalt-60. The initial product cost was very high (cobalt-60 and shielding costs). The units worked well, but after the expiry of the first half-life, they were not replaced.

Radioisotopes

Throughout the fifties and until the mid-sixties, CPD had established itself as a major supplier of bulk radioisotope products to much of the world, including Canada, the United States, parts of Europe, and Japan. The future looked bright.

In the late sixties, however, cancer-treatment clinics began using the many newer radioisotopes that were proving to be more suitable for the diagnosis and treatment of cancer. CPD was not ready for this market and, as a consequence, was relegated to being a bulk supplier of the older radioisotopes.

CPD'S (RCC) RESURGENCE PROGRAM

With annual revenues virtually flat at $10 million from 1970 to 1975, and the high internal expenses to pay for the new facilities plus the expensive LINAC program, the AECL board in 1975-76 had to decide whether to sell CPD to interested American buyers or to support CPD in a recovery program. The AECL board acknowledged that CPD had been good for Canada, that it had been profitable in the sixties and that, foreign ownership of part of Canada's nuclear capability was not acceptable. The board, therefore, decided to support a resurgence program.

Medical Products

Early acceptance of CPD's theratrons in the sixties had given it a leadership role in world teletherapy manufacture. When sales of theratrons began to decrease in the seventies, due, in part, to LINACs, CPD realized that if their market share was to improve, their theratrons would require conceptual changes (simplicity to reduce production expense, a design to meet competition on a feature-by-feature basis, and affordable pricing for smaller clinics).

A second series of major revisions, based on the best features of the series produced in the seventies, culminated in the dropping of vertical units. The new models comprised: the Phoenix – a basic, manually operated, no-frills theratron, with upgradable options; the Theratron 780C – a higher-capacity radiation unit, with computerized options; and the larger Theratron 1000. These reached the market in 1986, and a promising market recovery began.

Accessory devices for theratrons were always part of CPD's program. A patient pre-setup unit introduced in 1977 was the Simulator (now trademarked as Therasim). This device is smaller but similar to the theratron unit. X rays are used to assist in accurate tumour location, and to establish the required cobalt-60 dosage from the theratron. The Therasim minimizes patient stress by reducing treatment setup procedures beneath the large therapy unit, and permits the clinic to handle a larger patient load. Originally, the Therasim was considered a luxury item for clinics. However, in 1982, the 100th Therasim produced was shipped to the Saskatoon Cancer Clinic.

Coincident with the Therasim, a computerized "Theraplan" system was developed as a valuable accessory for the planning of ongoing theratron treatments. Theraplan consists of a tumour-registry system logged into a personal-computer database. It comprises a patient-dose compensating device, simulated with a water phantom and a film-dosimetry recording system, to measure therapy unit outputs and the transfer of these data to the computerized treatment plan for subsequent treatments.

The Montreal Neurological Institute (MNI) became interested in the development of a body-scanning device similar to a rudimentary unit that they had observed at the Bethesda Maryland Hospital. They developed a prototype for study and evaluation. MNI saw that a fully developed device using positron emitters (obtained from cyclotrons) would be useful in brain-scanning research. It asked CPD in 1978 to develop a 3-D Positron Emission Tomography unit (PET) for commercial sale. With a $2 million budget and two scientific teams (from CPD and Chalk River), a prototype unit, the Therescan 3128, was built. The computerized software and the unit were developed by CPD while the resonance detectors were developed by the Chalk River team. The unit was sold as a prototype for about $1 million to

MNI in 1980. In 1980, TRIUMF, of the University of British Columbia, placed an order for a similar unit for delivery in 1981. However, their order was cancelled the following year, due to a conflict of specifications. After some usage of the prototype, MNI requested CPD to develop a more sophisticated unit. CPD realized that few institutes would have the expensive infrastructure (cyclotron, radioactive processing facility) to operate such an assembly, as it was basically a medical research device. Few sales were expected and CPD was not able to finance additional development funds. The project came to an end in 1982. There are five PET units in Canada today.

The unit sold to MNI operated reliably for seven years, and was eventually resold to a Scandinavian group. Various other medical treatment devices were investigated, and prototypes built. Most were technically successful, but due to their expense or lack of market interest all were abandoned. Such devices were the Brachytron (a body intercavity unit for internal radiation techniques), a neutron-therapy unit (see chapter nine), and a cesium-137 source unit.

Industrial Products

The expected approval from the FDA to permit food irradiation in the United States failed to materialize, reflecting public reluctance to accept any food associated with irradiation. This delay and no other available market obligated CPD to accept a large inventory write-off of power-reactor cobalt-60 in 1974, thereby incurring a loss of more than $1 million that year.

While continuing the struggle to obtain FDA approval of food irradiation, CPD believed that a resurgence in this field would come from improved sales of the variety of cobalt-60 irradiation devices that they had developed during the mid-fifties – in particular, medical-supplies sterilization plants plus the sales of laboratory radiation devices, backed up by a program of cobalt-60 resupply to counterbalance decay.

Following the breakthrough in obtaining the medical sterilization contract from Johnson & Johnson in 1962, emphasis continued on the development and sales of larger-capacity plants. These were initially designed to contain as much as 3.7×10^7 GBq (1 million Ci) of cobalt-60, but soon the average design capacity far exceeded that amount.

An interesting adjunct to this growth was that the large quantity of power-reactor cobalt-60 that CPD was required to write off in 1974 had been sold before 1976. By 1974, a total of twenty-seven such plants had been installed by CPD, a few being installed in the United States for several medical suppliers.

Radioisotopes

To support a resurgence in the sales of radioisotope products, NRU operations were modified to permit easier entry/exit of targets to the high-flux regions of the reactor. Chalk River installed a loop in NRU for the irradiation of circulating xenon gas, which permitted selective tapping off of carrier-free[5] iodine-125 in good quantity, and in a more efficient way than the low-yield process CPD had established earlier. Economical quantity production of this radionuclide processed to a usable radiochemical had been a long-time CPD objective. It is used in quantity for radioimmunoassay studies.

Iodine-131, a carrier-free standard product since 1948, is used for thyroid imaging and therapy. A lengthy wet-chemical process was required to separate the iodine-131 from irradiated tellurium metal. It was messy, replete with costly process losses, and was uneconomical to produce. A dry-vacuum-distillation method of iodine-131 separation from irradiated tellurium dioxide had been successfully tested by CPD's R&D, but it required a larger target weight. The availability of the high NRU fluxes overcame this disadvantage without loss of radioactive quantity production. A full-scale extraction system was built that gave excellent and reliable results in a fraction of the usual processing time, and at reduced cost. The tellurium metal method was immediately abandoned.

The new product that CPD most wanted to produce in quantity was fission-product molybdenum-99, having a 66.7-hour

[5] Carrier-free: All atoms of the radioisotope exist in a pure form. It is not diluted, nor contaminated with any other radioisotope, or non-radioactive material, a desirable feature in the irradiation of human tissue.

half-life. Molybdenum-99 decays to technetium-99m, a nuclide that emits a single gamma photon of 140 keV, and has a six-hour half-life. Biological compounds targeted for selective absorption by body organs can be readily labeled with technetium-99m for scanning body functions. Technetium-99m is used in perfusion (blood flow), brain and heart scans, and to detect bone fractures. The energy level is easily collimated, resulting in sharp images by gamma cameras. Technetium-99m has been considered as the almost perfect diagnostic tool and has gained rapid market acceptance.

The product would be very profitable to any major supplier who was able to supply large and routine quantities of carrier-free molybdenum-99. In 1972, NRU was to undergo a lengthy service shutdown and would not be available for irradiations for an extended period. McMaster University (Hamilton, Ontario) assisted CPD to gain a foothold in the technetium-99m market by offering their small reactor (2 MW) to irradiate targets of enriched uranium that CPD would process. A three-year agreement was signed. The reactor power level was raised to 5 MW to increase production. However, yields were irregular, at less than 3.7×10^3 GBq (100 Ci) per week. CPD knew that, to be effective in the marketplace, routine quantities in the order of ten times that amount were needed. The availability of the McMaster reactor, however, allowed CPD to develop its molybdenum-99 process.

As soon as NRU was again operational, CPD switched its irradiation of uranium targets to NRU. Molybdenum-99 production quantities from NRU were easily sufficient to meet the large market demand. CPD began the fission-product separation process at the Tunney's Pasture laboratory. As the demand increased, the fission-product separation process quantities became too great for Tunney's Pasture. Also, the arrangement involved daily shipment of irradiated uranium targets from Chalk River to CPD for processing, and then the return of fission-product residues and the unused uranium-235 to Chalk River. It was decided to switch the first stage of chemical separation to Chalk River's universal cells. Chalk River went back into routine radiochemical processing for CPD. The unrefined fission-product molybdenum-99 was separated there and shipped in bulk to CPD for refinement and quality control.

To ensure reliable availability of molybdenum-99, CPD and Chalk River organized smooth production and transportation schedules. Rapid Customs clearance and the use of chartered aircraft were arranged. Moly-99, as it is called in the trade, provided substantial profits from 1978 on, with sales approaching $20 million annually by the mid-eighties.

Another product resulting from the molybdenum-99 program was the collection of xenon-133 fission product at the molybdenum-99 separation stage. This became a new product, useful in lung scanning.

In time, the accomplishments of CPD (RCC) became respected throughout the scientific world. Over the years, Australia, India, Pakistan, Belgium, Mexico, Thailand and Taiwan have consulted CPD in the development of their own radioisotope products.

Cyclotron-Produced Radioisotopes

Certain cyclotron-produced radioisotopes have special medical uses and have characteristics that cannot be met with reactor-produced nuclides. These are converted by pharmaceutical firms into aseptic products called radiopharmaceuticals. Such products command high selling prices.

In 1978, the CPD (RCC) product-development group proposed entering this growing segment of the market, to which they had no access. The large TRIUMF cyclotron at the University of British Columbia was made available on a contract basis. In 1980, CPD (RCC) constructed a four-hot-cell processing laboratory at TRIUMF, and ordered a 42 MeV-type CP42 cyclotron. Meanwhile, it shared the use of the TRIUMF 500 and 70 MeV beams. Production of thirteen-hour iodine-123 from enriched xenon-124 in bulk quantities was planned initially. It has been used extensively in heart and brain studies. In 1982-83, the first shipments of radiopharmaceuticals were made to hospitals and clinics in Canada.

The resurgence program for radioisotopes became a total success. CPD (RCC) had become a major supplier of both the older and the newer medical radioisotopes throughout the world.

FINAL RELOCATION FOR CPD/RCC

To cope with the production of the many new products and the higher radioactivity, it was obvious that the Tunney's Pasture laboratories had to be closed down. In 1979, final consolidation of CPD (RCC) at Kanata was started. By 1984, new offices and laboratories were completed (see figure 5.3). All Tunney's Pasture operations and some 180 personnel joined the 600 already at the Kanata site in the $35 million, 12,540 square-metre Kanata Isotope Production Facility (KIPF). KIPF contained twenty new hot cells constructed of lead and iron, instead of high-density concrete, reducing them to about one-third of their former size but having the same radiation-shielding capacity. Each radioisotope was to be processed in dedicated and isolated hot cells. These included cells for iodine-131, iodine-125, carbon-14, xenon-133, iridium-192, and yttrium-90 for radiography sources, and well-equipped facilities for the molybdenum-99/technetium-99m line. Cells were planned to handle special irradiations upon demand, for the R&D group, and to handle waste products. KIPF was built as an integral part of the 1971 cobalt-60 facility. Today it represents the largest and most modern radioisotope-production facility in North America.

FIGURE 5.3
Cobalt and Radioisotope Site, Kanata, 1980

BIBLIOGRAPHY

Bothwell, Robert. Nucleus*: The History of Atomic Energy of Canada Limited*. Toronto: University of Toronto Press, 1988.

Canada. Parliament. House of Commons. Special Committee on Research. *Minutes of Proceedings and Evidence*, no.1-9, 30 April-5 June 1956.

Harbert, J.C., and H.A. Ziessman. "Therapy with Intraarterial Microspheres." *Nuclear Medicine Annual* 1987, 295-320.

Johns, H.E., and T.A. Watson. "The Cobalt-60 Story." *Cancer in Ontario* 1982, 2-21.

Nordion International Inc. *Therasphere*. 91132BRO-06. Kanata, Ont.: Nordion Promotional Services.

Sims, Gordon H.E. "The Evolution of AECL." Master's thesis, Carleton University, Institute of Canadian Studies, 1979.

BASIC RESEARCH

Chapter Six

Basic Research: An Introduction

G.C. HANNA

This introductory chapter explains what is meant by basic research, outlines its historical foundation, and summarizes subsequent developments from an organizational point of view. The research itself is described in the three subsequent chapters.

A very complex terminology has grown up to describe various types of scientific research, much of it redundant and disputed. In a 1972 monograph on "Forty-five varieties of research (and development)", Lord Rothschild argued that "pure", "basic", and "fundamental" are synonyms, all describing research "done solely to increase knowledge without any practical application in view", and that all other research, which may be described as "underlying", "strategic", "mission-oriented" etc., is simply applied research. However, it is difficult in practice to maintain a completely clear distinction. Applied research can be directed to very long-term objectives, and its practitioners can be as curiosity-motivated as pure-research scientists; they in turn are only too happy if their contributions to pure knowledge are promptly applied to practical problems.

This section on basic research is not confined to pure research, but includes some applied programs, generally of a long-term nature, that are not covered elsewhere in this book. It does not include research in the life sciences, which is described in chapter three.

In the early years of the Canada/United Kingdom project, particularly in 1944 and 1945, almost all the research was directed to the design and operation of NRX and the chemical extraction plants. As the immediate problems were dealt with, at least provisionally, research activities began to separate into two conceptually distinguishable types. The first met the requirements of existing major projects and formally-approved new ones (e.g., the NRU reactor). It was greatly expanded, starting in the mid-fifties, to meet the needs of the power-reactor program, and it is described in other sections.

The second type of research was aimed at obtaining a more fundamental and detailed understanding of the various processes involved in nuclear science and its applications. The question of whether it was pure or applied research was not initially of much concern. Pure research had discovered fission and its application had come very quickly. It was therefore not unreasonable, in the early days of the project, to think that further research in this new area of science could lead to discoveries having some practical importance in the near future. The scientists were eager to develop the tools of their trade, and as they did, their researches, in a rapidly expanding field, did not appear to them as obviously pure or applied. This early period is characterized by the large number of research scientists who were simultaneously engaged in activities that, in retrospect, are identifiable as pure research, applied research, and instrument development. Items such as mass spectrometers, particle detectors, and electronic systems were designed and built by the experimenters themselves; nowadays, corresponding items are readily available from commercial sources, regrettably almost all outside Canada.

Unlike the applied research programs which, as noted above, were greatly expanded, pure research in the physical sciences has been maintained at a relatively constant level. With access to advanced facilities, many of them unique in Canada, the scientists have successfully exploited their opportunities to carry out forefront research of international quality, and AECL has done its best to update the facilities on which the research depends.

That AECL has a strong program in basic research is ultimately attributable to C.J. Mackenzie, who was acting president of the NRC when the atomic energy project was set up. He recognized it as an opportunity, in a prestigious area, for the advancement of Canadian science, which he saw as vital to the future of the country. J.D. Cockcroft, director of the project, was also a strong supporter of broadly based research, as would be amply demonstrated later at Harwell. E.W.R. Steacie, the deputy director, emphasized the importance of fundamental research in an August 17, 1944 memorandum to Mackenzie, an emphasis that characterized his entire career, which culminated in the presidency of NRC (1952-62).

In September 1944, Cockcroft and Steacie proposed that NRC should build an atomic energy research centre in Ottawa to take over from the Montreal laboratory when it was closed down. This centre would be additional to the major facilities and associated laboratories at Chalk River. Mackenzie was sympathetic, but G.C. Laurence, the senior Canadian physicist, argued strongly against dividing up what would be a relatively small research group after the United Kingdom staff returned home. In July 1945 it was decided, perhaps mainly for financial reasons, not to proceed with the Ottawa centre, but to transfer all the Montreal laboratory activities to Chalk River.

NRX, the world's best research reactor at that time, would of course be the centrepiece of the program, but there were other elements. The construction of a Van de Graaff accelerator had been initiated by Cockcroft in February 1945, and this laid the foundation for what became AECL's largest pure research program, nuclear-structure physics. This accelerator was to have been located at the proposed Ottawa centre, and Cockcroft justified it as "a necessary tool for the investigation of nuclear fission and absorption levels, since it can provide a monokinetic neutron source of controllable energy. Such a generator has been in constant use in the American project and it is our intention to build one in England." Not exactly pure research.

The early program also included high-energy physics, using cosmic rays, and this expertise made possible the study, described in chapter nine, of neutron production by high-energy particles. There were even plans, by B. Pontecorvo in 1946, for an underground experiment to detect solar neutrinos. This would have anticipated by some twenty years Davis's celebrated experiment, using the technique proposed by Pontecorvo, in the deep Homestake Gold Mine, South Dakota.

W.B. Lewis, who succeeded Cockcroft in 1946, was a strong and effective supporter of basic research. He demanded high standards, not least in matters of presentation and publication, but left the scientific direction of pure research to the scientists while he concentrated his formidable abilities on applied-research activities.

When AECL took over from NRC in 1952, the commitment to basic research was maintained. Nor was there any abrupt change in 1955 when the power program was approved, giving a prime focus to what had been an imprecisely defined mission. The reactor-based research continued to flourish, with NRU (1957) displacing NRX for most of the work. The Van de Graaff was replaced by the world's first tandem accelerator in 1959, which was succeeded by a larger one in 1967 (see chapter eight). Research in chemistry and materials science (see chapter seven) was sustained with a variety of advanced equipment, including mass spectrometers, a mass separator, electron microscopes, and electron and positive-ion accelerators. The basic-research program would have been greatly expanded had Lewis obtained approval for his Intense Neutron Generator (ING) project, but this was cancelled by the government in 1968 (see chapter nine).

In 1976, anticipating demands for more detailed accountability, AECL set up, under an R&D program committee, six steering committees to plan, coordinate, and report on R&D activities in i) power reactor systems, ii) the nuclear fuel cycle, iii) environmental protection and radioactive waste management, iv) heavy water processes, v) assessments and new applications, and vi) underlying and advanced systems research.

The purview of this last steering committee was essentially the area covered in the present section (and the R&D described in chapter three); i.e., "underlying" included pure research, and "advanced systems" referred to the accelerator and fusion-related work described in chapter nine. The steering committee justified its research activities, which accounted for almost one-third of the Research Company's budget, as:

- providing AECL's current mission with better fundamental understanding of the processes involved in nuclear energy systems, with a knowledgeable consultancy service, and with a mobile reserve of versatile individuals for dealing with emergencies;
- meeting AECL's need to be aware of, and to be able to respond to, developments elsewhere, by exploiting the "subculture of science" (meaning personal contacts with researchers in other institutions), which could provide intelligence when formal channels were restricted. It was considered particularly important to be able to evaluate without delay the potential of plausible schemes that could otherwise lead AECL into expensive blind alleys;
- providing the basis for future advance.

In retrospect, this rationale did not adequately emphasize the prime purpose of pure research, which is the advancement of human knowledge, and the particular responsibility that developed countries and large laboratories have for pure research – an activity whose benefits to those who fund it are seldom direct and immediate. It could also have been pointed out that achievements in this area added to AECL's overall prestige.

The rationale had some success. The underlying and advanced systems research continued to receive about the same share of the total R&D budget, and the nuclear physicists' superconducting cyclotron project, which was started in 1974, was allowed to proceed to the construction phase in 1978 (see chapter nine).

By this time, most of the more applied programs in underlying research (other than those described in chapter three) were in chemistry and materials science (see chapter seven). They included work on radiation chemistry, high-temperature-solution chemistry, isotope separation, radiation-induced creep and growth, hydrogen uptake by zirconium, and zirconium corrosion.

The pure research program had three major, world-class components. The largest was the accelerator-based nuclear-physics program, which concentrated on the use of heavy ions to study the structure of nuclei (see chapter eight).

The second one, at NRU, investigated a wide range of phenomena in condensed matter physics by neutron scattering. It exploited, in particular, the triple-axis-spectrometer technique that had been pioneered at Chalk River thirty years earlier. Thanks to excellent instrumentation, this program remained fully competitive with programs based on newer higher-flux reactors in Europe and the United States (see chapter eight).

The third research program arose out of the 1962-63 discovery at Chalk River of channeling, which is the ability of energetic positively-charged particles to penetrate further into crystalline material by following preferred directions (i.e., channels) between the regularly spaced atoms in the crystal. Of the three pure research programs, it provided the most immediate applications, and its use in solid-state studies, especially of semiconductors, was energetically pursued, and attracted visiting scientists from around the world (see chapter seven).

Unfortunately, budgetary pressures were becoming increasingly severe and, rather than continue to squeeze all three programs, AECL decided in 1985 to transfer the channeling program, recognizing that it could, unlike the other two, be successfully continued in a Canadian university environment – which it was, thanks to considerable initial help from AECL.

The neutron-scattering program was continued because Chalk River is one of the few places in the world where such research can be done, and the research represents outstanding value for money, making minimal demands on a facility (NRU) already funded for other purposes. For nuclear physics, the cyclotron was just being commissioned, which would ensure the future of this Tandem Accelerator Superconducting Cyclotron (TASCC) laboratory as one of the world leaders in nuclear-structure research. Both TASCC and the neutron-scattering installations at NRU constitute national facilities used by Canadian universities, who have supplied major pieces of equipment. This was no doubt a significant consideration in 1985 and will remain an important justification for the future survival of these programs.

In the early eighties AECL started to change the way its programs were planned, performed and reported to government. By February 1986, a new organization was in place (see chapter two, figure 10) with four program responsibility centres: PRC 1, for reactor development; PRC 2, for waste management; PRC 3, for physics and health sciences; and PRC 4, for radiation applications and isotopes. Of the R&D activities previously classed as underlying and advanced systems research, those described in chapters three, eight, and nine all went to PRC 3, except for accelerator physics (chapter nine) and the environmental research work at Whiteshell, which went to PRC 4 and PRC 2, respectively. The chemistry and materials-science programs of chapter seven that remained after the channeling program disappeared were considered to be applicable to specific near-term objectives. The materials-science programs went to PRC 1, chemistry at Whiteshell to PRC 2, and chemistry at Chalk River to either PRC 1 or PRC 4.

ACKNOWLEDGMENTS

The author is much indebted to R.M. McNamara and her staff at Chalk River Central Records for their unfailing patience and efficiency in searching early files, and to L. McNally at the National Archives of Canada for invaluable assistance in locating other relevant material.

BIBLIOGRAPHY

Bothwell, Robert. *Nucleus: The History of Atomic Energy of Canada Limited.* Toronto: University of Toronto Press, 1988: 64-66.

Chalk River Files:

1500/AECL-Montreal Labs, Vols. 3-6, September 1944 - April 1946.

1510/CRNL, Vol. 1, October 1944 - October 1954.

9026-1, Vols. 1 and 2, October 1947 - March 1958.

1615/R10, May 1976.

1510-1, Memo from R.G. Hart, 4 September 1985.

Hanna, G.C. "The Origin of the Chalk River Van de Graaff Accelerator." *Physics in Canada* 50 (1994): 219-22.

Mackenzie, C.J. Papers. NA, MG30 B122. Vol. 2, Diaries, October 1944 - January 1946.

National Research Council Records, NA, RG77.

Vol. 283, Atomic Energy, September 1945 - November 1947.

Vol. 283, Radiological Research Misc., October 1944.

Vol. 284, Radiological Research Policy, August 1944 - August 1946.

File 41-1-30 in 85-86/178 Box 4, Van de Graaff Generator, May 1945 - May 1948.

Lord Rothschild. "Forty-five Varieties of Research (and Development)." *Nature* 239 (1972): 373-8.

Chapter Seven

Chemistry and Materials Science

T.A. EASTWOOD

INTRODUCTION

The basic research programs in chemistry had their origins in the Montreal Laboratory. The task of the laboratory was the exploration of an alternative to the plutonium production route being followed in the Manhattan Project. The American program was based on graphite-moderated reactors and, as a backup, the Anglo-Canadian team pursued the heavy-water-reactor option. A major contribution of the chemists was the successful development of the process used to extract plutonium from uranium irradiated in the reactor. Working independently, they duplicated one of the main achievements of the chemists in the much larger Manhattan Project. In addition, they developed the process used to separate uranium-233 from irradiated thorium carbonate, a secondary objective of the Montreal Laboratory. This work aroused the scientific curiosity of the staff and led quite naturally to the research described here under the headings: Nuclear Chemistry and Heavy-Element Chemistry.

The third main component of the early chemistry-research programs arose directly from the commitment to heavy-water-reactor technology. It was expected that the nuclear radiations in the reactor would decompose the heavy-water moderator. The details of the chemical steps leading to decomposition were not well understood, however, and were of significant scientific interest. Research on this topic in radiation chemistry was launched in Montreal and continued throughout the period of this history.

As the research potential of NRX became apparent, research activities broadened. The reactor was used to make radiotracers for application in a variety of new fields, and it was shown to be a powerful tool for neutron-activation analysis. The effects of reactor irradiation on solids soon attracted attention. These activities are summarized under the headings: Radiotracer Studies, Irradiation Effects, and Analytical Methods. As the nuclear power program advanced, the research programs developed in tandem and are described later in this chapter.

NUCLEAR CHEMISTRY

There seems to be no generally accepted definition of nuclear chemistry. Within the Canadian project, it was considered to be the study of nuclear processes, interdisciplinary with physics, in which chemistry plays an important role. One such process that occurs in nuclear reactors is the synthesis of new chemical elements: the production of plutonium (element 94) from uranium (element 92), the aim of the Montreal Laboratory, is a well-known example.

Shortly after NRX startup, elements of even higher atomic number were made in the reactor. Plutonium was irradiated to yield americium, and it, in turn, was irradiated to make curium (atomic numbers 95 and 96, respectively). These elements had first been synthesized about five years earlier at the University of California. The nuclear properties of the americium and curium isotopes produced in NRX were investigated in the early days at Chalk River.

The question arises: is there an end to the new elements that can be made this way? Indeed, there is. With the progression to heavier nuclides, the half-lives tend to be shorter and, an additional process, spontaneous fission, opens another route by which heavy nuclides can decay. Production then becomes a competition between the processes for formation versus those leading to destruction by decay. The key factor is the neutron flux, and, in this feature, NRX was superior to other reactors for a number of years. That is why it was in demand by scientists active in this line of research, not only at Chalk River but also at other laboratories.

Eventually, NRX was surpassed in neutron flux by the Materials Testing Reactor (MTR) in the United States. This reactor opened fresh possibilities for heavy- element synthesis. In the mid-fifties, AECL chemists cooperated with scientists at the Knolls Atomic Power Laboratory at Schenectady, New York, to investigate the heavy elements in samples of plutonium the latter had irradiated in the MTR. In addition to isotopes of americium and curium, they were looking for isotopes of berkelium, californium, einsteinium, and fermium, elements that had been discovered a few years earlier, with atomic numbers 97, 98, 99, and 100. They thought that isotopes of even heavier elements might be produced, but they knew that they would have to work quickly to study them because the half-lives would be short. The sample that had received the longest irradiation was processed as rapidly as possible after the end of the irradiation, and all the trans-plutonium elements up to, and including, fermium were indeed found. Using the techniques that had been developed at Chalk River, the nuclear properties of the isotopes of these elements were investigated in the co-operative program. The results confirmed and, in some instances, refined the information in the scientific literature. No sign of elements beyond fermium was found. Heavier elements have since been synthesized elsewhere, but in accelerator irradiations, not in reactors.

A major interest of the nuclear chemists in the Montreal Laboratory was the fission process. From a practical standpoint, this is the most important nuclear process on earth, and from the scientific viewpoint, its complexities attracted the attention of experimentalists and theorists alike. In order to study any one of the several hundred different fission-product nuclides, available in only small amounts in the early days, it was first necessary to purify and identify them chemically. This was an extraordinary challenge, because more than thirty different elements are produced in fission. In fact, this chemical hurdle had been an obstacle on the road to the discovery of fission.

The yield of products from the fission of uranium-235 by reactor neutrons when plotted against the mass of the product gives a double-humped curve, with maxima at masses of about 97 and 137. One of the earliest publications of this curve resulted from research on the radioactive fission products in the Montreal Laboratory. A similar curve for the fission of uranium-233 was also published from there a short time later. These were the beginnings of a stream of publications from the Chalk River Laboratories, refining and extending the early work. Measurements of the radioactive products were augmented by mass-spectrometric determinations of stable or long-lived nuclides made at McMaster University (Hamilton, Ontario), some of which were supported by AECL contracts. The yields of many nuclides were put on an absolute basis, as compared with the early relative yields, and the radioactive decay characteristics of the products were also investigated. Measurements made in the fifties revealed small irregularities in the otherwise-smooth yield curve. The Chalk River chemists studied them with special care because it was thought that they could shed light on the mechanism of fission. This turned out to be correct and the irregularities are now generally considered to reflect structure in the arrangement of neutrons and protons in the fissioning nucleus.

Expertise developed from research on the radioactive-decay properties of the fission products was applied later to the wide variety of other radionuclides that could be produced in NRX. A noteworthy example was the study of the radioactive decay of calcium-41 made at Chalk River around 1950. This long-lived isotope, produced by irradiating calcium oxide or carbonate with neutrons, decays by electron-capture, a rare mode of decay in light elements. Rigorous chemical purification to rid the calcium-41 of radioactive impurities and sophisticated counting techniques were required for the successful conclusion of the work. In another early example, the half-life of carbon-14, the fundamental factor in radiocarbon dating of archaeological artifacts, was determined with greater precision than reported previously from elsewhere.

Measurements of the cross-sections of radionuclides in reactors received special attention at Chalk River as the power reactor program advanced through the fifties. The term cross-section is defined properly in chapter one; in the present context, it can be thought of as a measure of the probability for a specified reaction to occur in the reactor. Cross-sections vary over a range of more than a billion, so yields from nuclear reactions can range from the abundant to the meagre. Cross-sections cannot be calculated accurately from theory; consequently, measurements are required. The cross-sections of the uranium and plutonium isotopes were the foci of a wide variety of measurements, both chemical and physical, for they are of fundamental importance in CANDU power-reactor design, as well as of scientific interest. The cross-sections and yields of the fission products were also determined to calculate the loss of neutrons to the fission products as they accumulate in the fuel with increasing irradiation.

Neutrons abound in nuclear reactors, yet AECL chemists were able to make radionuclides in NRX that are neutron-deficient; i.e., their nuclei contain fewer neutrons than naturally occurring isotopes of the same element. One way was to use nuclear reactions initiated by the energetic neutrons born in fission. When a fast neutron is absorbed into a nucleus, the excess energy may be carried off by two, or in rare instances three, neutrons, yielding a neutron-deficient product. A reaction of the (n,2n) type led to the production of bismuth-208, a long-lived radioisotope of bismuth not known previously.

Another approach was to use secondary reactions: neutron reactions with a light element produced charged particles, which subsequently reacted with other nuclei. For example, carbon-11, a neutron-deficient isotope of carbon, was produced by irradiating lithium borate in NRX. Tritons, produced by the interaction of reactor neutrons with lithium-6 nuclei, react with nearby boron-10 nuclei to produce carbon-11 and two neutrons. Other radionuclides, including beryllium-7, nitrogen-13, fluorine-18, and sodium-22, were obtained this way.

By the early sixties, the nuclear chemists were turning their attention from the reactors to the tandem accelerator. Their main interest was the measurement of cross-sections for reactions between neutrons of precisely known energies, ranging from about 2 to 20 MeV, and a wide variety of target nuclides. The neutrons were produced in the reaction between deuterons accelerated by the tandem and tritium, and the tandem provided the precise control of deuteron energy required for the measurements. The results were of fundamental interest and provided useful information for the development of neutron dosimeters and in radiation damage studies.

Basic research in nuclear chemistry at AECL gradually wound down in the late sixties, as a result of staff changes and the emergence of programs of higher priority.

HEAVY-ELEMENT CHEMISTRY

In parallel with their work on methods for separating plutonium from irradiated uranium, the chemists in the Montreal Laboratory and later at Chalk River, sought to determine the families to which these elements belong. Scientists order the elements in sequence by their atomic number and place them in families, the members of which have similar properties. This arrangement is called the Periodic Table of the Elements. The table was essentially complete at the turn of the century, but the familial assignments of the heaviest elements were somewhat controversial. Few specialists pursued the matter, for, apart from radium, the heavy elements attracted little notice. With the advent of the nuclear age, interest increased greatly and all the nuclear establishments became active in research on the heavy elements, including the Montreal Laboratory.

G.T. Seaborg, a co-discoverer of plutonium at the University of California, was a strong supporter of the view that all the elements from actinium, atomic number 89, to the undiscovered element 104, belonged to one family. This would include all the synthetic elements mentioned earlier.

He argued that in this actinide family, all members had the same valence (outermost) electron shell, and that an inner shell, with little influence on valence, was filled in sequence as the atomic number increased. In this respect, the actinide family members resembled the lanthanide (or rare earth) family, found near the middle of the Periodic Table, where inner-shell filling was well established. With the family assignment properly established, the properties of new elements could be predicted from the chemistry of other members of their family. Many of the actinide elements existed in such small amounts, however, that proof of Seaborg's actinide hypothesis was elusive.

Uranium was known to exist in several valence states, and the initial research in Montreal explored relationships among the states using electrochemical methods. These experiments were extended to plutonium when it became available. A wide variety of oxidizing and reducing agents was used in related research. Plutonium was found in positive valence states of three, four, and six, with a preference for four in mild oxidizing or reducing solutions. The solubilities of many plutonium salts were measured, partly with an eye to methods for separation and purification. Precipitates containing other elements with which plutonium would co-precipitate quantitatively at tracer levels were sought, and lanthanum fluoride was recommended. Plutonium was found to readily form a complex with inorganic anions, such as sulphate, as well as with a multiplicity of organic molecules. The optical absorption spectra of aqueous solutions of plutonium showed sharp bands comparable to those of uranium, and analogous to those of the lanthanide elements. Scientifically, the knowledge gained through this research was consistent with the actinide hypothesis. From a practical standpoint, it built a firm understanding of methods for separating uranium and plutonium from the fission products and from each other, for purifying plutonium, and for developing analytical methods.

A noteworthy characteristic of this work is that the stock of plutonium amounted to only a few micrograms at first, and slowly increased to milligrams as irradiated uranium became available. The plutonium was recycled through the numerous experiments, and if a test-tube broke or other accidents scattered the sample, the blotting paper placed on purpose to catch the spills was processed to recover the plutonium. Many hours were spent on these tedious recovery operations.

Metallurgical research provided further tests of the actinide hypothesis. As discussed in chapter two, plutonium-aluminum-alloy fuel elements were fabricated at Chalk River and used in NRX in the early years. In research experiments throughout the fifties and early sixties, using the same basic process, many intermetallic compounds were prepared of uranium, neptunium, plutonium, or americium with aluminum and other elements. Some of these compounds make excellent small sources of neutrons; alpha particles from decay of the heavy element produce neutrons when they interact with the light element in the compound. The efficiencies of the various compounds for this purpose were determined. Similarities in the alloying properties and crystal structures of the intermetallic compounds of these heavy elements supported the view that they belonged in one family.

Experiments aimed at reducing plutonium fluoride to the metal, on the scale of hundreds of milligrams, were made using barium and related metals in a high vacuum. The physical properties of the products, such as density, crystal structure, and melting point, were then determined. The techniques were later extended to the investigation of methods for separating plutonium from irradiated uranium using molten silver, silver-gold alloy, certain of the lanthanide metals, and bismuth.

In sum, the AECL results in the forties and fifties were consistent with Seaborg's actinide hypothesis, but did not prove it. Since then, further results from other laboratories have firmly established the validity of the actinide hypothesis. It formed the basis for separating and identifying new synthetic elements heavier than curium and, for example, was invoked in the separation of these elements for the research mentioned earlier. Work on heavy-element chemistry lapsed in AECL laboratories for several years, but resumed at Whiteshell around 1970.

RADIATION CHEMISTRY — EARLY DAYS

Radiation chemistry embraces the study of the chemical changes produced by high-energy radiation from natural and artificial sources. It has its roots in Mme. Curie's research on radioactivity, and grew rapidly in step with the development of nuclear energy.

In the fledgling Canadian project, radiation chemistry was particularly important, because the heavy-water moderator in NRX was expected to decompose into deuterium and oxygen. NRX was equipped to recombine the gases and return the heavy water to the reactor. About the time of NRX startup, however, research at the Brookhaven National Laboratory in the United States showed that, if water is pure, the decomposition is negligibly small. Steps were taken to ensure the purity of the moderator, and the recombiner unit was not often needed. Although there was no net decomposition, a lot of radiation chemistry was going on and water radiolysis was a prominent program of research at Chalk River.

The radiolysis experiments were aimed at testing a contemporary hypothesis. It held that gamma rays dissociate water molecules, at widely spaced intervals along their paths, into positive ions and electrons. The ion-electron pairs quickly recombine to form water molecules again. These re-formed molecules are in higher-energy states than normal water molecules, and eventually shed the excess energy by decomposing into reactive intermediate species. Thus, clusters of reactive species are formed at intervals along each gamma-ray track. The clusters were estimated to be a few molecular diameters across, separated by much greater distances. The intermediates were thought to be neutral atoms and free radicals.

Atoms and free radicals are reactive by virtue of unpaired electrons in their outer electron shells; they are usually formed by the decomposition of energy-rich compounds, and are well known in other fields of chemistry. The intermediates from the radiolysis of ordinary water would be hydrogen atoms and hydroxyl free radicals, written as H• and OH•, respectively, typically three of each in each cluster.

In some clusters, the atoms and radicals would react to produce H_2, H_2O_2, and H_2O, all of which would eventually disperse in the bulk water. If reaction failed to occur in the clusters, the intermediates themselves would diffuse into the bulk water. There, if no impurities were present, they would react with dissolved hydrogen and hydrogen peroxide to re-form water. This explained why there is no net decomposition of the NRX moderator when it is pure. If solutes were present (metal ions, for example), the atoms and radicals would react preferentially with them; deuterium and oxygen (from the decomposition of the peroxide) would accumulate, and eventually escape from solution.

Most free radicals are too short-lived to have been observed directly with the techniques available at the time, so their identities and roles were hypothetical, based on knowledge from other fields of chemistry. Calculations to test the hypothesis had been reported in the literature, but to make them manageable, the authors had been forced to assume only one reactive species, a mathematical hybrid of H• and OH•, to which reaction and diffusion rates were assigned. To include two species, as required by the hypothesis, was beyond the mathematical techniques employed. For the new breed of computer mathematician at Chalk River, this chemistry problem presented a nice challenge. Using numerical techniques rather than classical analytical mathematics, and the Datatron, the first, and newly commissioned, electronic computer at the laboratory, the Chalk River researchers quickly made the calculations for two species reacting and diffusing as hypothesized.[1] The agreement between the calculations and the experimental results gave encouraging support for the hypothesis.

The calculations did not, however, reproduce all the experimental results. In particular, there was a notable lack of accord

---

[1] Each calculation took longer (eight hours or more) than any done up to that time. The computer staff wrote a special sub-program to start the calculations on the Datatron at night, unattended, to run to the next morning. This, of course, became standard practice.

with experiments at Chalk River in which deuterium or tritium was dissolved in the ordinary water being irradiated. The hydrogen product, containing both the light isotope, H-1, and either H-2 or H-3, depending on the additive, was then mass analyzed to determine the hydrogen isotope ratios. The mass spectrometers used at that time were home-made, temperamental machines, but were worth the trouble they gave because isotope experiments were very effective tests of theory.

Shortly thereafter, in the late fifties, it was shown from experiments elsewhere that an early step in water radiolysis is the formation of solvated electrons. Some electrons from the interaction of the gamma rays with water molecules were not recaptured by their parent ions; instead, they became associated with several water molecules. Similar hydration spheres are commonly associated with ions in water; sodium and chloride ions are examples. The solvated electrons turned out to have almost the same chemical properties as hydrogen atoms. When the computer calculations were done with reaction and diffusion rates appropriate to solvated electrons in place of hydrogen atoms, the agreement with experiment was much better than that obtained formerly, and confidence in the theory increased accordingly.

AECL radiation chemists and computer scientists have continued to cooperate in developing sophisticated programs to predict the course and outcome of chemical reactions, and the programs have found numerous applications inside and outside AECL.

In the early sixties, the radiation chemists at Chalk River turned their attention to the hydrocarbons. Several aspects of the radiolytic behaviour of these compounds of hydrogen and carbon made them interesting subjects for research. For instance, it had been known for about thirty years that benzene, C_6H_6, is more stable to ionizing radiation than its near-relative, cyclohexane, C_6H_{12}, by a factor of about one hundred. Both molecules have ring structures; benzene has alternating single and double bonds between the carbon atoms, while cyclohexane has only single bonds. It is this stability of benzene-type structure, coupled with their high boiling points, that make the terphenyls, three benzene rings bonded together in a string, effective reactor coolants. It was also known that as little as three percent of benzene in cyclohexane reduced the decomposition of cyclohexane by as much as fifty percent. This intrigued the chemists and appeared to be a form of radiation protection, a matter of special interest to radiation biologists.

The Chalk River chemists concentrated on cyclohexane solutions of benzene and other molecules with similar chemical properties. As in their research on water, they substituted deuterium for ordinary hydrogen in the compounds used in the experiments. Initially, they mass-analyzed only the hydrogen produced in the reactions, but later they were able to account for all the participating atoms. This advance was made possible by gas chromatography, a powerful technique then in its infancy, and the Chalk River chemists pioneered the application of the mass spectrometer as a detector in chromatography.

The experiments were among the first to show that positive ions and electrons are prominent reacting species in the early stages of hydrocarbon radiolysis, as elucidated in a dozen published Chalk River papers. Radiation protection in benzene-cyclohexane mixtures was shown to be due, at least in part, to a process called energy transfer. Benzene accepts energy from excited cyclohexane molecules which otherwise would decompose; benzene molecules, by virtue of their structure, accommodate the energy internally without decomposing. Thus, for a given dose of ionizing radiation, the decomposition of a mixture is less than that observed with pure cyclohexane.

Whiteshell joined Chalk River in research on radiation chemistry starting in the mid-sixties. By that time, instrumentation had advanced to the point that direct observation of the transient species in radiolysis, solvated electrons, free radicals, and so on, became possible.

RADIOTRACER STUDIES

With an excellent reactor in which to make radionuclides and the techniques for handling them well understood, it was natural for the chemists in the early days at Chalk River to apply radiotracers to their research interests. Among the tracer studies were pioneering experiments in which the rates of reaction of organic compounds enriched in the radioactive isotope carbon-14 were compared with those containing stable carbon-12 or carbon-13; i.e., the so-called isotope effect. Mass-spectrometry techniques were later employed in such experiments and were also used in research on isotope effects in chlorine reactions. The results were used to elucidate what chemical bonds were broken or formed in the reactions (i.e., the mechanisms of the reactions), and used for comparison with theory.

Radiotracer studies with bromine-82 showed that bromine and bromate ion in perchloric acid solution undergo isotopic exchange. The data implied the occurrence of secondary reactions leading to the formation of hypobromous acid. The mechanisms of the reactions revealed in the tracer experiments were later deduced from more conventional experiments. Isotope-exchange methods were also used to compare the properties of complex ions as a function of the position of their parent elements in the lanthanide family. A similar dependence of thermodynamic properties among the alkali metals gave results, unobtainable by prior methods, for comparison with theory. Radioactive isotopes of sodium, of mass 22 and 24, were used to investigate isotope separation using an organic cation exchange resin.

In research on the effects of fast-neutron irradiations on materials, discussed below, there is often a need to know the distance travelled by an atom recoiling from a collision with an energetic neutron. Because the distances are very small and, therefore, difficult to measure, such information was very scarce. In response to this need, a method for measuring these ranges was developed from research that started at Chalk River in the late fifties.

The key was a method for stripping away a series of thin, uniform layers from a metal sheet using a "chemical milling" process. It was first worked out with aluminum and depended on forming a layer of oxide on an aluminum anode in an electrolytic bath. By controlling the conditions, the thickness of the anodic oxide could be regulated precisely and, after removing the sample from the bath, the oxide was dissolved without attacking the metal. The sample could then be returned to the bath to repeat the process as often as needed. In this way, layers only a few atoms thick could be stripped away one after the other. The method was later extended to tantalum, tungsten, silicon and gold.

For the range measurements, radiotracer ions were accelerated to selected energies in the kilovolt region using a mass separator as an ion gun. Mass separators are similar in principle to mass spectrometers, but have no beam-defining slits, so the transmission is very high. The instrument for this program was built in Sweden in 1958, and was the first step to many valuable and enduring cooperative programs with Scandinavian laboratories.

The accelerated ions were directed at the target foil and, after the bombardment, their distribution in the target was determined by measuring the radioactivity remaining in the target after successive layers were stripped away using the technique just described. To ascertain the dependence on ion and target mass, ion masses were varied from 24 to 222 and the target elements ranged in atomic number from aluminum at 13 to gold at 79. Thus, data on ion ranges were obtained for the whole region of interest. The results agreed with a theory developed at about the same time at the University of Aarhus, Denmark, and the common interest formed a strong bond between researchers there and at Chalk River.

In the course of these experiments, it was found that a few ions penetrated much farther than most into aluminum foil. This result was completely unexpected and to demonstrate that it was not an experimental artifact, work on a second method for determining ion ranges was initiated at the Nobel Institute of Physics at Stockholm during a visit by the lead Chalk River scientist in 1962. Radon-222, an isotope of radon that decays with the emission of alpha particles, was implanted into aluminum foil and the energy spectrum of the alpha particles emerging from the foil was measured. Since alpha particles lose energy in proportion to the thickness of matter through which they pass, the spectrum could be interpreted to give the depth distribution of the implanted radon. Here again, it was found that a few atoms penetrated to a greater depth than the average, thus confirming results from the electrochemical stripping experiments. The spectroscopic technique was refined at Chalk

River over the next few years and yielded similar results. These experiments and the range measurements furnished the first experimental evidence for ion channeling, a quite unexpected phenomenon that was studied intensively at Chalk River and elsewhere.

A different application of the nuclear spectroscopic technique was the measurement of the relative mobilities of oxygen and metal ions within a growing anodic oxide. Radioactive marker atoms were implanted in a thin surface oxide layer on the metal specimen using the mass separator. A marker with nearly ideal properties was xenon-135. Xenon is quite inert chemically, and the isotope of mass 135 yields a mono-energetic electron when it decays. The oxide was then grown electrolytically to produce a coat many atom-layers thick, and the position of the marker was determined at several stages during the growth by measuring the electron energy loss. If oxygen is the mobile species, the oxide forms at the metal-oxide interface and the xenon marker remains near the outermost surface. If, on the other hand, the metal ions move, growth occurs at the oxide-electrolyte interface and buries the marker. Zirconium oxide was found to grow mainly by oxygen migration, for example, while in tantalum oxide, both the species are mobile, oxygen with about twice the mobility of the metal. A number of elements in several different families in the Periodic Table were investigated, and the experiments were soon accepted internationally as yielding the first reliable solid-state marker results. The method was brought to a high degree of refinement with tantalum.

IRRADIATION EFFECTS

Reactor irradiation produces physical, as well as chemical, changes, and fundamental research on the effects of irradiation on crystalline solids was started at Chalk River about 1951. It was known that many of the effects are caused by fast neutrons and energetic nuclei, such as fission fragments, which knock atoms out of their normal crystal-lattice positions. This results in the creation of a pair of what are called point defects: the displaced or interstitial atom that becomes wedged among normal lattice atoms, and the vacancy, the hole in the lattice where the interstitial atom used to reside. Defects may migrate through the crystal and will disappear should they arrive at a sink. The important types of sink are: linear defects in the otherwise regular crystal lattice; voids, small holes in the crystal; and grain boundaries. These last are absent in perfect crystals, but some solids are composed of small crystals (e.g., ordinary metals), and the boundaries between the crystal grains are common sinks for point defects.

In the first experiments, ordered alloys of copper-gold were irradiated in NRX. Ordered alloys are characterized by regions in which the atoms of the two elements are in a fixed arrangement. Two competing effects were resolved: disordering by fast neutrons, and what appeared to be an ordering effect of thermal neutrons complicated by the transmutation of gold-198 to mercury-198.

Interstitial atoms are at a higher energy level than those on normal lattice sites, and form a reservoir of energy in the solid. Heating allows the interstitials to return to their normal positions and releases the stored energy. Measurements of the stored energy in irradiated graphite were started in 1953, and, a few years later, measurements were made on irradiated lithium fluoride and several oxides of uranium. In the latter two cases, alpha particles and tritons, or fission fragments, respectively, are responsible for displacing host atoms from their normal lattice sites.

Around 1955, Chalk River scientists began two types of experiments to measure the energy required to form a defect pair, or to move defects in the crystal lattice. In one approach, the specimen was irradiated with fast neutrons at a temperature low enough to freeze the defects. The electrical resistivity of the specimen, a very sensitive monitor of point defects, was measured as the temperature was slowly raised, thus releasing the defects. From an analysis of the resistivity changes, the energy required for a vacancy to migrate in the lattice could be deduced, and the effects of neutron irradiation could be compared with other methods for introducing point defects. In the second type of experiment, the rate of diffusion of a radioactive tracer was measured as a function of temperature of the specimen. In the simplest experiments, the tracer was an isotope of the element being studied. Here again, the vacancy migration energy could be deduced. The first experiments of both types were made with platinum and later extended to other metals and semiconductors.

When zirconium alloys became the choice for power reactor pressure tubes in about 1957, the expertise developed in the early research activities was used to investigate irradiation effects in zirconium alloys.

ANALYTICAL METHODS — EARLY DAYS

Much of the research just described depended on measurements of the decay rates of radionuclides. In the early days, alpha- and beta-emitting nuclides were generally used in the experiments. Although it was relatively easy to compare decay rates of similar sources of a given radionuclide, comparisons under different conditions, or of different nuclides, were more difficult. Methods for improving the accuracy of these more exacting measurements received a good deal of attention. Experimenters switched to gamma-ray measurements with the introduction of sodium iodide scintillation spectrometers in the fifties, and of solid-state spectrometers about a decade later. These made possible not only the determination of the amounts of radionuclide present, but also identified them by their gamma-ray spectra.

Chalk River chemists analyzed the radionuclides in the debris from weapons tests, beginning with the first USSR nuclear explosion in 1949. The amounts were much smaller than used in the chemistry laboratories and new approaches had to be developed. The keys to success were the elimination of extraneous radioactive material from the laboratories to avoid contaminating the samples, and low background techniques; i.e., methods for reducing the radiation detected by the counters in the absence of the samples. Soon they were able to determine the type of weapon tested, and its approximate yield. Since such information was valuable to the Defence Research Board of Canada, this aspect of the work was later transferred there.

Radionuclides in the fallout entered the food chain, and the techniques were adapted to analyze dairy products, vegetables, etc., from across Canada. When procedures became standardized, this program was moved to Federal Department of Health and Welfare laboratories.

Low-background techniques were equally valuable in AECL programs and became highly developed. Building materials for the rooms in which the measurements were to be made, such as cement, sand, and paint, were chosen to minimize the radioactive content. Even the electric light bulbs used in these rooms were taken apart to remove the small amount of slightly radioactive cement used to join the glass to the metal base. The low counter-backgrounds facilitated measurements of radionuclides accidentally assimilated into the organs and tissues of AECL employees, and in environmental research.

The high neutron flux in NRX was exploited in research on activation analysis during the late fifties. In this procedure, the sample is irradiated with neutrons and the radioactive products are measured, often by comparison with standard samples of known composition processed in the same way as the sample being analyzed. High sensitivity is possible for many elements, depending mainly on their nuclear properties. Resolution of the gamma-ray spectrum of the activated sample, obtained with a sodium iodide spectrometer, for example, often permits several elements to be determined simultaneously, speedily, and without destruction of the sample. It is an especially valuable method for establishing whether samples collected in different places are from the same source. For example, by comparing the trace elements in samples of human hair, the Royal Canadian Mounted Police, in cooperation with scientists at Chalk River, obtained evidence that resulted in a criminal conviction in a Canadian court. This was the first judicial recognition of activation analysis in Canada and, so far as we know, in the world.

The Canadian nuclear program placed heavy demands on conventional methods of chemical analysis. Reactor materials required analysis for neutron absorbers, and methods for analyzing uranium fuel and aluminum sheathing for cadmium, boron, and other strong neutron absorbers had to be developed from first principles, since standard methods did not exist. The assay of plutonium solutions was an early challenge.[2]

Several mass spectrometers for the precise measurement of atom ratios were designed and built for chemistry research at Chalk River. Work on the second instrument, specifically for the demanding work with elements of high atomic mass, was started in the spring of 1950 and commissioned two years later. Typical of such ventures, not all the applications of the instrument were clearly foreseen when the design was started. It was not until a few years after commissioning that its full value was realized. By then, it was indispensable in both the fuel-development and reactor-physics programs. In the former, the uranium-isotope ratios before and after irradiation were used to calculate the energy produced in experimental fuel. In the latter program, the accuracy of reactor-physics calculations could be tested by comparing the results with the uranium-isotope ratios actually found in discharged fuel. Plutonium mass analyses were also needed in these programs and a third machine with safety features for work with this element was built a few years later. The many high-quality fuel analyses made with these instruments were major contributions in the evolution of CANDU power reactors.

CHANNELING AND ITS APPLICATIONS

In parallel with the experimental measurements of heavy ion ranges at Chalk River, scientists at the Oak Ridge National Laboratory, United States, studied the trajectories of atoms injected into crystalline targets by computer simulation. They found that a few of their simulated atoms travelled to surprising depths in the target, apparently confined to the open spaces (channels) between the rows of atoms in their model crystal by the electrostatic forces they had included in their simulation. They suggested that, in the Chalk River experiments, some crystal grains in the polycrystalline target foils were fortuitously aligned with their channels parallel to the incident ion beam, thus allowing the ions to penetrate further into the target than those hitting randomly oriented grains. At Chalk River, the foil experiments were repeated with single crystals of aluminum, and the results quickly confirmed this hypothesis. Independent verification followed rapidly from other laboratories.

The theoretical framework for the process, which came to be called ion channeling, was developed at the University of Aarhus, Denmark, in 1964. A feature of ion channeling that surprised scientists at first, and one that makes it applicable to many fields of research, is that the yields of nuclear processes that depend on close encounters between the incident ion and the target atom fall almost to zero for channeled ions. It was surprising because, before the discovery of channeling, one would not have expected the geometrical arrangement of atoms in the target, on a scale so much greater than nuclear dimensions, to have any effect on the yield of nuclear reactions. Preliminary tests made at Aarhus in 1964, in which the lead Chalk River scientist participated, and also at AERE, Harwell, United Kingdom, were the first to demonstrate the predicted reduction in nuclear reactions induced by protons. Further experimental confirmation of the theory was obtained over the next few years.

To summarize, channeling occurs whenever a beam of positively charged nuclear particles is incident on a crystal in a direction parallel to prominent rows of atoms forming the crystal lattice; in crystallographic terminology this is called a low-index direction. The particles are steered on helical paths through the channels between the rows by the

[2] These analyses were of more than academic interest. AECL sold the irradiated fuel rods from NRU to the United States for a price based on the plutonium content at about $100 per gram. The rods were dissolved in the United States, and samples of the solutions were analyzed there and at Chalk River. The American analysts used a method based on measuring the plutonium disintegration rate by counting the alpha particles. At Chalk River, the absorption of light by the solution, which is proportional to plutonium concentration, was measured spectrophotometrically. The latter were consistently higher, and the Canadian chemists managed to convince their American counterparts that their method was the more accurate one. Winning the debate was worth about $9 million to the Canadian program.

concerted effects of the electrostatic fields of hundreds of atoms forming the channels.[3]

As understanding of the channeling process improved in the late sixties, possibilities for applying it to other fields of research began to receive attention. At about the same time, a 2-MV Van de Graaff accelerator was installed at Chalk River for use in this program and in radiation chemistry. The first application was in research on ion implantation of semiconductor materials, and was launched during a visit from the head of the implantation group at Hughes Research Laboratories, Malibu, California.

In the joint experiments, each process used to make semiconductor electronic devices was investigated. Ions of the element required to dope the semiconductor and give it the desired electrical properties were implanted using the mass separator. During this process, the ions knock the atoms of the crystalline semiconductor from their normal lattice positions and produce point defects. If the implantation is continued, the concentration of defects becomes so great that they merge and form an amorphous layer. This damage to the perfection of the crystal structure impairs the electrical properties of electronic devices and is therefore undesirable. Point defects and amorphous regions in a crystal also adversely affect the channeling process. In the joint experiments, it was shown that channeling techniques could give a quantitative measure of the damage produced in semiconductor crystals during ion implantation.

The next stage in device fabrication is to heat the implanted crystal for a period of time sufficient for the damaged region to re-crystallize using the undamaged portion as a template. Joint experiments showed that it was also possible to follow this mechanism by channeling measurements. In addition, they gave the ratio of dopant atoms on crystal lattice sites to those in interstitial spaces in annealed devices.

The experiments resolved many questions about the use of ion implantation for fabricating semiconductor devices, and proved to be an ideal subject for testing the applicability of channeling in solid-state research. This valuable collaboration continued for nearly twenty years, although the Hughes scientist changed affiliations during the period. Other laboratories quickly embraced the channeling technique in related research, and by 1973, its use was reported from around the world in over one-thousand technical papers.

Chalk River scientists were the first to show that channeling is a powerful technique for the study of point-defect behaviour in alloys. Crystal-lattice vacancies are often found in association with the dilute, or solute, component of an alloy, and may influence the positions assumed by solute atoms in the crystal lattice. When these atoms are in, or encroach on, interstitial spaces, channeling techniques can often reveal the configuration of the solute atom-defect cluster. Numerous alloys were studied throughout the seventies, and several types of cluster were investigated, the most complex example of which is probably the tetrahedral arrangement of vacancies around interstitial indium atoms in aluminum crystals.

Beginning in 1974, channeling was used to augment more usual surface-science techniques in research on metal catalysts of the type used for hydrogen isotope exchange (see figure 7.1). Noteworthy results include information about the arrangement of atoms on catalyst surfaces, and correlations between surface structure and the extent to which the surface is covered by adsorbed gases.

In addition to the above, channeling has been applied at Chalk River in several areas of research of somewhat narrower scope. These topics include the measurement of nuclear lifetimes;

[3] In the late sixties, the federal cabinet minister responsible for AECL at the time, the Honourable Jean-Luc Pépin, paid an official visit to Chalk River and was briefed by the staff about program highlights. For a while he listened patiently to an account of the discovery of channeling, which was described with the aid of a "ball-and-stick" model of a simple crystal. Soon he interrupted, however, to exclaim: "Isn't that exactly what one would expect from looking at the model? Why are you so excited about it?" In such crystal models the open spaces between rows of atoms are quite obvious. The preferential penetration of protons in these low-index directions had, in fact, been predicted as early as 1912. The discovery of X-ray diffraction at about that time and the revolution in physics then in progress diverted attention from simple electrostatic steering responsible for channeling. The difficulty of measuring the short ranges of ions in solids was an added impediment.

FIGURE 7.1
Scientists at CRL used this ultra-high-vacuum equipment on a beam line from an ion accelerator for research on the structure of layers of gases adsorbed on catalyst surfaces.

the systematics of heavy-ion stopping powers; inner-shell ionization arising from ion-atom collisions; and the steering of GeV protons by bent single crystals.

MATERIALS SCIENCE

Materials science is at the boundaries of physics, chemistry and, when metals are studied, metallurgy. As the name implies, the subjects for research have practical applications. In 1957, when Zircaloy-2 was first considered for pressure tubes, zirconium alloys became the structural materials of greatest importance to AECL. Materials science covers the spectrum from fundamental studies at one end to applied research at the other. Work of the latter type in AECL is described in chapters twelve and thirteen, while the former is covered herein, although the division is not sharply drawn here, nor was it in the laboratories in which the research was done.

Zirconium was a "new metal" in 1957, attractive to reactor designers because its neutron-capture cross-section is lower than the cross-sections of other strong, corrosion-resistant metals, and there was much to learn about its fundamental properties. It was known that reactor irradiation produces point defects, and the subsequent behaviour of the defects modifies the physical properties of the material. Initially, therefore, much of the effort was aimed at establishing techniques and adapting them to investigate point-defect behaviour in this new metal.

The first experiments used resistivity and tracer diffusion methods to explore defect properties. By the early sixties at Chalk River, and a few years later at Whiteshell, transmission-electron-microscopy techniques were being used. Scanning-electron-microscopy was added in 1970.

The interaction between point defects and dislocations has been studied at Whiteshell since 1970 using the internal-friction technique. This technique depends on very precise measurements of oscillatory vibrations set up in the specimen (see figure 7.2). An example of its application is mentioned later, when zirconium hydride is discussed. Theoretical and computational methods have also been developed at Whiteshell to predict the precise arrangement of the atoms around defects and dislocations.

The use of channeling to determine the configuration of point defects in the neighbourhood of solute atoms has already been touched on in the previous section. Yet another technique introduced at Chalk River in the early eighties depends on the positron-electron annihilation phenomenon, discussed in chapter eight. Positrons tend to survive longer at vacancies than in the perfect crystal, because the electron density is usually lower there. Thus, lifetime measurements reflect vacancy behaviour.

FIGURE 7.2
The extremely small amount of energy required to move point defects in a crystal of reactor-irradiated magnesium oxide (shown at the bottom of the photograph) was deduced by scientists at WL from measurements made with this oscillating pendulum device.

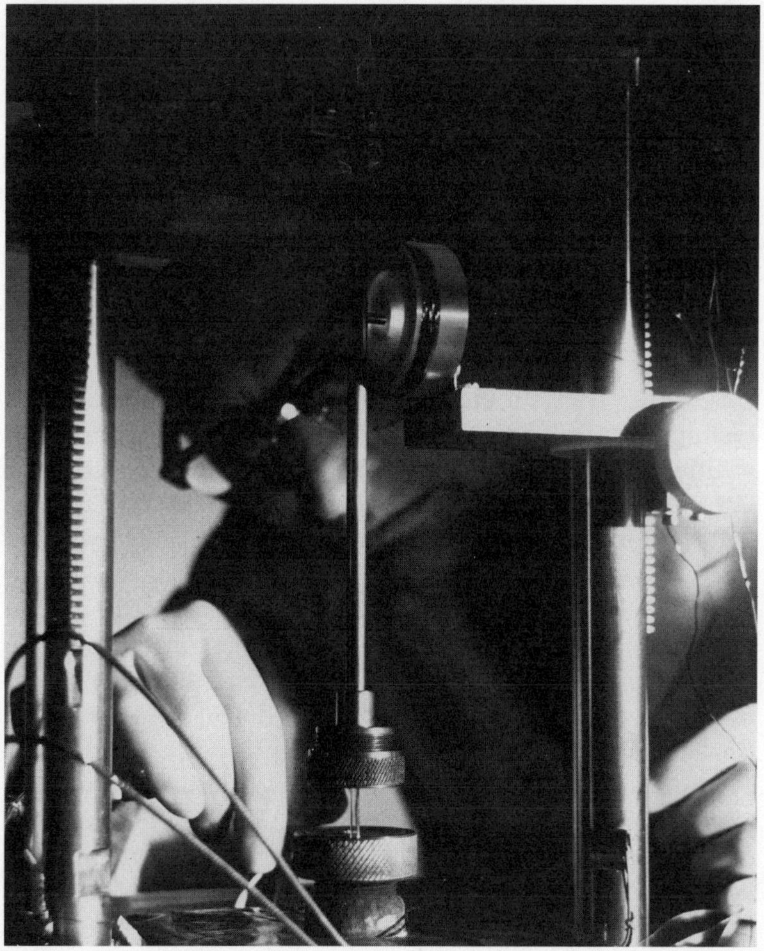

The results accruing from these studies are presented in some two hundred technical papers. Over fifty are directly related to zirconium and its alloys, and contributed to the international store of knowledge about the basic effects of irradiation on these materials.

Pressure tubes in the CANDU reactors sag and elongate in service, as described in chapter twelve. As discussed there, experimental measurements in NRX and NRU were difficult and time-consuming, and the specimens became highly radioactive. Around 1967, complementary approaches to the study of the processes fundamental to these shape changes were started at Chalk River and Whiteshell. Over the next two decades, the effects of electron and heavy-ion, as well as fast-neutron, irradiations were investigated. In addition, the theory of the processes was explored at both sites. By drawing on the experience and knowledge of irradiation effects accumulated over this period, AECL scientists and engineers have, to a steadily increasing extent, based predictions of pressure-tube-shape changes on sound physical principles.

Tests showed that Zircaloy-2 has good resistance to corrosion by water at the operating temperatures of power reactors over a period of a few years. The pressure tubes in CANDU reactors, however, were expected to last much longer, and to try to anticipate deleterious long-term effects, and understand them should any arise, fundamental research on the mechanism of corrosion was done at Chalk River from 1963 onwards.

It was known that in the early stage of zirconium corrosion, oxygen from the water migrates through the surface oxide film to the underlying metal; electrical neutrality is maintained by a counter-flow of electrons. The process was explored in detail using oxygen-17 as a tracer, and the $^{17}O(^{3}He,^{4}He)^{16}O$ nuclear reaction to follow the migration of the tracer during the corrosion process. With polycrystalline specimens, it was found that the rate of diffusion was a thousand times faster through the boundaries between oxide crystallites growing on adjacent crystallites in the metal than through the main body of the oxide. Incidentally, the novel tracer technique used in these experiments is an example of the type of cooperation possible at Chalk River. The chemists were able to satisfy the physicists' requirement for a stable target of oxygen-17 for experiments with the tandem accelerator, and the physicists provided the helium-3 beam from the tandem for the diffusion studies.

The counter-flow of electrons during corrosion was found to depend on the type of alloy and local environmental conditions. A hypothesis embracing both the oxygen and electron observations, and explaining the effects of reactor irradiation on the oxidation rate, has been developed. Additional results indicate that the later stages of corrosion proceed by oxygen migration through pores in the oxide layer, and factors affecting these processes were also established.

Zirconium and its alloys absorb hydrogen, and brittle platelets of zirconium hydride form in the metal if the amount exceeds solubility limits. The formation and orientation of such platelets through stress-induced diffusion resulted in the cracks in some pressure tubes, first observed in the Pickering-3 reactor in 1974. It turned out these tubes had been installed incorrectly. The behaviour of hydrogen in zirconium had been investigated in AECL laboratories since about 1958, and scientists were quickly able to provide a qualitative understanding of the fundamental mechanism by which these cracks developed in service. Within a few weeks, the mechanism was placed on a sound quantitative basis, but the importance of the topic has meant that much additional refinement of the theory has been considered essential to the CANDU program.

The cladding on CANDU reactor fuel is also based on zirconium. In addition to the environmental factors already mentioned, cladding may experience stress-corrosion cracking (SCC): i.e., attack by the fission products that accumulate in the fuel elements, coupled with mechanical stress. The intense investigation organized to find the cause of fuel-element failures, first appearing at Douglas Point in 1971 and later at Pickering, showed that SCC was responsible for many of the failures. Fundamental research on the process started at Chalk River in 1971 and contributed to the team investigation reported in chapter thirteen. As described there, the evidence points to fission-product iodine as the active agent, and a remedy was found to counteract its attack.

Although zirconium is no longer the new metal it was in the fifties, there remains a good deal to learn about the fundamentals of its alloys. Since they are extensively used in CANDU reactors, this area of materials science is particularly important to AECL.

RADIATION CHEMISTRY — LATER PERIOD

The first of the instrumental advances for research on radiation chemistry anticipated earlier in this chapter was the electron-spin-resonance (esr) technique introduced at Chalk River around 1963. As already discussed, the free radicals and neutral atoms, so common in chemical reactions induced by high-energy radiations, possess an unpaired electron. The electron spins as it orbits its parent nucleus, in roughly the same way the earth turns on its axis as it circles the sun. The esr spectrometer searches out resonances between the energy states of the spinning electron and an applied electric field that can be scanned across the microwave region, in the presence of a strong magnetic field. Research in other laboratories had shown that the resonance spectrum can, under favourable circumstances, be used to identify the chemical species containing the unpaired electron.

A wide variety of compounds was studied at Chalk River, including olefins, acetone, several different alcohols, aqueous solutions, nitrous oxide, and amines. From the esr spectra, the intermediate species produced by radiolysis could often be deduced, including free radicals, neutral atoms, and trapped electrons. The results confirmed the production of reactive intermediates postulated to be present on the basis of less direct methods, or introduced new and more explicit evidence. With knowledge of the intermediates, understanding of the course of events leading from the initial absorption of radiation to the final chemical products was greatly improved.

The esr technique was introduced at Whiteshell about 1967 and continued very successfully thereafter. In 1969, the program at Chalk River was terminated in favour of research on isotope-separation processes. More information on the Whiteshell program is provided later in this section.

The first accelerator at Chalk River for pulse radiolysis research was commissioned in 1968. In this machine, each pulse was formed by charging a bank of capacitors in parallel and discharging it in series. The resultant pulse was applied to the field emission cathode of a vacuum tube. The resulting electrons were directed through a thin window at currents of about

4000 A and energies up to 1.6 MeV. The radiation dose rate delivered to the gas in a cell placed in front of the window was up to two-hundred billion times the dose rate from conventional gamma-ray sources. From a radiation chemistry viewpoint, there are several advantages to accelerators of this type. They can produce concentrations of free radicals approaching in strength those of the reagents commonly used in chemical research, and reactions between radicals that do not occur at low dose rates are frequently observed. Furthermore, reactions of ions, commonly encountered in the radiolysis of gases, are usually much simpler than at low dose rates. These factors simplify the interpretation of the results.

Chalk River was one of the few laboratories to use such an accelerator and the scientists made many contributions to the understanding of the radiation chemistry of di- and tri-atomic gases, including oxygen, nitrogen, water, carbon dioxide, carbon monoxide, hydrogen chloride, and others. In addition, they deduced the rate constants for many of the reactions of ions in which the electrons are in different states of excitation. These data are useful in other fields, such as upper-atmosphere chemistry, and pulse radiolysis is a fruitful method for obtaining them.

A pulse radiolysis technique of a more conventional type was introduced at Chalk River in 1969, and at Whiteshell in 1972, using electrons from accelerators of the Van de Graaff type (see figure 7.3). Although the pulses were less intense than those from accelerators in the capacitor-discharge class, the electron energies were higher, and with greater penetrating power, experiments with liquids could be made.

At Chalk River, the major interest was in the radiation chemistry of amine solutions of the alkali metals. Of special note was the chemistry of three remarkable species that exist in equilibrium in such solutions: the solvated electron, the aggregate molecule, and the metal anion, M^-, where M stands for lithium, sodium, potassium, or cesium. The aggregate molecule was thought to be a complex of the metal cation, M^+, and the solvated electron, and hence is electrically neutral, but the structure was controversial. The metal anion, it was generally agreed, is formed by the addition of a solvated electron to the aggregate molecule. The chemistry is similar to the solution chemistry of the hydrogen-amine process for heavy-water production, and the research enlarged the community of chemists at Chalk River familiar with the field. The advantage of radiolysis is that it produces higher concentrations of these species than occur in ordinary solutions and the work considerably expanded the basic understanding of their chemistry.

At Whiteshell, the focus was mainly on the chemistry of the free radicals and solvated electrons produced in radiolysis, and a wide variety of compounds was studied. Of particular note

FIGURE 7.3
The lifetimes of transient chemical species were measured at WL using optical spectroscopy synchronized with pulses from the Van de Graaff accelerator in the background.

were the direct observations of several of these species using a combination of esr and pulse-radiolysis techniques. The beam tube of the accelerator was extended through the massive pole piece of the electromagnet of the esr spectrometer, and a flow system was used to continuously refresh the liquid being irradiated and flush out the products. In this way, it was possible to characterize the radicals produced in various alcohols and aqueous solutions. Information about the solvated electron in water was also refined in experiments in which each electron lived only ten microseconds on average.

A fundamental aim in all these pulse radiolysis experiments was to push the understanding of the chemical processes closer and closer in time to the primary interaction of radiation with matter. Research at Whiteshell provided the culminating AECL example. The experiments were made to test a theoretical prediction of the effects of changes in magnet field strength on the light emitted by the molecular fragments in liquids during their brief existence – less than a microsecond – following irradiation pulses. The theoretician, from the University of Sheffield, and the experimentalists cooperated during a summer appointment of the former in 1974, and after refinements to both the theory and the experiments, the sought-for agreement was found. Further investigation of the effect and its applications have been made at Whiteshell and other laboratories around the world.

HIGH-TEMPERATURE SOLUTION CHEMISTRY

High-temperature water is an important constituent of CANDU reactors and heavy-water plants. Its presence is also invoked in the analysis of potential reactor accidents and in predictive studies of how radionuclides might migrate from geological disposal vaults. Fundamental data about aqueous solutions at high temperature are scarce, however, because the measurements are difficult to make. Not only are high pressures required to keep water liquid at high temperatures, hot water is very corrosive. Chemists at Whiteshell began to contribute to knowledge in this field in the early seventies, and have made many contributions during the period of this review.

Their program had two complementary parts: experimental on the one hand, and predictive or theoretical on the other. The approach was based on thermodynamics, the science of heat energy, and on chemical thermodynamics, in particular. This branch of the science deals with heats of chemical reactions, heats of solution, and other measures of the forces driving chemical changes.

The initial focus was on the thermodynamic variables of compounds of the metals present in process streams of reactors and heavy water plants; e.g., iron, cobalt, and nickel. Several methods for extrapolating room-temperature data to temperatures up to 300°C were advanced. Electrochemical measurements were used to determine some of the basic data and to confirm some of the extrapolated values. The compounds of interest dissociate into pairs of oppositely charged ions when they dissolve. From the experimental results, the researchers could deduce characteristic properties for both members of the pair in many of the compounds. These data are useful in their own right and for comparison with theory. Furthermore, the chemists used them to deduce the properties of additional single ions from conventional thermodynamic measurements.

As the waste-management program advanced, the techniques were used to predict the high-temperature properties of about one-hundred different species of the actinide elements in solids and high-temperature water solutions. The solubility of uranium dioxide was measured up to 300°C and the behaviour of typical glass-waste forms in high temperature water was studied. Chapter nineteen describes how such basic information contributes to the waste-management program.

The techniques were also used to determine the properties of compounds of some fission-product elements at elevated temperatures. The work on technetium compounds was of special scientific interest. This element, atomic number 43, does not occur naturally, but is produced in good yield by fission of uranium. Thus, the Whiteshell chemists were able to study compounds of an element not available to many others. Primarily, however, the fission products were studied for their relevance to waste management and reactor-safety analysis. In addition to technetium, these included iodine, selenium and palladium. In reactor safety analysis, it is postulated that some reactor accidents may result in the release of fission products and high-temperature water into a reactor's containment building. Research on the fission-product elements, particularly iodine, which has complex chemistry, contributed to the knowledge necessary to anticipate the chemical consequences of such hypothetical accidents.

Thus the research results contributed to basic knowledge of a wide selection of chemical elements in high-temperature water, and were quickly applied in several AECL programs. They are also applicable to other Canadian industries, such as mining, and pulp and paper.

ISOTOPE CHEMISTRY

Chapter seventeen discusses the alternatives to the Girdler Sulfide (GS) process that were investigated for making heavy water. In parallel with the applied research-and-development work described there, some basic studies on two of the alternatives, the amine-hydrogen and the hydrogen-water processes, were undertaken beginning in the late sixties. For example, fundamental to the latter process is the deuterium-isotope separation factor between hydrogen and water. This factor was measured at Chalk River over a wider temperature range and with greater precision than reported by experimentalists in other laboratories. The results not only met AECL needs, but led to a refinement of the theory for the calculation of the equilibrium constant for the reaction from first principles that resolved a long-standing discrepancy with experimental determinations.

For practical application, both processes require a catalyst. In the amine-hydrogen process, the catalyst is dissolved in the amine and the chemistry is quite complex. The mechanism by which the catalyst acts, and the nature of unwanted side reactions that remove catalyst, were investigated. In the hydrogen-water process, the reactants pass over a solid catalyst, and both water and hydrogen molecules must be adsorbed on the surface of the catalyst for isotope exchange to occur. Other molecules compete for space on the surface and may deactivate the catalyst. Research done throughout the seventies, using some very advanced surface-chemistry techniques, showed how these reactions happen. Towards the end of the period and continuing to 1985, ion channeling was used to complement more conventional surface-science techniques. At Whiteshell, water-soluble catalysts for the hydrogen-water process were investigated from a fundamental point-of-view as alternatives to the surface catalysts developed at Chalk River.

A heavy-water upgrading plant operated at Chalk River based on deuterium enrichment by water electrolysis. To provide scientific support for the process, the effects of electrochemical variables were studied in the laboratory. With the discovery of a promising surface catalyst for the hydrogen-water process, a natural extension was to combine the two processes, and this led to the Combined Electrolytic-Catalytic Exchange process.

By around 1975, worldwide advances in laser technology, and the understanding of the interaction of laser light with chemical compounds, had advanced to the stage that isotope separation using lasers might be considered possible. To stay abreast of these developments, a research program on laser isotope separation, initially centred on deuterium, was launched at Chalk River; it continued during the balance of this history.

High-powered lasers can pump many quanta of light into target molecules almost simultaneously (see figure 7.4). Because the light-absorption spectra of isotopes of elements in the target differ, the large amount of energy deposited by the laser may be associated preferentially with a particular isotope. This concentration of energy may, in turn, lead to dissociation of the excited molecule at the site of the selected isotope. Following separation, the isotope can be swept up by chemical reaction or other means.

Hydrogen fluoride and carbon-dioxide lasers were used with quite a variety of simple molecules in vapour form: formic acid, methanol, methylamine, fluoroform, and related compounds. The effects of collisions between the excited target molecule and unexcited molecules, which tend to deactivate the target molecule, were important issues. The work led to AECL

FIGURE 7.4
Research on isotope separation using high-powered lasers at CRL required the complex equipment shown here, and stringent safety measures.

research contracts with Canadian universities, and cooperative experiments in which expertise was shared with scientists at several laboratories in Canada and the United States. Possibilities for separating isotopes of other elements were assessed: e.g., removal from zirconium of the isotope that is the main contributor to the neutron cross-section of the natural element, for use in pressure tubes; uranium-235 for enriched fuel; and isotopes for medical applications. Although a laser-based heavy-water process remains an elusive economic goal, a laser was used to enrich the deuterium in a few drops of water nearly 2,000-fold (i.e., from 0.015 percent to 30 percent) in a laboratory tour de force.

ANALYTICAL METHODS — LATER PERIOD

As programs progressed and new ones arose, analytical methods advanced in parallel to meet new requirements at Whiteshell and Chalk River. With the growth of the fuel-development program, faster methods evolved to meet the increased demand for analyses to help determine the energy output from test specimens and, in mixed fuels, apportion shares to the appropriate isotopes of uranium, plutonium and thorium.

Water and organic coolant in the circuits of operating reactors were analyzed for the trace elements responsible for the generation of radioisotopes that are then carried outside the reactor shielding by the coolant. These methods were also used to analyze water in waste-management studies. Special attention was given to the determination of several elements simultaneously in the same sample or, where practicable, continuously in liquid or vapour streams. In a similar vein, the deuterium content of process streams at heavy-water plants and in the research laboratories was measured on-line.

Advanced techniques were used to analyze the surfaces and near-surface layers encountered in research on catalysts and corrosion. The solutions of the organic chemicals used to decontaminate reactor circuits required analyses to maintain them at efficient and economic operating levels. The amine-hydrogen process for heavy water raised unusual requirements for the analysis of organic-nitrogen compounds. Investigation of pressure-tube hydriding required the determination of low levels of deuterium in many highly radioactive specimens cut from tubes removed from the Pickering reactors.

Requirements for the analysis of one-of-a-kind samples from physics, engineering, or environmental programs were not uncommon and often required small research programs for their completion. An unusual example of this capability was the work done at Whiteshell on the fragments of the Russian

Cosmos nuclear-powered satellite, which fell to earth in northern Canada in 1978. The whole repertoire of chemical, isotope and surface analytical techniques was applied to determine the hazard from the reactor parts and help the authorities decide on the Canadian response.

EPILOGUE

The eighties was a period of transition and change in AECL. In materials science and chemistry, high-priority work for the Pickering reactors, waste management, and commercial activities demanded attention from some of the scientists normally engaged in basic research. Also, budgetary pressures increased, and funding for the sophisticated equipment for work at the frontiers of these fields became scarce. These factors combined to make it a difficult time for basic research in chemistry and materials science.

Several steps were taken to alleviate the difficulties. Although always a feature of the laboratories, special emphasis was placed on the development of cooperative programs with other institutions. Thus, visiting-scientist appointments at Chalk River and Whiteshell were advantageous, and part-time and sabbatical arrangements for AECL staff at other laboratories were also helpful. To meet the needs of materials-science experiments, irradiation facilities were rented in a British reactor. These measures helped to keep the programs thriving, but the squeeze on basic research continued.

In late 1985, management concluded that the pressures could be sustained no longer. It was decided that underlying research in chemistry and materials science should be discontinued as discrete programs. AECL made arrangements for the continuation of some research arising from the channeling phenomenon at a Canadian university. These included the transfer of staff and necessary accelerator equipment, and the provision of financial support for several years. The balance of the research staff transferred to related applied fields or to commercial activities.

Rapid change continued, and by the end of the decade conditions were such that some projects in the underlying research category could be launched. At the time of writing, they continue at both the Whiteshell and Chalk River laboratories.

ACKNOWLEDGMENTS

B. Cox, J.A. Davies, P.J. Dyne and M. Tomlinson contributed major building blocks to the foundations of this chapter. Assistance with AECL documents and records was courteously provided by Chalk River staff. To all these people I extend my sincere thanks.

BIBLIOGRAPHY

Since space does not permit a complete list of references, a selection of review articles and summary papers is given here. For more detail, the reader is referred to the "List of Publications" issued by AECL, mentioned in chapter one. In addition to the record in the technical literature, progress reports on the work discussed in this chapter were made four times each year, starting in 1949. Although access to some is restricted, many of these reports are available through AECL and those dating from 1967 at CRL are in the AECL "List of Publications".

Atrens, A., I.G. Ritchie, and K.W. Sprungmann. "Time and Amplitude Dependent Damping in a Single Crystal of Zirconium." In *Proceedings of the Sixth International Conference on Internal Friction and Ultrasonic Attenuation in Solids, 4-7 July 1977*, 633-7. University of Tokyo Press, 1977. (Also AECL-5951.)

Cassidy, R.M., and S. Elchuk. "Application of High Performance Liquid Chromatography to the Analysis of Nuclear Materials." *Journal of Liquid Chromatography* 4 (1981): 379-88. (Also AECL-7129.)

Cox, B. *Towards an Understanding of Zirconium Alloy Corrosion.* AECL-5548. 1976.

Cox, B., R. Dutton, T.A. Eastwood, and I.V. Mitchell. *AECL Research Programmes in Materials Science.* AECL-7072. 1980.

Crocker, I.H., T.A. Eastwood, D.R. Smith, R.B. Stewart, M. Tomlinson, and D.F. Torgerson. *AECL Research Programs in Chemistry.* AECL-7071. 1980.

Davies, J.A. "Electrochemistry and Nuclear Science – A Fruitful Combination." *Chemistry in Canada* 17 (1965): 53-4. (Also AECL-2329.)

Davies, J.A., G. Amsel, and J.W. Mayer. "Reflections and Reminiscences from the Early History of RBS, NRA and Channeling." *Nuclear Instruments and Methods in Physics Research. Section B* 64 (1992): 12-28.

Dyne, P.J., D.R. Smith, and J.A. Stone. "Radiation Chemistry." *Annual Reviews of Physical Chemistry* 14 (1963): 313-34. (AECL-1747.)

Evans, D.K., and R.D. McAlpine. "The Importance of Collisions in Laser Isotope Separation." In *Proceedings of the International Conference on Lasers '83, 12-16 December 1983*, 603-8 (1983). (Also AECL-8344.)

Garisto, F., R.J. Lemire, J. Paquette, P.P.S. Saluja, S. Sunder, D.F. Torgerson, A.C. Vikis, D.J. Wren and J.C. Wren. "Radioiodine Chemistry in Reactor Accident Conditions. In *Proceedings of an International Symposium on Source Term Evaluation for Accident Conditions, Columbus, Ohio, 28 October – 1 November 1985*, 501-14. Paper IAEA/SM/00281/046. Vienna: International Atomic Energy Agency, 1986. (Also AECL-9143.)

Grummitt, W.E., and G. Wilkinson. "Fission Products of U^{235}." *Nature* 158 (1946): 163, letter to editor. (Also issued as National Research Council of Canada report NRC-1755.)

Harvey, B.G. "The Actinide Elements and the Chemistry of Plutonium." *Nucleonics* 2 (April 1948): 30-40. (Also issued as National Research Council of Canada report NRC-1756.)

Jackman, T.E., J.A. Davies, D.P. Jackson, W.N. Unertl, and P.R. Norton. "The Pt(110) Phase Transition: a Study by Rutherford Backscattering, Nuclear Microanalysis, LEED and Thermal Desorption Spectroscopy." *Surface Science* 120 (1982): 389-412. (Also AECL-7733.)

Jervis, R.E. and W.D. Mackintosh. "Activation Analyses of Interest to Atomic-Energy Programs." *Proceedings of the Second International Conference on the Peaceful Uses of Atomic Energy, Geneva, 1-13 September 1958*, vol. 28, 470-77. Paper A/CONF.15/P/189. Geneva: United Nations, 1958. (Also AECL-599.)

Jones, M., R.P. Schuman, J.P. Butler, G. Cowper, T.A. Eastwood, and H.G. Jackson. "Isotopes of Einsteinium and Fermium Produced by Neutron Irradiation of Plutonium." *Physical Review* 102 (1956): 203-7. (Also AECL-287.)

MacEwen, S.R., and G.J.C. Carpenter. "Calculations of Irradiation Growth in Zirconium." *Journal of Nuclear Materials* 90 (1980): 108-32. (Also AECL-6762.)

Puls, M.P., and R. Dutton. "Thermodynamics of Crack Growth." In *Micro and Macro Mechanics of Crack Growth*, 19-28, K. Sadananda, B.B. Rath, and D.J. Michel, eds. Metallurgical Society of the AIME, 1981. (Also AECL-7846.)

Rolston, J.H. "Research on the Separation of Hydrogen Isotopes by Catalytic Exchange." In *Proceedings of an International Symposium on Synthesis and Applications of Isotopically Labeled Compounds." Kansas City, Missouri, June 1982*, 385-90, 1983. (Also AECL-7703.)

Seddon, W.A., and J.W. Fletcher. "The Contribution of Pulse Radiolysis to the Chemistry of Alkali Metal Solutions." *Radiation Physics and Chemistry* 15 (1980): 247-54. (Also AECL-6631.)

Torgerson, D.F., N.H. Sagert, D.W. Shoesmith, and P. Taylor. *Underlying Chemistry Research for the Nuclear Waste Management Program*. AECL-7786. 1984.

Wikjord, A.G., T.E. Rummery, F.E. Doern, and D.G. Owen. "Corrosion and Deposition During the Exposure of Carbon Steel to Hydrogen Sulphide-Water Solutions." *Corrosion Science* 20 (1980): 651-71. (Also AECL-6724.)

Chapter Eight

Physics

J.C.D. MILTON

Nuclear Physics

INTRODUCTION

The atomic nucleus is unbelievably small yet it contains all but the tiniest fraction of the mass of every object that surrounds us. To gain some idea of how small, consider atoms. If it were possible to line them up single file, it would take tens of millions of them to stretch over one centimetre. If one could expand the diameter of one atom until it were the size of a football field, the nucleus would be represented by a pea at the centre of the field. And yet what wonders a nucleus contains. We have been studying it for 100 years, yet the deeper we look into it, the more surprises we find. We do know that it is composed of neutrons and protons,[1] which taken together are called nucleons. The nucleus can emit very many things, especially when it is disturbed (see Box A). Because β particles, which are electrons, come out of the nucleus, it was once thought that it must contain electrons; but it does not. As someone has said, "Because barks come out of dogs does not mean that dogs are composed of barks". How do we study the nucleus? No microscope can see anything so small. There is already a hint in what has been said above, but if we had to wait to see what came out of the nuclei that nature gave us, we would know very little about their make-up; yet that is all that scientists in the early years of this century could do. Most of what we know has been learned by perturbing the nucleus, usually by hurling other particles at it, such as nucleons and combinations of nucleons, and observing what comes out. This may sound crude, but it is really the only method we have. If the projectile is charged, such as a proton, electron or a particle, the bombarding is done by machines called particle accelerators, or more commonly, just accelerators. If the projectile is uncharged, such as a neutron or a neutrino, other means are employed.

The nucleus is an important part of all matter. Its study can help us understand the origin and behaviour of our universe; it can be used as a laboratory to study the nature of the fundamental particles themselves. A nucleus can exhibit a wide range of complex and beautiful phenomena; it can contain anywhere from one to more than 250 nucleons. It therefore covers the mid-ground between simple systems of one, two or three particles that we can describe with "simple" single-particle theories, and those with many thousands of particles,

---

[1] It is true that other particles can exist in the nucleus with names such as meson and hyperon, and they do play a role, but their existence is transitory.

BOX A

Radioactive nuclei, both natural and man-made, and stable nuclei when they are disturbed, can emit radiations, both particle and electromagnetic. Many particles can be emitted by a nucleus, including other nuclei if it is sufficiently excited; however, common radioactive nuclei emit predominantly radiations of three types: α particles, β particles, and the nucleons (i.e., protons and neutrons). α particles are the nuclei of the helium atom; they have two neutrons and two protons tightly bound together, but no surrounding electrons. The β particles are electrons, either negatively or positively charged. In the latter case, they are usually referred to as positrons. But how can electrons come out of the nucleus when they are not there? The answer is that the weak force can transform a neutron to a proton and an electron, producing at the same time an exceedingly light (possibly massless) particle called a neutrino, symbol ν. (Actually it is an antineutrino.) The weak force can also transform a proton to a neutron plus a positive electron and a neutrino, if energy is available from an external source, such as the rest of the nucleus. In both cases, the electron is referred to as a β ray, or as β^+ or β^- if there is a need to differentiate. Electrons are only referred to as β rays when they arise from the action of the weak force.

The electromagnetic radiations are called γ rays; they are the same as the visible light and the X rays that are emitted by excited atoms, except that they have a higher energy, usually much higher. Because of their greater energy they have a shorter wave length than the visible light emitted by atoms, and they are much more penetrating, even more than X rays. When a γ ray does strike an atom, a high-energy electron is frequently ejected. As it has a charge, this high-energy electron can ionize other atoms; i.e., separate other electrons from their parent atoms. These secondary, or free, electrons may be collected in a detector such as an ionization chamber, a proportional counter, a Geiger counter, or a germanium detector. After the atom has been ionized, it is left in an excited state; it frequently deexcites by emitting visible light. If the material is a scintillator, such as sodium iodide or bismuth germanate, this light can escape from the material and be picked up in a photomultiplier. The amount of light, or ionization, depends on the energy of the γ ray; the more energy the more ionization, so the detectors mentioned above can give us information on the energy of the γ ray (the Geiger counter is an exception; it gives no information on the energy, but only the number of γ rays).

such as a gas, that can be well described by statistical theories. The most compelling reason for studying the nucleus is that it is one of the few ways in which we can study the strong force, the strongest of the four forces in nature,[2] and the one that holds nuclei together by overcoming the mutual repulsion of the positive charges on the protons.

Thermal neutrons[3] have several advantages as projectiles with which to bombard the nucleus. First of all, they carry no charge, so they can easily enter the nucleus; the result is a nucleus in a narrow range of excited states (see Box B). A charged particle usually must have considerable energy in order to overcome the electrostatic repulsion between it and the positively charged nucleus. The result can be either a single nucleus in a highly excited state, or two nuclei (the target and the projectile) in various stages of excitation. The resulting system is thus considerably disrupted and much less well defined. Of course, negatively charged electrons have no such problem, and they are indeed used extensively with nuclei to study the weak force, but they cannot investigate the strong force.

Because the neutron is a useful particle for probing the nucleus, we can begin to see how Chalk River became an important contributor to knowledge in nuclear physics, and in particular

2 The four forces are, in order of their strength, the strong force, the electromagnetic force, the weak force and gravity.

3 Thermal neutrons are those whose energies are in thermal equilibrium with the molecules of the reactor moderator.

BOX B

A nucleus with a given number of protons and neutrons, usually referred to as a nuclide,[1] can exist in a number of states. The one of lowest energy is called the ground state; all the others have higher energy and are called excited states. Near the ground state the excited states are few and usually well separated, but at higher and higher energies the states become very numerous, closer together, broaden, and eventually, overlap.

[1] Footnote 4.

to what physicists call nuclear structure, by which they mean the way that nucleons are organized inside the nucleus. Research reactors in the forties and fifties provided hitherto undreamed-of numbers of neutrons, and hence were at the forefront of many branches of science, nuclear physics being one of them. Any of the small number of laboratories then fortunate enough to have such a facility could enjoy a free run into essentially untouched, seemingly limitless fields of ground-breaking research in nuclear physics. Chalk River, with its high-flux reactors, NRX and NRU, and strong research policies combining fundamental and applied science, was well positioned to take a commanding lead in those early years. This favourable position ultimately declined as more laboratories entered the field with research reactors surpassing NRU, and so, with the passage of time, the unique opportunities in reactor-based nuclear physics at Chalk River diminished. In contrast, as we shall see later, the accelerator-based nuclear physics, though subject to the same obsolescence of equipment, moved from strength to strength and remains competitive to this day. In 1958, the small, vertical electrostatic accelerator of the type often called a Van de Graaff accelerator was replaced by the world's first tandem accelerator; it was subsequently replaced in 1967 with the new and larger Emperor, or MP tandem, and more recently, with the addition of a superconducting cyclotron, the present powerful Tandem Accelerator SuperConducting Cyclotron (TASCC) facility was created (see figure 8.1).

Nor was this preeminence in equipment restricted to large facilities such as reactors and accelerators. In the early days, Chalk River was unusually well endowed with excellent resources in electronics – not simply the best low-noise fast amplifiers, kicksorters (pulse height analyzers), discriminators and coincidence circuits (most of them designed and built in-house), but with technologists who could rapidly design and build custom electronic systems. In the sixties, the necessity of providing one's own electronics became less necessary, but by then the first small transistorized digital computers were coming on the market, and Chalk River had the knowledge necessary to use them for data acquisition. In addition, the laboratory had outstanding expertise in detectors. We shall learn later of some of these activities, but for now will mention an application having nothing to do with reactors or accelerators: the measurement of the half-life of the muon. A small building at Chalk River was outfitted with a large cosmic-ray telescope and (for those days) some very fast and complex electronics. The result was the first accurate measurement of the lifetime of the muon (see Box G), at that time a mysterious new particle. Today its accurately known lifetime, 2.19703 ± 0.00004 microseconds, is in good agreement with Chalk River's 1951 value of 2.22 ± 0.02.

NUCLEAR PHYSICS AT THE REACTORS

Neutron Capture Gamma Rays

We have already seen that the neutron should be a useful probe of the nucleus, and so it was with the study of the γ rays from neutron capture; that is, those that are emitted promptly upon the absorption of a neutron. For nuclear physics, the ability to resolve and determine the precise energies, intensities and other basic properties of individual γ rays was a high priority.

BOX C

According to the Shorter Oxford Dictionary, the word "spectrometer" was first used in 1874 and was at the time applied to an instrument used to measure the index of refraction of a material. However, the word was soon applied to instruments separating light into various colours and measuring their intensities. The colour of light is determined by the wavelength. When in the twentieth century it became known that the wavelength and the energy were simply related, the word spectrometer came to mean any instrument that separated and measured a range of entities, that is, a spectrum, into its components. In nuclear physics, the component is usually the energy, the momentum or the mass. Thus a γ spectrometer usually separates and measures the energies and intensities in a spectrum of γ rays, a β spectrometer the momentum of β rays, and a mass spectrometer the mass of particles.

Spectrometers come in many types and sizes: they may be as small as a germanium crystal weighing less than a few grams, to measure the spectrum of charged particles, such as α particles, or be several stories high, weighing hundreds of tons, as at a high-energy accelerator such as at CERN or Fermilab, to measure the masses and energies of very high-energy particles. Spectrometers may make use of single or multiple electric or magnetic fields, of refraction or diffraction. A pair spectrometer measures the momenta of both the electron and the positron created when a high-energy γ ray is annihilated. Unlike other particle spectrometers, which strive for uniform fields, a lens spectrometer (see for example figure 8.2,4) exploits the properties of a highly non-uniform field.

See Box H for further information on a particular type of spectrometer.

One of the first programs in nuclear physics at the startup of NRX in 1947 investigated the prompt γ rays arising from the capture of a thermal neutron. The record-breaking neutron flux of NRX enabled Chalk River scientists to make the first high-resolution measurements of these γ rays. The first published Chalk River measurements on an important fundamental physical property – the energy released when a neutron and a proton are bound together to form a deuteron – attracted worldwide attention; it resolved a discrepancy in the results of other laboratories that were pursuing the same problem using older methods.

For several years after NRX startup, the high flux of the reactor, combined with the excellent resolution and precision of a new type of spectrometer (called a pair spectrometer (see Box C), resulted in an abundance of measurements that were unmatched at the time. They earned the scientific team a reputation for excellence, and added significantly to the prestige of the whole laboratory. To cope with the flood of data generated by a reactor running day and night, the team installed one of the earliest fully automated control and data-recording systems for a spectrometer in such an environment. Among contributions to physics gleaned from these early measurements were: 1) details of γ emission and the densities of states in nuclei throughout the periodic table of the elements, 2) the first elucidation of the energy dependence of the intensities of γ rays of a specific electromagnetic type emitted from highly excited nuclei, and 3) the first demonstration of simple high-energy spectra implying unexpectedly simple structures in certain heavy elements, a result that lent strong support to the then-emerging nuclear shell model (see Box D).

Subsequently, at both NRX and NRU reactors, an improved pair spectrometer (figure 8.3,2), a novel high-resolution diffraction spectrometer, and later (1964) the first large lithium drifted germanium detector (both of the latter two devices being first developed at Chalk River) were employed to extend the capture γ-ray studies. Included were a few special measurements from which could be extracted certain otherwise inaccessible but necessary cross-sections for thermal-neutron capture.

By 1959, through the use of ever more diverse techniques, and neutrons with energies well above the average of the thermal

BOX D

In the late forties a new nuclear model appeared, called the shell model, in analogy with the very successful atomic shell model. Previously the only models available were the liquid drop model, which assumed that the nucleus was simply a fluid drop, and the statistical model, which assumed that the nucleus was a kind of soup of neutrons and protons. Since we know that the nucleus contains a relatively small number of nucleons, such models are clearly inadequate for many purposes. Just as for the electrons in the atomic shell model, the nuclear shell model supposes that the nucleons are arranged in a series of shells, and again as for atoms, these shells were named s, p, d, f, h, i, etc. The nuclear shell model is, however, a good deal more complicated than its atomic predecessor, because in the case of the atom, the electrons in the shell move under the influence of a potential created by an external body, the nucleus. In the nucleus, the nucleons move in a potential that they themselves create. The situation is further complicated by the fact that the nucleus contains two types of particles – neutrons and protons – and, except in the lightest nuclei, in unequal numbers. Thus the protons may be filling one shell, such as the s shell, while the neutrons are filling the d shell, giving rise to nomenclature such as s-d shell nuclei. This makes the calculations considerably more difficult.

When a shell is filled, or closed (i.e., it contains all the particles that it is allowed to), the resulting nuclei are more stable than their neighbours, just as with atoms for which the closed shell atoms are the rare gases. The nucleons in this case are tightly bound together, so that the formation of these nuclei releases more energy (or requires less) than the formation of their neighbours; they are also spherical and resist deformation. Such nuclei are often called magic; if both the neutron and proton shells are closed, they are called doubly magic.

In the fifties, a somewhat different nuclear model appeared, the collective model. The shell model is an independent particle model, but in the collective model, all the nucleons behave as a group. Two such group modes are vibrational and rotational. The latter has turned out to be very important in nuclei. Any non-spherical nucleus may exhibit rotational properties, and these give rise to a particularly regular γ-ray spectrum, the rotational spectrum. Determination of these rotations has turned out to be a most fruitful way of studying nuclei.

Subsequently, the nuclear shell model and the collective model were united to create the unified model. It is the best we have today, and it is very good. For example, it accounts for the fact that many nuclei, if not most, are not spherical but are spheroidal, or even ellipsoidal.

---

spectrum, many other detailed nuclear properties were determined. Some of the new techniques involved collaboration with other laboratories, such as the Brookhaven National Laboratory in the United States, which designed and built equipment to be installed at NRU. The combined data obtained from these studies remain today at the core of empirical codes for shielding design and assessment for both fission and fusion reactors, while the formalism established to account for the results has been adopted into modern theories of γ-ray intensity distributions.

In a collaboration unusual at the height of the cold war in the sixties, Chalk River scientists and a team from the Kurchatov Institute in Moscow, which had employed a different method to measure capture γ rays, produced a comprehensive joint compendium of all thermal neutron capture data then available, the majority being provided by their own laboratories. This compilation was the definitive repository of information on γ rays from thermal-neutron capture for many years.

In recent times, the worldwide use of the unique, element-specific signatures provided by prompt γ rays listed in the above compilations and in a more extensive recent tabulation published by Chalk River have grown rapidly. Because these radiations are emitted promptly on capture of the neutron, and because of their relatively high energy, they have certain advantages over the delayed, lower energy γ rays detected in thermal-neutron activation analysis. A disadvantage is that they require an on-the-spot neutron source of high intensity. As a result, they have found use in restricted but diverse applications, such as on-line monitoring of process streams, determination of sulphur in coal and oil, analysis of biological and environmental materials, prospecting, and determination of nitrogenous explosives in luggage.

Early Beta and Gamma Spectroscopy

It was realized that a new high-flux reactor (such as NRX) would provide an abundance of new radioactivities not previously available. L.G. Elliott and colleagues began construction of a magnetic lens β-ray spectrometer (see Box C) at the Montreal Laboratories in 1944. It was to be used for the study of β and γ activity of ^{233}Pa, and of the γ rays of fission products, especially ^{140}La. By June 1946, two lens spectrometers had been placed end-to-end for coincidence studies at Montreal, and a high-transmission spectrometer had been built for shipment to Chalk River. At that time the only accurate way to determine γ-ray energies was to measure the energy of conversion electrons and make the necessary simple corrections.

By the end of 1947 a large magnetic spectrometer had been set up and aligned at the south thermal column of NRX, and studies of ^{113}Cd(n,γ) ^{114}Cd (see Box E) were in progress. A variety of experiments followed. Of particular interest was the first observation of Doppler-shift broadening and its use in determining short nuclear lifetimes. The external conversion line of the 479 keV γ ray emitted following the reaction ^{10}B(n,α)^{7}Li showed Doppler broadening (from α recoil) and analysis of the shape showed that the mean lifetime of the ^{7}Li excited state is less than 3×10^{-13} seconds.

By 1950 there were three β-spectrometer installations. J.M. Robson had one at the NRX reactor for determining the lifetime of the neutron (see figure 8.2,4). It is energetically possible for a neutron to decay to a proton, and evidence for this had been obtained at the Oak Ridge National Laboratory in the United States. In an ingenious experiment, Robson was able to measure the spectrum of β rays in coincidence with the proton, and to show that it had the expected shape and end point. His careful measurements led to publication, in 1951, of the first accurate half-life of the neutron, 12.8 ± 2.5 min. Subsequently, Robson attempted an even-more daring experiment. Beta decay of a nucleus is mediated by the weak interaction; the neutron is the simplest system that undergoes β decay and is therefore well suited to calculation. The probability of the β particle and the ν being emitted with a given angle depends on the fundamental nature of the weak interaction. Robson attempted to measure the probability distribution of the angle between the proton and

BOX E

Since the early days of nuclear physics, a standard nomenclature has been used to specify the reaction taking place, viz. X(x,y)Y. X is always the target nucleus, and x the projectile, which may or may not be a nucleus. The products of the reaction are y and Y. By convention, Y is the heavy product, and y one or more light products. For example, in the reaction given below, ^{113}Cd(n,γ)^{114}Cd, neutrons bombard the isotope ^{113}Cd to produce a γ ray and ^{114}Cd. If multiple light products are produced, y will be written as 2p, nα, 2n2p, etc., with no comma between the light products. A slight exception occurs in the case of fission, in which, for example, the symbol ^{235}U(d,pf) means that a deuteron (d) gives up its neutron (n) to form the nucleus ^{236}U, which subsequently undergoes fission to produce a very wide range of product nuclei (the fission products). The proton (p) continues largely unscathed, except for a change in direction, and is frequently referred to as a spectator.

the electron, from which the correlation of the β and the ν can be deduced. This extremely difficult experiment met with only limited success.

A second spectrometer with a double-lens was set up at Chalk River. This spectrometer had considerably sharper resolution than the earlier arrangements. It was used for studies of many β spectra, including ^{198}Au, ^{170}Tm, and ^{181}Hf. The third setup was an end-to-end pair of lens spectrometers in the NRX annex. The members of the pair could be operated independently, so that β rays from a source at the centre could be measured in one spectrometer, and coincident conversion electrons in the other, an arrangement that proved most useful in the development of the fast-slow coincidence technique to be described later.

By 1950, scintillation counters began to be used in β and γ studies; they comprised small anthracene crystals for speed, and larger crystals of sodium iodide doped with thallium, NaI(Tl), for high efficiency. While the energy resolution of NaI(Tl) detectors was modest (e.g., 5 percent at 1 MeV), they proved useful in many γ-ray studies and in particular coincidence experiments used to determine a sequence of γ rays. One experiment was the spin-parity sequence of energy levels in the doubly magic ^{208}Pb nucleus (see Box D) thought to be 0+(ground), 2+, 4+ ... as in other even-even nuclei. When the angular distribution between the γ rays from the two lowest states was measured, the spin-parity sequence proved to be 0+, 3-, 5-, an exception to the general rule.

Another interesting technique was the development of very high gain proportional counters, initially for the purpose of detecting the 2.8 keV X rays from the decay of ^{37}Ar. The motivation was an experiment proposed by Pontecorvo to detect neutrinos, which would produce ^{37}Ar from ^{37}Cl in a large tank of carbon tetrachloride. The experiment was never carried out, but the counters were used to measure the β spectrum of tritium (see Box A). The analysis of the shape of the spectrum near its 18.6 keV endpoint showed that the mass of the light neutrino was less than 500 eV, an upper limit that was not reduced for many years. The most recent limit (1991) is 7.3 eV.

Fission

In the late forties, knowledge of the fission process was meagre. The theory of fission, based on the liquid drop model of heavy nuclei, was able to predict neutron-induced fission threshold energies and the total energy release, but not much more. It was therefore appropriate that the fundamental physical process on which all fission reactors was based should be studied at Chalk River.

Soon after the initial operation of the NRX reactor in 1947, the energies of both fragments were measured in coincidence in a double ionization chamber. From such measurements the prompt[4] mass yield, was measured and compared with the final mass yield, as previously determined at the Montreal Laboratory and in the United States by chemical means. About the same time, a series of experiments was begun to determine the properties of the prompt fission neutrons. Although it is these neutrons that make a sustained chain reaction possible, very little was known about their origin and properties. The first experiment used a beam of thermal neutrons from one of the two graphite thermal columns, a gridded ionization chamber, to register the fragment energy in fission of ^{233}U, ^{235}U, and ^{239}Pu, and a high-pressure methane proportional counter for detecting the fast neutrons.

The results verified that most of the neutrons are emitted from the full-velocity fragments. Moreover, it was found that about 30 percent more neutrons came from the light fragment than from the heavy. It was well known then that fission by thermal neutrons is asymmetrical in the mass yield, the most probable mass numbers in ^{235}U(n,f) being 94 and 140. It was clear from the energetics of the process that the neutrons must come from a higher average excitation energy

4 The prompt masses are those existing immediately after the fragments separate, but before the prompt neutrons are emitted. Although an average of 2.5 neutrons is emitted in the fission of uranium, any number between 0 and about 5 may be emitted, with each emitted neutron lowering the fragment mass by one unit. The resulting mass after neutron emission is the final mass.

in the light than in the heavy fragment. The answer to why the excitation energies should be so different in the light and heavy fragments would only emerge years later, with the development of the shell model of nuclear structure (see Box D).

In 1951, a second series of experiments was begun to learn more about neutron emission from the fission fragments. In the first of the more detailed experiments, the single ion chamber used earlier was replaced with a back-to-back chamber similar to the one mentioned above, which could measure the energies of both fission fragments simultaneously. Two neutron detectors were used, one on each side of the ion chamber. The fragment energies were recorded on two slowly moving, synchronized strip-chart recorders. The amplitude of the stroke on the chart was a measure of the fragment energy; the detection of a neutron in the same direction as the fragment was noted by a tick on the side of the chart. Needless to say, the total number of events analyzed by this state-of-the-art, two-dimensional analyzer was limited by the courage and stamina of the reader. Simple relations between the energies and momenta of the fragments allowed the calculation of the masses in each event. The neutron yield could now be correlated with the mass of the fragment moving in the direction of a particular neutron counter.

A surprising result of this experiment was that the lightest fragments in both the light and heavy groups emitted very few neutrons, but near the symmetric mass region, the light fragment emitted many more neutrons than the paired heavy fragment, although the latter was only very slightly heavier. This sawtooth variation of the number of neutrons with the fragment mass, initially believed impossible by many, has since been corroborated for every case investigated, including spontaneous, thermal-neutron and proton-induced fission.

In the mid-fifties, the velocities of paired fission fragments were measured by time of flight, and an improved mass resolution of 2.5 to 3.0 mass units was obtained. The data recording was automated to the extent then possible – coded pulse heights recorded on punched paper tape, subsequently transferred to punched cards for reading by the Datatron, a vacuum tube electronic computer with a 4,000-word memory (8 kilobytes in modern terminology). The experiment yielded the average total kinetic energy in the spontaneous fission of ^{252}Cf. The energies of the prompt γ rays were also measured and showed a negative correlation with the total kinetic energy of the fragments, as expected.

With the advent of the NRU reactor in 1958, the time-of-flight measurements (figure 8.2,5) were extended to the thermal-neutron fission of ^{233}U, ^{235}U, and ^{239}Pu, and with improved fragment detectors, the mass resolution was reduced to 2.0 units. With this high-resolution fine structure was seen in the prompt mass yields. It is most pronounced at high total kinetic energies, where the fragments have little excitation energy, and most noticeable in the fission of ^{235}U and ^{239}Pu. Another feature clearly shown was the large drop in total kinetic energy when the masses are nearly equal. This drop is now believed to result from an increased deformability of the light fragment near the symmetric mass region and is related to the sawtooth variation of neutron emission discussed earlier. Near the closed shells, such as at N=50, 82 and Z=50 (N and Z are the number of neutrons and protons), the fragments are rigid, whereas those far from closed shells, such as those at masses 110 and 160, are softer (see Box D).

In the mid-sixties, attention was directed to charged particle-induced fission, taking advantage of the newly installed MP Tandem Accelerator, and of solid-state detectors for measuring particle energies. The fissioning nuclei studied were ^{236}U and ^{240}Pu. Whereas these nuclei had been studied earlier in neutron capture-induced fission, in the new experiments they were produced by the reactions ^{235}U(d,pf) (see Box E) and ^{239}Pu(d,pf). Particular attention was paid to the excitation energy range in the product nuclei from 5 MeV to 10 MeV, a range that brackets the excitation energy produced in neutron capture in ^{235}U and ^{239}Pu.

To characterize the transition state (fission saddle-point states), the angular distribution of the fragments with respect to the classical recoil axis was measured. In both cases studied, a large anisotropy was observed at excitation energies near the fission threshold. These results were interpreted in terms of the angular momentum quantum numbers of the transition states populated in the (d,pf) reaction.

The (d,pf) work was subsequently extended to lower excitation energies with a resolution of 100 keV in ^{239}Pu(d,pf). In a tour de force requiring continuous data taking for over 300 hours, a resolved level structure was found at excitation energies between 4.8 and 5.2 MeV in ^{240}Pu. The fragment angular distributions for all seven observed levels were found to be the same.

In the fifties and sixties, Chalk River was a leading institution in the field of fission physics, the NRX and NRU reactors providing exceptional facilities in the field. By the late sixties, it became clear that further fruitful research in the fission process with thermal neutrons was unlikely. On the other hand, it was becoming clear that, for example, the fission transition state properties could be approached through heavy ion collisions of a variety of projectiles and targets.

Two-photon decay

In the late winter of 1974, a preprint was brought from the Institute Laue-Langevin (ILL) in Grenoble, reporting the observation of coincident γ rays adding up to 2.22 MeV (in addition to the well-known single γ ray of that energy) following the capture of neutrons by a water target. The branching ratio for this two-photon branch, compared to the single-photon branch, was reported to be much larger than any theoretical prediction. This was the beginning of a most exciting period for at least four physicists at Chalk River. The experiment was quite easy to set up. It took about a week to get the beam line ready and the electronics working. To quote one of the experimenters:

> We took data for about a week. Every physicist at Chalk River knew what we were doing during that week. It took about twelve hours to do a measurement so each day we would set up some configuration, run it overnight, see the results on the analyzer when we came in the next morning, discuss the result during the day and, in particular, at lunch with everyone else, reconfigure the equipment and run again overnight.
>
> In the first measurement with the detectors opposite each other, we saw the same high branching ratio. Then we placed the detectors at 90° to each other and shielded from each other. The branching ratio dropped dramatically to less than 10^{-4} after an overnight run. We then returned the detectors to the 180° position and placed lead plates in front of each one. This latter measurement confirmed our suspicions. It was not two-photon decay with both γ rays coming from the target, but a single γ ray from the target that scattered out of one detector into the other, leaving some energy in both detectors.
>
> During the third week, we wrote up a note for Physical Review Letters. The fourth week we rested. Needless to say getting the note into print took much longer – over three months.
>
> We continued, in a more mundane way, to improve the sensitivity and statistics of the measurement. We finally got the branching ratio limit down to less than $5x10^{-6}$, which was still a factor of 20 above the more conventional theoretical predictions. We also looked for but failed to find doubly radiative neutron capture in ^{2}H and ^{16}O.[5]

NUCLEAR PHYSICS AT THE ACCELERATORS

The Chalk River Electrostatic Accelerators

The superconducting cyclotron is but the latest of a series of accelerators at Chalk River. The first, the Van de Graaff (1952-60) (figure 8.1,1), was a "home-made" vertical machine with a maximum voltage of 3 million volts (MV) on a terminal at the top. It could accelerate several light ions,[6] but was mainly used with protons and ^{3}He. It was superseded by an EN tandem Van de Graaff (1958-67) (figure 8.1,3,4,5) made by High Voltage Engineering Corporation. Originally, the central terminal was rated at a voltage of 5 MV, but later it and the approximately sixty subsequent ENs ran successfully at 7 MV. The Chalk River EN was the first tandem in the world and could accelerate useful beams

---

5 E.D. Earle, private communication.

6 An ion is an atom with a non-zero net electric charge. The terms "light ion" and "heavy ion" are not precisely defined in common usage. Here we will consider an ion of helium or lighter as "light", and carbon or heavier as "heavy". This leaves ions of lithium and boron as neither light nor heavy.

FIGURE 8.1
ELECTROSTATIC ACCELERATORS AT CRL

1. The vertical 3 MV Van de Graaff was first in the series of CRL's electrostatic accelerators, and its first "high" energy accelerator. It is shown with its gas-holding tank removed being inspected before operation by Nuclear Physics Branch staff.
2. Physicists discuss results in the "control room" of the 3 MV machine.
3. The arrival of the pressure vessel for the world's first tandem accelerator, HVEC's 5 MV EN Van de Graaff arrives at CRL and is greeted with great interest by Nuclear Physics Branch staff.

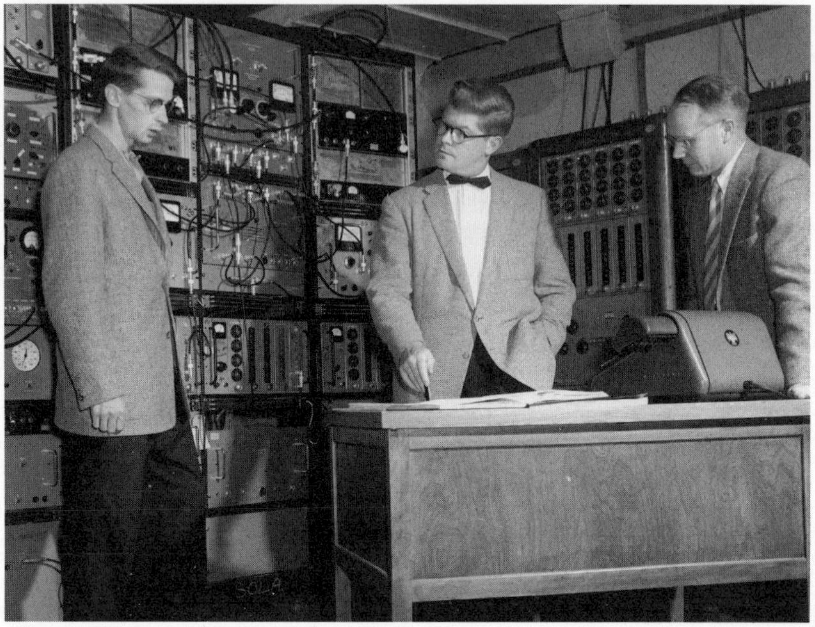

2

3

1

4. AECL commissioning staff at the EN tandem with the 90° analyzing magnet in the foreground and the switching magnet with its 3 alternative beam lines on the left.
5. The EN tandem control room shortly after startup.
6. The MP tandem Accelerator.

5

4

6

of heavy ions (e.g., carbon and oxygen). This technical development caused a revolution in nuclear physics, especially when coupled to the further development of silicon crystal heavy-ion counters, germanium crystal γ-ray counters, and transistorized multichannel analyzers. These devices were all built at Chalk River, which for a time held a lead in this area of nuclear physics. The EN tandem was in turn replaced by the present MP tandem with 10 MV on the central terminal. Modifications in 1972 upgraded the MP tandem (figure 8.1,6) to 13 MV, and subsequently in 1986 to 15 MV. Planning for a superconducting cyclotron energy booster was begun in 1972. Plans called for designing and building the first superconducting cyclotron in the world. Its installation occurred in two phases: a proof of principle stage completed in 1987, and the total facility stage (TASCC), fully commissioned in 1991.

Light Particle Reactions

In 1948 there was a 150 kV Cockcroft-Walton accelerator available, capable of accelerating deuterons. It had been built a short time before for research on very light nuclei, but was proving insufficient, and a larger accelerator of 250 kV was being built. It was known that when deuterons bombarded tritium, there was a prolific production of energetic (14 MeV) neutrons. The plan was to bombard various elements with these fast neutrons and look for the resulting radioactivities. At that time there was in existence a fairly good, semi-quantitative treatment of nuclear reactions, called the statistical model (see Box D). According to it, one would expect an energetic neutron to be captured on collision and then share its energy with all the protons and neutrons of that nucleus. Then, much later on a nuclear timescale, a proton, or one or two neutrons, would by chance gain enough energy to escape. For the inelastic reactions, the expected yields were (n,2n), (n,p), and (n,α) (see Box E), in that order.

The theory here was quite clear: the last two reactions on this list would become increasingly rare to the point of extinction, as the target nucleus became heavier. The proposal was to look for these reactions in some of the convenient light nuclei. Where the reaction product was known, the reaction cross-section could be quantitatively measured from the radioactivity of the product. Adequately accurate measurements were possible with solid samples, without chemical separation, thanks to the careful calibration of β-counting efficiencies that had been made by the chemists. The necessary tritium targets for the 150 kV accelerator were also prepared by them.

In fact, everything was ready: accelerator, targets, counting equipment, supplies, and theory. The accelerator was, however, rather primitive, and the counting equipment was crude by the standards of even a few years later. The laboratory was a great, empty, draughty barn of a building. Nevertheless, the experiments went well. The first results, on the lightest elements, were pretty much as expected, and for the heavier nuclei, the (n,2n) results continued broadly to follow the expected trends. However, reactions giving charged particles, protons and alpha particles began to show astonishingly large probabilities. The mere fact that they were observable at all ran strongly against the statistical theory.

At this point, some of the theorists became interested. To them it was quite clear that we were seeing a predicted, but hitherto unseen reaction: direct interaction in which an energetic particle comes in so swiftly that it bumps out one of the particles in the target nucleus, while the others act merely as spectators. Such a reaction, in a heavy target, when charged particles are to be emitted, is orders of magnitude more probable than the statistical heating-up kind of reaction described by the earlier theory. This, then, was a new phenomenon and the first experimental confirmation of the predictions of a model based on the independent existence of separate particles in nuclear matter.

These and other direct interactions arising from proton, deuteron and alpha bombardment, including the excitation of nuclear rotations and vibrations, were soon seen in other laboratories. The theory was worked out in detail, so that the nuclear structure and reaction mechanisms were well understood. Some years later, others became interested in this work, and several examples of rotational excitation by means of 14 MeV neutrons were studied. The cross-sections were found to be in satisfactory agreement with the predictions based on the γ decay of these states.

Although the production of 14 MeV neutrons from the deuterium on tritium (DT) reaction continues to be of interest in many other fields, its use for the study of nuclear structure and reactions was limited. Even with the best advances in technology and computing, it finally proved to be just too difficult to get further useful information. One other measurement was made, nonetheless: a precise determination of the fast-fission cross-section of uranium. By 1968, 14 MeV neutrons had run their course as a principal research tool in nuclear physics, and the 250 kV Cockcroft-Walton accelerator was shut down.

Nuclear Molecules

The arrival of the world's first tandem accelerator at Chalk River in 1957 presented the nuclear physicists with the opportunity of studying nuclear reactions with heavy ions in detail never possible before. The EN tandem accelerator (figure 8.1,4) first accelerates negative ions to a 5-million volt positive electrode, where several electrons are removed from the fast-moving ion. Afterwards, the now multiply charged positive ion again acquires energy by further acceleration to ground potential. For example, a beam of carbon atoms with 20 MeV can readily be produced in microampere ion quantities.

Such strong beams of carbon ions made possible new types of experiments, often with completely unexpected results. For example, in the bombardment of carbon by carbon, it was found that the reaction yield varied with ion energy in such a way as to suggest that the two carbon nuclei did not immediately fuse, but formed a transient nuclear molecule. This discovery created much interest and was later shown to be present in other cases, with indications that, in addition to shell structure and rotational behaviour like spheroidal liquid drops, nuclei could show strong cluster structure, making them sometimes appear like molecules.

This discovery would have been very difficult to make without the development of the tandem accelerator and silicon heavy ion counters. The studies also had indirect implications for the creation of the new field of accelerator mass spectrometry in 1977, because of the discovery that some negative atomic ions could not be formed for subsequent tandem acceleration. Attempts were made in the early sixties to accelerate negative nitrogen ions to study their nuclear reactions. This was found to be impossible and negative nitrogen hydride ions had to be used instead. It is now known that there is no stable negative nitrogen ion.

In the early sixties, it was realized that this could be used to separate ^{14}C and ^{14}N for radiocarbon dating, but it was technically not possible until the development of powerful new ion sources with special properties some fifteen years later. In the meantime, one of the first experiments in the new subject was done identifying boron negative ions by accelerating them in the tandem accelerator (figure 8.1,4) and subsequently stripping all five electrons off the boron, prior to counting the ions after magnetic analysis. Radiocarbon dating by accelerator mass spectrometry, with applications in archaeology, storage site selection for nuclear waste, oceanography, climatology, and many other disciplines, grew out of these early experiments, illustrating once again the unexpected applications from basic research.

Nuclear Spins from Angular Distributions

We have seen (in Box B) that nuclei can be excited into an enormous variety of states. Each state has a different set of quantum numbers, one of which is the spin (a measure of the angular momentum). One method for an excited nucleus to shed excess energy is by the emission of γ rays. It was predicted, and soon found, that the number of γ rays detected following nuclear reactions varied with the angle between the direction of the γ rays and the direction of the particles (see figure 8.2 for instruments developed to measure these angular correlations) such as protons or alpha particles initiating the nuclear reaction. These are the so-called angular distributions of nuclear radiations, and in the fifties the laboratories at Chalk River became well known for pioneering work in this field. The shape of the angular distribution is determined by the spins (and parities[7]) of the initial

[7] Parity in physics is a purely quantum mechanical concept. It refers to the symmetry properties of the wave function under the action of reflection in a mirror, and may be positive or negative.

and final states of the emitting nucleus. Today there is a wealth of information on the excited states of nuclei, much of it stemming from the early work at Chalk River.

The theory of γ-ray angular distributions is quite complicated and in the sixties experimental nuclear physicists had difficulty with it, because of the knowledge of quantum mechanics it required. Fortunately, Chalk River had recently hired several young, enthusiastic theoretical physicists who collaborated with the experimental physicists to revolutionize the subject. The theorists educated the experimentalists in the intricacies of angular distribution theory.

During the mid-fifties, γ-ray angular distributions and more complicated correlations were exploited to measure the spins and other properties of nuclear states in an ever-more sophisticated fashion. Eventually, two special simplified geometrical arrangements, which became widely known as the Litherland geometries I and II,[8] became standard. A theoretical paper was published on this subject and led to multitudinous measurements of nuclear spins around the world. Such measurements were essential precursors to the discovery of collective motion in light nuclei (discussed later) and to, for example, the measurements that showed that the neon observed in stars must be made at much higher temperatures than previously supposed.

This successful fusion of experiment and theory would have been of much lesser importance if, simultaneously at Chalk River, imaginative work in electronics had not also been in progress; multichannel analyzers enabled the complicated γ-ray spectra to be unravelled quickly and the presence of nearby computers, primitive by today's standards, made the rapid analysis of the data possible.

Collective Motion in Light Nuclei

In contrast to heavy nuclei, where the liquid drop model reigned supreme in its increasingly successful description of the collective motion, the nuclear shell model (see Box D), was applied with some success to the lighter nuclei, especially those lighter than oxygen. These two views of nuclear structure were first thought to be mutually exclusive, but with the help of some of the work with the Van de Graaff, it was shown that the two views were complementary.

The work on collective motion of nucleons in light nuclei started in 1954 with the decision to study $^{24}Mg(p,\gamma)^{25}Al$, because it was expected to show a simple γ-ray spectrum suitable for analyzing with one of the all-vacuum-tube, thirty-channel analyzers built by Moody at Chalk River in the late forties. The angular distributions were also expected to be easier to analyze. In fact, the situation proved to be complicated and the γ-ray decay scheme and angular distributions were only obtained after much work. The data were not understood at first, and several abortive attempts to explain it on the basis of the shell model bogged down in impossible complications. Finally, on the way by train to the Washington meeting of the American Physical Society in 1955, to present a paper on the subject, one of the group, while reading Bohr and Mottleson's papers on rotational bands in heavy nuclei, realized that the data for ^{25}Al could all be explained by assuming that three rotational bands were present in the low quantum states of ^{25}Al.

This discovery was followed by a lot of exciting activity, as it was possible to predict the existence and properties of other quantum states in ^{25}Al. For example, the third member of the ground state rotational band was predicted to be a resonance in the nuclear reaction with an angular momentum of four units of $\hbar$ (Planck's Constant). This is a high angular momentum for a 1.2 MeV proton and explains why this quantum state had not been seen previously. It was found after a short search, its energy and other properties having been predicted.

Several years of work on ^{25}Al followed this discovery, and during the work one of the team had the good fortune to visit Bohr, Mottleson and Nilsson in Copenhagen, who explained the latest calculations of Nilsson on the shell model calculations of the quantum states in a spheroidal potential. On his way back to Chalk River in the fall of 1955, this emissary, armed with the galley proofs of Nilsson's paper, complete with corrections in Danish, was able to predict the results of other experiments that

[8] Named after A.E. Litherland.

were later carried out. ^{25}Al is still one of the classic examples of a light nucleus that clearly demonstrates the relevance of the collective model in light nuclei.

The discovery of strong Coulomb excitation (i.e., excitation by the transient electric field of a passing heavy ion), led to the suggestion that heavy nuclei, such as tungsten and tantalum, had large electric quadrupole moments and so were football-shaped. Such nuclei could show rotational bands of states (see Box D) associated with the collective motion of the nucleons in the nucleus. This motion was not the rigid rotation of nucleus originally envisaged by Rutherford, but a new type of collective flow of the nucleons suggested by Bohr and Mottleson, which is closer to the motions of the atoms in a liquid drop.

Coulomb excitation was studied with the EN tandem at Chalk River, after its discovery elsewhere, and the topic of collective motion of the nucleus was actively pursued by researchers using the ion beams from the tandem and using radioactive nuclei prepared by reactor irradiations as sources in β-ray spectrometers.

The Nuclei in the s-d Shell

The Chalk River nuclear physics group has made very important contributions to the understanding of nuclei of the s-d shell (Box D); i.e., nuclei from ^{16}O to ^{40}Ca. The s-d shell nuclei provided crucial data for the development of nuclear-shell-model theory. Here the sophisticated calculations were manageable and realistic, even with many-body systems of considerable complexity. Strong collaboration with Canadian university scientists developed and continues to the present time. About 1970, a very important collaboration resulted in a systematic study of the γ-ray transition rates and α-particle transition probabilities for the decay of the levels in the ground-state band of ^{20}Ne. The results of this important study at Chalk River were consistent with the microscopic approach of the shell-model predictions, and not with the simpler collective-model prediction. These experiments were recognized as crucial to nuclear theory, since the properties of ^{20}Ne could be calculated with precision sufficient to make stringent tests of the shell model.

In the same period, major contributions to the study of the nuclear moments of s-d shell nuclei were made through development in our understanding of the Coulomb reorientation effect and magnetic hyperfine interactions induced when heavy ions collide and separate (see Box F).

The group made substantial contributions to the methods of measuring radiative lifetimes of nuclear levels during the seventies and eighties. These techniques allowed the group, again collaborating with university scientists, to make a systematic study of transition strengths in s-d shell nuclei, and led to a better understanding of the underlying nuclear structure. It also led to systematic measurements of selected transitions in mirror nuclei (the mirror of a nucleus with Z protons and N neutrons has N protons and Z neutrons). These experiments have resulted in the determination of the polarization charges carried

BOX F

When a nucleus spins in a magnetic field, the direction of the axis about which it is spinning can only assume certain (quantized) directions, and each orientation has a different energy, the size of which is larger the larger the magnetic field. A γ ray emitted by such a spinning nucleus will have its energy modified by the energy associated with the orientation. This modification, called hyperfine splitting, will be very small unless the magnetic field is very large. It happens that the field at the nucleus created by the electrons of the atom can be very large indeed. The transient magnetic field at the nucleus created by a passing high-speed heavy ion can also be large, large enough to change the direction of the axis of spin. This is called the Coulomb reorientation effect.

by protons and neutrons in the nucleus, and contributed to the impressive advances in nuclear structure theory that have taken place. These discoveries and subsequent recognition by the international physics and technological communities enhanced AECL's reputation in nuclear endeavours.

Doppler-shift Attenuation Method for Measuring Short Lifetimes

In the Doppler-Shift Attenuation method (DSAM), the lifetime of an excited level emitting a γ ray is compared with the slowing down time (approximately 3×10^{-13} s in our case) of recoiling ions in a solid. The Doppler effect on the frequency or energy of the emitted radiation is used to determine the velocity of the nucleus at the time of its decay.

The first experiment measuring the lifetime of an excited nuclear state by DSAM was carried out at Chalk River in 1948. (See the section entitled "Early Beta and Gamma Spectroscopy".) Later, the power of the method was greatly enhanced when used in conjunction with heavy-ion beams from the tandem accelerators. Later still, another huge leap in sensitivity was obtained through the use of the high-resolution Ge(Li) detectors newly developed at Chalk River, and of the electronic computer signal processing systems discussed above. The method was subsequently refined, improved and applied to the study of nuclei throughout the periodic table. Today, the DSAM technique is used to measure lifetimes as short as 10^{-15} s to as long as 10^{-11} s.

An important advance in DSAM developed at Chalk River was the use of targets made by implanting 30-40 keV helium ions into thin metal foils. The technique became known as the inverse reaction method, because it reversed the previous method of bombarding a heavy target nucleus with a light ion. The much higher momentum of the heavy ion colliding with the helium target meant that higher recoil velocities of the interaction product could be achieved and hence the accuracy of the lifetime measurement could be increased. The inverse reaction method is advantageous in all Doppler-shift lifetime measurements.

The accuracy of DSAM is primarily dependent on our knowledge of the deceleration of the ions in solids. Thus comprehensive systematic experiments on the stopping powers of various materials for many ions were carried out at Chalk River.

The Recoil Distance Method

In the recoil distance method (RDM), the nuclei excited in a reaction recoil freely in a vacuum, with a velocity of a few percent of that of the velocity of light, until they hit a moveable stopper that reduces their velocity to zero in less than one picosecond (10^{-12} s). Because of the Doppler effect, the energies of the γ rays emitted by the stopped and moving nuclei are different. Knowing the velocity of the recoiling ions from the Doppler-shift, the distance between the target and stopper, and the ratio of the stopped to moving recoils, it is possible to obtain lifetimes in the range 10^{-12} s to 10^{-9} s. In 1970, a capacitance method of measuring very small distances between the very flat, thin target foil and the stopper was introduced, extending the technique to the few-picosecond region, and thereby allowing its use with heavier nuclei.

Later, it was discovered that the radiation patterns from highly stripped excited ions recoiling in vacuum were greatly perturbed by magnetic hyperfine interactions between the nucleus and its remaining tightly bound electrons. These effects allowed magnetic moments of excited states with lifetimes accessible to RDM – moments previously inaccessible to measurement – to be determined. An understanding of this phenomenon developed at Chalk River and elsewhere, largely based on experiments at Chalk River, made it possible for Chalk River physicists to exploit the method fully to improve our knowledge of nuclear structure.

DSAM and RDM, both largely pioneered at Chalk River, are now widely used in nuclear structure studies at accelerator laboratories throughout the world. They have given us detailed knowledge of the electromagnetic transitions in nuclei, much of it generated at Chalk River.

Investigations of the Weak Force

Superallowed Beta Decay

Nuclear properties are usually studied so that scientists can understand the nucleus itself. As our understanding and techniques have improved, however, it has been possible to use the nucleus as a laboratory in which the fundamental forces of nature can be probed. Beginning in the early seventies, one such opportunity was recognized at the tandem accelerator, and later at TASCC. This group was to spend more than twenty years studying a particular type of β decay called superallowed in a series of selected nuclei.

The weak force (see Box A) is responsible for beta decay, the process by which an isotope of one element transforms into the isotope of another, via emission of an electron (or positron) and an anti-neutrino (or neutrino). It had been hypothesized in the late fifties that a certain conservation principle (the conservation of vector current – CVC) should apply to the weak force and, over the next decade, increasingly demanding tests of CVC had been pursued. One of these tests was to study the β decay of spin-zero, positive-parity nuclei that decay to analogue states in their daughters; i.e., to states with exactly the same quantum numbers as the parents. Such transitions are called superallowed. Under these very special conditions, the transition strengths are almost a direct measure of the weak force unadulterated by uncertainties in the nuclear structure of the states involved. A series of such measurements in different nuclei can then test CVC conservation with substantial precision.

A unique isotope separator was designed and installed; it was used to prepare pure samples of sub-second radioactivities for precise half-life measurements. Helium-jet and gas-transfer systems were developed for branching-ratio measurements. A new technique using the Q3D[9] magnetic spectrometer was invented for measuring the total energy released in the transition. Finally, a variety of theoretical techniques was applied to the calculation of small correction terms necessary in extracting fundamental information from the measurements. In this way, unprecedented levels of accuracy and precision were attained.

By 1992, CVC conservation had been demonstrated to 3 parts in 10,000. These results are being used to put constraints on the Standard Model (Box G), a model that unifies the description of the weak and electromagnetic forces.

The isotope separator mentioned above has also been used to study "exotic" nuclei; that is, those rather far away from the region of nuclear stability, particularly those approaching the

[9] A large, high-precision, high-transmission magnetic spectrometer, so-called because it consists of one quadrupole magnet followed by three dipole magnets.

BOX G

A comprehensive description of the Standard Model is beyond the scope of this work. For our purposes it is sufficient to say that according to the model, particles such as nucleons are not fundamental; the fundamental particles are quarks and leptons. The nucleons are formed from three quarks held together by gluons, the carriers of the strong force and so named in analogy with the photon, the carrier of the electromagnetic force. The leptons are light, weakly interacting particles (see Box A), of which the electron, the muon and the neutrino are examples. The particles formed of quarks are divided into two categories: the baryons (heavy particles) formed of three quarks (e.g., the nucleons); and mesons (particles of mass intermediate between that of the leptons and the baryons), formed of two quarks (e.g., the pion). In early days the muon was frequently referred to as the μ meson. However today it is known not to be meson but a lepton.

proton drip line, beyond which the nucleus becomes unstable to proton emission. The half-lives and masses have been determined for many such nuclei, the masses both from measuring α-decay energies and decay chains, and by direct measurement of the mass by the separator. This is possible because of the exceptionally high mass resolution of the separator.

Parity Violation in the Nucleon-Nucleon Interaction

Measurements of parity violation in the nucleon-nucleon system have been pursued for many years to determine the role, if any, of the weak interaction. It is characteristic of the weak interaction, unlike the strong, that it violates conservation of parity. Early measurements on a two-nucleon system reported in 1972 by a Russian group on the reaction $n + p \longrightarrow d + \gamma$ indicated a parity-violating component about 200 hundred times greater than the theoretical limit. This discrepancy was considered very significant, since the two-nucleon calculation is simple and straight-forward.

Around 1980, an experiment to measure the same parity-violating component via the inverse reaction, $d + \gamma \longrightarrow n + p$, was begun at Chalk River. An intense beam of polarized electrons[10] was required, as well as a score of high-efficiency, high-speed neutron detectors. Electrons were generated by a gallium arsenide photoemission source and accelerated by the high-current Chalk River Electron Test Accelerator (ETA).

To reach the required sensitivity of 1 in 10^8, it was necessary to count at least 10^{16} neutrons. The source, the accelerator, and the detectors were equal to this monumental task, but the positional stability of the electron beam under electron polarization reversals was not good enough, and only an upper limit that scarcely eliminated the Russian value was reached. The experiment was thus disappointing, but the gallium arsenide source won its developer an Institute of Electrical and Electronics Engineers prize.

Accelerator Mass Spectrometry

Immediately following the 1977 publication of the first reports of the detection of ^{14}C at natural levels with a tandem accelerator, by the groups at Toronto-Rochester and Simon Fraser-McMaster, a team at Chalk River began similar measurements. The first successful detection of ^{14}C at natural concentrations was achieved in January 1978. The use of time-of-flight discrimination through the Q3D spectrometer gave exceptionally clean detection of ^{14}C relative to ^{13}C backgrounds – a serious problem elsewhere at the time. In the period 1978-82 the Chalk River program concentrated on ^{14}C with a highly sensitive system based on two sequential magnetic analyses followed by a velocity filter. During this time, the ages of about 100 different items were determined. The accuracy was better than 5 percent, the age limit 45,000 years. Among the major topics studied were Antarctic meteorites, ground water samples related to the waste management program, mammal bones from western Canada and most notably, the Taber Child. Its geological setting suggested an age of ~35,000 years, which would have made it one of the earliest evidences for human occupation of the central plains of North America. From a 1.6 gram sample, an age of $4{,}100 \pm 750$ years was obtained, showing the skeleton to be post-glacial. The ^{14}C program was suspended in 1982 when the tandem was shut down for installation of the new TASCC facility. The program was resumed in 1988 after the completion of the TASCC installation project. It is presently devoted to ^{36}Cl measurements of interest to the Nuclear Waste Management Program, as well as to hydrology, environmental tracing and glaciology.

OTHER NUCLEAR PHYSICS ACTIVITIES

Theoretical Physics

The theoretical physics group at Chalk River has always been small, less than a dozen physicists, but it has lent invaluable support to the nuclear physics and the condensed matter programs. In the early years it was the backbone of the reactor physics program, and continued to contribute in a smaller way up to 1984. In addition, it provided considerable assistance to other non-physics programs in the laboratory. Because of this,

[10] The polarized electrons produced polarized gamma rays by bremsstrahlung on a tantalum target.

much of its work has been discussed under the particular program; for example, the work on two-photon decay, angular correlations of γ rays, superallowed β decay, β spectrometers, liquid helium, neutron-scattering cross-sections, and others. However, some work does not fit neatly under the programs discussed so far. One example is the early work of A.G.W. Cameron on the primordial formation of nuclei and their level densities. This was a massive empirical study, using for the most part IBM punched card computers, and was for many years the standard reference work on both subjects. Another was the work begun in the early seventies and extending into the eighties on quarks in nuclei. It grew out of work in particle physics, such as that on the weak interactions. Although the non-relativistic quark model had had astounding success in baryon spectroscopy, its ideas had not been incorporated into nuclear models. The research at Chalk River, in collaboration with groups at the Université de Montreal and Julich/Bochum in Germany, were fruitful first steps in understanding the role played by quantum chromodynamics (that is, quarks and gluons) in the strong interaction between nucleons in the nucleus. More recently, Chalk River theorists have made contributions, particularly in the realm of string theory, towards the attainment of the creation of a unified field theory.

Development and Exploitation of Ge(Li) Detectors at Chalk River

For many years, nuclear physicists dreamed of having γ- ray detectors with energy resolution comparable to that obtainable with the internal conversion technique. The introduction of NaI(Tl) scintillation detectors in 1948 was a great step forward. While they had high detection efficiency, the energy resolution was modest. In the late fifties, I.L. Fowler had developed a number of thin solid-state detectors, mainly for heavy-particle detection. Attempts were begun to produce larger (thicker) detectors of germanium crystals by drifting lithium ions into them, to compensate for impurities, and create the equivalent of an impurity-free germanium volume that would act as a solid-state ion chamber.

By June 1963, photopeak resolutions for ^{137}Cs of 1 percent had been measured from a device 15 mm in diameter having a 1 mm thick layer. By September 1963, G.T. Ewan and A. Tavendale reported measurements of γ-ray spectra from an 18 mm diameter, 8 mm deep flat Ge(Li) detector with energy resolutions of 4.5 keV at 122 keV and 6.0 keV at 1333 keV. By June 1964, progress on a fast-drift method for making lithium-drifted germanium detectors was reported, and by September a drift of 13 mm was achieved. A contract was given to RCA Victor Co., in Montreal, to develop these detectors commercially, and by December 7 encapsulated detectors of 5 mm depth had been received from them. Subsequently, work was begun on an open-ended coaxial diode, a method that was a significant step towards higher-efficiency, higher-resolution γ-ray spectrometers. Ewan and Tavendale were awarded a Radiation Industry prize by the American Nuclear Society in 1967.

In the years that followed, other laboratories began constructing Ge(Li) detectors and the shortage of good germanium crystals from Belgium became a limiting factor. RCA's venture as a possible commercial supplier of Ge(Li) detectors was limited by this shortage, as well as by the headquarters decision that all foreign sales of Ge(Li) detectors must be handled through the New York office. Ultimately, a spin-off company from Oak Ridge National Laboratory (United States) called ORTEC, began growing its own germanium detectors, and perfected a method of growing intrinsic germanium crystals in which impurity compensation by drifting in lithium ions was no longer needed. This removed the hardship of having to keep Ge(Li) detectors at liquid nitrogen temperature lest the lithium ions drift out of position and the detector become useless.

Instrument Development at Chalk River

Double-Focussing Pi-Root-Two Beta Spectrometer at Chalk River

First thoughts on an iron-free, pi-root-2, β-ray spectrometer with a planned resolution of 0.015 percent in momentum at a maximum solid angle of 1.5 percent of 4π steradians were articulated in March 1955. By March 1956, the design was complete; three pairs of coaxial coils were specified, the largest being 4 m in diameter.

The theoretical calculations showed that great care in manufacturing the coils would be needed. It was also necessary to eliminate any interference from the earth's and man-made magnetic fields. A suitable site was found and a non-magnetic building designed to accommodate the nine-metre diameter coils that would nullify any stray magnetic fields. The building construction was completed by the summer of 1957, and Canadair, being successful in meeting the required stringent tolerances, delivered the all-aluminum coils and support structure to Chalk River in November 1957. By March 1958, the spectrometer was operational and experiments had begun. The spectrometer performed exactly as predicted, and momentum resolutions as sharp as 0.01 percent in momentum (at reduced transmission) proved possible.

In the decade that followed, many nuclear decay schemes were investigated. Techniques for preparing very thin sources were developed, because the energy loss of electrons leaving the source material was a limiting factor on the sharpness of internal conversion lines, as indeed was the natural width of the lines themselves.

The spectrometer was copied in a number of other laboratories. A 3/4 size version was built in Tokyo at the Institute for Nuclear Studies, and 1/3 size versions at the California Institute of Technology, Idaho Falls, Oak Ridge National Laboratories, NRC in Ottawa and the National Physical Laboratories in Teddington (United Kingdom).

In 1962, a technique for absolute calibration was developed; the energy of the 411 keV external conversion line of gold was measured very precisely against positron annihilation radiation. Thus with the rest mass of the electron as a basis, several other γ-transition energies were determined with high accuracy. These provided the most accurately available γ-ray energy standards for the next decade, and were particularly useful in calibrating Ge(Li) γ detectors in the late sixties and early seventies.

The latest experiment with this spectrometer took place in 1983-86 in a search for the rest mass of the neutrino, following a report of observation of the influence of a heavy neutrino with a mass of 17 keV in the β spectrum of tritium. The β spectrum of ^{63}Ni, another low-energy β emitter, was examined, but the expected kink from the heavy neutrino failed to appear. The question of whether there is real evidence for 17 keV heavy neutrinos remained a hot topic until 1992. The evidence is now against the existence of a 17 keV neutrino.

Fast-Slow Coincidence Technique

A new concept for measuring very short (for those days) nuclear lifetimes occurred to R.E. Bell about 1948. The technique, which became known as the fast-slow coincidence technique, separated the measurement of time from that of pulse size and selected only those events that corresponded to the desired energies. Bell used two detectors of fast organic phosphors coupled to the new photomultipliers, and a new tube developed for television to deliver the resulting pulses to a system of cables that gave an output only when the two pulses were within two nanoseconds of each other. The time spectrum was scanned by varying the lengths of the cables and the successful pulses were passed to one input of a slow triple-coincidence circuit. Thus an output occurred only when fast-time selection and the two slow pulse height selection limits were satisfied.

In 1949, the end-to-end β-spectrometer, described earlier, was equipped with two stilbene detectors. The first use of this technique showed that the 83 keV β-ray transition in ^{170}Yb had a half-life of 1.60 nanoseconds. Many nuclear lifetime measurements followed using sources of ^{198}Au ^{181}Hf, ^{60}Co, and ^{239}Np, and much new information was obtained on the decay schemes of these nuclei.

By the early fifties, the fission- and β-ray groups replaced single-point time selection with time-sorting by using time-to-amplitude converters (TAC). This technique was widely copied and is now in standard use in most nuclear physics laboratories throughout the world using off-the-shelf electronics.

On-line Data Handling at the Accelerators

In 1962, Chalk River was one of the first laboratories to purchase a digital computer for on-line data acquisition. The computer, a Digital Equipment Corporation PDP-1, was designed to operate with many types of input-output devices. Specialized interfaces were designed and built at Chalk River to store single-parameter spectra directly in the memory of PDP-1 and to acquire, monitor and output multiparameter data to magnetic tape. Because of its versatility and reliability, the PDP-1 computer served the needs of tandem-based experiments for nearly twenty years.

One of the first time-sharing (multi-user) computers, a DEC PDP-10, was purchased in the late sixties. Once again, specialized hardware was developed for data acquisition/analysis. However, this computer was used primarily for off-line analysis, because the time-sharing operating system could not be optimized for real-time applications.

In 1983, when the MP Tandem was shut down for the installation of the superconducting cyclotron, the PDP-1 and the PDP-10 were replaced with a CAMAC (Computer Automated Measurement and Control) based data acquisition system interfaced to a Perkin-Elmer model 3230, 32-bit minicomputer. This computer had a speed twenty to fifty times greater than that of the PDP-1, and an on-line storage capacity 160,000 times greater. Subsequently, a second, even faster, computer with considerably greater storage capacity was added. Software developed at Oak Ridge was acquired and adapted to meet local needs. Specialized CAMAC hardware was developed in-house, and front-end microprocessors were obtained to control and acquire data from complex experimental devices, such as the 8π spectrometer (figure 8.2,3), a $5 million apparatus with 92 independent detectors and data rates of up to 400 kilobytes/s.

These superb data-handling facilities, for many years unique in the field, were crucial in creating and maintaining the reputation for excellence enjoyed by Chalk River in the nuclear physics community, and for establishing the many firsts attained by the tandem physicists.

Cosmic-Ray Neutron Monitors

In 1957, Hugh Carmichael and Ivan (Dick) Fowler developed some very large BF_3 proportional counters, 15 cm in diameter by 180 cm long. An array of several of these counters was mounted in paraffin wax in Deep River in 1964, where, with several subsequent modifications, they operated until 1995, producing a continuous hour-by-hour record of the atmospheric neutron flux that results primarily from cosmic rays originating in the sun. Solar flares create a rapid increase in particles emitted from the sun and are responsible for the magnetic storms that frequently wreak havoc with radio communications and with large electric power networks. The cosmic rays reach the earth before the magnetic disturbances, and hence the Deep River neutron monitor provides an early warning of such storms. Consequently, the results were relayed to NASA and to Strategic Air Command headquarters in Colorado until 1972. These detectors were subsequently manufactured in Canada and shipped to about a dozen monitoring stations around the globe in time for the International Year of the Quiet Sun (1964-65). Many of these are still operating.

Large Arrays of Gamma Detectors

Gamma spectroscopy has benefitted greatly from the development at Chalk River of large arrays of detectors. The first was the Ferguson Goniometer in 1959, originally referred to as "the heavy angular distribution table" (see figure 8.2,1). It consisted of a large NaI(Tl) detector (12 cm diam by 12 cm), moveable in a horizontal plane, and a second similar detector moveable in a vertical plane on an arm that could itself rotate about a vertical axis. It occupied an entire small room. The second was the Icosahedron, a device that could locate 20 NaI(Tl) detectors at fixed points in the icosahedron. The Icosahedron was little used, but served as a prototype for the Lotus (1966) (figure 8.2,2), an elaboration of the Ferguson Goniometer, which could locate seven large NaI(Tl) detectors at arbitrary polar and azimuthal angles, limited only by the constraints imposed by the massive arms necessary to position the heavy detectors accurately. In a modified form, the Lotus still exists today, but was superseded in 1986 by a much more powerful spectrometer called the 8π (see figure 8.2,3).

Other Innovative Technical Developments

Many other innovative devices and techniques were developed at Chalk River. One of these was the Double Flat-Crystal Transmission Spectrometer for γ rays (1957-79)(figure 8.3,2). This ingenious device was developed to the stage of making

FIGURE 8.2
BETA, GAMMA, AND FISSION SPECTROMETERS AT CRL

1. The large Goniometer with two experimenters
2. The Lotus with its creator

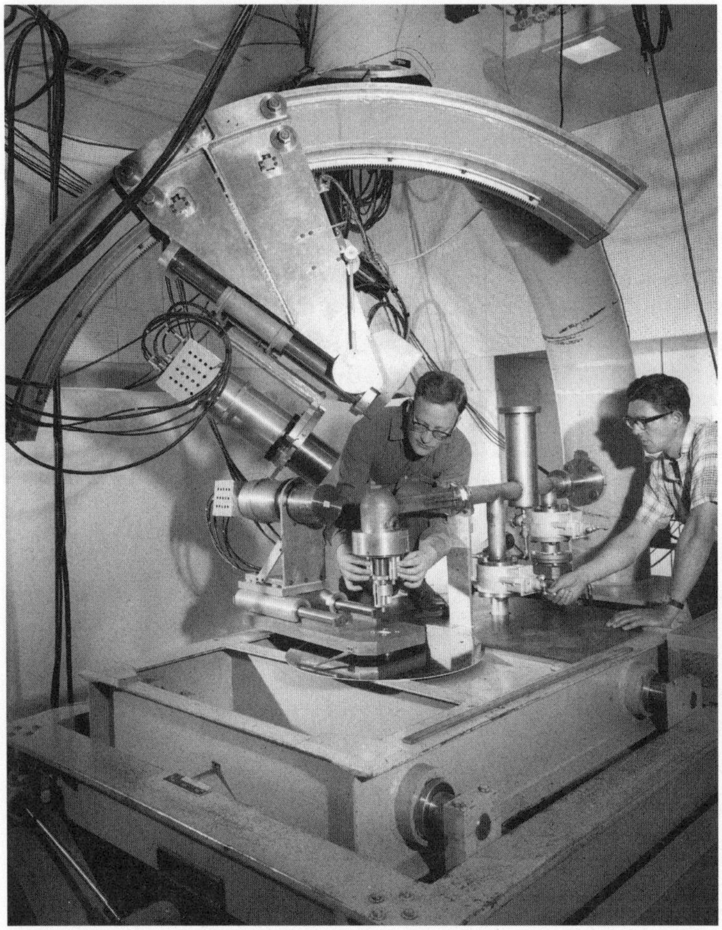

1

2

3

4

5

3. The 8 Spectrometer with the proposers and users of this exceptional instrument.
4. John Robson examines the apparatus, at NRX, with which he measured the angular correlation between the β particle and the ν in the decay of the neutron.
5. The fission group makes some adjustments to the fission-fragment fission-neutron time-of-flight apparatus at NRU.

very precise measurements of the wavelength of the 511 keV annihilation radiation and of the 411 keV γ ray in ^{198}Au. Another was the Compton Monochromator (1966-76), which made use of a beam of high-energy γ rays from neutron capture in ^{59}Ni falling on an aluminum crystal to produce monochromatic γ rays with energies between 2 and 8.3 MeV. These were used to make very high resolution measurements of the photofission cross-sections of ^{233}U, ^{235}U, and ^{239}Pu. The measurements were subsequently refined even further using a Bremsstrahlung Monochromator (1974-83), installed first at the electron linear accelerator at the University of Toronto and later at the University of Illinois.

Radioactive Standards Laboratory

From the earliest days, it was necessary to have standards against which the strengths of α-, β-, and γ-emitting nuclides could be measured, and a group was set up at Chalk River to provide them. The provision of standards is a legal responsibility of NRC, of which at that time Chalk River was a part. In 1952, when AECL was formed, the responsibility for producing radioactive standards remained largely at Chalk River and remains there to this day, although a standards group was set up at NRC in Ottawa. A radioactive standard is somewhat different from other standards, since a given source decreases more or less rapidly with time. Thus it is necessary to know at least three things: the source's radioactive half-life, its emission rate, and the time at which it was measured. It is the job of a radiometrology group to produce radioactive sources for which these three things are accurately known, yet the only one that is easy to measure is the time of the measurement. The difficulties with β emitters are particularly acute. The Chalk River group realized early that it was necessary to do absolute β counting. This they did by developing a 4π counter and several ingenious techniques based on it. All standardization is a comparison; however, in the case of radioactivity there is no ultimate standard, such as a platinum-iridium kilogram in Paris or a cesium clock. This difficulty is "solved" by a round-robin of international comparisons in which a proposed standard is measured by each of about twenty national standards laboratories. It is a matter of tremendous pride that Chalk River has consistently shone in these comparisons, although many nations have much larger radiometrology groups.

Condensed Matter Physics

INTRODUCTION

According to the duality principle in quantum mechanics, particles exhibit wave-like qualities, and waves exhibit particle-like qualities. Thus the neutron, normally considered a particle, has wave-like properties. It so happens that the wavelength of a neutron at thermal energies is comparable to the spacing of atoms in a crystal. This fortunate property means that neutrons can be diffracted by the regularly spaced atoms in a crystal, just as light is diffracted by the regular lines on a grating. Such diffraction studies either by neutrons or X rays give information on the position of atoms. In certain cases, neutrons have considerable advantage over X rays: they can "see" light atoms; they can differentiate between atoms of nearly equal atomic number; they can, with a few exceptions, penetrate deeply into matter; and, very importantly, they are sensitive to the magnetic properties of atoms.[11] However, most importantly, the fact that the neutron energies are thermal means that their energies are comparable to the excitation energies of the crystal. Thermal neutrons are thus particularly useful for studying these excitations, and this gives them a large advantage over X rays.

Condensed Matter Physics at NRX

Exactly the same considerations as those at the beginning of the nuclear physics section apply to studies of bulk matter. NRX, with its high flux of thermal neutrons, was ideal for the exploitation of neutron scattering techniques, which had been initiated at the Argonne (ANL) and Oak Ridge (ORNL) National

[11] The magnetic interaction of light waves is extremely weak so that only with the most intense laser beams can the magnetic properties of matter be investigated.

Laboratories in the United States, and it was clear that the field was simply waiting for Chalk River scientists to make exciting discoveries – which they did. By 1950, the use of beams of thermal neutrons from NRX to study the structure of condensed matter (solids and liquids) had been established (figure 8.3,1). Initially the program concentrated on gases, but by the end of this period, the emphasis had shifted, with tremendous success and far-reaching significance, to condensed matter.

Thermal neutrons from the core of a reactor emerge with a wide range of energies. For useful experiments it is necessary to have a neutron beam with a well-defined energy impinge on the sample under study. This can be done with filters, choppers, or by diffraction from single crystals. If changes in the neutron energy must also be measured (e.g., in inelastic scattering), similar measurements must be made on the scattered neutron.

The early workers undertook the development of spectrometers that exploited each of these techniques. One requirement was the ability to move heavy shielding with precision; the use of anti-aircraft gun mounts proved to be a more acceptable solution than the earlier one of physicists with crowbars. Large, fist-sized, single crystals were desirable; by 1954, suitable crystals of several metals had been grown. Large, low-noise,[12] reliable, boron trifluoride detectors were needed and available on-site. Appropriate electronic (vacuum tube) instruments were also developed on-site, as few were available commercially.

From the conception of the program, certainly as early as 1949, experiments on liquid helium were envisaged. By 1952, a Collins liquifier was installed not far from the reactor face at NRX, and a container suitable for holding cryogenic liquids for many hours, even days, in a neutron beam, was built.

The early instruments provided a monochromatic beam; i.e., neutrons of a single wavelength or energy, suitable for diffraction experiments. Results on urea and ammonium chloride were published in 1951 as a proof of the technique, while on a more practical level, studies on the texture of uranium and rolled aluminum commenced. The first experiments on cryogenic liquids were undertaken in 1952, with preliminary data on helium, argon, nitrogen and oxygen. After NRX restarted in 1954, this program was expanded and more detailed and accurate results obtained; those for liquid helium in particular attracted wide international attention.

These experiments on structure gave qualitatively the same type of information as that from X rays, although exploiting the advantages of neutrons over X rays. But neutrons could do much more, a fact quickly realized by physicists everywhere. It is well known that the energy changes that take place on scattering reflect the motions of the atoms; the trick was to find a feasible way to measure these changes. As early as 1951, B.N. Brockhouse and his colleagues were developing methods for doing this, and over the decade demonstrated them in a remarkable series of experiments. Initially, Brockhouse and D.G. Hurst used the known dependence on energy of the high neutron absorption of certain elements to obtain information on neutron energy transfer. The results were interesting, but could not reveal the exciting details they knew were contained there; these revelations had to await the restored NRX and the development of new instruments, notably the triple-axis spectrometer (see figures 8.3,3 and 8.4).

Triple-axis Spectrometer

The neutron scattering experiments described in the previous section gave qualitatively the same information as could be obtained with X rays; i.e., the *positions* of atoms from measurements of the changes in direction (and hence momentum) of the scattered neutron. But to exploit the ability of neutrons to measure the *motions* of atoms, it was necessary to find a feasible way of measuring the changes in neutron energy that take place on scattering. By the mid-fifties, Brockhouse was developing several methods for doing this, and over the next few years, he demonstrated them in a remarkable series of experiments.

Competition was strong with Brookhaven (United States), Harwell (United Kingdom) and Saclay (France), among others,

[12] That is, free of the radioactive contaminants and unwanted electrical breakdown that give rise to spurious pulses.

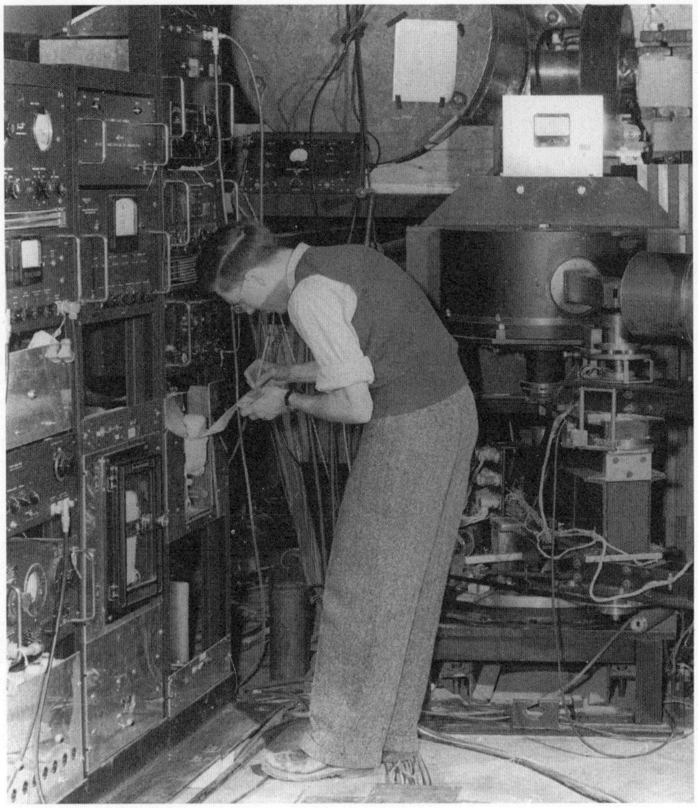

1

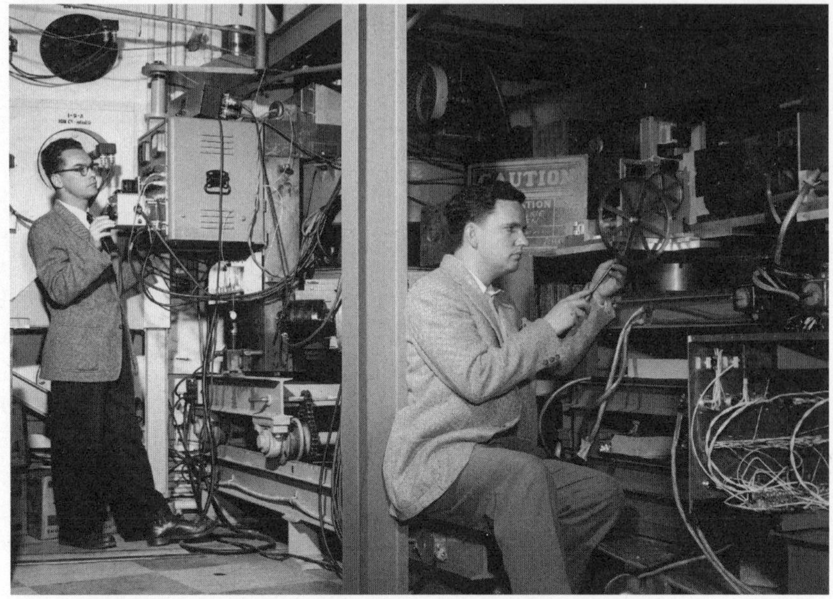

2

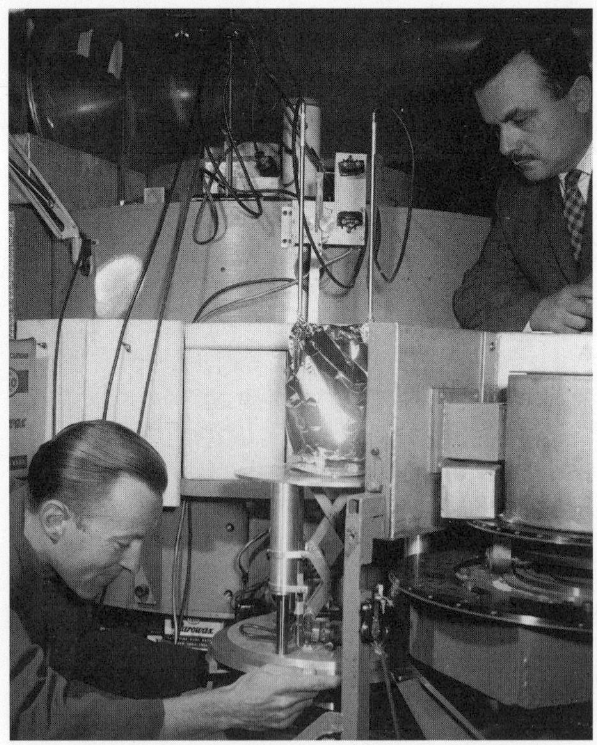

3

FIGURE 8.3
NEUTRON SPECTROMETERS AT NRX

1. Researcher Norm Alcock examines some output from the first neutron diffraction spectrometer (designed by Don Hurst) at the NRX reactor (circa 1947).
2. On the left, a member of the capture gamma-ray group at one face of the NRU reactor with the pair spectrometer designed and built by the group and used to measure neutron capture gamma rays. Another staff member adjusts the positioning device on the flat crystal spectrometer used to make exceedingly precise measurements of gamma-ray energies.
3. Technologist Ed Glaser makes adjustments to the system he designed to move all parts of the Constant Q Triple Axis Spectrometer in the way required to achieve constant Q motion. Bert Brockhouse looks on.

FIGURE 8.4
Schematic diagram of the first constant Q triple axis spectrometer which was mounted at NRU in 1958. M_1 is a neutron filter, X_1 is the monochromating crystal, S is the sample under study and X_2 is the analyzing crystal.

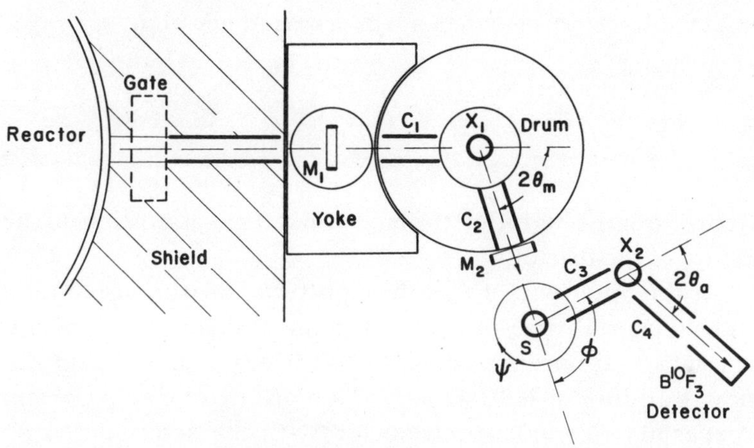

FIGURE 8.5
The first photon dispersion curves ever measured. They were obtained for aluminum at the NRX reactor and described in a letter to the editor – Physical Review 100 (1955) 756 by B.N. Brockhouse and A.T. Stewart.

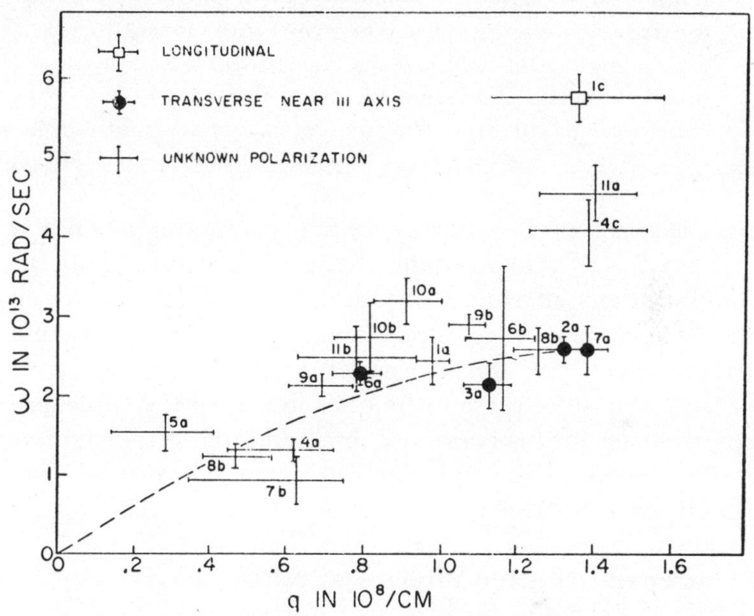

all striving to establish firsts. That Brockhouse achieved so much so quickly was the result of a synergy of his instinct for good physics, his foresight in choice of technique, the local skills that produced large single crystals and state-of-the-art neutron detectors, an excellent technical staff, the superb reactor facilities and support services at Chalk River, and of course, a lot of hard work.

Brockhouse's experiments attracted international attention and, along with later achievements, led to various honours, including the Oliver E. Buckley Prize of the American Physical Society, and Fellowship in the Royal Society of London, and later, the Canadian Association of Physicists' Medal for Achievement in Physics, the Order of Canada, and, as a final crowning glory, the joint award of the Nobel Prize for Physics in 1994 (see below). The outstanding achievement of this period was the first measurement of the energy vs. momentum (or the frequency/wavelength) relationship for lattice waves in crystals, first for aluminum (see figure 8.5) and later for germanium. Closely related but different measurements on vanadium and water resulted in other highlights from this period.

As stated previously, such experiments require knowledge of the direction and energy of the neutron before and after scattering. Determining the beam direction is straightforward, and there are several practical ways to determine energy. In the time-of-flight method, rotating choppers[13] (at least two) placed in the beam provide sharp pulses of neutrons with a well-defined speed, which after scattering are timed over a known distance. In

[13] In this method, a beam of neutrons emerging from the reactor is chopped in segments, whose length in time is short compared to the time between segments. The chopper is a wheel or drum made of some material opaque to neutrons, with a hole or slot through it, to allow neutrons to pass when the hole is lined up along the beam. The neutrons in the bunches have different velocities and hence different flight times over a long (a few metres) path between the chopper and the sample, and between the sample and the detectors.

BOX H

A triple-axis spectrometer (figures 8.3,3 and 8.4) is a device that produces monoenergetic neutrons from the broad spectrum of thermal neutrons emanating from a reactor, and then measures the energy of these neutrons with respect to the scattering angle after they have interacted with a sample. It performs these tasks with the aid of three crystals mounted over three axes of rotation. The first crystal selects neutrons in a narrow band of energies by means of Bragg scattering (see Box I). The second is the sample under study; it need not be a crystal, but to exploit the full power of the instrument, it usually is. The third selects monoenergetic scattered neutrons, again by Bragg scattering. In this case the energy is determined by the angle of the neutron detector.

a triple-axis crystal spectrometer, Bragg scattering (see Box H and Box I) is used to determine the speed of neutrons both before and after scattering. A third method uses filters to transmit neutrons of some energies preferentially; it is less precise but has often been used. Brockhouse developed variations and combinations of all of these techniques, but used the triple-axis spectrometer for his best work, including the very extensive measurements on lead and the alkali halides begun at NRX but completed at NRU.

Condensed Matter Physics at NRU

With the restart of NRU in late 1958, Brockhouse's triple-axis spectrometer was ready to operate there with a beam of neutrons from one of the high-flux through-tubes that passed close to the centre of the reactor.

At the other end of this through-tube, Brockhouse built a hybrid crystal/time-of-flight instrument called the rotating-crystal spectrometer, a device he had tried out at NRX and was now building at NRU as part of a more extensive program. A special triple-axis spectrometer (with characteristics especially chosen for experiments on liquid helium) was being built at one of the very large (30 cm diameter) holes in the NRU thermal column. A third neutron scattering facility at NRU was a high-precision multi-chopper time-of-flight spectrometer being built under the Canada-United Kingdom cooperation agreement with Harwell, primarily for studies of reactor-moderator materials.

BOX I

The scattered intensity of a beam striking a crystal is greatly enhanced when the distance along the beam direction from one plane of atoms to the next is equal to the wavelength of the beam, and when the angle of incidence is equal to the angle of scattering. Such enhanced scattering is called Bragg scattering. It can occur for different angles of incidence when the distance is 2, 3, 4 and so on, wavelengths, and from differing planes of atoms in a complex crystal, giving rise to a number of Bragg peaks. The phenomenon results from constructive interference of the incident and scattered waves. Bragg scattering from crystals of low neutron absorption, such as aluminum or silicon, can be very efficient. The width of the Bragg peak depends on the perfection of the crystal, the more regular or impurity-free the crystal, the narrower (and higher) the Bragg peak. Thus a Bragg peak from a good silicon crystal can have a very narrow spread in neutron energies (0.1 percent is routinely available).

By the mid-seventies, the neutron scattering instrumentation at NRU had undergone significant changes. Three new computer-controlled triple-axis spectrometers were operated by Chalk River staff, often with the collaboration of visitors from Canadian and foreign universities. Both the universities of McMaster and Guelph had their own spectrometers at NRU.

The advance that made the triple-axis spectrometer so powerful for studying lattice vibrations in solids was the invention by Brockhouse in 1959 of the Constant-Momentum-Transfer (or Constant Q) method of making measurements. For accurate interpretation, these studies require the measurement of the distribution of scattered-neutron energies at single, preselected values of the neutron momentum transfer. However, kinematic considerations easily show that if the energy transfer varies, then the momentum transfer must also vary. Brockhouse realized that one could change various spectrometer angles (for example, the crystal orientation and scattering angle) in such a way that the magnitude and directions of the momentum transfer would remain the same at each point of the energy distribution.

The practical realization of this idea was first achieved in experiments on lead with banks of cleverly constructed manual switches; the results helped make Chalk River the undisputed leader in this field. At an international conference where this technique was reported, it was jokingly described as a triumph of experiment over mathematics. Crystal spectrometers operated in this way collect data at a much slower rate than pulsed-beam, time-of-flight spectrometers, but the latter measure complete time distributions in many detectors simultaneously. By contrast, the triple-axis spectrometer has usually been restricted to point-by-point measurement in a single detector. The advantages of the triple-axis spectrometer are two: its specificity and its flexibility. First, its specificity: the time-of-flight method is a scatter-gun approach, in which the interesting data have to be gleaned from a large body of data, whereas the triple-axis spectrometer can be arranged to measure exactly the quantities of interest.[14] Second, its flexibility: to be efficient, time-of-flight requires large banks of detectors, which are not easy to modify or relocate. Time-of-flight is well suited to liquids, gases and amorphous solids, whereas the triple-axis spectrometer is the method of choice for crystalline materials. The big successes of the sixties were measurements on lead, sodium iodide, liquid helium, the magnetic materials magnetite and cobalt, and the ferroelectric material strontium titanate.

The extraordinary importance for condensed matter science of these two developments, the triple-axis crystal spectrometer and the Method of Constant Q, can scarcely be overestimated. Their importance, along with all the pioneering neutron-scattering research performed with them, was recognized in 1994 with the award of the Nobel Price for Physics to Dr. Brockhouse and to an American colleague, Dr. C.G. Shull, who was honoured for his pioneering work in neutron diffraction (elastic scattering).

In lead, detailed measurements were made of the relationship between the energy and momentum for the lattice vibrations travelling in different directions in the crystal (the so-called phonon[15] dispersion curves). Analysis of these curves showed that the forces holding the lead atoms in the crystal were of extremely long range. These measurements were sufficiently precise to demonstrate a direct influence of the electrons in the metal on the lattice vibrations. This influence had been predicted on theoretical grounds (the so-called Kohn effect), but this was its first reported observation. Measurements at higher temperatures showed that the forces between the lead atoms were apparently temperature-dependent, and for the first time demonstrated that the lifetime of the phonons gets shorter as the temperature increases.

In sodium iodide, an alkali halide crystal similar in structure to common salt, the measurements were the first for a crystal with more than one kind of atom, and yielded results that had extensive implications for both theory and experiment.

---

[14] Modern high-speed, low-cost electronics might be able to overcome this disadvantage of the time-of-flight method.

[15] The atomic vibrations of a crystal lattice may be quantized. The quantum of these excitations is called a phonon, in analogy with the photon of electromagnetic radiation.

In liquid helium, the most accurate and extensive measurements to date, and also the first at elevated pressures, were made with the new spectrometer. In magnetite, the measurements resulted in the first-ever direct observation of spin waves (i.e., magnetic disturbances that propagate in the crystal in a wave-like manner) in any magnetic material, and in cobalt the first in a metal.

The lead, sodium iodide, liquid helium and magnetite/cobalt measurements were followed by extensive programs at NRU on both simpler and more complicated metals, insulators and molecular crystals of increasing complexity, liquid helium under a wide range of temperatures and pressures and, later in the sixties, an extensive program on magnetic materials.

The metals program ranged from measurements on sodium (figure 8.6), perhaps the simplest of all metals from a theoretical standpoint, to complicated alloys such as brass (a mixture of copper and zinc) and alloys of niobium and molybdenum. Measurements on pure niobium and molybdenum reflected the diverse superconducting characteristics of these metals and spawned considerable international activity on the underlying theory.

The insulators and molecular crystals program included extensive measurements on strontium titanate and provided a dramatic demonstration of the connection between ferroelectricity and the extreme temperature dependence of the frequencies of certain types of atomic vibration. This program later expanded to include complicated semiconductors, such as gallium arsenide, and complex molecular crystals such as hexamethylenetetramine, some of which had to be grown at the temperature of liquid nitrogen.

The liquid helium measurements continued under increasingly better experimental conditions. In the thirties, it was postulated that when liquid helium was superfluid, at temperatures less than about two degrees above absolute zero, about 10 percent of the atoms essentially did not move (the so-called zero momentum state). New methods of analysis, combined with high-precision neutron scattering measurements, gave the first convincing direct experimental evidence that this theoretical concept was correct. Other results were completely unexpected; for example, at large scattering angles the width of the scattered peak did not increase monotonically with the transferred momentum, but oscillated. Furthermore, below the critical point

FIGURE 8.6
The first accurate determination of the phonon frequency distribution from neutron scattering. (A.D.B. Woods, B.N. Brockhouse, A.T. Stewart, and R. Bowers; Phys. Rev. 128 (1962) 1112)

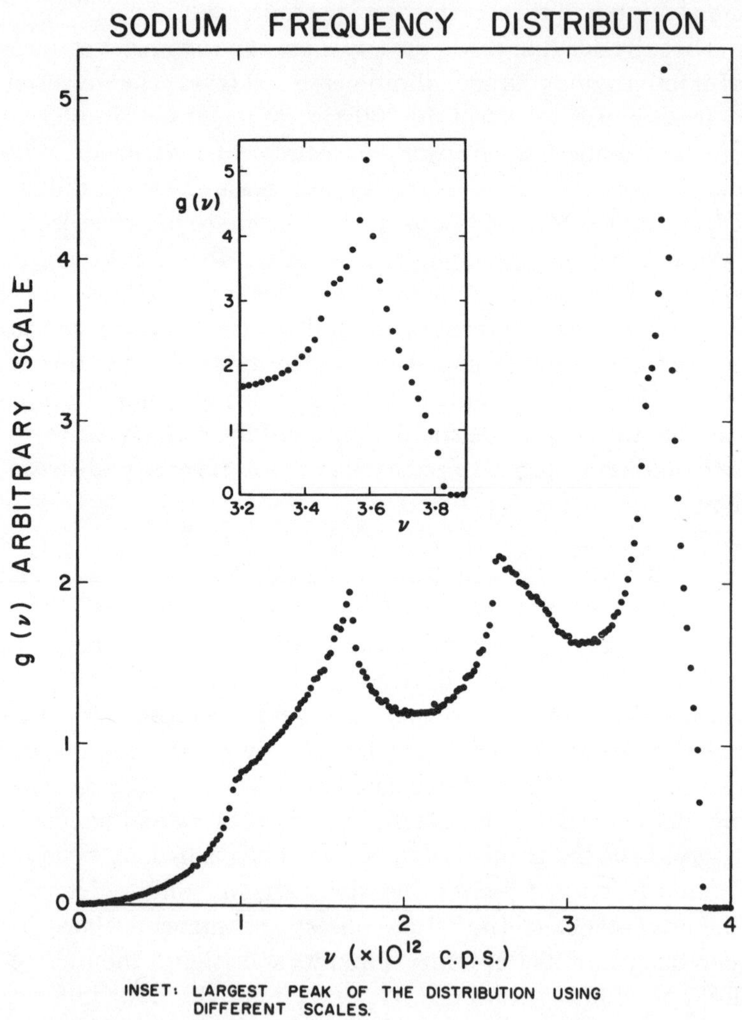

(the lambda point), the width did not increase with increasing momentum transfer, but decreased.

The program on magnetic materials covered measurements on crystals ranging from rare earth metals, such as erbium and holmium, to antiferromagnetic salts, such as potassium manganese fluoride. Also included were mixtures of materials that allowed the effects of the variations of certain magnetic parameters to be investigated.

Other notable work included experiments on water, liquid lead, liquid tin, liquid neon, many exotic magnetic and nonmagnetic crystals, and biological materials. Since the sixties, collaboration with Canadian and foreign universities has been steadily increasing, and has now been formalized to the point where a joint organization, the Canadian Institute for Neutron Scattering (CINS), has been established. CINS has obtained grants from the Natural Science and Engineering Research Council that have been used for major equipment and its installation at NRU, e.g., DUALSPEC, a triple-axis and diffraction spectrometer combined. CINS is active in promoting new neutron scattering facilities in Canada.

The eighties saw the development of a technique to measure strains in materials caused in the manufacture of components (e.g., by welding or forming). Materials tend to corrode or fail in regions of high strain, and this technique, commercialized as Applied Neutron Diffraction for Industry (ANDI), has proved to be valuable in improving manufacturing processes in items as diverse as heat-exchanger tubes and Saturn rockets.[16] The development of ANDI is a fascinating story, but occurred almost entirely after the cut-off date for this history.

The neutron scattering experimental program has been supported throughout by a small but active theory effort. Before the early sixties, the mathematical analysis needed for the interpretation of the experiments on liquid lead was created. In 1959, a model, the Cochran "shell model", was developed for the interpretation of the sodium iodide and related experiments. This model subsequently led to a new theory to describe the onset of ferroelectricity. Significant contributions were made to the theory describing the effects of increasing temperature on vibrating atoms in crystals. Since 1965, theoretical achievements varying from new, powerful ways to analyze experimental results to a recently published book, entitled Neutron Optics, have been made. Some of this analysis was crucial to the interpretation of liquid helium measurements, while the treatise on neutron optics has been recognized as a comprehensive review indispensable for people working in this field.

Positrons at Chalk River

Chalk River scientists were responsible for the first two experiments to show that positron annihilation could be a useful tool in solid-state research. In the first experiment, R.E. Bell and R.L. Graham, using a new high-resolution electronic timing circuit developed for nuclear physics, discovered that some positrons in certain solids have a lifetime much longer than the rest. These long-lived positrons were found only in insulating solids such as plastics or fused quartz, which had vacant lattice sites. They correctly postulated that some positrons formed positronium[17] in these little cavities, and that orthopositronium, because it has one unit of angular momentum, cannot self-annihilate quickly into two γ rays, as can parapositronium with spin 0. It lives much longer, eventually annihilating with an electron picked off another molecule. This postulate was confirmed a year or so later in an experiment that measured the three γ-annihilation rates in these same materials. The rate confirmed the existence of orthopositronium. An interesting problem arose in these three γ-coincidence experiments: the coincidence rate for the three counters in the plane of the specimen – which is the condition to detect the three-photon decay of orthopositronium – was less than the coincidence rate observed when one counter was lifted out of the plane to record the background rate. It was some time before these scientists realized that the gain of a photomultiplier tube can

[16] A full-sized, 3.6 m-diameter section of the rocket was mounted at a spectrometer in NRU for this work.

[17] Positronium is like a hydrogen atom with a positive electron replacing the proton.

change slightly when that tube is tilted in the earth's magnetic field!

The other experiment was the careful measurement of angular correlation of the two γ rays from the electron positron annihilation. If the electron and positron are not moving, the angle must be exactly 180°. The slight departure from 180° that is observed is a direct measure of the speed of the electron. The first experiment illustrating that this method could be used to observe electron velocities was done by a Chalk River physicist on leave at Dalhousie University following the NRX accident. Following this preliminary work, many other experiments with much better resolution were done at Chalk River, proving the usefulness of this new technique for solid-state physics. Others had previously attempted these measurements but failed, either from inadequate angular resolution or an inadequate time scale. In the first one, the time scale was too short to see the long lifetime, and in the second, the angular resolution was too poor to see any structure in the velocity spectrum of electrons in solids.

These two experimental techniques, pioneered and first used successfully at Chalk River, have now spread around the world. Hundreds of scientists are using them in studies ranging from many-body effects in electron gases to the determination of defects near the surface of implanted semi-conductor materials.

ACKNOWLEDGMENTS

Essential assistance in writing this chapter was provided by T.K. Alexander, G.C. Ball, G.A. Bartholomew, R.L. Clarke, E.D. Earle, J.S. Fraser, R.L. Graham, J.C. Hardy, J.W. Knowles, A.E. Litherland, A.T. Stewart, J.G.V. Taylor, and A.D.B. Woods.

BIBLIOGRAPHY

Bartholomew, G.A., G. Dolling, M. Harvey, and J.C.D. Milton. *AECL Programs in Basic Physics Research*. AECL-7073. 1982.

Bartholomew, G.A., and J.S. Fraser. *AECL Programs in Advanced Systems Research*. AECL-7074. 1981.

Brockhouse, B.N., M.A. Clark, J.W. Knowles, J.C.D. Milton, and R.N. Sinclair. "Physics Research at the Chalk River Reactors." In *Pile Neutron Research in Physics: Proceedings of the Symposium on Pile Neutron Research in Physics, Vienna 17-21 October 1960*, 25-48. Vienna: International Atomic Energy Agency, 1962. (Also AECL-1103.)

Brockhouse, B.N., A.D.B. Woods, G. Dolling, and I.M. Thorson. "Research with Inelastic Neutron Scattering at the NRU Reactor." In *Proceedings of the Third International Conference on the Peaceful Uses of Atomic Energy, Geneva, 31 August-9 September 1964*, vol. 7, 419-29. Paper A/CONF.28/P/12. Geneva: United Nations, 1965. (Also AECL-2012.)

Hanna, G.C. "The Origin of the Chalk River Van de Graaff Accelerator." *Physics in Canada* 50 (1994): 219-22.

Laurence, G.C. *Early Years of Nuclear Energy Research in Canada*. Chalk River: Atomic Energy of Canada Limited, 1980.

Milton, J.C.D. *After Dinner Talk to the International Nuclear Target Development Society, 18 September 1986*. Unpublished report CRNL-4178. Chalk River: Atomic Energy of Canada Limited, 1987.

Preston, M.A. *Physics of the Nucleus*. Reading, Mass.: Addison-Wesley, 1962.

Research Facilities at CRNL. 2d ed. AECL-3555. 1970.

Sargent, B.W. "Nuclear Physics in Canada." In *History and Philosophy of Science*, Montreal: McGill University, 1979.

Sargent, B.W. "Nuclear Physics in Canada in the 1930s." *In Otto Hahn and the Rise of Nuclear Physics*, 221-40. Edited by William R. Shea. Dordrecht, Holland: D. Reidel, 1983.

Sears, V.F. "Fifty Years of Theoretical Physics at AECL: 1943-1993." *Physics in Canada* 49 (1993): 209-15. (Also AECL-10930.)

Woods, A.D.B. *Research and Development in the AECL Research Company*. AECL-9708. 1988.

Chapter Nine

Accelerators and Fusion

G.C. HANNA

INTRODUCTION

Unlike chapters seven and eight in this section, chapter nine is concerned with applied research. This research is included in the section on Basic Research because it lay outside the mainstream of CANDU development and because much of it was carried out by scientists from the basic research area. It was concerned, primarily, with the large-scale production of neutrons by means other than nuclear fission, i.e., from high-energy nuclear reactions using charged-particle accelerators or from controlled thermonuclear fusion.

Two applications for accelerator-produced neutrons were explored. The first, electronuclear breeding, and the second, a high-flux research facility, are described in the next two sections of this chapter. The accelerator work led to extensive activities in other applications of accelerators which are described in the third section. The fourth section gives a brief account of fusion and AECL's activities in this field, and the chapter concludes with an overall assessment of the programs.

ELECTRONUCLEAR BREEDING

The concept of electronuclear breeding arose out of W.B. Lewis's pioneering analyses of how nuclear power could best contribute to global energy requirements. In a series of reports written between August 1951 and May 1952, he developed the idea that the Chalk River approach to atomic power should be the long irradiation of natural uranium. His economic analysis indicated that it could compete with coal for the generation of electricity, provided that the uranium fuel could be irradiated long enough to produce a thermal energy output of 3,000 megawatt days (MWd) per tonne before having to be removed from the reactor. This had already been achieved in NRX; CANDU reactors currently yield about 7,000 MWd per tonne.

However, if *all* the natural uranium could be fissioned it would yield almost a million MWd per tonne. In principle, a large fraction of the uranium can be fissioned through the use of fast neutron reactors to "breed" the fissile material plutonium-239 from the "fertile" uranium-238 and produce power at the same time (see chapter one and chapter eighteen). Similarly, thorium-232 can be fissioned by breeding uranium-233, and this is theoretically possible in thermal neutron reactors.

The third report of the series (DR-22) states that "it is appreciated that the large-scale exploitation of atomic power involves the operation of "power breeder" reactors...". The report provides an extensive analysis of plutonium breeding and argues that "the exploitation of the long irradiation of natural uranium is not likely to be limited by cost competition from plutonium power breeders, but only by the total amount of fresh natural uranium which can be obtained at a cheap enough price." It goes on: "The possibility of a cheaper production of power from, for example, artificially generated fast neutrons and thorium, and some remoter possibilities are discussed in an accompanying highly speculative paper DR-23". Lewis's ideas were further developed over the next few months and presented in DR-24.

The essential idea, which later became known as electronuclear breeding, was to use electrically generated neutrons to convert fertile to fissile material. Evidently the electrical energy required should be as small a fraction as possible of the energy that can be obtained from the fissile material produced. The most promising method for generating the neutrons appeared to be the bombardment of a heavy-element target, uranium or thorium, with an intense beam of protons or deuterons accelerated to an energy of several hundred MeV. The high energy is important; a larger fraction of the beam energy goes into neutron-producing reactions than at low energy, where most of it is dissipated in ionization processes.

High-energy particles can break off chips or splinters of nuclear material from heavy nuclei. This phenomenon is called "spallation," a term borrowed from geology and mining. Most of the neutrons produced in the high-energy bombardment of heavy nuclei are not associated with the spallation process – they are just "boiled off" from excited nuclei – but they are nevertheless generally referred to as "spallation neutrons".

Producing enough spallation neutrons for electronuclear breeding on a significant scale called for accelerator beam currents much larger than were then available. Lewis was attracted to the challenge of developing high-power accelerators, perhaps because of his earlier work on the radiofrequency (rf) system of the Cambridge University cyclotron, and his subsequent involvement with high-power radar systems during the war. Only limited resources were available at Chalk River in the early fifties, and Lewis was looking to developments in the United Kingdom and the United States to provide the necessary advances in accelerator technology, though he was also encouraging NRC and Canadian universities to embark on accelerator programs.

In July 1952, during the initial development of these ideas, a future systems organization was set up. It included a high-energy systems panel, which considered the theoretical calculation and experimental measurement of neutron yields, and also the requisite accelerator technology. This interest was communicated to the atomic energy authorities in the United Kingdom and the United States on the assumption that information bearing on it would go under joint American-British-Canadian classification. The United Kingdom agreed; the United States apparently found that the query was too embarrassing to make any reply at all, but agreed with the maintenance of security.

In fact, the United States had just cancelled its own project, which went by the cover name of the materials testing accelerator (MTA). It had been started in early 1950, by Lawrence and Alvarez at Berkeley, to produce fissile material (and/or tritium) for the American weapons program by using an enormous linear accelerator. The plan was to generate a 500 mA beam of 350 MeV deuterons, in a structure 18 m in diameter and 450 m long. This beam, with a power of 175 MW, would bombard a complex depleted-uranium target assembly and produce 564 kg of plutonium per year, approximately equal to the output of a 1,500 MW (thermal) production reactor. Its advantage over a reactor was that it did not depend on natural uranium, which was thought to be in short supply. A prototype "front end" (the first 27 m of the accelerator) was built at the abandoned Livermore naval air station, and in the summer of 1952 it was beginning to show some promise. By this time, however, the growing supply of uranium and the improving efficiency of production reactors obviated the need for production accelerators. On 7 August 1952, the USAEC deferred all plans for the full accelerator and left the prototype to die a natural death. Reports on the project were declassified in February 1957.

The existence of a high-current accelerator project at Livermore had been revealed, though not its power or purpose, in a list of USAEC contracts published in the July 1952 issue of the journal *Nucleonics*. In an assessment made earlier that month, Lewis had noted that the Berkeley scientists had been publishing a great deal of information bearing on neutron yields at high energy, but had been silent for the past three years; he now considered (July 28) that the Livermore accelerator suggested that the United States might already be interested in the subject.

The MTA accelerator employed primitive technology, but an improved design was developed using a higher frequency (48 MHz rather than 12 MHz) with a corresponding reduction in tank diameter. A front end was built at Berkeley and operated more satisfactorily; it produced 30 mA of 7.5 MeV deuterons.

None of this American information was available to the Canadians, who were dependent on United Kingdom cooperation, especially on a group at Harwell that was just starting to design a 600 MeV linear accelerator for particle-physics research. This machine was not intended to produce large average currents, but much of the technology was directly relevant.

Most of the Chalk River effort was directed to the question of neutron yields. Calculations were made, but the underlying nuclear theory was considered to be unreliable at the energies of interest (several hundred MeV), and attention was quickly focused on the possibility of experimental measurements. The highest energy accelerator in Canada was the McGill cyclotron and some measurements were made there using 77 MeV protons, but they were obviously of limited value. Accelerators under construction in the United Kingdom would eventually be suitable but the indefinite delay was not considered to be acceptable. In the absence of American cooperation, it was decided to attempt to use the high-energy proton component of cosmic rays.

Protons in the energy range 250-900 MeV make up only about 0.5 percent of the total cosmic-ray particle flux at sea level, and about 2 percent of it at an altitude of 3,000 m. Equipment was designed and built to select individual protons from the large background of unwanted particles (mostly muons), measure their energies, and measure the number of neutrons they produced in slabs of tin, tungsten, lead and uranium.

Although it seemed possible to do the experiment at Chalk River (altitude 150 m), it was clear that a higher altitude would have the double advantage of an increased proton flux and an increased proton/muon ratio; comparisons of data at two altitudes would check that muon effects were being correctly allowed for. The equipment was therefore assembled in a nine-metre-long trailer, which, it was originally planned, would be taken to a site near Sunshine Lodge, Banff, at an altitude of 2,200 m. It subsequently turned out to be politically possible to use the 3,260 m Echo Lake site of the Inter-University High Altitude Laboratory in Colorado, where the proton flux in the energy range of interest is more than a factor 3 greater than at Sunshine (and a factor 11 greater than at Chalk River).

The United States agreement that Chalk River scientists could use their cosmic rays came in the spring of 1953. They had decided that the research could be regarded as unclassified high-energy physics, so long as measurements were made only on simple targets and not on complex target configurations, which had received a lot of attention during the MTA project. The experiment was carried out on this basis. The American authorities were very cooperative, and the Echo Lake laboratory was most accommodating – $2 a night per person for lodging and $1 a day for the experiment. Food and electrical power were extra.

No request was made to use an American accelerator, in particular the recently commissioned 3 GeV Cosmotron at the Brookhaven National Laboratory (so named because it was to take over the job previously done by cosmic rays). By the time it might have been politically possible to use the Cosmotron, the Chalk River group felt that their cosmic-ray apparatus would give satisfactory results; using an accelerator might give data more quickly, but new apparatus would have to be built to operate in an unfamiliar environment.

However, the cosmic-ray experiment turned out to be much more difficult than expected, and diagnosing and rectifying problems with the apparatus took time. Final modifications were made and the equipment recommissioned at Echo Lake by two physicists in December 1955. They worked twenty-five-hour days and were back on solar time when they went home for Christmas. A satisfactory series of measurements was made during the next three months. The trailer laboratory returned to Chalk River in April 1956, a long series of runs followed (in Deep River), and the results were published in 1960. In spite of the delay, they constituted the only experimental data on neutron production in heavy elements by protons above 200 MeV. Within the experimental uncertainties (typically 20 percent) they confirmed theoretical predictions which had by then become more reliable.

Their practical usefulness was in abeyance. Economic studies at Chalk River had concluded, in January 1955, that developments in power reactors had postponed the day when high-energy systems might be economical producers of neutrons. As described later in this chapter, the economics of electronuclear breeding were re-examined from time to time. The use of spallation neutrons to produce a very high flux of thermal neutrons for experimental purposes was the basis of the ING proposal, which used the Echo Lake data until measurements were made on the Cosmotron in 1965.

ING: THE INTENSE NEUTRON GENERATOR

Early research at Chalk River had benefited greatly from the NRX reactor, which had the world's highest thermal-neutron flux,[1] 6×10^{13}, until it was overtaken in 1952 by the United States Materials Testing Reactor (MTR) with a thermal flux of 4×10^{14} (and a fast flux of 2×10^{14}). NRU, commissioned in 1957, has a thermal flux of 3×10^{14}. It was designed as a general purpose reactor for support of a wide range of basic and applied research activities and not as a high-flux reactor per se.

For a given configuration of fuel and moderator the flux is simply proportional to reactor power, which is limited by the rate at which heat can be extracted from the fuel. Specially configured reactors can provide regions where the thermal flux for a given power is substantially higher than in conventionally configured reactors, like NRU (and NRX). Widespread interest in such high-flux reactors was developing in the fifties, and the High Flux Beam Reactor (HFBR) at the Brookhaven National Laboratory, which has a flux of 10^{15}, was started up in 1965. Its prime purpose was to provide beams of thermal neutrons for the type of research already under way at NRU and other reactors.

It was in this context that the director of the physics division met with three of his staff in the spring of 1963 and asked them to consider what Chalk River should do in the high-flux field. By June 1963, they had made a preliminary study of three potentially useful sources of neutrons; fission, spallation and fusion. Recognizing that flux is limited by heat removal, they rated these sources according to the energy released in the production of one useful neutron – 130 MeV for fission, 40 MeV for spallation, and 17 MeV for fusion. They dropped fusion, conceptually the best, because its practical realization was unpredictable, and their study of contemporary accelerator technology convinced them that the spallation route was feasible. Following consultation with the neutron users at Chalk River, a high neutron flux research facility study committee (HNFC) was set up in July 1963 to give further detailed consideration to the possibility of a high-intensity spallation neutron source based on a proton accelerator. For convenience it will be referred to as ING, the intense neutron generator, although it only received that name at a later stage.

The conceptual design of ING proceeded rapidly, with valuable assistance from accelerator groups in Europe and the United States who were developing proposals for "meson factories." These required proton beams in the relevant energy range (0.5-1 GeV) at an intensity of 0.5-1 mA. ING needed much higher intensity, but there did not seem to be any fundamental reason why this could not be achieved.

The ING accelerator[2] was designed to provide a 65 mA proton beam of energy 970 MeV, along with some low-intensity beams for meson production and other purposes. The main beam would be bent downwards so as to impinge vertically on a liquid-bismuth target, 10 cm in diameter and 60 cm deep, surrounded by a tank filled with heavy water 250 cm in diameter

---

[1] In this section the numerical values that are quoted refer to fluxes of neutrons per square centimetre per second, but the units are omitted for the sake of brevity.

[2] Figure 9.1 shows ING in its 1967 version, when a 1.5 km-long linear accelerator had replaced the originally proposed separated orbit cyclotrons, and lead-bismuth had replaced bismuth (figure 9.2).

and 300 cm deep (see figures 9.1 and 9.2). As in a reactor like NRU, there would be internal irradiation positions and beam tubes leading out of the tank. The maximum thermal flux would be 10^{16}, scaled back from the originally hoped-for value of a thousand times that of NRU, but still a factor of ten larger than is available today at the world's three leading research reactors (HFBR at Brookhaven, HFIR at Oak Ridge and the ILL reactor at Grenoble). The estimated capital cost was $60 million, in 1964 dollars, including $42 million for the accelerator, but excluding engineering design costs and the cost of the bismuth target for which there were no design studies. The predicted operating cost was $7 million per year.

Meanwhile, other proposals had been developed under the aegis of a future systems committee, set up by Lewis in the fall of 1963 to consider possible future initiatives for Chalk River. These proposals were for programs in fusion, fast reactors, the long-distance transmission of energy, fuel cells, magnetohydrodynamic energy conversion, and the direct conversion of heat to electricity. The various subcommittees that developed these proposals reported in the spring of 1964. Only in the case of fusion was a recommendation made for a significant program (which was not accepted); for the rest the recommendations were, at most, for a modest activity that would preserve some awareness of outside developments.

Lewis presented the ING project to J.L. Gray and the AECL board of directors in May 1964. He proposed a period of further study to establish practicability, with expenditures small at first but growing rapidly into the range of $4-10 million a year. It was agreed that the work should continue.

In his presentation, Lewis stressed not only ING's scientific importance but also its potential for generating revenue from isotope production, and the role ING could play in developing profitable high-technology capabilities in Canadian industry. While these considerations were always being advanced in support of ING, they did raise two problems: the extent to which the research capabilities of ING should be compromised to increase isotope production, and the extra cost and delay that would inevitably be required to vitalize an initially unqualified and therefore largely unresponsive industry.

Another difficulty, which was to become very serious, also arose at this stage. The problem was the siting of ING. For Lewis, one of the principal elements in the rationale for ING was its importance for the scientific future of Chalk River, and he naturally insisted that it be located there.

FIGURE 9.1
ING – Thermal Neutron Facility. Cutaway view of the target building.

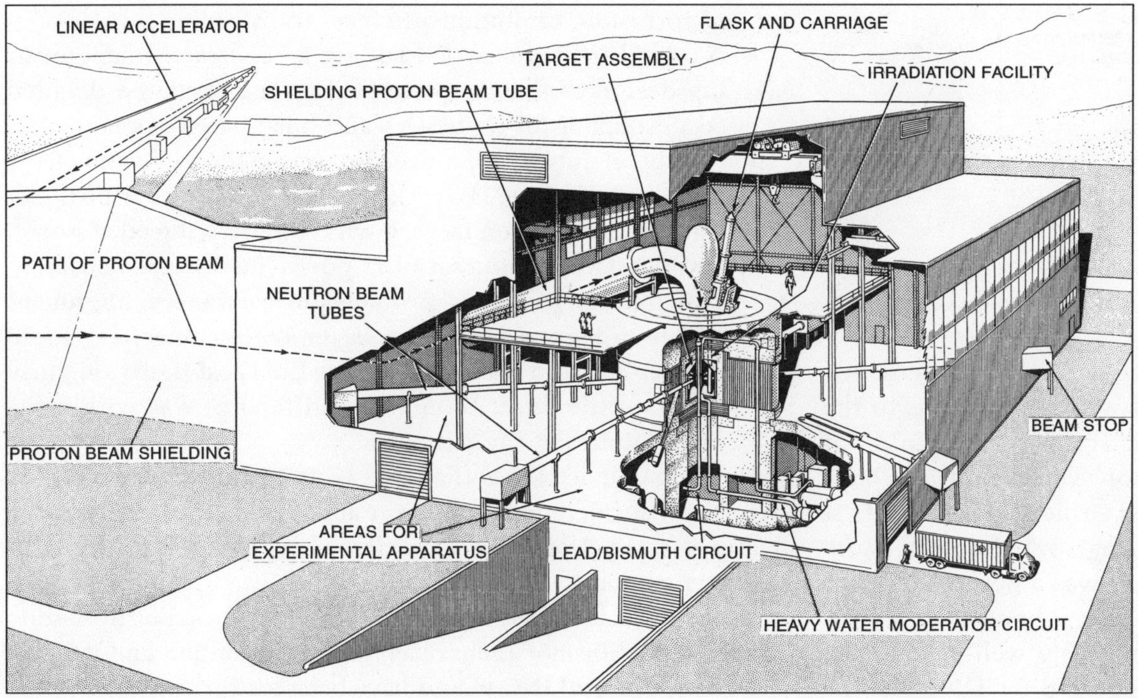

FIGURE 9.2
The ING Target/Moderator Assembly Schematic

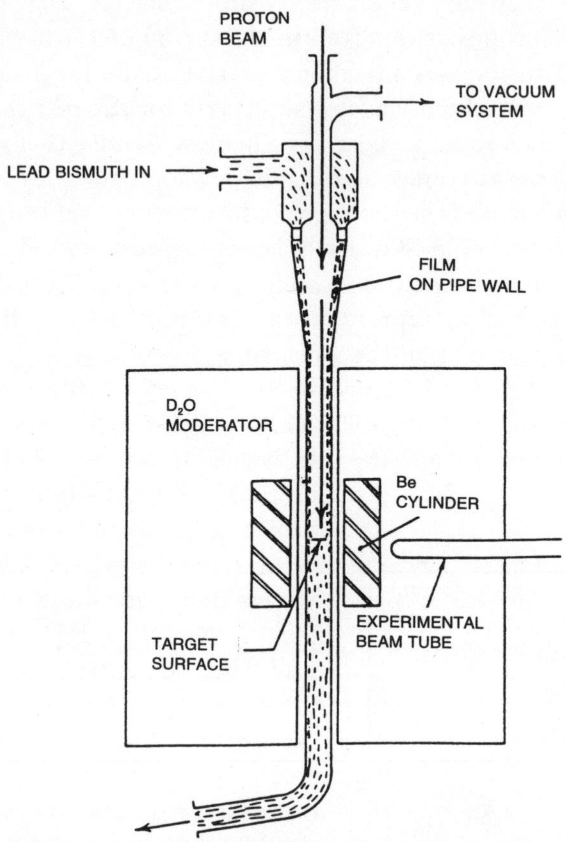

However, Gray had qualms regarding the expansion of the laboratory that would be necessary.

The immediate problem was the cost, and Lewis returned from his presentation to the board with instructions to the ING group that the cost of the accelerator itself should not exceed $25 million. The accelerator system that had been proposed comprised three Separated Orbit Cyclotrons (a first stage injecting into a larger second stage, which injected into the still-larger final stage). The SOC was a new accelerator concept and the only alternative was a linear accelerator (linac) system. The linac was a relatively well-established device, though not on the scale required for ING, but the SOC was judged to be less expensive to build, because of its smaller size and more efficient radiofrequency (rf) cavity system. This latter feature would also reduce the operating costs below those of a linac. Without some unforeseeable new development, there did not seem to be much prospect for halving the cost of the accelerator.

In any event, a substantial R&D program was initiated to establish confidence in the design and to develop more reliable cost estimates. Some of the achievements of the next two-year period were:

- Measurements were made of the neutron yields from thick targets of beryllium, tin, lead, and natural uranium bombarded with protons in the energy range 0.5 to 2 GeV. They were carried out, in cooperation with an Oak Ridge group, on the Brookhaven Cosmotron in February 1965, and they provided more accurate data than had previously been available.
- An engineering design for the liquid-metal target was started and experiments were underway to study problems of corrosion, cavitation and mass transfer.
- A consulting firm carried out, under contract, an engineering design of the complete facility and prepared detailed cost estimates for both SOC- and linac-based systems.
- Finally, in July 1966, a decision was made in favour of the linac over the SOC. Los Alamos had developed a linac design for their meson factory with greatly reduced rf losses, and satisfactory sources of rf power for it were becoming available. Engineering problems of tolerances, alignment and foundations would be easier with a linac, as would reparability and realignment. The final cost figures slightly favoured the linac, though the difference was within the uncertainties of the estimates.

The research activities that ING would support also received further consideration. They were almost all in the basic research area; e.g., nuclear physics, nuclear chemistry, and condensed-matter studies by neutron scattering. Research in the materials-science area included radiation-damage studies, but providing special facilities for them received little attention and it is not clear how useful ING would have been for such work.

The list of justifications for ING included electronuclear breeding, but it was not a strong argument, especially since an economic analysis included in the proposal showed that this prospect still lay in the indefinite future. It was intended that a low-intensity beam would be steered to a "nuclear power technology facility," where relevant parameters could be measured; some measurements of this type were carried out later under contract at the TRIUMF accelerator.

All of this work, the design of the ING facility and the planning of its research program, was carried out with very little input from the Canadian research community outside Chalk River. Prior to the authorization of the design study in May 1964, this was perhaps understandable, but as the study progressed it became clear that outside communities had to be brought into the picture. Accordingly, a symposium on ING was held in Chalk River in April 1965, to which people from government, industry, and universities were invited.

The reactions of the university scientists were the most pronounced. They deplored the idea of AECL unilaterally designing a major scientific facility that they would then be expected to support and use – they should have been involved at the outset. In response, Lewis set up and chaired an ING study advisory committee, made up almost entirely of university physicists, to address their concerns.

The committee had a difficult job, compounded by the increases in the estimated costs for ING, which grew rapidly, reaching $155 million by June 1966. No progress was made in resolving the difficult questions of where ING should be sited, and of the organizational arrangements e.g., between AECL and a possible multi-university institute.

Nevertheless, the time had come, two years into the design study, for a decision on whether to proceed with ING. The proposal that Lewis took to the executive committee of the AECL board of directors in August 1966 had Gray's acquiescence rather than his full support. Gray had written two memoranda to Lewis in May expressing his concern about the high cost of ING and his opposition to Lewis's position on siting and organization. The executive committee decided that the government had to be consulted before the project went any further, and Lewis made another presentation later that month to a subcommittee of Privy Council members and other government representatives. They in turn referred it (on 4 October 1966) to the newly created Science Council, which set up an ad-hoc committee of eight men, drawn from universities, industry, and the Economic and Science Councils, to review the proposal. The committee engaged two Canadian-born accelerator experts from the United States as consultants.

On 17 March 1967, the committee reported to the Science Council, which accepted its findings. The principal ones were:

- In the context of probable future national expenditures on R&D, the financial and manpower demands of ING were not excessive.
- A number of technical issues, particularly rf power supplies, ion-source life, system reliability and control, and liquid-metal target design, called for further feasibility studies, probably lasting two years before a commitment to construction could be justified. (The Science Council recommended that up to $7.5 million be provided for this purpose.)
- Maximizing the benefits of ING would probably require a site more generally accessible than Chalk River. To safeguard the interests of universities and industry, there should be an immediate change in the project organization, with the setting up of a senior managing board incorporating university and industry members. Questions such as siting could be studied during the two years of further feasibility studies.
- A careful review should be undertaken towards the end of the feasibility studies.

For more than a year, the government took no action. R&D work continued at Chalk River and, under contract, in industry and universities, though with less financial support than the Science Council had recommended. Meanwhile, opposition to ING from outside AECL grew even stronger, with deans of engineering expressing particular hostility.

At last, following the general election in June, the Trudeau government cancelled the ING project on 20 September 1968, because of its "very high cost".

It was probably the right decision. ING's proponents were pushing to extreme limits technologies with which they had no experience, and estimated costs ($750 million in 1993 dollars) were correspondingly unreliable. Given that public support for science was waning, the government could not be expected to make the open-ended commitment that ING demanded.

AECL's failure to include Canadian universities in the planning of ING is often cited as a major cause of its demise. In fact, only a handful of ING supporters could have been recruited, since there was little relevant technical experience at the universities and limited interest in the research program proposed for ING. Most of the university researchers saw the high cost of ING as putting their own research funding at risk, and more prior consultation could hardly have persuaded them otherwise.

For its proponents, ING was the project that would assure Chalk River's future as a major research laboratory with the mandate "to conduct fundamental nuclear research and to promote the practical applications of atomic energy."[3] They also had the resources to develop proposals such as ING to an advanced stage, resources that were far larger than those available to universities, and this funding imbalance underlay the conflicts that arose.

The cancellation of ING was a great disappointment for Lewis. In November 1968 he was appointed to a staff position in AECL's head office organization, without any direct responsibility for research at Chalk River, though he was still located there. He continued to take a keen interest in alternative accelerator systems, but his attention turned increasingly to the economics of advanced fuel cycles, particularly the Valubreeder reactor (see chapter eighteen).

ACCELERATOR PHYSICS

The Post-ING Transition

Following the cancellation of ING, Chalk River moved quickly to establish a program that would exploit the capabilities developed during the ING project, and it was approved by Gray on 26 November 1968, just two months after the project was cancelled.

The new program had two parts. The first continued work on lead-bismuth, with emphasis on its possible usefulness as an in-reactor coolant. It soon became clear, however, that it would be difficult to find compatible materials for fuel cladding and coolant channels that had low neutron absorption. Resources for an expanded program were not available and the work was stopped in September 1969.

The second part was the development of accelerator technology with emphasis on high-current, high-efficiency devices. Electronuclear breeding would be a long-term objective, but some near-term applications were identified, notably the use of high-current electron accelerators for industrial radiation processing, and the use of cheap, large rf power sources in industrial heating applications.

Apart from the lead-bismuth target, ING would have faced three main technical challenges:

1 providing an intense source of protons and holding them together, against the disruptive influence of their mutual electrostatic repulsion, while they were given an initial acceleration;

2 accelerating them to their final energy as efficiently as possible; and

3 reducing the costs of generating the large amounts of rf power that would be required.

Two major experiments had been planned to verify the ING design: the proton injection experiment (PIE), and the electron test accelerator (ETA), which, respectively, addressed items (1) and (2), fast electrons being a cheap alternative to fast protons for (2). These experiments, described below, formed a major part of the post-ING program.

Work on (3) continued at a reduced level and several experimental magnetrons were built and tested. In 1970, a promising industrial heating application was identified and a contract was given to Varian Associates of Canada Ltd. to develop a commercial magnetron. The application evaporated, but work continued at low priority, leading in 1974 to a pre-production

[3] AECL's *Annual Report* for 1961-62, page 7.

magnetron that generated 140 kW of 915 MHz power with a very gratifying efficiency of 80 percent. In the absence of any firm industrial application, this work was then set aside.

During the ING study a number of possible alternative accelerator systems had been considered. One of the most interesting was the electron ring accelerator, which had been invented in the USSR and demonstrated there on a very small scale. The basic idea was to produce a stable ring of relativistic electrons, and then load it with protons at a concentration of about one proton per 200 electrons. The ring is then accelerated, and it drags the protons with it. Because of their high effective charge, the protons could, in principle, be accelerated to 1 GeV over a distance of just a few metres. Experiments were started in 1970 on generating the very short intense bursts of electrons that would be required, but they were terminated early in 1975, to release effort for the superconducting cyclotron.

Accelerators for Electronuclear Breeding

An analysis made during the ING study concluded that electronuclear breeding using high-energy protons could become economic only if the price of uranium were to increase, as high-grade ores were used up, by a factor of about three.

The economics of an "accelerator breeder" (AB) were re-examined on a number of occasions, with similar results. The last review, in 1981, considered an AB producing one tonne of fissile material per year, which would service about six 2,000 MWe CANDU stations using a thorium fuel cycle with a conversion-efficiency factor of 0.9 and operated at 80 percent availability. The estimated capital cost in 1981 dollars was $1.5 billion, 30 percent of it for the 300 mA 1 GeV accelerator and its power supplies, and the rest for the target and its surrounding blanket, where the fissile material is produced. This cost was too high; the price of uranium-235, taken as $48 a gram, would have to increase by a factor of at least three before the AB system could become competitive.

Thus electronuclear breeding remained a distant prospect throughout this fifteen-year period, and one could argue that the AB ought to have been abandoned earlier. However, the AB did provide a technically feasible way of putting an upper limit on any future rise in the cost of nuclear energy, more feasible, during this period at least, than the alternative of fusion.

It should also be recognized that the relevant activities were carried out at relatively low power, requiring only modest expenditures, that they were making important contributions to accelerator technology, and that they were providing the foundations for devices with near-term applications. Furthermore, the development of high-current accelerators for neutron production and other purposes was a major interest at several other leading laboratories. Links established during the ING program were maintained, with great benefit to Chalk River, especially with Los Alamos, Brookhaven, and Oak Ridge.

The high current test facility (HCTF) built for the PIE experiment (see figure 9.3) followed the design proposed for the first stage of ING. Hydrogen gas is ionized by an arc discharge and protons are extracted. This ion source is inside the high-voltage dome, which is connected to the positive terminal of a 750 kV DC power supply, and the protons accelerate towards ground potential along an accelerating column. They are then steered by two magnets into an Alvarez tank, named for its inventor, which accelerates them to an energy of 3 MeV. Radiofrequency power (168 kW at 268 MHz) fed into the tank sets up oscillating voltages between hollow electrodes located along the tank axis; they have entrance and exit holes (not shown in figure 9.3) to let the protons through. The distances between successive electrodes are so arranged that the protons are in the interelectrode gaps when the voltages are in the right direction to accelerate them, and are inside the electrodes, and thus unaffected, when the voltages reverse. As the protons accelerate they travel further in each rf cycle; to allow for this the electrode lengths and spacings are increased along the tank axis.

In an AB the protons would be further accelerated in a series of Alvarez tanks to an energy of 200 MeV and then to 1 GeV in a "coupled-cavity" linac.

ETA (see figure 9.4) is a coupled-cavity linac of the type, developed at Los Alamos, that was chosen for ING. Radiofrequency power (805 MHz) from two 100 kW klystrons is fed to a series of resonant cavities coupled together with off-axis cavities. Electrons are accelerated by the oscillating electric fields in the on-axis cavities, which are each 180 degrees out of phase with their neighbours. The lengths of the cavities are chosen so that the electrons take one half cycle to traverse them, and are therefore always in an accelerating field. As in the Alvarez tank, distances have to match velocity, but v, the electron velocity, rapidly approaches c, the velocity of light, so that the initial "graded-β section" (β being v/c) is quite short.

ETA was designed to accelerate a 20 mA electron beam to an energy of 4 MeV. The work went well and by 1979 had established that much higher currents could be accelerated given more rf power; e.g., 100 mA with 500 kW, 80 percent of it going into the beam, and the rest into heating the cavities.

The Alvarez tank had problems with the rf supply and heating effects in the tank. It first accelerated protons in 1980 and during a final series of runs in 1985 a 5 mA beam, limited by the output of the 750 keV accelerator, was accelerated to an energy 10 percent higher than the 3 MeV design value. The experience gained established confidence that accelerators of this type could meet AB requirements.

FIGURE 9.3
High Current Test Facility

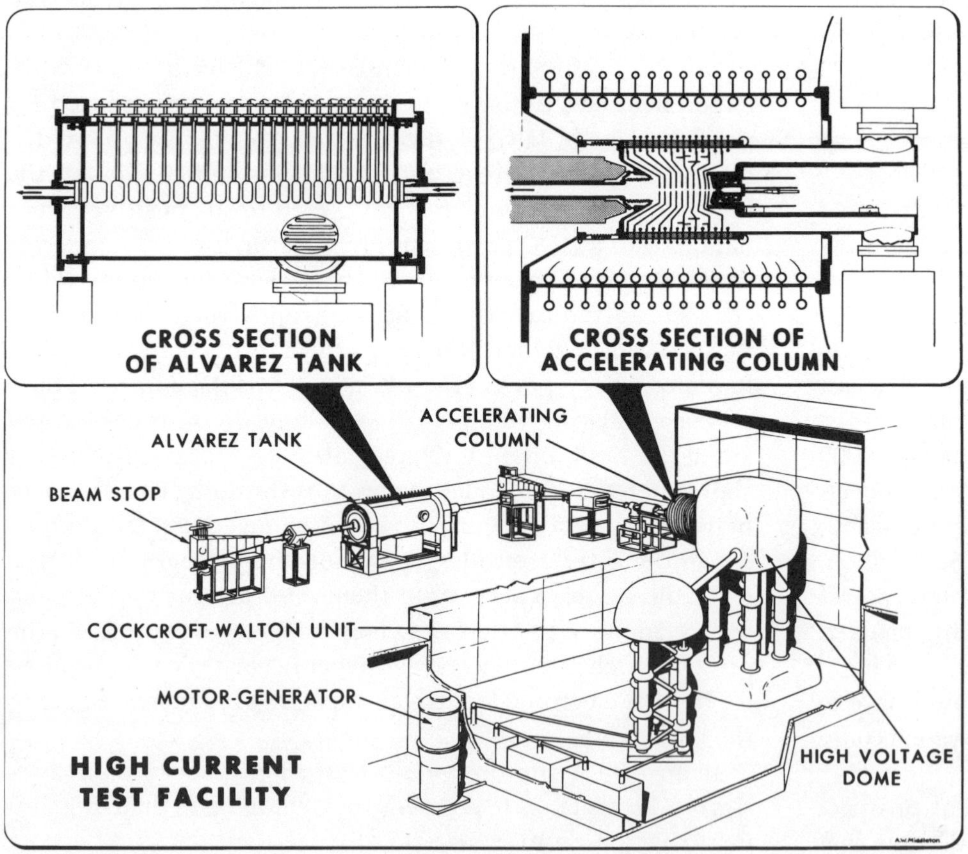

Such was not the case with the 750 keV accelerator. It was always recognized that initial acceleration would be difficult. The protons must be accelerated as fast as possible to minimize the spreading of the beam caused by their mutual repulsion, which means that the accelerating column must be operated with the maximum possible voltage gradient. This maximum is reduced when a beam is being accelerated, because of a multitude of secondary effects that the beam can produce. The accelerator behaved well at low current (5 mA), but a long struggle to improve performance at 40 mA met with little success. Nevertheless, it worked at least as well as any other similar accelerator.

Fortunately, by this time a new device had appeared: the radiofrequency quadrupole (RFQ) accelerator, which was invented in the USSR and further developed at Los Alamos. The principle is illustrated in figure 9.5. An rf voltage is applied to the top and bottom electrodes, which are connected together, and an equal voltage of opposite phase to the

FIGURE 9.4
The Electron Test Accelerator

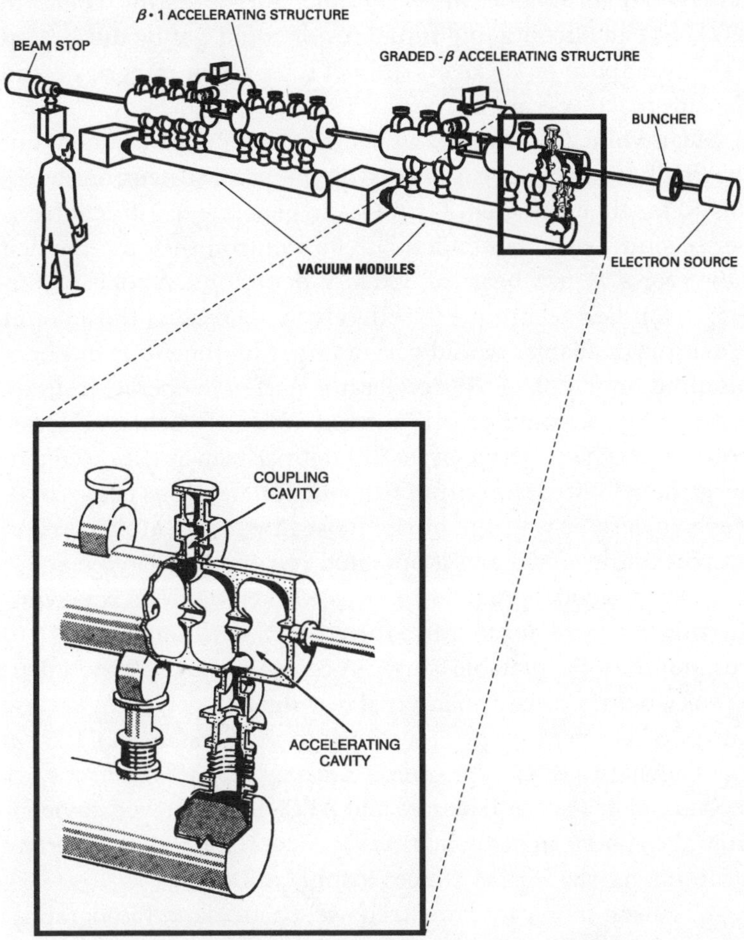

FIGURE 9.5
The RFQ Accelerator Schematic

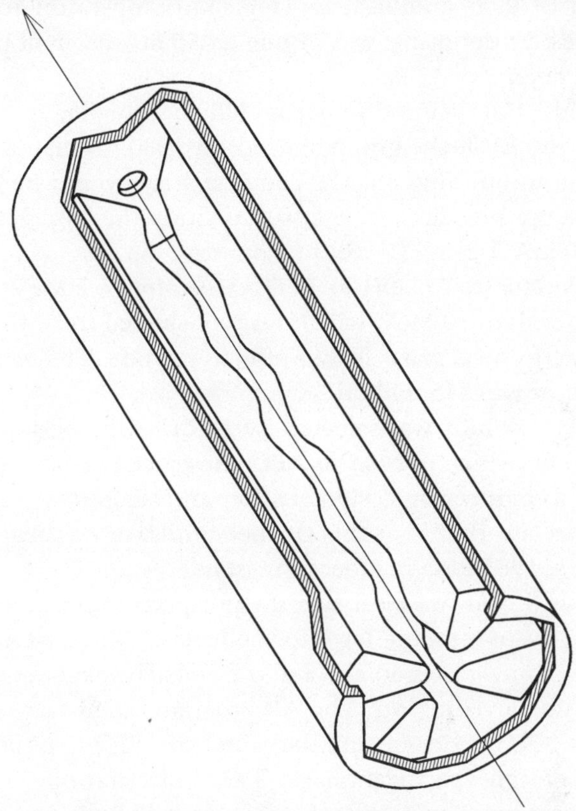

(similarly connected) right and left electrodes. The resulting rf quadrupole field has a strong-focusing effect on particles travelling in a more or less axial direction. If the spacing between top and bottom electrodes varies periodically along their length, and there is a similar but out-of-phase variation between the left and right electrodes,[4] an axially directed component of the electric field is produced. The RFQ thereby becomes an accelerator, but one with strong-focusing forces to hold the beam together. Satisfactory performance demands tight dimensional tolerances and extremely smooth surfaces.

The RFQ can accept protons at low energy (e.g., 75 keV), which greatly eases the initial acceleration problem, and accelerate them to an energy suitable for injection into an Alvarez tank (e.g., 2 MeV). Chalk River started RFQ work in 1981, maintaining close cooperation with Los Alamos.

[4] The top and bottom electrodes are closest together when the left and right ones are farthest apart, and vice versa. The spatial period of the interelectrode distance is increased to match the increasing distance travelled by the particles in one rf cycle as they accelerate.

An ion source suitable for an AB was available, thanks to a development program that had been under way since the ING days. In 1979, a "duoPIGatron" ion source, operating in an Injector Test Experiment, was giving a 450 mA beam of protons with an energy of 42 keV.

Confident in their ion source and linac technology, and with faith in the RFQ, the group started, in 1980, to plan a staged approach to building an AB. The first stage would be ZEBRA (zero energy breeder accelerator), to study the launching of a full 300 mA beam. It would comprise an ion source and preaccelerator (to 75 keV), an RFQ accelerator (to 2 MeV) and an Alvarez tank to 10 MeV, which was considered to be the minimum energy necessary for an effective study. The estimated capital cost was $15 million.

In 1981, ZEBRA was selected by AECL as the best available program on which to base an AECL Research presence in Quebec. Civil engineering design of office and laboratory buildings started in late 1982, but this Quebec initiative was abandoned in August 1983 for a number of reasons.

By this time it was clear that the prospects for electronuclear breeding were receding far into the future – there were ample supplies of low-cost uranium, and the global demand for nuclear energy was levelling off. The AB program quietly lapsed, but work on RFQ continued, and the prototype, RFQ 1, successfully met its design specifications in 1990, accelerating 75 mA of 50 keV protons to an energy of 600 keV. Chalk River was a world leader in this RFQ work, and received substantial funding for it from Los Alamos. Unfortunately, the financial constraints within AECL led to the decision in 1992 to terminate this program, and the equipment was transferred to Los Alamos in March 1993.

Smaller Accelerators for Neutron Production

In 1970, AECL's commercial products division (CPD) expressed interest in cancer therapy using neutrons. Suitable neutron sources were not available and Chalk River agreed to develop a strong 14 MeV neutron source. An accelerator conceptually similar to the 750 keV HCTF accelerator would provide a beam of 300 keV deuterons, which would bombard a titanium tritide target. This target would be sandwiched between thin layers of aluminum oxide, to prevent the tritium escaping, and thereby achieve an acceptable target lifetime. Target testing began in 1971, but an encouraging initial result could not be duplicated – the aluminum-oxide layers did not retain their integrity – and this approach was abandoned in 1973.

Meanwhile, CPD interest in neutron therapy had waned, but the Chalk River biologists needed an intense fast-neutron source (FINS) for their research. FINS was to generate 4×10^{12} neutrons per second – just about adequate for neutron therapy – which called for a 25 mA beam of 300 keV deuterons. A rotating target, a standard technique for effectively increasing the amount of tritium available, would give a target lifetime of about one-hundred hours. The FINS accelerator started to operate in 1976, but stable operation at full current was only achieved three years later. The rotating target did not perform as expected and outgassing limited the output to about 10^{12} neutrons per second. Nevertheless, it was one of the most intense 14 MeV neutron sources in the world and it operated reliably for many years.

In 1973, a conceptual design was worked out for a superconducting cyclotron for neutron therapy that was small enough to be mounted on a rotatable gantry. A 25 mA beam of 30 MeV deuterons would bombard an internal beryllium target, giving a dose rate of 60 Gy per minute at a source-to-skin distance of 125 cm (equivalent to a FINS-type source of strength 3×10^{13} neutrons per second). The idea was patented and AECL later received payment from the American manufacturer (Medcyc) of a neutron-therapy cyclotron for the Harper Grace Hospital in Detroit.

A neutron source for industrial use, particularly radiography, was designed in 1982 and named CANUTRON. A 50 mA beam of 2.5 MeV protons from an RFQ accelerator impinging on a liquid lithium target would produce 4×10^{13} neutrons per second, giving a thermal flux of 10^{12} neutrons per cm^2 per second in a light-water moderator. The idea was not taken up.

Electron Accelerators for Therapy

As described in chapter five, AECL's CPD had outstanding success with cobalt-therapy machines, but by the mid-sixties electron accelerators were recognized as potential competitors. Accelerators can produce X rays that are of higher energy, and consequently more penetrating, than the gamma rays from cobalt-60. This greater penetration is advantageous in the treatment of deep-seated tumours, because it reduces the unwanted dose to intervening tissue relative to the dose that has to be delivered to the tumour. The X rays, produced when the electrons hit a metal target, have lower energies than the electrons; e.g., the X rays from a 3 MeV accelerator have about the same depth-dose distribution as the cobalt-60 gamma rays (1.17 and 1.33 MeV). The electrons can also be used directly (i.e., with the target swung out of the way) to treat more accessible tumours.

High-energy electron accelerators had been used for cancer treatment before cobalt-60 became available – in particular, 24 MeV betatrons, one of which had been installed at the University of Saskatchewan in 1948. These were inconveniently large machines, and the era of compact accelerators mounted on rotatable gantries began with the Clinac 4, launched by Varian Associates of Palo Alto, California, in 1967 (CPD knew of their plans two years earlier). This 4 MeV linear accelerator (linac) gave a depth-dose distribution only marginally better than a cobalt-60 machine, and it was more expensive and less reliable. Nevertheless, it rapidly became very popular, mainly because the Blue Cross medical insurance plan in the United States had established a fee of $10 per accelerator treatment versus $5 for cobalt-60.

CPD initially considered cooperating with an established accelerator manufacturer – including Varian, who had made an offer – but by December 1969 decided to enter the market independently. The first accelerator would be a low-energy, 4-6 MeV, medical linac (LEML) to be developed at Chalk River, and this would be followed by a 25-35 MeV microtron, to be developed at the University of Western Ontario (London, Ontario) with funding from CPD. In both cases, the accelerators would be incorporated into therapy units to be designed by CPD.

LEML would provide electrons of the same energy as ETA, but only one hundredth the current, and it would be physically much smaller. A pulsed magnetron would supply rf power (approximately 2.5 MW peak, 2 kW average) at a frequency of 3 GHz. LEML would operate, like ETA, on the standing-wave principle, in contrast with traditional travelling-wave linacs, which require more rf power to produce the same electron energy. (A microtron also makes efficient use of rf power. The electron beam, accelerated in an rf structure, is repetitively steered, by magnetic fields, back into the structure, where it receives further accelerations.)

The original plan was not followed. What happened was very complicated, and only the main features are described here. The first complication was due to the Chalk River group who, welcoming a program that promised near-term benefits, moved quickly to explore a range of possibilities. These included, by September 1970, a high-energy medical linac of 25-35 MeV energy (HEML) – actually a two-linac system – that could be powered by the same magnetron that LEML would use. Clearly, this was in competition with the microtron and it was agreed that a choice would have to be made at some future time. (CPD finally abandoned the microtron possibility in 1975). The next development was CPD's April 1971 decision to drop the LEML in favour of an intermediate-energy (8 MeV) linac, IEML.

Plans changed dramatically in 1972 when CPD negotiated an agreement with the Paris-based Compagnie Générale de Radiologie (CGR), which was AECL's agent for medical product sales in many European and South American countries. CGR was marketing two therapy units, the 6 MeV Neptune and the 32 MeV Sagittaire – both using travelling wave linacs – and wanted CPD's help to market them in North America. CGR had no interest in collaborating on LEML, HEML or the microtron, and, since they were developing a 16-20 MeV unit called Saturne, CPD's need for any Canadian accelerator was far from clear. In any case, CPD's resources were almost completely committed to the unexpectedly big job of modifying Neptune to meet North American standards. (Sagittaire already did, but it was very expensive and there would be few sales.) The modified Neptune, marketed as Therac 6, was in fact a CPD-designed therapy unit incorporating a CGR accelerator (eventually, CPD built this also). The first unit was delivered to a hospital in 1975. Similarly, the Saturne accelerator in a CPD-designed therapy unit was marketed as Therac 20. Altogether, CPD sold thirty-six Therac 6s and twenty Therac 20s. They performed well, but the enterprise was unprofitable for reasons described in chapter five.

The 1972 CGR/CPD agreement coincided with a breakthrough at Chalk River – the concept of a single "turn-around" linac for HEML[5] (see figure 9.6). The electron beam is launched from an annular gun and accelerated through a series of coupled cavities energized from a 3 GHz 2.6 MW pulsed magnetron. The on-axis coupled system, designed at Chalk River, is cheaper to manufacture than the side-coupled type used in ETA. The electrons leave the accelerating structure with an energy of 12.5 MeV, are then turned back into it, where they pick up another 12.5 MeV, and then pass through the hole in the annular gun to the extraction magnet. Full acceleration on the second pass is obtained only if the electrons reenter in phase with the rf field; by moving the magnet this phase can be changed and the output energy varied down to 5 MeV.

Many problems had to be solved before the bright idea of 1972 became the Therac 25. Most were solved in a logical and

[5] Production models differed somewhat from the version shown in figure 9.6. One of the focusing elements, the gap lens, was replaced by a quadrupole triplet, and the annular gun was replaced by an off-axis electron emitter and electrostatic inflection system.

FIGURE 9.6
Therac 25

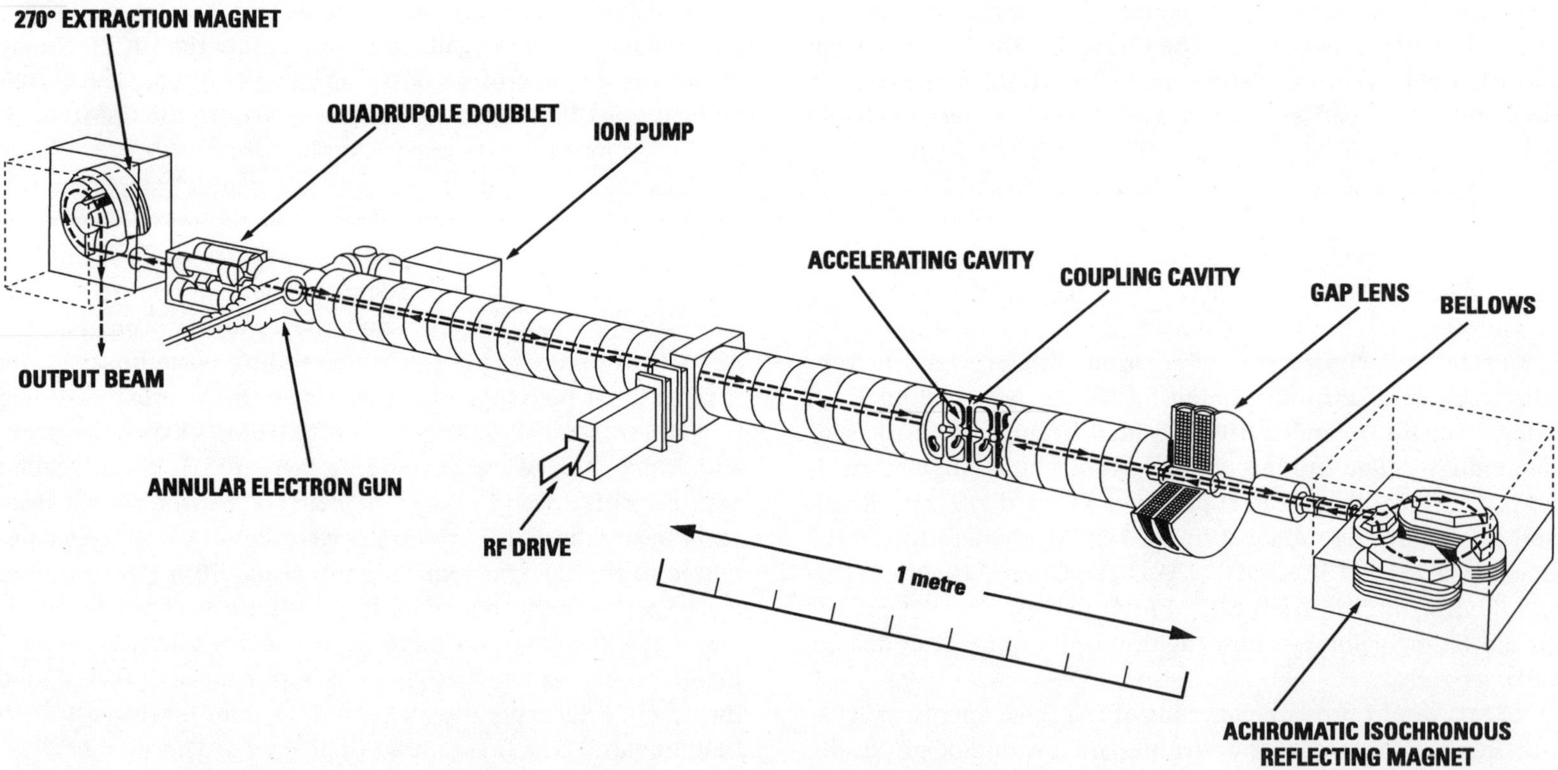

coherent way (e.g., meeting the very exacting performance requirements of the reflecting magnet), but developing a satisfactory electron gun involved as much art as science, and it took a long time. One problem was never solved, however, and as a result Therac 25 cannot be used for low-energy X-ray therapy. Achieving the required dose rate demands a much larger electron current than at high energy, and this large current has to be accelerated to 12.5 MeV and then decelerated. During this process some beam is spilled and unwanted X rays and neutrons are produced. Their intensity exceeded regulatory standards, even with external shielding. When the electrons themselves are used for therapy, much smaller currents are required and standards are met.

Therac 25 deliveries started in December 1982, and after some commissioning difficulties the first unit entered service in October 1983. By July 1986 twelve were operating – seven in Canada and five in the United States.

Between July 1985 and January 1987 there were six serious accidents involving overdoses to patients. They were traced to deficiencies in the software of the computer control system. Appropriate changes were made and the machines were allowed to resume operation. One of the American clinics that had experienced an accident returned its machine, and one has been retired, but the other ten were operating reliably at the time of writing (1993).

In January 1985, AECL decided to stop making medical accelerators. The Therac 6 and Therac 20 programs had lost money and the Therac 25 was also unprofitable. The Therac 25 was a complicated machine and its development was a long and costly process. Many customers were lost because of delayed market entry, and others because the expected low-energy photon capability turned out to be unavailable.

There were many reasons for the failure of AECL's medical accelerator program. The Chalk River group overestimated the advantages of the new standing-wave technology, and underestimated the time and effort required to convert a laboratory demonstration into a saleable product. This was hardly surprising in a research-oriented group of people, full of bright ideas but short of practical experience, who were determined to demonstrate that their long-term program could provide a near-term benefit. CPD was not exactly blameless either, but the fatal weakness was that the whole medical accelerator program lacked strong central direction. One might have expected AECL, at this stage in its maturity, to have provided an adequate management structure.

The Superconducting Cyclotron

In 1972, the Chalk River nuclear physicists put forward a research program that called for an additional accelerator to increase the energies of the heavy ion beams available from their tandem accelerator (see chapter eight). After studying several possibilities, the accelerator physics branch proposed a superconducting cyclotron (SCC) as the best choice. The advantage of using superconducting coils was the very high magnetic field that could be produced (5 tesla), which reduced the size and hence the cost relative to conventional cyclotrons. Advances in superconducting technology, especially the United Kingdom development of twisted conductors, had made it possible to produce very high and very stable magnetic fields over large volumes for the bubble chambers then being used in high-energy physics. The Chalk River scientists were the first to recognize that the technology was good enough for use in cyclotrons. After some persuasion by Chalk River, the idea was accepted by Michigan State University, and later by Milan. Contact with those groups was close and mutually beneficial.

Figure 9.7 shows the basic features of the SCC. Heavy ions from the tandem pass through a thin carbon foil, which strips off more of their electrons; this increases their net charge and decreases their radius of curvature in the magnetic field. They are then accelerated by rf fields in the eight gaps between the four hollow "dees" and the four top-and-bottom pairs of "flutter poles" (early cyclotrons had two D-shaped rf electrodes, and the name has stuck). As they speed up, the ions spiral outwards until, in their outermost orbit, radius 65 cm, they are electrostatically deflected and extracted from the cyclotron. The flutter poles produce an azimuthally varying magnetic field, which provides vertical focusing, and each contains a set of thirteen "trim rods" (a Chalk River innovation) for adjusting the magnetic field. This field has to increase with increasing radius to compensate for the relativistic increase in ion mass and so keep the ion motion in phase with the rf.

By adjusting the magnetic field and rf conditions, the SCC can be made to accelerate all ions, from lithium to uranium, to energies of at least 10 MeV per nucleon and with an energy spread of less than 1 part in 2,000. In its first run, for example, 71 MeV iodine-127 ions from the tandem, charge 7+, were stripped to charge 23+ and then accelerated to 1,270 MeV.

It was realized at the outset that the SCC faced a number of technical challenges – in particular, the stability and reproducibility of the magnetic field, and the behaviour of the rf acceleration system. They were not amenable to study using small-scale models, and approval was obtained from the AECL board of directors in June 1974 to construct full-scale versions of the magnetic and rf systems. The work went well; satisfactory performance of the two systems was demonstrated, and in July 1978 the board approved the first phase of the project – the construction of the tandem cyclotron facility (TASCC) with a single experimental target location. (The second phase, which added beam transport lines and associated equipment for nine additional target locations, was approved in February 1986).

FIGURE 9.7
Superconducting Cyclotron

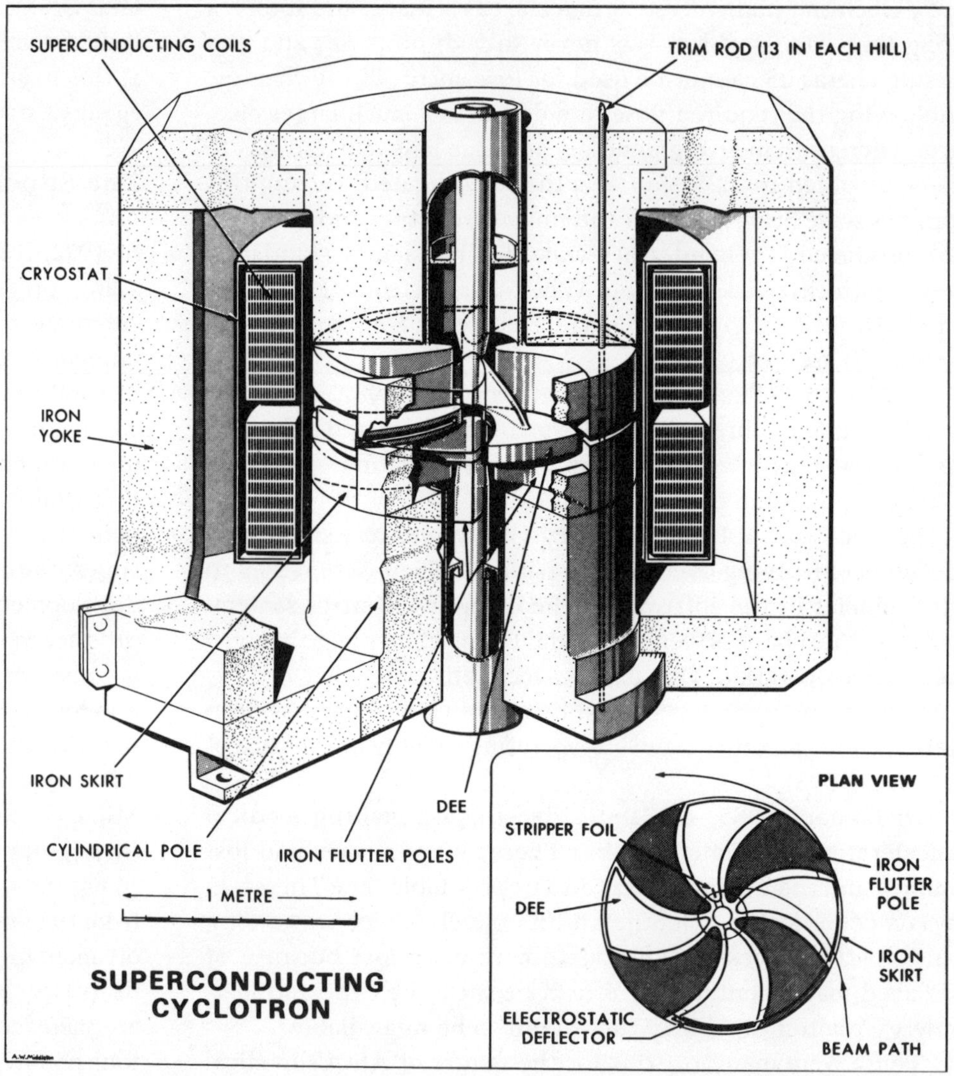

The inner wall of the magnet cryostat was then modified to accommodate the rf accelerating structure and the two major components of the cyclotron operated together in 1982. It was then dismantled and moved to the tandem laboratory. First beam was extracted on 19 November 1985.

It is not possible in this brief account to describe the many elegant and innovative features that distinguish this cyclotron. The design has proved extremely successful, enabling a comparatively small group of accelerator physicists to provide the Canadian nuclear physics community with a powerful and versatile accelerator at modest cost.

Other Activities

The development of accelerators for industrial radiation processing headed the list of items that had been proposed for the post-ING program. Following the successful operation of the electron test accelerator, careful consideration was given to possible applications for such 100 percent-duty-cycle linacs. Early in 1984, a conceptual design was put forward for RAPELA which could produce a 50 mA beam of 10 MeV electrons; i.e., 500 kW of beam power compared with 80 kW in ETA. However, industrial interest was lacking.

A new idea was born the following year: IMPELA, a long-pulse, variable-duty-cycle linac that would operate efficiently at lower beam power. Construction of a 10 MeV 50 kW prototype began in 1986 (see figure 9.8). It gave first beam on 14 April 1989 and full power on 8 November 1989. It won an American award as one of the 100 most significant technical products introduced in 1990. Three units have been sold, and the prototype continues to operate, very reliably, at Chalk River, carrying out irradiations for commercial customers (see also chapter twenty).

In October 1982, the leader of the ETA project was invited by the Canadian high-energy physics community to chair a committee that would recommend what Canada should contribute to a large electron-proton collider (HERA) planned for Hamburg, Germany. By December 1983, it had been decided that Chalk River would design and build a coupled-cavity rf system (2.5 MW) for beam bunching in the main ring. The cost, about $7 million, would be met by the Natural Sciences and Engineering Research Council, the agency responsible for funding university research in Canada. It was not a large contribution in the HERA context ($500 million), but it was the first firm offer from outside Germany, and it earned Canada considerable credit.

Unfortunately, a design change at HERA in May 1984 obviated the need for the proposed system, and a new Canadian package had to be organized. Eventually it comprised two smaller accelerating systems from Chalk River, plus a beam transport line from the TRIUMF laboratory, at a total cost of about $6 million. The accelerating systems, though based on Fermilab (United States) designs, required the development of a new technique to swing cavity frequency and a very fast feedback system to control it. The work took longer than expected, but the systems arrived in

FIGURE 9.8
IMPELA Prototype

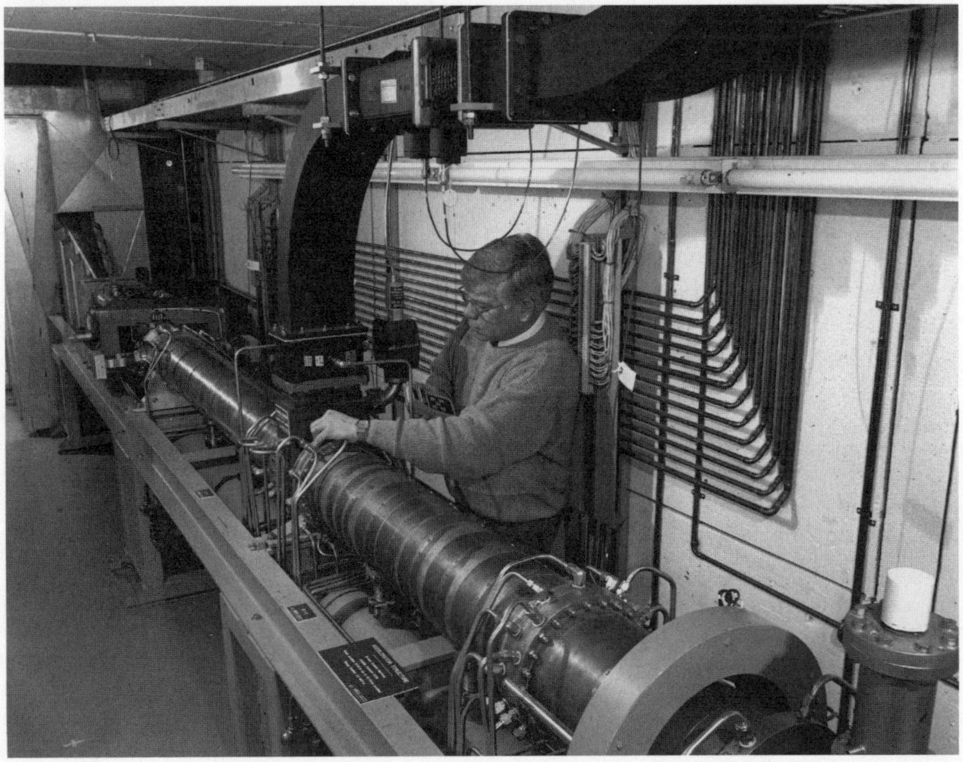

time, thanks to some slippage in the HERA schedule, and are operating very satisfactorily (see also chapter twenty).

The development of ion sources to produce intense proton beams had been extended to other ions in 1979. By 1984, contract work was being undertaken for two American manufacturers of semiconductor equipment, and a licensing agreement was concluded with one of them (Eaton/Nova) for the use of the Chalk River duoPIGatron in ion implanters; it realized about $500,000.

An activity very different from those just discussed began in 1984: the study of particle acceleration by lasers. Harnessing the very high fields that lasers can produce in plasmas could lead to a new generation of accelerators with much higher energy than is practicable with current technology. The Chalk River laser accelerator has generated a field of 1.7 GeV/m, accelerating an injected electron beam of energy 12 MeV to 29 MeV in a distance of 1 cm.

FUSION

One might have expected AECL, the Canadian agency responsible for the peaceful utilization of atomic energy, to have embarked at an early stage on a program to develop fusion as a source of energy. In fact it did not. This section gives an elementary description of the fusion process and outlines AECL's involvement.

Nuclear fusion refers to reactions between light nuclei that produce heavier nuclei, e.g., the DT reaction, in which a collision between a deuterium and a tritium nucleus produces an alpha particle, a neutron and 17.6 MeV of energy. To be able to react, the nuclei must have enough kinetic energy to overcome their mutual electrostatic repulsion; in thermonuclear fusion, this energy is provided by heating the DT mixture to a temperature of about 100 million degrees. At such temperatures, the atoms lose their electrons and the reacting medium becomes a plasma. Fusion reactions take place spontaneously in stars, but on earth the problem is to create a sufficiently dense hot plasma and hold it together long enough to achieve break-even; i.e., for the thermonuclear reactions to produce as much energy as that used to create the plasma. Until very recently this had been done only in nuclear weapons, where the reaction is so fast that a lot of energy is produced before the plasma blows up; this is called inertial confinement fusion (ICF) and is the principle employed in laser fusion, which is described below.

Plasmas can be held together by specially shaped magnetic fields, and this magnetic confinement (MCF) approach to controlled thermonuclear fusion is being vigorously pursued around the world, in particular with the construction of successively larger versions of the tokamak machine, a USSR invention. By the late eighties, large tokamak experiments were close to break-even, and in 1991 the European JET (Joint European Torus) experiment exceeded break-even by 14 percent.

The total world expenditure on fusion research is currently about $2 billion a year, and its proponents estimate that fusion will be established as a practical source of energy some time in the second quarter of the next century. Comparing it with fission, they claim that with fusion there is an unlimited supply of cheap fuel (the deuterium in the oceans), there is no danger of a divergent chain reaction, and the radioactive wastes are much more manageable, since there is no need to mine uranium and no fission products or plutonium are produced. No attempt will be made here to assess the force of these safety and environmental arguments, but it should be noted that adequate supplies of fission fuels will be available for a very long time.[6]

Research into the possibility of controlled thermonuclear fusion in the United States, the United Kingdom and the USSR

[6] In his 1971 paper "Energy in the Future: the Role of Nuclear Fission and Fusion", Lewis estimated that supplying 15 billion people with 5-50 thermal kilowatts each (i.e., 5 to 50 times the average per capita consumption in the world today) "...would cause no severe pressure on world resources of uranium and thorium for hundreds of centuries...". For DT fusion the tritium is made by bombarding lithium with the neutrons produced in the DT reaction, so that the available energy is set by the availability of lithium. This limit is very roughly comparable with what is available from uranium and thorium. However it is reasonable to suppose that, long before there is a shortage of any of these materials, the much more difficult DD reaction will have been harnessed, which would provide 15 billion people with 15 thermal kilowatts each for a billion years. The cost of deuterium is not important; the current price of $300 for a kilogram of heavy water translates into 0.00004 cents per electrical kWh.

was initially secret, but in 1956 the Russians disclosed some of their results, and on 24 January 1958, the United States and the United Kingdom announced the declassification of results obtained in their programs. On the previous day, the British held a press conference claiming – wrongly as it turned out – that controlled thermonuclear fusion had been achieved in the ZETA toroid at Harwell.

A few days later Lewis wrote an appraisal of the situation pointing out that there was still a long way to go – the number of neutrons produced by ZETA, assuming they were of thermonuclear origin, indicated a fusion power output of a millionth of a watt. He also discounted the advantages claimed for fusion regarding fuel supply and radioactivity.

Consistent with Lewis's scepticism, AECL did not set up a fusion research program, though the possibility was reviewed on three subsequent occasions, as described below. AECL kept in touch with outside developments through a group of in-house scientists, some of whom had a background in fusion research. This fusion awareness activity worked well, thanks to the excellent relations maintained with leading fusion-research laboratories, some of which were facilitated by exchanges of technical information in areas common to fusion and accelerator technology.

The first serious consideration of an AECL fusion program was in 1964 as part of the future systems study already mentioned in connection with ING. The proposal that was submitted recommended assembling a group of fifteen professionals who would decide what type of major, $1-million machine should be built and then proceed to use it for research. The proposal did not identify any particular Canadian expertise or focus that would make the program more than a small addition to the existing world effort, and it was not funded.

In January 1973, following the partial declassification by the United States of weapons-related research, Chalk River set up a laser fusion working party (LFWP) to review the prospects for laser fusion. This approach looked promising for Canada because two laboratories, the Defence Research Establishment, Valcartier (DREV), and NRC, were at the forefront of high-power carbon-dioxide laser development, and laser-plasma interaction studies were underway, notably at NRC.

In laser fusion, a small frozen DT pellet is fired into an evacuated reaction chamber that is surrounded by a battery of high-power lasers, all aimed at the centre of the chamber. When the pellet reaches the centre, the lasers are fired and their radiation compresses and heats the pellet. The resulting fusion reaction is so rapid that a substantial amount of energy is generated before the pellet expands and disperses, which is the ICF principle.

Members of the LFWP visited many laboratories in Canada and the United States, including Livermore and Los Alamos, and consulted their experts, who were very helpful. It became clear that there were serious problems: the pellet compression process was not well understood, and much of this research would remain classified because of its relevance to weapons; the optimum wavelength was unknown, though it was almost certain to be much smaller than that of carbon dioxide lasers, where Canadian high-power laser expertise was concentrated; laser performance and cost would have to improve dramatically; many engineering problems would have to be solved (e.g., pellet fabrication and delivery, and materials damage, including damage to lasers).

The LFWP concluded that any Canadian initiative would have little effect on the rate of progress, and its recommendation, made in October 1973, was that AECL should await developments and reassess the situation when scientific feasibility had been demonstrated.

In fact, progress has been slower than expected and ICF, including the possible use of particle beams instead of lasers, is still much less advanced than MCF. The current world effort is about 90 percent MCF and 10 percent ICF with most of the latter contributed by the American program, which is largely defence-oriented.

The LFWP recommendation was accepted, and AECL continued its fusion-awareness activity at about the pre-1973 level. However, to safeguard AECL's potential future role, an agreement was reached in 1975 with NRC, which had been designated lead agency for fusion, that AECL would be the logical agency to take responsibility if and when a major commitment to a Canadian fusion reactor project was warranted.

The introduction to the LFWP report had stated that, because of the smaller energy release per neutron in fusion than in fission, fusion could be considered a better source of neutrons than of energy. This suggested replacing the idea of using fusion to produce power directly with a scheme in which most of the power would come from a near-breeder fission reactor, with fusion supplying a relatively small number of make-up neutrons, probably in the form of fissile material. In this role, fusion would be in direct competition with accelerators.

In 1981 AECL conducted a major study of the economics of fusion as a source of neutrons for fission-energy systems. A DT fusion reactor would be surrounded by a blanket, containing uranium and/or thorium and lithium, in which the 14.1 MeV neutrons from the DT reaction would be multiplied, by a variety of nuclear reactions, and produce additional energy, fissile material, and enough tritium to fuel the fusion reaction. Typically, a uranium-lithium blanket would produce 1.6 atoms of plutonium-239 and 130 MeV of energy per deuterium atom consumed, in addition to the mandatory atom of tritium. However, blanket operation, including the extraction of fissile material and tritium, represented a formidable engineering challenge, and the estimated thirty-year-lifetime cost of the system exceeded that of currently available stand-alone reactor systems by about a factor of three. As noted earlier, the accelerator breeder was found to be similarly uneconomic, though the accelerator technology was considered to be closer to realization than controlled thermonuclear fusion. Interest in either approach waned as the prospect of thorium fuel cycles receded (see chapter eighteen).

Although AECL did not embark on a fusion program of its own during this period, it played a role in the establishment of a national fusion program (NFP). This was a complicated business and only a brief outline can be given here. It started with Project FC (Fusion Canada), a federally funded study by five Quebec institutions that, in November 1974, recommended programs in ICF with carbon-dioxide lasers, MCF with a small tokamak, fusion-reactor materials with a strong neutron source, and the design and evaluation of conceptual fusion reactors. An advisory committee, which included an AECL representative, reported on this proposal in March 1975. It supported the ICF and materials work, but not the building of a tokamak, and recommended only a very small reactor design effort.

Response from the federal government came slowly. An interdepartmental panel on energy R&D (PERD), set up in January 1974, had designated NRC as the lead agency for fusion in June 1974, but it was only in June 1978 that NRC was able to announce an NFP. Its principal components were: participation in foreign projects, an ICF program at NRC, and a tokamak facility at IREQ, Hydro-Quebec's research institute in Varennes. The limited funds available in the first few years of the program were used to finance detailed studies of the feasibility of projects in these areas.

In 1981, the federal and Quebec governments concluded an agreement to share the cost of building the Varennes tokamak, roughly $40 million over a five-year period. It produced its first plasma in 1987 and has since made significant contributions to fusion research.

In 1982, the federal and Ontario governments and Ontario Hydro agreed to set up a Canadian fusion fuels technology project (CFFTP) as part of the NFP. CFFTP would exploit the tritium technology that had been developed in the CANDU reactor program, and tackle the process-engineering and materials problems associated with the tritium breeding blanket and the first wall of a fusion reactor. AECL undertakes significant tritium-related research, some of which predated the setting up of CFFTP (see chapter fourteen).

The intended ICF component of the NFP has not materialized. NRC's own program took a different direction, and attempts by the University of Alberta to obtain funding for a large laser-fusion program have been unsuccessful.

AECL took over the management of the NFP from NRC in a phased operation that was completed by 1 April 1987. With a funding level of some $25 million per year, the NFP is modest compared to the fusion programs of some other countries, but its tightly focused research is highly regarded and Canada plays an effective role in international fusion activities. Strategically the NFP is designed to ensure that Canada can make the best possible use of the ongoing opportunities offered by fusion R&D around the world.

EPILOGUE

An assessment of the programs described in this chapter inevitably focuses on Dr. W.B. Lewis, who was responsible for Chalk River having an accelerator program and not having a fusion program.

He proposed the use of accelerators to product neutrons for "breeding" fissile material at a time when the perceived shortage of uranium provided strategic, if not immediately economic, justification and he continued to believe, throughout his life, that accelerator-produced neutrons could have a role in meeting the world's eventual energy requirements. It is far from certain that he was wrong.

His championship of ING is no surprise – neutrons for research were of incalculable value – and he found it hard to accept that the Canadian government was capable of making a substantial commitment to CANDU but had nothing for what he saw as an important component of Canada's scientific and technological future.

After the ING program was stopped, Lewis continued to take an interest in futuristic (e.g., ion-drag) accelerators, but he had no direct responsibility for the post-ING accelerator program. Perhaps he would have supplied the firm direction it needed – there were always too many projects and too few people.

Regarding fusion, this was a simple matter for Lewis. He knew, and he was right, that its realization would take much longer than its proponents claimed. In that sense it was no threat to AECL's programs. Nor was it an opportunity for Chalk River or for Canada; there were no home-grown bright ideas, like CANDU, but only the possibility of making a small addition to the research already going on in other countries. As might be expected, some of the Canadian fusion proponents regarded him – a senior scientist whose views were sought at the highest level – as their implacable enemy. In fact, his pronouncements were always fair, and founded on a knowledge base at least as extensive as theirs. Altogether, in the area of fusion, as in so many other areas, he served his laboratory and his country well.

ACKNOWLEDGMENTS

The author has made extensive use of Technical Information Services at Chalk River and is very appreciative of the help received, particularly from B.A. Barrie in the Scientific Documents Distribution Office, M.R. Charbonneau in the Library, and R.M. McNamara and her staff in Central Records. He is also greatly indebted to former colleagues, especially to C.B. Bigham and J.H. Ormrod for accelerator physics, to D.P. Jackson for fusion, and to G.A. Bartholomew for ING and the post-ING program generally. Valuable comments from outside AECL were provided by R.M. Fawcett and D.S. Hanna.

BIBLIOGRAPHY

AECL Laser Fusion Working Party. *A Review of the Prospects for Laser Induced Thermonuclear Fusion.* AECL-4840. 1974.

Bartholomew, G.A., and J.S. Fraser. *AECL Programs in Advanced Systems Research.* AECL-7074. 1981.

Bartholomew, G.A., and P.R. Tunnicliffe. eds. *The AECL Study for an Intense Neutron Generator: Technical Details.* AECL-2600. 1966.

Bercovitch, M., H. Carmichael, G.C. Hanna, and E.P. Hincks. "Yields of Neutrons per Interaction in U, Pb, W, and Sn by Protons of Six Energies between 250 and 900 MeV Selected from Cosmic Radiation." *Physical Review* 119 (1960): 412-31.

Bigham, C.B., W.G. Davies, E.A. Heighway, J.D. Hepburn, C.R.J. Hoffmann, J.A. Hulbert, J.H. Ormrod, and H.R. Schneider, "First Operation of the Chalk River Superconducting Cyclotron." *Nuclear Instruments and Methods in Physics Research* A254 (1987): 237-51. (Also AECL brochure, *TASCC,* 1987).

Church, T.G. ed. *The AECL Study for an Intense Neutron Generator: ING Status Report, July 1967.* AECL-2750. 1967.

Doern, G. Bruce. *Science and Politics in Canada.* Montreal: McGill-Queen's University Press. 1972.

Ebrahim, N.A. "Laser Particle Acceleration Research at Chalk River Laboratories." *Research Trends in Physics* (1992): 310-51.

———. "Lasers, Plasmas, and Particle Accelerators." *Physics in Canada* 45 (1989): 178-88.

Fawcett, Ruth. *Nuclear Pursuits: The Scientific Biography of Wilfrid Bennett Lewis*. Montreal: McGill-Queen's University Press, 1994.

Fraser, J.S., C.R. Hoffmann, S.O. Schriber, P.M. Garvey, and B.M. Townes. *A Review of Prospects for an Accelerator Breeder*. AECL-7260. 1981.

Geiger, J.S., and G.A. Bartholomew, eds. *A Review of the Prospects for Fusion Breeding of Fissile Material*. AECL-7259. 1981.

IMPELA. Kanata, Ont.: AECL Business Unit, n.d.

Lawrence, E.O. "High Current Accelerators." *Science* 122 (1955): 1127-32.

Lewis, W.B. *Atomic Energy Objectives for Applied Nuclear Physics Research*. Report DR-23. Chalk River: Atomic Energy of Canada Limited, 1952.

———. *Chalk River Prospects for Atomic Power*. Rev. 2. Report DR-22. Chalk River: Atomic Energy of Canada Limited, 1957. (Also AECL-395.)

———. "Energy in the Future – the Role of Nuclear Fission and Fusion." *Proceedings of the Royal Society of Edinburgh* A70 (1971/72): 219-23. (Also AECL-4409.)

———. *The Significance of the Yield of Neutrons from Heavy Nuclei Excited to High Energies*. Report DR-24. Chalk River: Atomic Energy of Canada Limited, 1952. (Also AECL-968.)

MACPHEE Technical Corporation. *Fusion: Energy for the Future*. Rev. ed. Chalk River: National Fusion Program, AECL Research, Chalk River Laboratories, 1991.

McMichael, G.E. "CW Linear Accelerator Developments at CRL: Status and Future." *Nuclear Instruments and Methods in Physics Research* B10/11 (1985): 825-30.

Schriber, S.O. *Canadian Accelerator Breeder System Development*. AECL-7840, 1982.

Science Council of Canada. *The Proposal for an Intense Neutron Generator: Initial Assessment and Recommendations*. Report, no. 2. Ottawa: Science Council, 1967.

Science Council of Canada. *The Proposal for an Intense Neutron Generator: Scientific and Economic Evaluation*. Special Study, no. 4. Ottawa: Science Secretariat, Privy Council Office, 1967.

Taylor, T., G. VanDyk, L.W. Funk, R.M. Hutcheon, and S.O. Schriber. "Therac 25, A New Medical Accelerator Concept." *IEEE Transactions on Nuclear Science* NS-30 (1983): 1768-71.

Ungrin, J., S.T. Craig, N.H. Drewell, N.A. Ebrahim, J-P. Labrie, C.B. Lawrence, V.A. Mason, and B.F. White. "IMPELA: An Industrial Accelerator Family." In *Accelerator Physics Branch, Chalk River Laboratories: Annual Technical Report, 1989,* 30-1. Ed. by J.A. Hulbert. AECL-10235. 1990.

York, Herbert F. *The Advisors: Oppenheimer, Teller, and the Superbomb*. San Francisco: W.H. Freeman, 1976: 122-5.

CANDU DEVELOPMENT

Chapter Ten

CANDU[1] and Its Evolution

H.K. RAE

Chapter one described the beginnings of the atomic-energy program in Canada, and how Chalk River became established as an internationally prominent research centre by the early fifties. In common with other countries where similar research centres had evolved in this period, notably the United States, the United Kingdom, France, and the USSR, there was at Chalk River much speculation and some preliminary conceptual studies of the prospects for nuclear power generation.

W.B. Lewis, the director of research at Chalk River, initially leaned toward breeder reactor concepts to maximize the energy obtainable from available uranium and thorium resources, which at that time were thought to be very limited.

G.C. Laurence, a senior member of Chalk River staff, began in 1950 to advocate the natural uranium-fueled heavy-water reactor for power production, a concept that made maximum use of existing Canadian nuclear experience.

Chalk River was aware of British and American developments through the United States-United Kingdom-Canada tripartite agreement. However, detailed information on nuclear power development was not provided by either the United Kingdom or the United States, because their power programs were interwoven with their military programs. American legislation prior to 1954 severely restricted release of information about its pressurized-water-reactor development. But Chalk River's NRX reactor was an essential testbed for the fuel for the United States navy's nuclear-submarine program. The navy was instrumental in having the first high-temperature, high-pressure water loop installed in NRX in 1951 (see chapter two).

By 1951, Canada's nuclear power program began to take shape. In August of that year, Lewis, drawing on input from many staff at Chalk River, prepared "An

---

[1] For readers unfamiliar with the CANDU power reactor a brief description is appended to this chapter, providing background for chapters ten through eighteen.

Atomic Power Proposal", a report that gave direction to this program and began what would become Lewis's missionary zeal for the natural-uranium-fueled, heavy-water-moderated power reactor. In his report he proposed that it might be possible for nuclear power to compete with coal as a source of electricity. He cited three achievements or near-achievements as the basis for his forecast. First was the achievement of irradiations of NRX fuel to a total heat output of 70 MWh/kg of uranium, making uranium three or four times cheaper than coal or oil as a heat source, and there were prospects for longer irradiations. Second, the design for the NRU reactor was in progress; it incorporated means for continuous operation with on-power fueling, and confirmed that high-power output should be attainable from a physically small reactor. Third, the experience with the United States navy loop experiments confirmed the practicality of operating a high-temperature, high-pressure water system with nuclear fuel, and zirconium seemed to be a candidate to combine adequate strength, good corrosion resistance and low neutron capture as a fuel-sheathing material.

Because of the presumed scarcity of uranium, Lewis stressed in his early proposal, and in later refinements of it, the need to chemically process the irradiated fuel and recycle the recovered uranium and plutonium to the reactor. By multiple recycles he foresaw a total consumption of as much as 3 percent of the initial uranium, ten times that consumed in a single irradiation of 70 MWh/kg. However, as it later transpired, uranium is a relatively abundant element and its oxide can be irradiated to 200 MWh/kg in CANDU, so recycling became unnecessary and uncompetitive. The story of fuel cycles and how their study was revived in the seventies and eighties as future options for CANDU is told in chapter eighteen.

By 1952, the atomic energy project at Chalk River was the largest NRC division and further growth was planned, with the construction of the NRU research reactor underway. If Lewis's proposal for nuclear power development was accepted, even greater growth might be necessary, and a greater degree of industrial orientation would be desirable.

In addition, Chalk River was beginning to generate commercial revenues from isotope sales and loop irradiations. These factors led to the formation of AECL as a crown corporation, and the transfer of the Chalk River laboratories to AECL on 1 April 1952.

By the early fifties, Ontario Hydro had become interested in the prospects of nuclear power as an alternative to coal for electricity production. In January 1952, the premier of Ontario Leslie Frost and the chief engineer of Ontario Hydro, Richard Hearn met with C.J. Mackenzie, president of NRC and soon-to-be president of AECL, and C.D. Howe, the federal minister responsible for NRC and AECL. Thus, the stage was set for the long and productive period of cooperation between AECL and Ontario Hydro that became one of the essential elements in the successful development of CANDU. The immediate results of the meeting were the appointment of Hearn to AECL's board of directors, and preliminary joint studies of potential reactor designs by AECL and Ontario Hydro engineers. Half of the eight members of AECL's first board of directors were representatives of Canadian electrical utilities, and this strong utility connection continued for nearly two decades throughout the formative years of CANDU development.

But progress was slow, in part due to the priority that had to be given to salvaging NRX after the December 1952 accident. In September 1953, Chalk River hosted AECL's first Atomic Power Symposium, attended by representatives of industry and the utilities to discuss the feasibility studies and supporting research and development. While heavy water was strongly favoured as the moderator, the choice of coolant was less clear; options considered included light water, heavy water, steam, gas, organic liquids and liquid metals. Lewis proposed the construction of a small demonstration power reactor. This symposium implemented a policy that originated with C.D. Howe, requiring the early involvement of utilities and industry in the nuclear power program, and limiting AECL's role to R&D.

THE NUCLEAR POWER GROUP AND NPD-1

Coincident with the appointment in November 1953 of W.J. Bennett to replace C.J. Mackenzie as AECL president, the power program began to move forward at an increasing pace. Bennett announced to AECL staff that nuclear power was to be their main objective. He reorganized Chalk River, appointing Lewis as vice-president for research and development (a job he was already doing) and Lorne Gray as vice-president for administration and operations. Early in 1954, utilities, engineering companies and manufacturers were invited to participate in a feasibility study of nuclear power. Representatives of Babcock-Wilcox and Goldie-McCulloch Co. Ltd., British Columbia Electric Company Ltd., Canadian-Brazilian Services Ltd., the Hydro-Electric Power Commission of Ontario (Ontario Hydro), Montreal Engineering Co. Ltd., and Shawinigan Water and Power Co. Ltd, nine in all, came to Chalk River and formed what became known as the nuclear power group (NPG). It functioned as a unit of Chalk River's engineering design division and was led by Harold Smith of Ontario Hydro. The group had access to everyone at Chalk River, and included as its tenth member an AECL reactor physicist.

The aim was to develop a nuclear power reactor that could compete economically with coal-fired stations. It was to be fueled with natural uranium and moderated by heavy water, because this concept appeared promising for power production, it was well-suited to Canadian resources and did not require enriched uranium, it was an extension of existing Canadian reactor experience, and it was perceived as requiring only a small development program. In retrospect, we know that the development program was large, but it was probably less than might have been required had a different reactor concept been chosen. Implicit in the above rationale was Lewis's conviction that low fueling costs would be crucial to success, since they would offset the anticipated high capital costs. Another factor must have been the knowledge that large-scale heavy-water production plants were operating in the United States, that the cost of production was not unreasonably high, and that this technology would be available to Canada (see chapter seventeen). Finally, this reactor concept appeared to fit in with Ontario Hydro's history of capital-intensive power stations that had low operating costs. Ontario Hydro also wanted a design that was consistent with the limited industrial resources available in Canada.

Although the reactor power was not specified, a maximum capital cost was set that limited the size to about 10 MW, thereby limiting the risk associated with this first step toward a new technology.

The fuel considered was uranium metal sheathed in Zircaloy-2, an alloy of zirconium being developed in the United States for its nuclear submarine program. Both ordinary water and heavy-water coolant were considered. Both steel pressure vessel and zirconium alloy pressure tube reactor designs were examined. An AECL committee, the power reactor group, chaired by W.B. Lewis, provided guidance and review as the study progressed. At the end of 1954, the NPG concluded that a power rating in the range of 10 to 20 MW would be within the capital cost target; that there was little difference in cost between a pressure vessel and pressure-tube design, but that the former was preferred since it was based on established technology; and that heavy water was the preferred coolant, because in larger reactors it would give lower unit-energy costs.

The design parameters for the prototype were agreed upon and proposals were invited for detailed design and construction. Seven bids were received. AECL selected Canadian General Electric in March 1955 as the prime contractor for NPD – the Nuclear Power Demonstration. CGE would contribute $2 million in engineering development in support of the design to be carried out in Peterborough, where they had decided to locate their Civilian Atomic Power Division (CAPD). The total cost of NPD was expected to be $20 million. AECL's annual operating budget for Chalk River was only about $10 million at this stage (excluding the construction of NRU). Ontario Hydro was confirmed as the utility to be involved in the project; they would supply the site and the conventional power equipment, buy steam from the nuclear system and operate the plant. AECL would pay for and own the nuclear system.

CGE supplied about a dozen of its own engineers to the initial staff of CAPD and its general manager, Ian McRae. They recruited about the same number of engineers from Chalk River, including Ian MacKay who was to manage engineering, and several of the members of the NPG including John Foster as manager of design. Both AECL and Ontario Hydro retained a strong influence on key decisions through regular review meetings.

With the design and development proceeding in parallel, it was inevitable that design changes would be needed as new information became available; this would lead to a better demonstration, but delay the reactor's completion and increase its cost. Charting a course through this minefield was the task of the NPD technical committee, chaired by Lewis; the NPD project committee, consisting of senior AECL, Ontario Hydro and CGE managers and chaired by Gray; and ultimately Bennett and AECL's board of directors. Lewis was frequently the champion of change.

Initially, NPD was to have an electrical output of 10 MW. However, by August 1955, AECL had decided to increase this to 20 MW, so that fuel performance information would be produced more quickly and the fuel rating in central channels would be closer to that expected in future large commercial reactors. The channel diameter had already been selected as 83 mm based on a combination of engineering design considerations and reactor-physics optimization studies. A seven-element fuel bundle had been selected, but with the increase in fuel rating it became necessary to further subdivide the fuel in the central channels and a 19-element bundle design was selected for use in these channels. This early decision on a 19-element bundle configuration, 80-mm in diameter, set the pattern for CANDU fuel development and it became the ancestor of all CANDU fuel.

At about the same time, the fuel material was also changed. Originally, Zircaloy-clad uranium metal had been chosen, although some priority was also given to developing a high-temperature aluminum alloy as sheathing, because of the high cost of zirconium. Work on uranium oxide as a fuel had also begun. Information exchange with the United States and the United Kingdom had greatly improved in 1954 with the relaxation of secrecy, in preparation for the first United Nations conference on the peaceful uses of atomic energy. Both the United States and the United Kingdom were experiencing problems with uranium metal. The advocates of uranium oxide were extolling its advantages. In the summer of 1955, Westinghouse's Bettis Atomic Power Laboratory (BAPL), one of the United States navy's reactor development contractors, tested an intentionally defected uranium dioxide fuel element in an NRX loop. Although CRL was not told what the fuel was, a routine X ray before irradiation showed it was only half as dense as uranium metal and the inference was clear. Its excellent performance and AECL's intimate knowledge of that performance convinced Lewis that oxide was the correct choice. In October he went to an NPD technical committee meeting at Peterborough and told CGE that the fuel would be uranium oxide.

While the design of NPD proceeded, the NPG was augmented, renamed the nuclear power branch, and given the task of studying a large uranium-heavy-water reactor, to determine a conceptual design and power output likely to compete with alternative power sources in Canada, and to define the development work required. This study took two years.

In the spring of 1957, the nuclear power branch reported its conclusions as follows: power output had the largest effect on unit energy cost and a unit of least 200 MW was required to compete with a coal-fired power station. A 200 MW station should be built, since this was the best fit to Ontario Hydro's anticipated needs in ten years' time. The size of pressure vessel required would be much larger than an equivalent enriched uranium light-water reactor vessel, and beyond current experience. Its manufacture would require a technology unlikely to be developed in Canada (in fact, the much smaller NPD vessel was being fabricated in Scotland). Zircaloy pressure-tube technology was being developed in the United States for the Hanford N Reactor[2] so a pressure-tube design was now considered to be practicable, and was recommended. Since a thirty-year amortization period would be needed to achieve a low unit energy cost, the station life should be at least thirty years. However, it would be premature to set this as the target for the reactor, and therefore the core should be replaceable; that would be possible with a pressure-tube design. A horizontal orientation of the pressure tubes simplified the support of the reactor and its fuel, and made bi-directional on-power fueling practicable. A major and essential innovation, one unique to CANDU, was this

[2] A graphite-moderated pressurized-water reactor being built for the United States government to produce both plutonium and electricity.

fueling system. Short fuel bundles would be inserted in one end of a fuel channel and would be moved progressively through the channel as each new bundle was added and an irradiated one removed from the other end. By fueling adjacent channels in opposite directions a more uniform neutron flux could be obtained throughout the reactor. The higher net neutron production from fresh fuel would offset the neutron losses by absorption in the fission products within fuel nearing the end of its irradiation in the adjacent channel. All the fuel would leave a channel at roughly the same total irradiation, which would be about twice that achievable with unidirectional fueling or with full-length fuel bundles. The higher average fuel irradiation achieved with this fueling scheme helped to compensate for the extra neutron loss by absorption in the pressure tubes. One disadvantage of the pressure-tube design was a higher capital cost due to an increase in heavy water volume in the moderator relative to that in a pressure-vessel design, since a wider spacing between channels was required to accommodate feeder pipes at each end of each channel and the head of the fueling machine.

Uranium oxide sheathed in Zircaloy was recommended as the fuel; even though it gave slightly higher unit-energy costs than uranium metal, its good performance seemed much more certain. It was predicted that the core design would permit an irradiation of 195 thermal MWh/kg with natural uranium. Uranium enrichment was found to reduce the unit energy cost somewhat, but only if a much higher fuel irradiation could be achieved. For any irradiation up to that attainable with natural uranium, the enriched fuel results in a substantially higher cost. Enrichment was not recommended, since fuel performance at high irradiation was unknown.

This study, led by Harold Smith, set the main conceptual design features of CANDU, and most of these have been retained throughout the program. The choice of a pressure-tube reactor design removed any limit on reactor size, so that the same basic design could be used as progressively larger units were required. However, now NPD, with its pressure vessel, was no longer a representative demonstration. And so two years into the design it would have to undergo a major change and literally be turned on its side. Bothwell, in *Nucleus*, his history of Atomic Energy of Canada Limited, describes how AECL, Ontario Hydro and CGE made this decision, which would double costs and delay construction by a year or more, but would provide essential early experience with the CANDU concept. Some orders, including that for the pressure vessel, had to be cancelled.

NPD-2

Major new features – the fuel channel, the horizontal calandria and the fueling machine – had to be designed, developed and tested for NPD-2. The nuclear power branch had proposed a fuel bundle length of one foot. The CGE team decided 50 cm was short enough, and this was a decision that Lewis endorsed immediately. The same length is still used for all CANDU fuel, even though it was recognized by about 1960 that a 200 cm long bundle would be adequate to achieve the desired fuel irradiation benefits of using the bi-directional, on-power fueling scheme. Having fewer end plates in the channel should permit a somewhat higher irradiation. However, the 50 cm length was recognized to be extremely convenient to make and handle, and has proven to be very reliable and sufficiently efficient.

The choice of fuel-sheath thickness is another prime example of Lewis's influence on the project. He was determined to minimize the non-fuel materials within the reactor core that waste neutrons and thereby limit the total heat output from the fuel (neutron economy is discussed in chapter eleven). His goal was a fuel irradiation of 240 MWh/kg, which would result in an attractively low fueling cost for commercial CANDUs. By late 1958, fuel-design details were being specified so that production could begin in 1959. The NPD technical committee reviewed the design. CGE and some of the AECL fuel developers recommended 0.6-mm-thick sheaths, 20 percent thinner than the manufacturers were already making for American reactors. Lewis proposed 0.25 mm and insisted on a sheath no thicker than 0.4 mm for the 19-element fuel, unless it could be proved why it had to be thicker. And he was determined that the performance of thin-sheathed fuel would be demonstrated in NPD from its initial startup. Since all agreed that the seven-element fuel for the outer channels should have 0.6 mm sheaths, this fuel was produced first, giving fuel developers and fabricators a year to investigate thin sheathing. By the beginning of 1960 it was clear that thin sheathing showed considerable promise, but irradiation experience was limited. Consequently, the decision was taken to produce two batches of 19-element fuel, one with 0.6 mm sheaths and the other with 0.4 mm. About 200 bundles of each formed the initial fuel loading for the central channels of NPD in early 1962. The thin sheaths performed well and, ever since, all CANDU fuel has been sheathed with 0.4-mm-thick Zircaloy tubing. The development of the NPD fuel and the evolution of this design to much higher power ratings in later reactors was a major multidisciplinary program (see chapter thirteen).

While American information on pressure tubes provided a starting point, the CANDU design conditions were different and the need to minimize neutron absorption was much greater. The development of fuel-channel materials became a large program, which continued to grow in size and breadth (see chapter twelve), and it remains a major development activity still.

Each of the 132 NPD fuel channels contained nine fuel bundles; however, the pressure-tube length was 4 m, so half of each bundle at the ends of the channel was outside the neutron flux.

An important decision in the design of NPD was the choice of carbon steel for the heat-transport-system piping, rather than the more expensive stainless steel being used in American reactors. It was recommended by CAPD for NPD-1, where the system piping surface was small, and was endorsed by the metallurgists and chemists at Chalk River. The choice was maintained for NPD-2, despite the much more extensive piping required for the pressure tube design, and carbon steel has been successfully used in all CANDU reactor heat transport systems (see chapter fourteen).

The carbon-steel inlet and outlet feeder pipes were to be connected to the fuel channels through end fittings, which would also provide access for the fueling machines to connect to the ends of each channel. To fasten the pressure tube to each end fitting, a rolled joint was selected as the best alternative. The end fitting was fabricated from a stainless-steel alloy having the required strength and a good match to the zirconium-alloy pressure tube in its coefficient of thermal expansion. Years of development and testing went into achieving an adequate design. The same rolled-joint concept has been used for all CANDU fuel channels, although it has evolved in details of design and fabrication. Inadequate control of rolling procedures was the cause of some pressure-tube failures in the Pickering and Bruce stations (see chapter twelve). As a result, closer tolerances and better inspection procedures were instituted.

An ingenious system was proposed for reactor control using moderator level combined with a means of rapidly dumping the moderator into a dump tank for reactor shutdown. This concept was applied to NPD, Douglas Point, RAPP and Pickering A. Moderator dump for shutdown was chosen for the early CANDUs because of shortcomings in the reliability of shut-off rods at that time. However, for Pickering and later CANDUs, moderator dump would not provide sufficiently rapid shutdown, so shut-off rods of demonstrated high reliability were adopted; moderator dump was retained in Pickering A to provide a greater depth of shutdown than shut-off rods alone, a function that was provided in all later CANDUs by the injection of a soluble neutron absorber (soluble poison, e.g., gadolinium nitrate) into the moderator. The main control of NPD was through moderator level, but this technique was of lesser importance in Douglas Point, RAPP and Pickering A. It was omitted from all later CANDUs. Reactor power was controlled by solid neutron absorbers in Douglas Point, and liquid ones in Pickering and later CANDUs (see chapter eleven).

A major concern with any heavy-water reactor is to avoid losses of this costly material ($60 per kilogram when NPD was being designed). Extensive leakage-collection systems were installed. However, these could not be applied to the many joints and valves in the high-pressure, heat-transport system, where most small leaks of heavy-water that might occur would immediately vapourize. The answer was to isolate the reactor vault and the boiler room so that large air driers could recover the bulk of this leakage. A preliminary version of this concept was installed at NPD. The commercial CANDU units have benefited from a substantial program to improve heavy-water containment and minimize losses (see chapter fourteen).

Lorne McConnell of Ontario Hydro created the NPD operating organization, which in turn produced the operational procedures that became the pattern for all Ontario Hydro nuclear stations, and for the other utilities that operate CANDUs. NPD was not only the first Canadian power reactor, it was also the training ground for designers, researchers and operators who would go on to manage the growth of the CANDU program over the next two decades.

Initial operation of NPD occurred in April 1962 and full power was reached on June 28. NPD successfully demonstrated the feasibility of the CANDU concept through reliable operation, safe refueling at full power and acceptably low heavy-water losses. NPD operated until 1985, when it became evident that its pressure tubes should be replaced (see chapter twelve); after a prolonged shutdown it was decided that the cost of retubing would far exceed the value of continuing to operate NPD, and it was taken out of service in 1987. It ran reasonably well throughout its operating life and provided early experience with CANDU components, systems and materials. NPD was also a testbed for new fuel-bundle designs, for alternative pressure-tube materials, new instruments and other components. However, its usefulness in this regard was rather limited by the fact that its neutron flux was considerably lower than that in large CANDUs. For a period, the heat-transport system was modified to permit coolant boiling; this provided confidence that the large commercial CANDU reactors could operate stably with net steam production in the more highly rated channels.

An experimental irradiation of cobalt in NPD in 1965-66 demonstrated the concept of cobalt-60 production in CANDU reactors (see chapter five). In NPD the cobalt was contained in three elements of a bundle, similar to a fuel bundle, with the other elements containing only helium. Three experimental bundles interspersed with natural-uranium fuel were loaded by the fueling machine into one channel. Due to the neutron absorption by the cobalt, the average burnup of all the fuel in NPD was reduced by about 3 percent during the eighteen-month experiment. Forty cobalt bundles of the NPD design were loaded into four channels of Douglas Point in 1966, reducing the fuel burnup by 12 percent. However, cobalt-60 production was low because of the poor early performance of Douglas Point (see below); when increased reactivity in the core was needed in 1969, the cobalt bundles were replaced by fuel. Arrangements were made for cobalt production in reactivity control rods (adjuster rods) in Pickering, and subsequently in other CANDUs (see chapter eleven). While awaiting the first output of cobalt-60 from Pickering, eighty bundles were loaded into NPD in 1971; in this case the reactor had to be partially fueled with slightly enriched uranium (1 percent uranium-235). Three-million curies were produced, but the cost was considerably higher than from adjuster rods in Pickering or later CANDUs.

The important contribution of NPD was in its early years. There were problems with components to be overcome – faulty valves, for instance. Heavy-water leakage and losses were high initially. Improvements to the recovery system and the elimination of many sources of leakage reduced average losses from 25 to 7 kg per day. The latter would be an acceptable loss rate for a large CANDU reactor.

The initial attempt at on-power fueling failed, resulting in the escape of heavy water from the heat-transport system. About 1,000 kg were lost, but the bulk of the escaped water (4,000 kg) was recovered and had to be purified and reconcentrated to reactor grade, since it had become mixed with ordinary water when the dousing system had been triggered by the pressure rise in the reactor vault. The fueling machines were repaired, but required considerable maintenance to retain adequate reliability. Machines of an improved design were installed in 1969.

DOUGLAS POINT

In 1957, while CGE got busy with NPD-2, AECL and Ontario Hydro decided to proceed with a detailed design assessment and development work for the 200 MW station, which would become known as Douglas Point. At the time, a power of 200 MW was considered to be a commercial station. In fact, when Douglas Point was built, its steam condenser would be the largest in Canada. However, it was soon surpassed by much larger CANDU units, and so for the last two decades Douglas Point has usually been referred to as a prototype reactor.

For the 200 MW design study AECL set up the nuclear power plant division (NPPD) in Toronto early in 1958 in space provided by Ontario Hydro. In essence, this decision was an arrangement imposed by Ontario Hydro, who would be the ultimate customer.

Ontario Hydro agreed to participate in the design study and supplied about fifteen engineers, including Harold Smith on a part-time basis to manage NPPD. The Toronto location provided for close association with Ontario Hydro and ready access to industry, but also meant less direct influence by Lewis and the R&D environment of Chalk River.[3] NPPD was given control of CGE's contract for NPD-2, so that this project would be coordinated with the 200 MW design study and its supporting engineering development work, also to be a part of NPPD.

J.L. Gray, who by May of 1958 would become president of AECL, outlined AECL policy in establishing NPPD in a report (AECL-561) entitled "A Statement of the Nuclear Power Development Program." Gray stated that arrangements were being made for participation of other utilities in addition to Ontario Hydro, and of manufacturing and engineering companies; that the design would be directed toward maximum use of Canadian manufacturing facilities; and that NPPD was envisaged as a transitional step until nuclear power had reached a stage of use at which utilities and industry could undertake further development on a normal commercial basis. However, NPPD grew into a permanent part of AECL (known as Power Projects, then CANDU Operations, and more recently AECL CANDU), providing engineering services for domestic CANDU stations and overall project execution and management for CANDU stations overseas.

John Foster joined NPPD as deputy manager and became manager after Smith became chief engineer at Ontario Hydro. Another wave of transfers from Chalk River to the CANDU design activities took place, and a half-dozen or so attachments from other companies to NPPD were obtained. Engineers were also recruited from the Avro Arrow project when it was cancelled in February 1959.

For a short time, NPPD was side-tracked into developing a conceptual design for a vertical reactor, but Lewis and Gray firmly insisted that the designers stick to the horizontal arrangement already chosen.

By the spring of 1959, the government, AECL and Ontario Hydro were anxious to advance the start of work on the 200 MW unit, mainly because of the rapid pace of development of nuclear power in the United States, Britain and France. AECL and Ontario Hydro reached an agreement whereby AECL would design and own the station, and Ontario Hydro would construct and operate and pay for the electricity. Ontario Hydro would also own the site and have an option to buy the station once it had demonstrated successful operation. The Douglas Point site, near Kincardine on Lake Huron, was selected as being suitably remote (in case of a major accident), yet not too far from Toronto. It was at this time that the acronym CANDU was selected, originally to designate the Douglas Point station. However, it was usual to call power stations by place names and CANDU became the generic name for the Canadian heavy-water power reactor. The Douglas Point project established the organizational pattern for Ontario Hydro's nuclear power program. It was a learning experience for NPPD, Ontario Hydro's construction division and a large number of Canadian manufacturers.

To achieve the ten-fold increase in power output from NPD to Douglas Point required a higher coolant temperature to increase steam-cycle efficiency, a doubling of the maximum power rating of the fuel, a higher average to maximum fuel power rating (flatter neutron flux distribution), a 5 m (rather than 4 m) channel length,[4] and 306 fuel channels instead of 132. There were many changes in the details of material selection, components and systems design from NPD. Zircaloy calandria tubes

[3] One senior member of NPPD recalled that the contribution of W.B. Lewis must never be underestimated. He was in the background but he was like Buddha or God. He had such an obvious grasp of nuclear energy.

[4] Although each fuel channel in Douglas Point contained twelve fuel bundles, only ten were within the flux.

and a stainless-steel calandria were selected for greater strength and better corrosion resistance than aluminum used in NPD, and became the norm for all future CANDUs. The reactor and its auxiliaries were housed in a concrete containment building with 1-m-thick walls, rather than in an underground vault, and a more elaborate dousing system was provided to control pressure in the event of a major release of steam. The evolution of Canadian reactor-safety philosophy and the design response to these requirements has already been described in chapter four. A major innovation was the use of extensive computer data processing and a limited degree of computer control; in this first application of computer control to a power reactor, the tasks of the computer were limited to those that could be done manually or by separate equipment if the computer was not working. Douglas Point was the first reactor in the world in which a reactivity-control element was positioned by a stored-program digital computer.

Although the location of NPPD in Toronto had diminished Lewis' direct influence on the design of Douglas Point, he continued to lead the supporting development work at CRL well into the sixties. As he had throughout the design of NPD, Lewis provided the stimulus, coordination and much of the direction of this program. He delved deeply into the details of many areas of research[5] to gain a detailed understanding so that he could assess problems, participate in the solutions and give more effective overall direction. He delighted in detailed technical participation, especially in the area of fission product movement in uranium dioxide (see chapter thirteen). But most importantly, he had a great talent for synthesizing disparate scientific, engineering and cost factors into a coherent picture that could lead to a solution, which he then pursued tenaciously. Lewis dominated CRL throughout this period and was an internationally recognized authority on nuclear energy. He fostered broad participation by AECL researchers in international meetings and collaboration with many other nuclear laboratories.

The Douglas Point project was stretched out to seven years by delays in completing the design and by late deliveries of equipment. Often, the equipment failed to meet specifications and had to be returned to manufacturers for upgrading. The Canadian content was 70 percent, supplied by some 600 companies.

Operation began late in 1966, and the early years were plagued by problems. It was not declared to be in-service[6] until September 1968. The main coolant pumps had to be modified; valves required frequent maintenance. Heavy-water leakage was excessive and much development work was required to bring it under control and keep losses within an acceptable range. The situation was further exacerbated by the difficulties of making repairs – space around equipment was constricted by over-zealous minimizing of heavy-water holdup in the heat-transport and moderator piping systems, and radiation fields were high. The latter problem was the result of poor water-chemistry control, resulting in the movement of radioactive corrosion products into the steam generators and other out-reactor components, a phenomenon discussed in chapter fourteen. This was remedied by developing a chemical process for removing the corrosion products, and improving the chemical control procedures.

Douglas Point was in some respects "over-designed" – it was too complex. It also demonstrated that the good, standard-engineering quality accepted for many conventional components was not good enough. This caused AECL to expand its R&D effort to take an in-depth look at many out-reactor components (for example, pump seals, valves, steam generators and heat exchangers), and to develop much more rigorous specifications for them (see chapter fourteen).

---

[5] In a speech to the Canadian Nuclear Association in 1985, John Foster recalled Dr. Lewis' style thus: "He [Dr. Lewis] walked into any intellectual space as if he didn't care who owned it. Trained as a physicist, he would debate on equal terms with metallurgists, biologists, economists and engineers in their own fields, but he was usually well prepared. His only measures of other people were their knowledge, intellectual ability and integrity. Depending on how they measured up in these respects, he treated his most junior subordinate in the same way he treated internationally renowned scientists. He could be acerbic at times, but he could also be very considerate."

[6] "In-service" commences when reactor operation has been demonstrated to be sufficiently reliable for commercial use.

These expensive lessons were learned in time to apply them to the large commercial reactor stations being built in the late sixties and early seventies.

As with NPD, the need to retube Douglas Point, combined with a growing list of other major modifications that were needed, led to the decision to take it out of service in 1984.

Douglas Point did not produce electricity as cheaply as expected, and Ontario Hydro never exercised its option to buy the station. However, during its eighteen years of operation, earned revenue paid for most direct operating costs. Ontario Hydro benefited considerably through the experience gained by its operating staff.

Unfortunately, the late sixties were a critical time in the world power-reactor scene. Douglas Point won no converts to the heavy-water design. Most countries opted for the American style of light-water reactor, and Canada was left largely alone as the developer of the heavy-water concept. Success was to be achieved, but too late to make a major impact on the world market.

PICKERING A

By 1964, studies by Ontario Hydro showed that a 500 MW reactor would be needed to compete with the newest coal-fired stations, which consisted of multiple 500 MW units. Also, the expected rate of expansion of the Ontario Hydro system required the addition of large blocks of power to serve Ontario's booming economy. Ontario Hydro decided to commit two 500 MW units. An agreement was worked out whereby Ontario Hydro would build, own and operate the units and AECL would do the nuclear design, as engineering consultant. Ontario Hydro would invest the equivalent of two coal-fired units, about 40 percent of the cost, and the Ontario and federal governments would provide 28 and 32 percent of the cost, respectively. The governments' investments would be repaid from the savings in the cost of electricity relative to that generated by Ontario Hydro's Lambton coal-fired station. In the federal case this was an important revenue for AECL for some years.

The timing of this commitment turned out to be extremely fortunate. It was made before the problems at Douglas Point had surfaced; otherwise, it might not have been approved so readily. But it was not made too early for the design to benefit from some of the lessons learned at NPD and Douglas Point.

At the same time, NPPD was confirmed as Canada's ongoing reactor design organization. A major factor leading to this decision was Ontario Hydro's aversion to being dependent on a single private-sector supplier. It left CGE with a design team, but no domestic market. NPPD changed its name to Power Projects and later moved to its own facilities at Sheridan Park (Mississauga, Ontario) in 1966.

Ontario Hydro wanted the new station, which would soon be expanded to four units, located close to the load centre. They recommended an available Ontario Hydro site at Pickering, 30 km east of downtown Toronto. This was approved after it was shown that the probability of an accident was very small and that the containment system would make the probability of any release of radioactivity of consequence to the public extremely remote.

The scale-up in power from Douglas Point was achieved by adding 27 percent more channels, increasing the channel diameter to 103 mm, with a corresponding 50 percent increase in allowable fuel bundle power, and achieving a flatter flux distribution to raise the average bundle power. Many at Chalk River opposed this change in channel diameter, in part because the 83 mm channel used at Douglas Point and NPD appeared to give somewhat lower power costs, but mainly because of the accumulated R&D experience with the smaller size. The debates on channel diameter at the power reactor development program evaluation committee (see chapter one) were long and acrimonious. Chalk River finally proposed a solution: use the 103 mm channel with a 28-element fuel bundle, but the elements must be the same size and operate at the same rating as those in Douglas Point, so that the existing body of fuel-development information would be clearly pertinent. In the longer term, the move to the larger channel facilitated the design of the larger reactors that followed Pickering.

The policy for Pickering was to select heat-transport system components only slightly larger than those at Douglas Point, to keep in step with Canadian manufacturing capabilities. The result was sixteen main coolant circulating pumps and twelve steam generators per reactor, 50 percent and 70 percent larger, respectively, than the ten pumps and eight steam generators installed at Douglas Point.

At Pickering the next step in automation of reactor control was taken. All major plant processes were to be controlled by the central computer. Since failure of the computer would shut the reactor down, a second identical computer was added on standby and able to take over at any time. This arrangement has worked well, and has been used in successive CANDUs with increasing levels of sophistication.

The Pickering reactors were an outstanding success in their first decade of operation. They demonstrated consistently high capacity factors, except for a period when about 10 percent of the tubes in two units had to be replaced because of cracking (see chapter twelve). The cost of electricity was less than from contemporary coal-fired stations. Heavy-water leakage and loss were well within the design targets. Radiation fields have remained low. In short, few of the problems experienced at NPD and Douglas Point were encountered. This good record has been clouded since 1983 by prolonged shutdowns to replace all the pressure tubes in all four units, for reasons explained in chapter twelve.

During early operation at Pickering, some fuel failures occurred after fuel-bundle power was raised rapidly in a large step as a result of refueling; similar failures were also occurring at Douglas Point. The immediate problem was controlled by restricting power changes during refueling, and a long-term solution was developed by modifying the fuel (see chapter thirteen). Subsequent CANDU fuel performance has been excellent. A recent exception occurred during the initial operation of Darlington (see below). A number of fuel bundles were severely damaged by vibration arising from pressure pulses generated by the heat-transport-system pumps. New pump impellers were installed that give a higher frequency of pressure pulse that is not transmitted to the fuel, thus restoring good fuel performance.

An important feature of on-power fueling is the ability to remove failed fuel bundles before they can deteriorate and seriously contaminate the heat-transport system. Thus, in CANDUs, fission products make only a minor contribution to radiation fields around steam generators or piping after shutdown (see chapter fourteen). Instruments that measure gaseous fission products and radioiodines (which are readily released from failed fuel) in the coolant are used to detect any failures. Two methods were developed to locate the channel containing the failed fuel bundles. One is to scan the gamma radiation from each of the outlet feeder pipes during shutdown, to detect accumulated solid fission products (e.g., zirconium-95, niobium-95 and lanthanum-140) that deposit on the walls. The other is to measure the amount of two short-lived volatile fission products, which emit delayed neutrons (bromine-87 and iodine-137), in a sample of coolant from each channel using an on-line delayed-neutron monitoring system. The latter are used at both Bruce stations and the CANDU-6 reactors, while gamma scanning is used at both Pickering stations and Darlington.

BRUCE AND LATER CANDUS

In 1968, Ontario Hydro committed four 745 MW units to be built adjacent to Douglas Point, and this became the Bruce A station. This was only two years after Douglas Point began operating and three years before Pickering started up. They were the first units to have large coolant circulating pumps, only four per unit. Steam generator size was more than doubled so that only eight per unit were required. The increased reactor power output was achieved by a combination of additional fuel channels and an increase in the fuel bundle power rating; the latter required a change to a 37-element bundle in which the diameter of the individual elements was reduced by 15 percent. Each Bruce A reactor was designed for a maximum total output equivalent to 905 MW electrical: a net electrical output of 745 MW, about 50 MW for unit services, such as the coolant circulating pumps, and additional steam for heavy-water production (see chapter seventeen) of up to the equivalent of 110 MW electrical.

Commencing operation in 1977, the Bruce A units continued the good performance of CANDU until steam-generator problems occurred in the nineties due to excessive deposits forming on the secondary side (see chapter fourteen). There were very few pressure-tube problems until the early nineties, when evidence was found of pressure-tube fretting by fuel-bundle vibration. This was also observed in Darlington.

Figure 10.1 illustrates the growth of the CANDU program over three decades. A second four-unit station was built at Pickering for initial operation in 1983. Figure 10.2 shows the eight Pickering reactors, all connected to a single vacuum building, and the two large turbine halls. A second four-unit station at Bruce was started up in 1984. The latter units were provided with larger turbines than in Bruce A and are each capable of a net output of 860 MW. Ontario Hydro's Darlington station comprises four 880 MW units. AECL helped India to build two CANDU units similar in design to Douglas Point; India has since built a series of similar reactors of their own modified design. CGE sold a 125 MW CANDU to Pakistan. Starting in 1970, AECL developed the single-unit station design, now known as CANDU 6. With an output of about 600 MW, it was based on the Pickering reactor, but uses the higher channel power possible with 37-element fuel bundles, large steam generators and pumps. Four units are in operation: Point Lepreau in New Brunswick, Embalse in Argentina, Gentilly-2 in Quebec and Wolsong I in Korea. In the late eighties, the first of five planned units was being built in Romania. In the early nineties Korea ordered three more units for the Wolsong site.

FIGURE 10.1
CANDU Power Stations

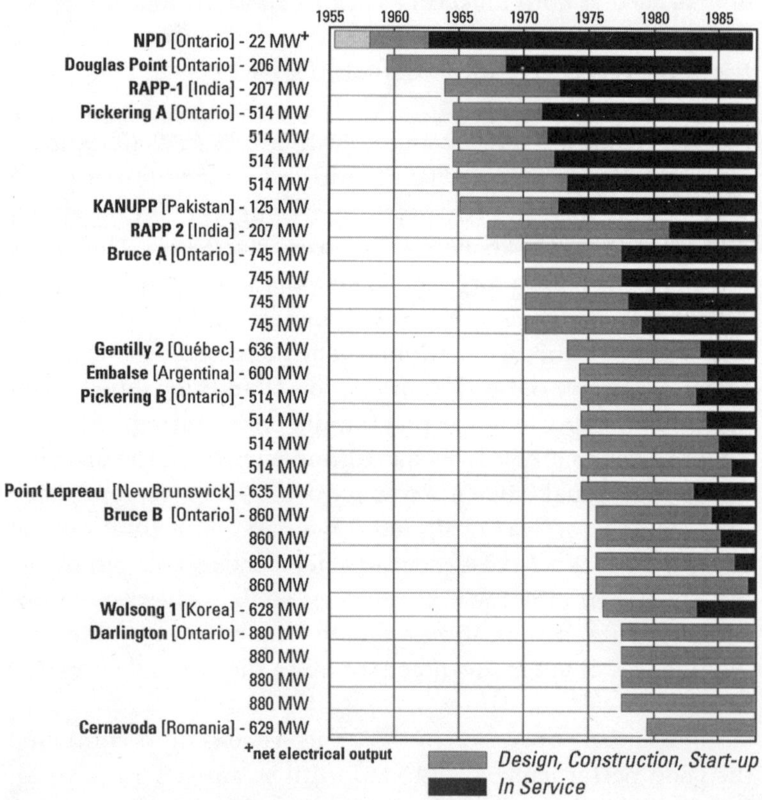

The twenty large commercial CANDU reactors in service by 1988 were designed and built over a period of nearly three decades, (1959-87). This may seem like a small program until it is realized that each reactor represents an investment equivalent to several billions of today's dollars. A large multi-unit CANDU station will produce during its lifetime a total amount of energy equivalent to a major oil discovery, such as Hibernia. The CANDU program is a major high-technology enterprise, and one with a high Canadian content.

Since the mid-eighties, AECL CANDU has been developing a smaller nuclear power station (450 MWe) which is designated CANDU 3. It is simplified in design, with fewer components and a layout to facilitate fast construction and ease of maintenance; a number of evolutionary improvements over the CANDU 6 have been incorporated to enhance safety, extend operating life and increase capacity factor. The overall target is a lower cost of electricity.

OTHER CANDU CONCEPTS — ORC AND BLW

Because the coolant and moderator are separate, the basic CANDU concept can accommodate coolants other than heavy water. Two that were investigated extensively and were used in demonstration or prototype reactors were an organic coolant and boiling light water; the corresponding power-reactor designs are referred to as CANDU-OCR and CANDU-BLW. The advantages perceived for these designs over the established heavy-water-cooled design were: lower capital cost, higher steam-cycle efficiency, lower heavy-water inventory, and very much lower heavy-water leakage.

FIGURE 10.2
Pickering Generating Stations A & B Showing the Vacuum Building to Left of Eight Reactor Units

Ian MacKay was attracted by the potential of an organic coolant to provide a high-outlet temperature, and thus a high steam cycle efficiency, at a relatively low coolant pressure. When he joined CGE he persuaded them to fund a small study. The organic-coolant concept attracted Lewis and AECL, and by the late fifties a substantial development program, split between Chalk River and Peterborough, was in progress. In 1959, when the decision was reached that AECL would build a second research centre in Manitoba, it would have as its focus a reactor designed for research and to demonstrate the organic-coolant concept. This reactor, WR-1, was brought into operation in 1965, and successfully demonstrated the feasibility of sustained operation at a coolant temperature of 400°C. The OCR program was stopped in 1972, despite this technical success and forecasts that the concept would perhaps produce electricity 15 percent cheaper than the standard CANDU (see chapter sixteen). There were four principal reasons: the cost of an OCR prototype was considered to be too large a risk for a relatively small and uncertain benefit; Pickering was a success and no utility was convinced of OCR's promise; the reliability of the fire suppression system (necessary with a combustible coolant) was challenged; and, OCR represented a distinct departure from the growing trend of world nuclear technology to water-cooled designs. Organic-cooled reactor programs in Europe and the United States, with which Canada had collaborated, had been abandoned by the seventies; the light-water reactor had won the race in these countries.

Development of the BLW concept began in the early sixties, and included a joint program with the British of studies in boiling water loops in NRX and NRU. Collaboration with Italy, Japan and the United States on studies of boiling-heat transfer was also important. When Hydro-Quebec showed an interest in CANDU, they sought a federal financing arrangement similar to that for Douglas Point, but this could only be justified for a prototype. AECL recommended the construction of a CANDU-BLW prototype, and a design group was set up in Toronto in 1965, and ultimately became part of Power Projects. This prototype became the Gentilly-1 station operated by Hydro-Quebec. It began operation in 1972, but only operated for short periods at full power. There were serious reactor-control problems requiring some re-design and modification; before this work was completed, and reflecting Hydro-Quebec's lack of interest in purchasing the station, the reactor was shut down permanently in 1978, and has since been decommissioned (see chapter sixteen). Heavy-water-moderated prototype reactors cooled with boiling-light water, but using enriched-uranium fuel, have operated successfully in Britain and Japan.

In retrospect, the choice of BLW for Gentilly-1 was perhaps unfortunate. A few years later, after WR-1 had operated successfully for a few years, the choice might have been an OCR. In any case, in a retrospective review in 1987, well after his retirement in 1974, J.L. Gray regretted the decision not to pursue the OCR concept. Although Lewis supported the choice of BLW for Gentilly-1 in 1965, he had become a strong advocate for OCR by 1970. But by the early seventies AECL had other priorities, like the supply of heavy water, that were straining its financial resources.

FIGURE 10.3
Point Lepreau Generating Station

CANDU PERFORMANCE

Year after year, CANDU reactors dominated the list of the world's top ten nuclear power stations in lifetime capacity factor.[7] For example, when the 285 stations throughout the world that are 500 MW or larger in size are ranked by lifetime capacity factor to the end of 1989, the top ten consist of six CANDUs, three German PWRs and one Belgian PWR. First was the Emsland unit in West Germany at 91 percent, followed by Point Lepreau at 89.6 percent. The other five CANDUs were Pickering units, 5, 6, and 7, and Bruce units 5 and 6. By the end of 1990, Point Lepreau, shown in figure 10.3, then entering its eighth year of operation, had the highest lifetime capacity factor in the world, 89.7 percent. However, some CANDU reactors have significantly lower lifetime capacity factors; e.g., Pickering units 1 to 4, because of long shutdowns for pressure-tube replacement, and some units at Bruce A and Darlington, because of recent steam generator and fuel problems. By 31 March 1994, Point Lepreau's lifetime capacity factor had increased to 91.4 percent, second in the world to Emsland at 91.5 percent.

In 1987, the Canadian Engineering Centennial recognized CANDU as one of ten outstanding achievements in engineering in Canada since 1887.

---

7 Capacity factor is the ratio of the actual energy produced during any period to the energy that could have been produced in the same period if the plant had operated continuously at its fully rated capacity.

Some of the factors that give CANDU an edge in capacity factor over other reactor types are on-power fueling, computer control, rigorous operator training and a strong R&D capability within AECL, which has worked closely with the utilities in developing improvements or correcting operating problems.

The role of R&D did not end with the successful operation of Pickering A or Bruce A. An essential continuing role has been detailed analysis and trouble shooting, which underlies continuing reliable performance. As described in chapter twenty, this R&D role has evolved to a dual one: a large generic program supported jointly by AECL and the CANDU Owners Group (utilities that operate CANDU reactors), and a variety of station-specific commercial contracts.

Of course, a higher capacity factor will reduce the cost of electricity. In 1984 Ontario Hydro compared the average cost of electricity for their CANDU stations with their forecast of the cost of electricity from light-water reactors if they were built and operated by Ontario Hydro. They concluded that electricity from light-water reactors would be 17 percent more expensive in Ontario. In 1993 the Korean Electric Power Corporation reported that its CANDU 6 unit produced electricity 15 percent cheaper than the average for their PWR units.

FIGURE 10.4
Projected Unit Energy Costs of Major Supply Options

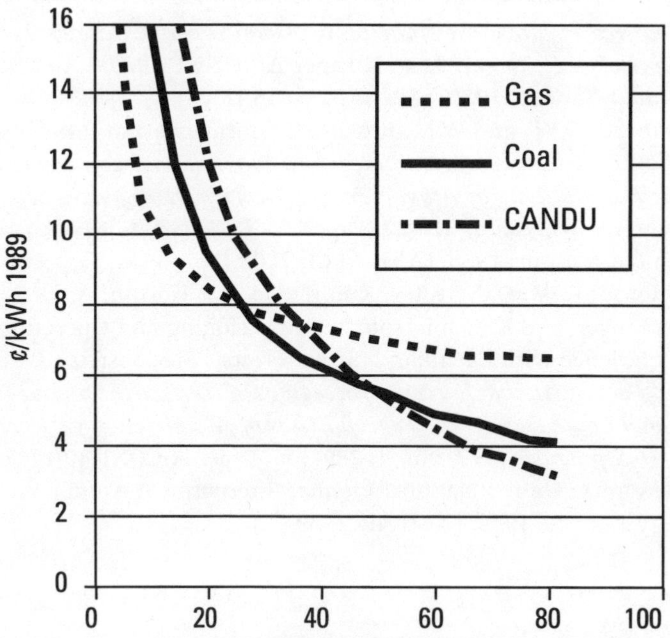

Inherent neutron economy, the use of natural uranium, reliable fuel performance and on-power fueling all contribute to achieving a low fueling cost. The cost of heavy-water loss, while meeting the design target, is not insignificant; however, it, together with the interest charges on the heavy-water inventory, are offset by a fueling cost that is considerably lower than that for light-water reactors.

The provision of closed-loop automatic reactivity control, combined with the inherently small change in reactivity with power, provides considerable flexibility in responding to demand power changes. Excellent load following performance has been reported for the CANDU station in Argentina, where conditions within the national power grid periodically require this mode of operation.

CANDUs have excellent environmental and safety records. Emissions of radioactive materials have been generally less than one percent of the regulatory limits. There has been no measurable dose to a member of the public from any acute release of radioactive material. The radiation dose to the majority of exposed workers has been well below the regulatory limit of 50 millisieverts per year. Health and Welfare Canada report[8] about fifty instances of overexposure throughout the operating history of CANDU power stations in Canada, fourteen of them in excess of twice the regulatory limit.

Ontario Hydro's experience with operating both nuclear and coal-fired base-load power stations has shown that the nuclear option has consistently provided electricity at a lower cost. Electricity from oil or gas is even more expensive than from either uranium or coal. Figure 10.4 shows Ontario Hydro's estimates of the future costs of power from major supply options – for capacity factors over 50 percent nuclear will continue to cost

[8] National Dose Registry, W.N. Sont, private communication. (1993)

less. If cost allowances for the health and environmental effects of each type of power production facility were to be included, nuclear energy's advantage would increase.

CANDUs produce one-fifth of Canada's electricity, with an outstanding record for safety, reliability and economy. This nuclear supply is concentrated in three provinces, accounting for more than half of the electricity produced in Ontario, one third of New Brunswick's and a small fraction of Quebec's. On a worldwide basis CANDUs account for nearly 5 percent of the total installed nuclear capacity of more than 300 GW (see Table 10.1 in the appendix).

BIBLIOGRAPHY

Atomic Energy of Canada Limited. Marketing Division. *CANDU-600 Station Design*. Report PP-28. Ottawa: Atomic Energy of Canada, 1976.

Bothwell, Robert. *Nucleus: The History of Atomic Energy of Canada Limited,* Toronto: University of Toronto Press, 1988.

Fawcett, Ruth. *Nuclear Pursuits: the Scientific Biography of Wilfrid Bennett Lewis.* Montreal: McGill-Queen's University Press, 1994.

Foster, J.S. "The First Twenty-Five Years." In *25th Annual Conference of the Canadian Nuclear Association: Proceedings*, 1-11. Toronto: Canadian Nuclear Association, 1985.

Gray, J.L. "Key Decisions." In *Canadian Engineering Centennial Conference: Proceedings of the Nuclear Power and Fusion Programs of the Canadian Nuclear Society, 18-22 May 1987, Montreal*, 10-16. Toronto: Canadian Nuclear Society, 1987.

Hart, R.S., and G.L. Brooks. "CANDU Reactors: Experience and Innovation." In *Good Performance in Nuclear Projects: Proceedings of an International Symposium, Tokyo, April 1989*, 266-75. Paris: Nuclear Energy Agency, Organisation for Economic Cooperation and Development, 1989. (Also AECL-9796.)

Lewis, W.B., and J.S. Foster. "Canadian Operating Experience with Heavy Water Power Reactors." In *Proceedings of the American Power Conference*, vol.32, 200-9. Chicago: Illinois Institute of Technology, 1970. (Also AECL-3569.)

MacKay, I.N. "The Canadian NPD-2 Nuclear Power Station." In *Proceedings of the Second United Nations International Conference on the Peaceful Uses of Atomic Energy, Geneva, 1-13 September 1958*, vol. 8, 313-21. Paper A/CONF.15/P/209. Geneva: United Nations, 1958. (Also AECL-619.)

McConnell, L.G. "Historical Overview." In *Canadian Engineering Centennial Conference: Proceedings of the Nuclear Power and Fusion Programs of the Canadian Nuclear Society, 18-22 May 1987, Montreal*, 3-9. Toronto: Canadian Nuclear Society, 1987.

Pon, G.A. *Evolution of CANDU Reactor Design*. AECL-6351. 1978.

Smith, H.A. *A Review of the Development and Status of CANDU Nuclear Power Plant*. Report NGD-5. Toronto: Ontario Hydro, 1975. (Prepared for the European Nuclear Conference, Paris, 1975, Plenary Session V, 23 April 1975.)

Smith, H.A., W.M. Walker, N.L. Williams, and E. Critoph. "A Study of a Full-Scale, Uranium and Heavy Water Nuclear Power Plant." In *Proceedings of the Second United Nations International Conference on the Peaceful Uses of Atomic Energy, Geneva, 1-13 September 1958*, vol. 9, 3-15. Paper A/CONF.15/P/208. Geneva: United Nations, 1958. (Also AECL-618.)

Woodhead, L.W., and W.M. Brown. "Performance and Problems of NPD." In *Proceedings of the Third International Conference on the Peaceful Uses of Atomic Energy, Geneva, 31 August-9 September 1964*, vol. 5, 313-22. Paper A/CONF.28/P/8. New York: United Nations, 1965. (Also AECL-2008.)

Woodhead, L.W., D.C. Milley, K.E. Elston, E.P. Horton, A. Dahlinger, and R.C. Johnson. "Commissioning and Operating Experience with Canadian Nuclear-electric Stations." *In Peaceful Uses of Atomic Energy: Proceedings of the Fourth International Conference on the Peaceful Uses of Atomic Energy, Geneva, 6-16 September 1971*, vol. 2, 289-306. Paper A/CONF.49/P/471. New York: United Nations; Vienna: International Atomic Energy Agency, 1972. (Also AECL-3972.)

APPENDIX

The CANDU Reactor

H.K. RAE

In a nuclear power plant heat from the fission process within the reactor is used to boil water and the resulting steam turns a turbine that drives a generator to produce electricity. A CANDU power plant is shown schematically in figure 10.5. The uranium fuel, in the form of bundles of elements, is placed in a number of fuel channels that pass through a low-pressure tank (called a calandria) containing the heavy-water moderator. This assembly of fuel, fuel channels, moderator and calandria is the reactor core. Heavy water coolant (separate from the moderator) flows through the fuel channels, around the fuel to remove the fission heat and is circulated through a steam generator (boiler) in a heat transport system. In the steam generator heat is transferred from the pressurized hot heavy water to ordinary water at a lower pressure which boils to produce steam to turn the turbine. After this steam has expanded through the turbine it is condensed, by a large flow

FIGURE 10.5
CANDU Power Plant Schematic

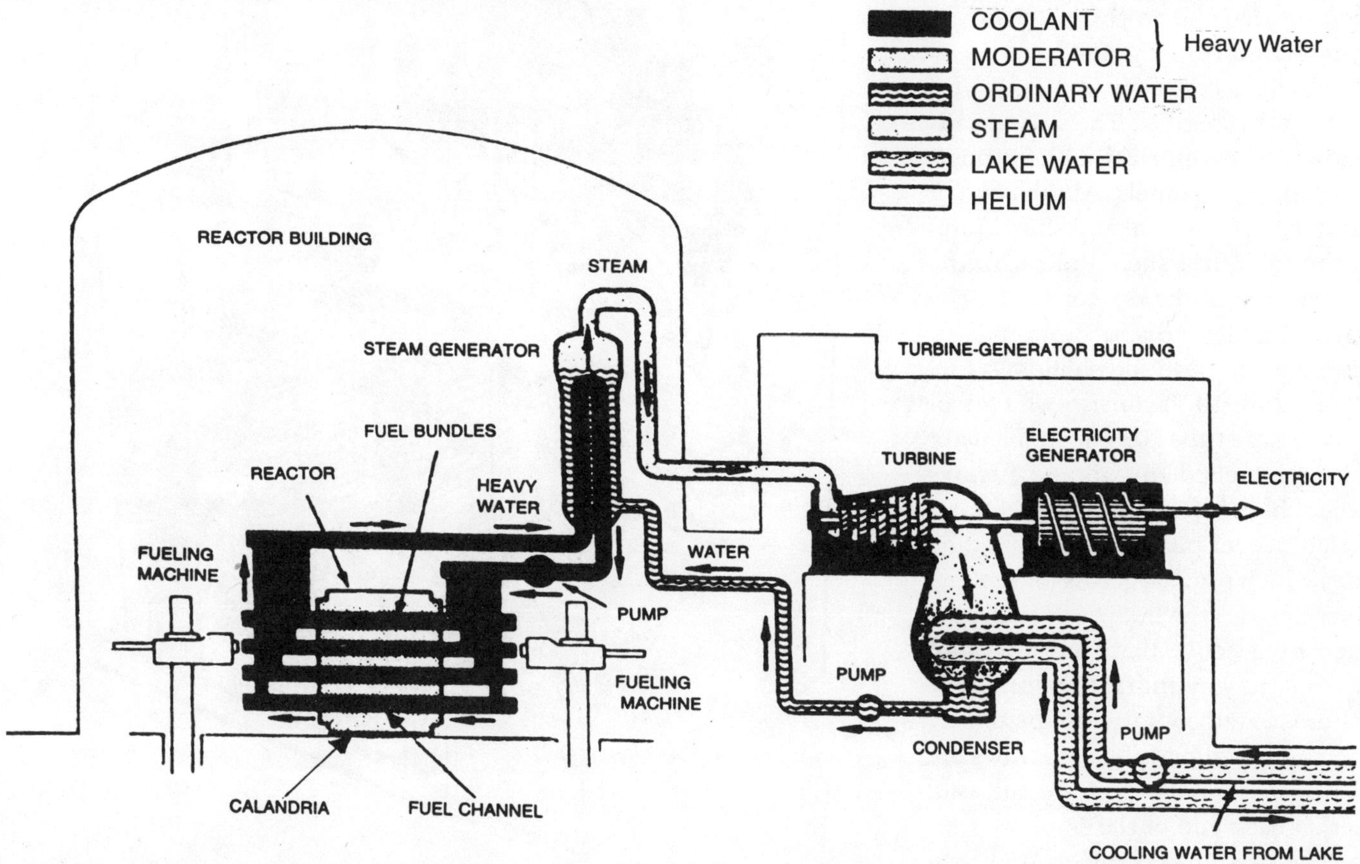

of cooling water, and the condensate is returned to the steam generator.

The CANDU concept can have other coolants than pressurized heavy water, and these alternative designs are discussed in chapter sixteen. The pressurized heavy water design is sometimes designated CANDU-PHW to distinguish it from these other CANDUs with different coolants, but in this book it is referred to simply as CANDU.

The following outline of a CANDU nuclear power station is based on the CANDU 6 design as exemplified by the 635 MW Point Lepreau station in New Brunswick (figure 10.3) which has led the world in operating reliability for a number of years.

Figure 10.6 is a cut-away drawing of a CANDU 6 reactor assembly. The calandria is a large horizontal cylindrical tank through which pass 380 fuel channels. At each end of the calandria is a cylindrical end shield integral with the calandria shell and set in the thick concrete wall of the reactor vault. The fuel channels are aligned and supported by lattice tubes within these end shields. A fuel channel (see figure 10.7) consists of a zirconium alloy pressure tube 103 mm in diameter and 6 m long attached to two end fittings which project beyond the end shields to attach to coolant feeder pipes and are sealed by end closures which provide access for fueling. Within the calandria the hot pressure tube is surrounded by a cool calandria tube, and separated from it by cylindrical spacers. The annulus thus created contains an insulating gas. The zirconium alloy calandria tubes are rolled into holes which penetrate the wall (tube sheet) at each end of the stainless steel calandria. The calandria is penetrated by a

FIGURE 10.6
CANDU Reactor Assembly

1. Calandria
2. Calandria shell
3. Calandria side tube sheet
4. Fueling machine side tube sheet
5. Lattice tubes
6. End fittings
7. Feeders
8. Calandria tubes
9. Steel ball shielding
10. Annular shielding slab
11. Pressure relief pipes
12. Rupture disc
13. Moderator inlets (4 each side)
14. Reactivity control nozzles
15. Reactivity control devices
16. Horizontal flux detectors (9)
17. Poison injector nozzles (6)
18. Ion chamber cooling piping
19. End shield cooling piping
20. Calandria vault (light water shield)

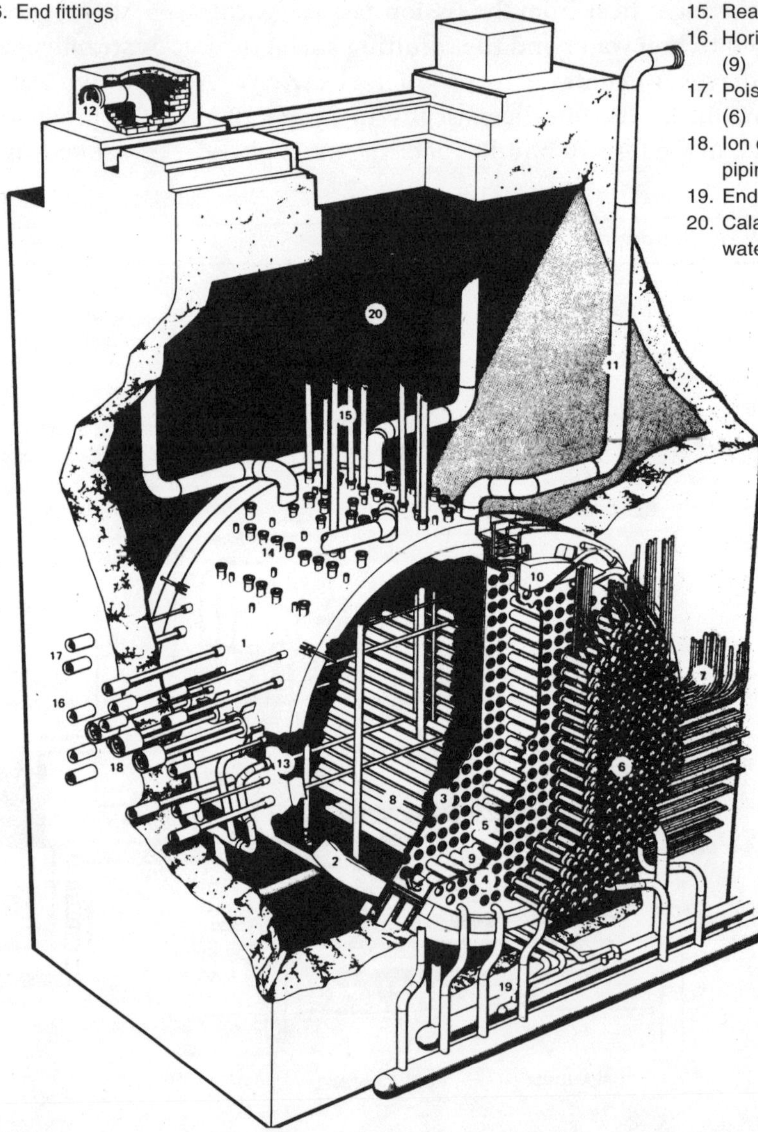

FIGURE 10.7
Schematic Arrangement of a Fuel Channel

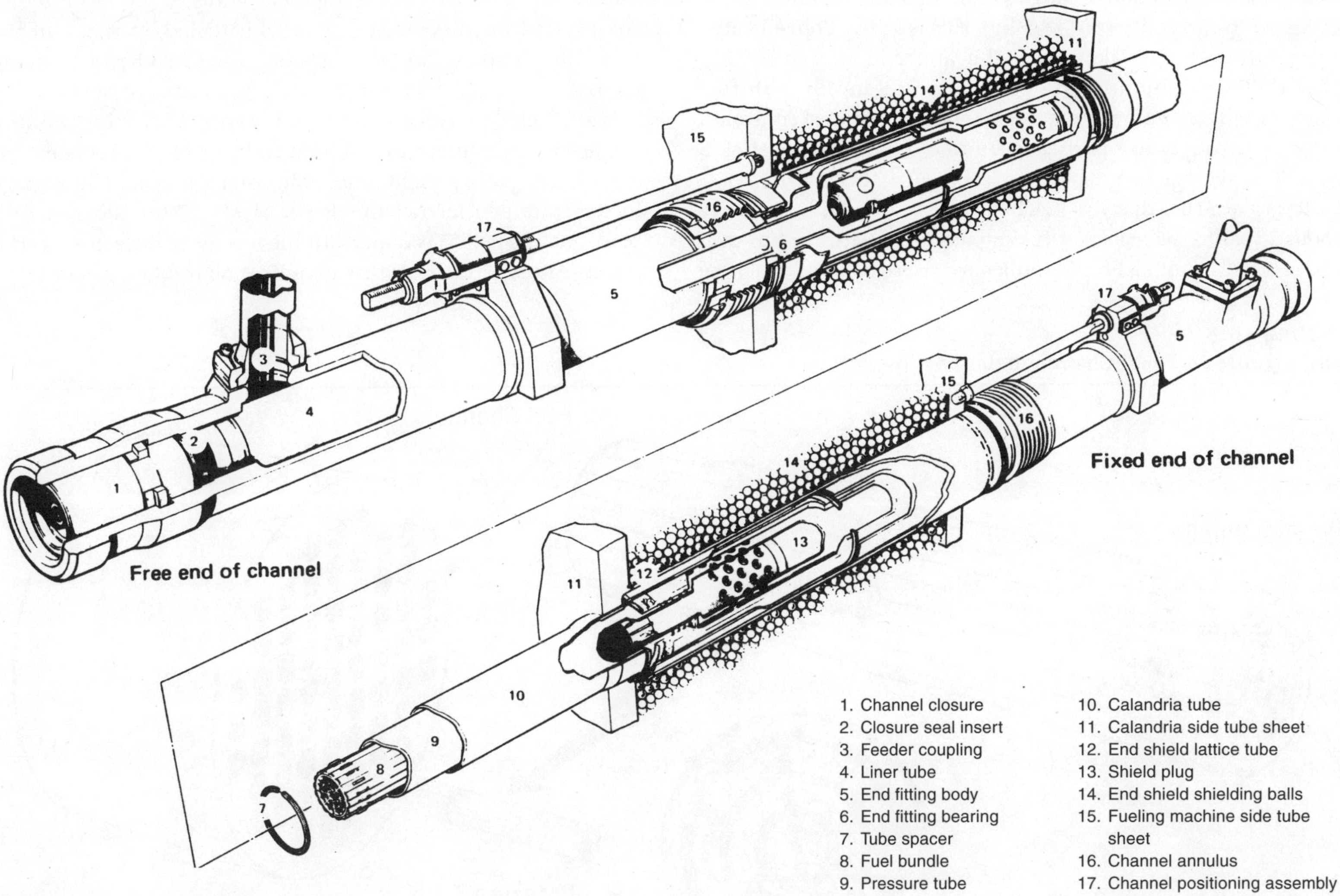

number of both horizontal and vertical tubes, at right angles to the pressure tubes, which contain reactor control mechanisms, reactor shut-off devices and neutron flux detectors.

Radiation shielding of the reactor is provided by ordinary water which fills the reactor vault, by the thick concrete walls, by a mixture of steel balls and water filling the end shields and by shielding plugs within the end fittings of each fuel channel.

Uranium ores are mined and milled and the concentrate is then refined to produce uranium oxide powder. This is pressed and fired to form hard, insoluble ceramic pellets (see figure 10.8). The pellets are sealed within zirconium alloy tubes to form fuel elements. A fuel bundle, consisting of 37 elements, is 500 mm long and 100 mm in diameter and weighs 21 kg (uranium content of 18.5 kg); this is a very concentrated energy source, equivalent in total heat production to 400 Mg of coal. Twelve bundles are placed end-to-end in each fuel channel where the heavy water coolant is pumped at a high velocity through and past each bundle to remove the fission heat. The average bundle power is 450 kW (thermal).

To replace fuel that has reached the end of its life, typically about a year in the reactor, a fueling machine is attached to each end of a fuel channel. Fresh bundles are inserted by one machine pushing an equal number of used bundles out into the other machine (see figure 10.9a). Subsequently, the used bundles are transferred to underwater storage. Generally, eight new bundles are inserted in pairs when a channel is fueled, so that half the fuel will occupy two positions along the fuel channel during its irradiation and half only a single position. Adjacent channels are fueled in opposite directions to provide a more uniform average fissile content throughout the reactor. This scheme increases the energy yield attainable from the fuel. The average fueling rate is fifteen channels per week (17 bundles per full-power day). Fueling is done with the reactor in operation, and it is this feature of CANDU that permits such frequent fueling.

FIGURE 10.8
Fuel Bundle and Fuel Channel Relationship

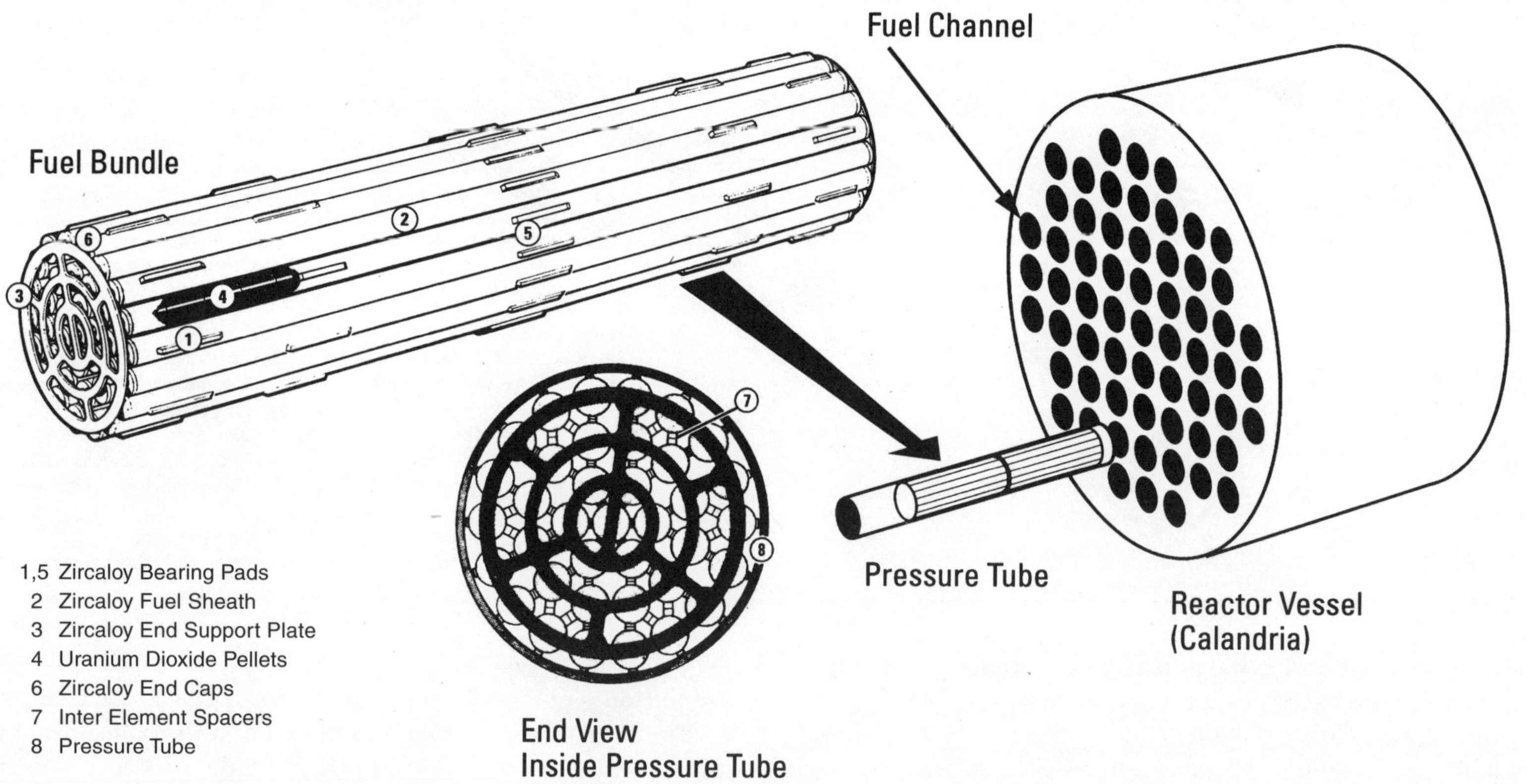

FIGURE 10.9
Design Features of the CANDU Reactor

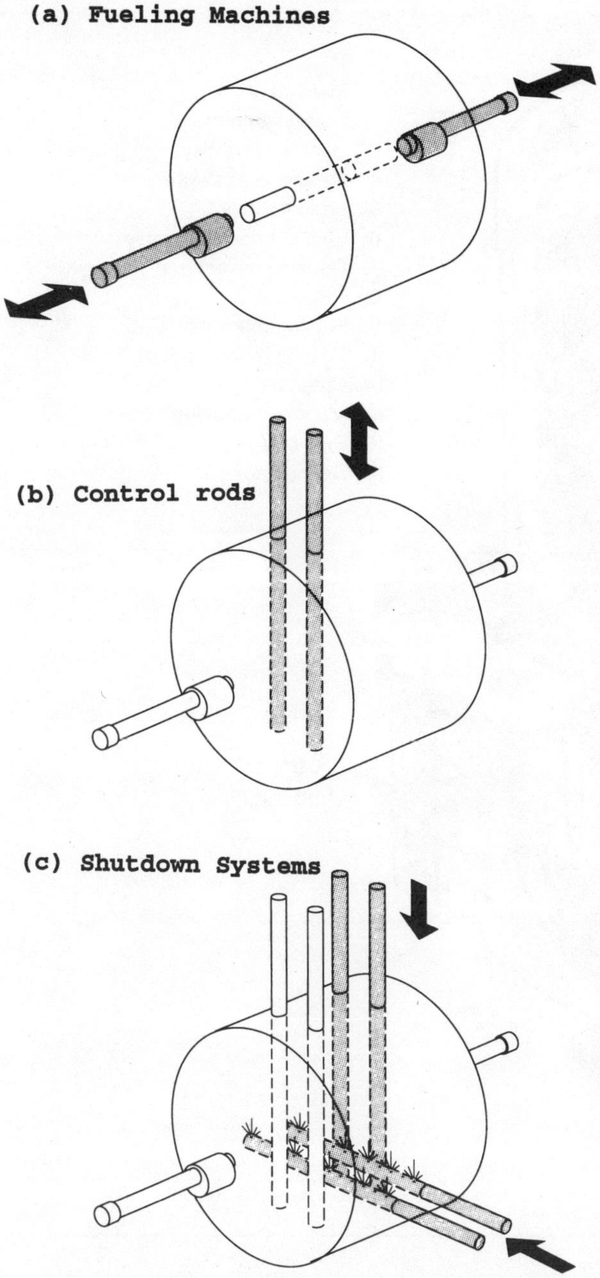

The selection of the coolant temperature is a compromise between higher steam cycle efficiency if the temperature is increased and thinner pressure tube walls if the temperature, and thus the required pressure, are lowered. Thinner tubes absorb fewer neutrons, allowing more energy to be produced from each kilogram of uranium entering the reactor. The outlet coolant conditions chosen are 310°C and 10 MPa (100 atmospheres), allowing the use of relatively thin tubes (4 mm wall thickness), and giving an overall net thermal efficiency of 29 percent.

The heat transport system consists of 760 feeder pipes, one from each end of each channel, which are attached to large headers (see figure 10.10). These connect to the four steam generators and the four main circulating pumps, and to a pressure control system.

The total quantity of heavy water in the moderator and coolant is 450 Mg (0.7 megagrams per net electrical megaWatt), and represents about a tenth of the power station's capital cost. Somewhat more than half of the heavy water is in the moderator system, mainly within the calandria. A cooling system maintains the moderator at an average temperature of 70°C.

On-power fueling provides the primary long-term control over the neutron flux distribution and power output. Short-term control of flux distribution and reactor power is provided by 14 zonal control units distributed through the reactor, each contains a quantity of ordinary water (a neutron absorber) that can be varied. Further long-term control of neutron flux and power is provided by neutron absorber rods (containing neutron-absorbing cobalt or stainless steel) which can be moved into and out of the reactor (see figure 10.9b).

A unique feature of CANDUs is the complete separation of the automatic safety systems from the normal process systems. Two independent shutdown systems, that are physically and functionally separate, can each rapidly and safely shut down the reactor for any failure or operating error that the control system cannot deal with (see figure 10.9c). One consists of a group of shut-off rods containing cadmium, an efficient neutron absorber, that can be fully inserted in two seconds. The other is a system for injecting a soluble neutron absorber,

FIGURE 10.10
CANDU 6 Reactor Building

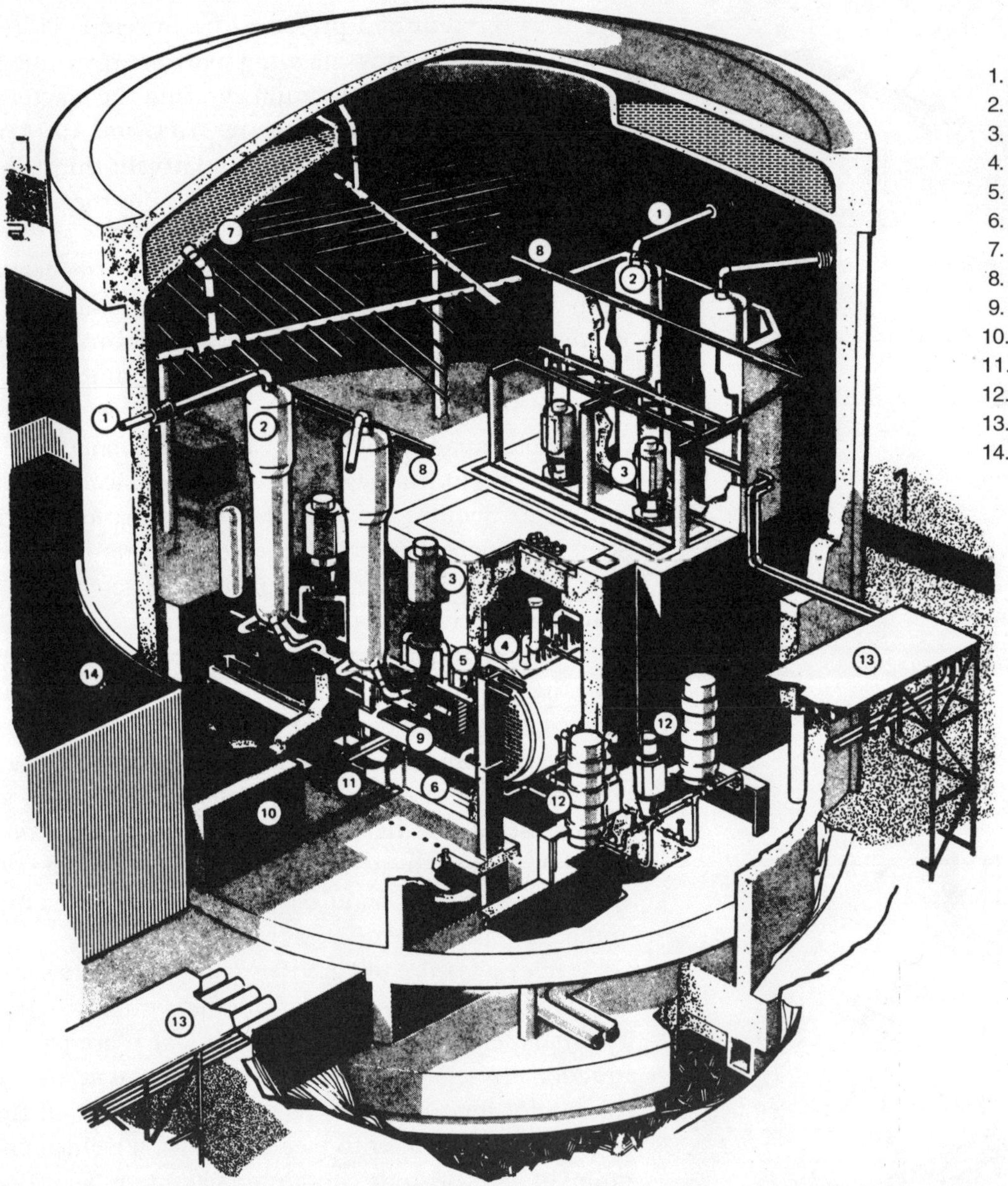

1. Main steam supply piping
2. Steam generators
3. Heat transport system pumps
4. Calandria assembly
5. Feeder pipes
6. Fuel channel assembly
7. Dousing water supply
8. Crane rails
9. Fueling machine
10. Fueling machine door
11. Catenary
12. Moderator cooling system
13. Pipe bridge
14. Service building

gadolinium nitrate, into the moderator through an array of nozzles. It can be removed later by ion exchange. The actuating circuits for each system are triplicated and actuation requires simultaneous signals in two channels; this permits regular testing of each individual channel without affecting operation of the reactor, as discussed in chapter four.

The emergency coolant injection system provides cooling to the core in the extremely unlikely event of a large rupture in the heat transport system. The injection system is designed to prevent or limit fuel failure and to remove the heat from radioactive decay of fission products that continues to be produced after the reactor is shut down.

Two types of containment are used for CANDUs. The CANDU 6 reactor and all its major coolant system components are contained within a concrete building (see figure 10.10) designed to contain all the steam that might be released in an accident; the pressure of this steam would be controlled by a spray of water from a large tank in the roof of the building – a dousing system. The dousing system is actuated by the rise of containment building pressure and is driven by gravity. In the Ontario Hydro multi-reactor stations, all reactor buildings are connected to a large building held at low pressure – a vacuum building (see figure 10.2). A pressure rise in a reactor building would open dead-weight relief valves leading to the vacuum building where a preset pressure rise would actuate a dousing system. Thus, pressure would be reduced to subatmospheric within containment very quickly after a coolant system rupture.

All major station processes are both monitored and controlled by a central computer. Since computer operation is essential for station operation, a second identical computer is on standby to provide the desired high degree of availability. Computers are also used in the safety systems.

By 1989, there were 417 power reactors operating and 121 being built in thirty-two countries around the world. These are summarized in Table 10.1 by reactor type. Two-thirds of the total capacity of 416 GW was supplied by pressurized water reactors (PWR), located in twenty-six countries. Boiling water reactors (BWR) are also widespread; they accounted for twenty percent of the total capacity, located in thirteen countries. CANDUs totalled 20 GW, or 86 percent of the pressurized heavy water reactor capacity, making up 5 percent of the world total, located in six countries. CANDUs, BWRs and PWRs are the only established candidates for the next generation of reactors.

No more gas-cooled reactors are being built.

The USSR light-water-cooled graphite-moderated reactors have been modified to improve safety standards since the Chernobyl accident. However, over the next decade they are likely to be shut down as new generating capacity is put in place. The USSR also has 30 PWRs in operation and being built.

The liquid metal fast breeder reactor (LMFBR) is at the demonstration phase. It is a possible contender for the next generation of reactors, but likely only when uranium becomes scarce.

The capacity of nuclear power plants worldwide in 1989 exceeded hydraulic capacity, and generated 17 percent of all electricity. Of the twenty-five countries that operated nuclear power plants in 1989, fourteen derived in excess of 20 percent of their electricity from this source; France had the highest nuclear contribution at 75 percent. In Canada the percentage was 16.

TABLE 10.1
World Nuclear Power Capacity*

REACTOR TYPE	OPERATING Number	OPERATING Capacity GW	BEING BUILT Number	BEING BUILT Capacity GW	TOTAL GW
Pressurized Water	235	196.9	81	73.9	270.8
Boiling Water	87	69.5	15	14.5	84.0
Heavy Water					
• Pressurized	27	14.6	17	8.9	23.5
• Other	2	0.2	-	-	0.2
Gas Cooled	43	15.7	-	-	15.7
Light Water					
• Graphite	20	14.2	4	3.7	17.9
Fast Breeder	3	1.2	4	2.5	3.7
	417	312.3	121	103.5	415.8

* From *Nuclear News 33* Vol. 2, 82 (1990)

The two types of light water reactors, PWRs and BWRs, require enriched uranium as fuel - typically a uranium-235 content of four or five times natural. Such a reactor has a smaller core than CANDU because there is considerably less space between fuel channels. The fuel and water are contained in a large pressure vessel, the water serving as both coolant and moderator. Fueling is done at infrequent intervals with the reactor shut down so that the vessel can be opened. The PWR has a coolant circuit similar to CANDU, operating at a high enough pressure that the coolant remains liquid. In the BWR the fuel is cooled by boiling water, and the steam produced within the reactor goes directly to the turbine.

Although quite different in reactor design, CANDUs and PWRs share many similarities in the overall nuclear steam supply system. Both use UO_2 and Zircaloy in the reactor core; material behaviour in natural water and heavy water are essentially identical; many coolant system components such as pumps and steam generators are similar; and most aspects of the steam cycle circuits are alike. There has been and continues to be good exchange of technical data and operating experience throughout the international water-cooled reactor community. Collaborative research programs with other countries, notably the United Kingdom and the United States, have enhanced Canada's access to new results at an early stage and have contributed to the evolution of PWRs in Europe and Japan. Thus, while CANDU is uniquely Canadian and based on the technology developed by AECL, Ontario Hydro and Canadian industry, it has been able to draw on ideas and experiences from nuclear programs in other countries, and has contributed to those programs.

Chapter Eleven

Reactor Physics and Control

M.F. DURET
H.K. RAE

Fuel Cycle Physics

M.F. DURET

INTRODUCTION

While the subject of nuclear reactor design is highly technical, only a few fundamental concepts are required to understand how a fission chain reaction can be established and maintained. In general, the materials in a nuclear reactor fall into three different categories: fissile material, fertile material and parasitic material. Fissions are often produced when neutrons are absorbed in fissile material (e.g., uranium-235), but can also occur when fast neutrons are absorbed in fertile material (e.g., uranium-238). However, in most cases when a neutron is absorbed in fertile material, it produces new fissile material (e.g., plutonium-239). These fertile and fissile materials are the major constituents of the fuel in the reactor. Parasitic materials absorb neutrons without producing either fissions or new fissile materials. Structural materials in the reactor fall into this category.

Consider the three categories at work in a nuclear reactor. Fissioning, during which the nucleus breaks up into smaller fragments called fission products (normally two but occasionally three), produces about 200 MeV of energy, highly penetrating gamma rays and two or three fast neutrons. These first-generation fast neutrons are slowed down to lower, or "thermal", energies by the reactor's moderator (see chapter one). If they escape capture in other materials in the reactor, and do not escape from the reactor they can produce further fissions in the fuel. If just enough second-generation neutrons are produced in these fissions to maintain the neutron population, the reactor is described as being "critical" and the power level remains constant. The ratio of neutron populations in successive generations is called the reactivity, k. If k for the system is less than 1, then the reactor is sub-critical and any existing neutron population will die away. If k is greater than 1, the neutron population (and consequently the reactor power) will increase. Excess reactivity can be removed from the system by adding neutron absorbers (control rods that absorb but do not produce neutrons, and so make a negative reactivity contribution to the system), or by increasing the escape (leakage) of neutrons from the core (by lowering the depth of the heavy-water moderator). The energy produced in fission is divided amongst the kinetic energy of the fragments, the kinetic energy of the neutrons, the energy of the prompt gamma rays, and the energy contained in all the decay products of the radioactive fragments. Most of this energy appears in the fuel, ultimately as heat.

The fission products fall into the category of parasitic neutron absorbers. As they accumulate in the fuel they cause a significant decrease in reactivity, which must be countered by raising the moderator level, withdrawing movable neutron absorbers (control rods) or adding fresh fuel.

The fundamental reactor-physics task is to determine (by calculation and/or experiment) the fates of all neutrons produced in the reactor, so that neutron absorption and neutron production, and hence reactivity, can be predicted for different reactor designs and operating situations.

At any position in the reactor core there are neutrons of many different energies. This neutron energy spectrum includes neutrons that are slowing down from their original high energies at fission and neutrons which have slowed down but have not yet been absorbed. The former are distributed in energy roughly as 1/E (E being the neutron energy) and the latter approximately as a Maxwellian distribution characterized by a neutron temperature close to that of the moderator. The rate of neutron absorption by a unit volume of any material is the product of neutron density, neutron velocity and the cross-section of the material. This cross-section (an area which is a measure of the probability that the neutron will be absorbed) varies with neutron energy, and varies widely from one nuclide to another. To simplify what can be a very complex calculation, and recognizing the lack of detailed nuclear properties in the early days, a simple model was developed at Chalk River which has been widely used. A representative cross-section was assigned for each type of neutron, those slowing down and those already slowed down to thermal energy. A compilation was prepared of these representative cross-sections for many nuclides as a function of temperature. To calculate neutron absorption in each lattice cell[1] of the reactor core, the composition of all materials present is required as well as the variation of temperature and the ratio of the two neutron spectral components with position within the cell. Measurements made in ZEEP with a hot pressure tube and coolant showed how these spectral components did vary with position in a lattice cell and with temperature, providing data to correlate with simple mathematical models. Methods such as this led to a description of the nuclear reactions taking place in the lattice cell, which used recipes and correlations with experimental measurements. The first computer program using this technique was called POOOF. A version of this code used by the designers was known as POWDERPUFFS. As more measurements became available and the semi-empirical models became more precise, the codes were improved. In the early years this development went on independently at Chalk River and Power Projects, resulting in different estimates for the same reactor at the two sites – much to the frustration of J.L. Gray at Head Office in Ottawa. Later, POWDERPUFFS became the accepted basic design code, continuing to evolve as more detailed analyses became possible. At Chalk River an improved semi-empirical lattice code known as LATREP (lattice recipes) was developed. As more powerful computers became available, more detailed calculations of the neutron spectrum became possible, and codes developed in other laboratories were in many cases used to study special situations. Much effort was devoted to adapting large multigroup transport codes, such as WIMS, to heavy-water reactor applications. Although POWDERPUFFS and LATREP have proven adequate for most CANDU basic design purposes, the more detailed codes have allowed more accurate analysis of safety margins in large multizone reactors or complex fuel-management schemes, thus achieving higher power ratings and lower-cost power.

With the powerful desktop computers and workstations available today, the computing environment has changed beyond recognition. Old recipe programs are now rarely used. Instead more fundamentally based codes have been developed. The cell code WIMS, originally obtained from the United Kingdom, has been modified in Canada to such an extent that it is now known as WIMS-AECL. Computing power is no longer a restriction on the complexity of models used to predict neutron spectra and the reactions occurring in the reactor core.

Although a great deal of data was known for many important nuclides in the early days of CANDU design, many other important cross-sections had not been measured. In addition, the as-manufactured nuclear characteristics of some possible

[1] A lattice cell consists of a fuel channel and its surrounding moderator.

structural materials were not very well known, because the amounts of impurities allowed were only loosely controlled in the manufacturing process. The problem of dealing with these complications was solved to some extent by using materials intended for the actual reactor construction in the experimental measurements, when possible.

Although the NRX reactor had already been designed, a small experimental reactor called ZEEP was built to investigate the reactor physics properties of a wide range of uranium lattices in heavy water moderator. ZEEP was a small, lightly shielded tank of heavy water in which an array of uranium rods could be suspended. Foils could be placed in the tank to measure the spatial distribution of the neutron flux and give some information on its energy spectrum. Criticality could be maintained by adjusting the height of the heavy-water moderator. Because of the meagre shielding, it operated at very low power. Experience with higher powers was obtained from the research reactor NRX, which was initially fuelled with uranium metal. A great deal of reactor physics data was provided by ZEEP and NRX.

Later, another low-power reactor, ZED-2, and the research reactor NRU, provided still more power and experimental facilities. As more powerful and versatile reactors were developed, methods of calculation became more advanced, progressing from laborious mathematical calculations done by hand to sophisticated electronic computer codes that performed very complicated reactor calculations, simulating an entire reactor core.

A focused approach

In the early fifties, a focused approach to reactor design was undertaken. The nuclear power group, described in chapter ten, began their studies of a nuclear power reactor, fueled with natural uranium and moderated with heavy water, that could compete economically with coal-fired stations. A large multidisciplinary effort was launched to study the overall plant design, the reactor design, the fuel design, the strategy for plant operation and fuel-cycle physics. Because these aspects are highly interrelated, an optimization procedure was developed to determine optimum and consistent specifications in each area. The procedure was based on minimizing the estimated total-unit-energy cost (TUEC) of electrical energy delivered by the plant, consistent with safe and reliable operation. This was an iterative procedure, since new ideas, knowledge and developments arose in each of the areas during the course of the investigation.

Initially, there were uncertainties in all areas of the project, perhaps the most serious of which arose in the physics area, because they tended to complicate possible solutions to major problems and uncertainties in other areas. Perhaps it was for this reason that it fell largely to those in the fuel-cycle physics area to initiate and coordinate the optimization function.

It was immediately apparent that a particularly important item in this optimization would be the total thermal energy extracted from the natural-uranium fuel before it had to be removed from the reactor (conventionally called the discharge fuel burnup); the larger the discharge burnup, the lower the fueling cost. Two reasons for removal of the fuel were foreseen: (1) physical deterioration and (2) nuclear deterioration (i.e., an inability to sustain a chain reaction). The challenge facing the fuel developers, which was successfully met, was to ensure that physical deterioration of the fuel would not be the limiting factor. This left the nuclear deterioration of the natural uranium fuel as a fuel-cycle physics problem.

The basic fuel-cycle problem

Experience with uranium-metal fuel in the NRX reactor had demonstrated that natural-uranium-fuelled, heavy-water-moderated reactors could be operated successfully at modest powers. However, burnups were quite low (72 MWh/kg or 3,000 MWd/te), because of distortions of the metallic fuel. Fuel processing and refabrication were initially envisaged as part of the natural-uranium fuel cycle. Later experience showed uranium oxide to be a much superior fuel, and burnups in the range of 240 MWh/kg (10,000 MWd/te) appeared to be possible without processing. However, the operating temperatures, pressures and power densities were all much lower in NRX than those contemplated for a power reactor. It became evident early in the engineering studies that the large core sizes required with heavy-water-moderated cores and the high coolant pressures required to provide high-temperature steam made the idea of a large containment vessel impracticable. It was therefore decided that the coolant should be contained in pressure tubes. These tubes were much larger than the tubes in NRX, to permit more complex fuel designs, having a greater heat-transfer area between the fuel and coolant. It was also recognized that, to minimize fueling costs, on-power fueling with short fuel bundles pushed through the reactor in opposite directions in alternate channels would be desirable. This fueling strategy (or variations of it) provided a relatively uniform discharge burnup of the fuel. In addition, it placed new and old fuel in close proximity in the reactor core, which reduced the effect of fuel irradiation on core reactivity and thereby reduced the need for reactivity control rods, which are required to compensate for the reactivity changes due to irradiation in most other types of reactors. The overall effect of this fueling scheme was to increase the achievable burnup by a factor of about two.

The basis for accurately predicting the nuclear characteristics of natural uranium/heavy-water lattices, particularly as a function of neutron irradiation, had not been established during the fifties. Thus the challenge in the fuel-cycle physics area was a two-pronged one:

- to obtain a detailed understanding of reactor lattices consisting of relatively large fuel channels with complex fuel geometries, and at relatively large channel spacings, to establish a basis for optimizing plant, reactor, fuel-channel design and operating strategy, and
- to show that such an optimized design and operating strategy would permit sufficient fuel burnup to make the system economically competitive.

This boiled down to a basic problem – that of being able to predict the "neutron balance" for power reactor designs under all proposed conditions and histories of operation. Here "neutron balance" means a detailed account of the sources of neutrons produced in the reactor and their fates in the operating reactor core, which of course must be critical. The difficulty with this problem is the particularly stringent need for accuracy. In a reactor using natural uranium fuel, neutrons are a precious commodity. As will be seen later, only a small fraction of the neutrons produced in a typical operating CANDU are available to provide for the deleterious effects of burnup as fission products accumulate in the fuel and progressively absorb more neutrons. Thus, errors much less than 1 percent are required in estimating neutron balances, if there is to be any confidence in calculated burnup predictions. For a typical CANDU reactor with a discharge burnup of 180 MWh/kg (7,500 MWd/te), a change of 0.001 in reactivity k, (commonly called a "milli-k") corresponds approximately to a change of 2.4 MWh/kg (100 MWd/te) in this discharge burnup. This sensitivity of fuel burnup to the availability of excess neutrons was recognized early in the project. As a result, neutron economy became a dominant guiding principle in all aspects of the design. It was a principle W.B. Lewis insisted on, and he spent much effort ensuring that neutron economy would be achieved.

In fact, as new pieces of data became available or were modified, Lewis would factor them into his estimate of burnup attainable in the current reactor design. Because of its many variations, both up and down, this estimate became popularly known as Lewis's "fever chart". He also used it to assess conceptual design changes. This estimate varied considerably in the early days (200-240 MWh/kg), but ended up around 180 MWh/kg (7,500 MWd/teU), which was quite satisfactory.

Accurate nuclear data were essential if reliable burnup predictions were to be achieved. These data would come from reactor-physics experiments, from work in nuclear chemistry (see chapter seven) and from atomic energy programs in other countries. In the early fifties, Lewis initiated the formation of a United Kingdom-United States-Canada tripartite nuclear data committee as a forum for evaluating all these data. This was superseded by much wider international collaboration as soon as that could be arranged.

Solving the fuel-cycle problem

At the beginning of the CANDU development program, computational capability was rudimentary and the nuclear data incomplete. Thus, as mentioned, experimental measurements were made on lattice assemblies resembling as closely as possible the actual power reactor lattice design.

The experimental program had two fundamentally different aspects:

- low-power experiments in lattice-test reactors using realistic fuel designs with pressure tubes and calandria tubes, and
- measurements of reactivity change with irradiation, which required reactors of higher power to obtain fuel irradiations of practical interest.

To measure lattices more characteristic of power reactors, ZED-2, a larger and more versatile version of ZEEP, was built. ZED-2 was probably the most useful reactor physics facility at Chalk River. Costing about $4 million to build (including the heavy water), it was a much more elaborate design than ZEEP, allowing a much wider range of measurements to be made, more accurately and more quickly. The research program in ZED-2 has required about ten professionals, technicians and support staff since it first went critical in September 1960.

ZED-2 was used to determine the nuclear characteristics of a variety of possible power-reactor lattices. In addition to the characteristics of the lattice itself, other experiments provided the basis for extrapolation to high-power conditions. These included measurements of the effect of changes in temperature, coolant density, and non-uniformities in the lattice. Initially, measurements were done on lattices that filled the calandria. Measurement techniques and computational capability improved with time to such an extent that later experiments could be done by substitution techniques, using as few as seven test assemblies. The substitution method consists of replacing some of the channels in a known lattice with those of the lattice whose parameters are to be measured. This capability considerably reduced the cost of the experiment, particularly for experiments using high-temperature, pressurized coolants.

The following list of fuel assemblies tested illustrates the increasing complexity of the fuel designs considered as the program proceeded:

- 7-element UO_2 and UC
- 19-element UO_2, U metal, $ThO_2/(^{235}U)O_2$
- 28-element UO_2
- 31-element PuO_2/UO_2 (simulating burnt-up fuel)
- 37-element UO_2,U_3Si, UC, U metal.

Measurements in these lattices gave information on the critical size of the lattice, the ratio of fissions in fertile nuclides (caused by high-energy neutrons) to fissions produced in fissile nuclides (caused by thermal neutrons), the ratio of fissile material produced (by neutron capture in fertile nuclides) to fissile material destroyed (by fission or neutron capture), and the shape of the neutron-energy spectrum. Measurements were also made of the spatial distribution of neutron flux within fuel assemblies and lattice cells, as well as in the lattice as a whole.

This information permitted derivation of a detailed neutron balance for the lattice. Table 11.1 lists the neutron balance for a typical cold, clean (without fission products), critical CANDU

TABLE 11.1
Neutron Balance Sheet
Typical Cold Clean Critical CANDU Lattice

Fates of 1000 fast neutrons			
Neutron Leakage			118
Non-fuel Absorptions			
Heavy Water		16	
Calandria Tubes		9	
Pressure Tubes		22	
Fuel Sheaths		6	
Sub-total		53	53
Fuel Absorptions			
U235		468	
U238			
Thermal	271		
Resonance	90		
Sub-total	361	361	
Sub-total		829	829
Total			**1000**
Sources of fast neutrons			
U235		974	
U238		26	
Total		**1000**	**1000**

lattice, derived from the ZED-2 experiments with 28-element uranium-dioxide fuel assemblies similar to the Pickering A fuel.

Criticality is achieved in ZED-2 by adjusting the level of the heavy-water moderator until operation is stable at a reactor power of a few tens of watts. The moderator level determines the effective core size and hence the neutron leakage, which is the variable used to obtain a neutron balance.

The core for the experiment listed in table 11.1 had an effective volume of twenty-two cubic metres. Power-reactor volumes are at least ten times larger than this, so most of the neutron leakage that occurred in ZED-2 would be available in a power reactor to compensate for other effects that waste neutrons. The availability of such excess neutrons is referred to as excess reactivity, and is measured in units of milli-k (mk). For all practical purposes, 1 mk corresponds to the availability of 1 excess neutron in 1,000.

Most of the factors involved in the extrapolation to high power could be investigated in the ZED-2 reactor. Heaters were used in the moderator to raise the lattice temperature uniformly by as much as 60°C. This simulated the temperature rise in a high-flux reactor due to the energy given up by neutrons decelerating in the moderator, as well as energy produced by gamma-ray absorption in the moderator. In addition, seven specially designed "hot channels" permitted fuel-channel temperatures to be raised to 300°C for some of the lattices. In a power reactor with boiling light-water coolant, boiling will occur at some point in the channel. To simulate this effect an array of hypodermic needles was used to bubble air through some of the channels.

As opposed to idealized lattices, power-reactor cores are not uniform. In addition to the inevitable changes arising from fuel irradiation and refuelling operations, control absorbers are introduced into the core to control the magnitude and shape of the flux. The size of these flux and reactivity perturbations was studied extensively in the ZED-2 reactor.

A factor in reactor fuel behaviour at high power that could not be studied in ZED-2 was the buildup of fission products. During reactor operation at high power, some fission products build up quite rapidly to a saturation level at which they are being destroyed by neutron absorption as rapidly as they are being produced in fission, and since both processes are proportional to the reactor flux they represent a more or less constant reactivity load. In a low-power reactor such as ZED-2, these reactivity loads are much too small to be measurable. Fortunately, these effects had become well understood from the operation of NRX.

Such specialized measurements allowed an estimate of the reactivity losses due to most of the factors involved in the extrapolation from measurements in ZED-2 to a typical power reactor operating at high power. A typical estimate is listed in table 11.2. The basis for the allowance for burnup is explained in more detail below.

A major uncertainty in the fuel-cycle physics was the reactivity change that occurs as the fuel is irradiated.

Initially, the only fuel nuclides of interest are uranium-235 and uranium-238. As the fuel is irradiated neutrons are absorbed in both nuclides. The principal nuclide transformations occurring in the fuel are illustrated in figure 11.1 below.

Two distinct series of transformations occur here. The uranium-235 captures a neutron and either fissions or becomes uranium-236. Since there is no production of uranium-235, its concentration in the fuel will gradually diminish. On the other hand, absorptions in uranium-238 produce a new fissile nuclide, plutonium-239, which will build up approaching an

TABLE 11.2

Allocation of Excess Reactivity in a Typical CANDU

REACTIVITY LOAD	EXCESS REACTIVITY ABOVE COLD, CLEAN CRITICAL (MK)
Full insertion of adjuster rods	20
Extra structural materials	3
Temperature effects	13
Saturating fission products	33
Control margin	2
Burnup	30
	101

FIGURE 11.1
Nuclear Transformations in the Fuel

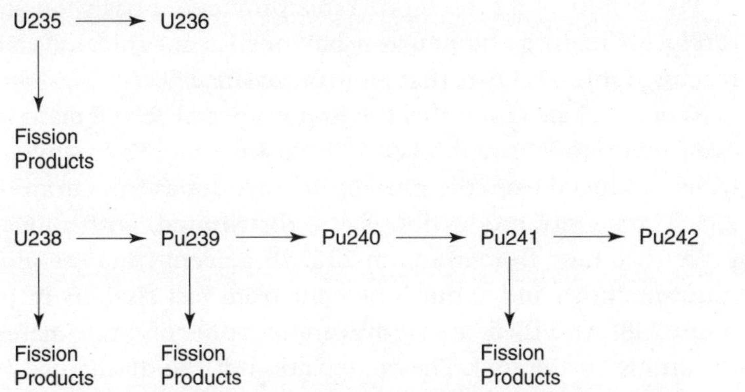

equilibrium, where it is destroyed by fission or by transformation to plutonium-240 as rapidly as it is being produced. In a similar way, plutonium-241, another fissile nuclide, will build up by neutron absorption in plutonium-240. However, for irradiations of interest, concentrations of plutonium-241 never reach an equilibrium level. For practical irradiations we do not need to consider the very low concentration of nuclides beyond plutonium-242 or uranium-236.

Thus, as the fuel is irradiated, competing reactivity effects occur. The loss of fissile uranium-235 is partially compensated by the growth of two fissile plutonium nuclides. Moreover, neutron absorption by the accumulating fission products leads to a reduction in the reactivity of the fuel. Although these changes in the composition of the fuel affect the reactivity significantly and produce large amounts of energy, the chemical changes are in fact very small. A mere 1 percent burnup of the uranium corresponds roughly to the production of 240 megawatt-hours of thermal energy per kilogram of uranium (MWh/kgU). To put this in familiar terms the electrical energy extracted from one kilogram of uranium with 1 percent burnup is enough to electrically heat an average home (70 kWh/day) for about 1,000 days.

The mathematical description of these nuclear transformations is simple to formulate in terms of the nuclear cross-sections of the materials involved. However, for many fission products the yields and cross-sections were unknown. In addition, all the cross-sections depend on the neutron energy spectrum, which depends on position in the reactor core and changes with time as the concentrations of various nuclides change. Thus the result may depend to some extent on the history of the fuel irradiation. It was therefore desirable to verify the mathematical predictions in a variety of different practical situations.

Three major experimental programs were undertaken with this problem in mind.

In the early fifties, the only fuel available with significant irradiation came from the operation of NRX. This led to cooperative measurement programs with both the United States and the United Kingdom.

Initially, measurements were made on samples from rod #683, one of the uranium metal rods originally installed in the reactor, which by September 1951 had attained a specific output of about 3,100 MWd/teU in its central region. Some of the samples cut from this highly radioactive uranium metal rod were sent to the Argonne National Laboratory, where measurements on reactivity and fuel composition as a function of irradiation were made.[2] Measurements on fuel composition were also made at Chalk River. In addition to providing valuable information about the cross-sections of the uranium and plutonium nuclides in the fuel, and the overall reactivity changes, these initial studies revealed the importance of such effects as the radial variation in the neutron spectrum within the rod and changes in the fuel density during irradiation.

In the mid-fifties, rods containing aluminum-clad uranium cylinders specifically designed to simplify measurements of this sort were irradiated in NRX. In a collaborative program with Harwell, samples of this fuel were sent to the United Kingdom, where their reactivity was measured in the GLEEP reactor, which had been specially built to make such measurements. They were then returned to Chalk River for chemical and mass-spectrometric analyses, to determine the amount of plutonium

[2] Because the neutron flux decreases toward the periphery of the reactor, samples from various positions along the rod correspond to differing total amounts of fuel burnup or irradiation.

in the fuel and the isotopic composition of both the uranium and plutonium.

To provide, at Chalk River, a facility similar to GLEEP, the PTR reactor was built, and started operation in 1957. This low-power reactor consisted of a highly-enriched uranium fuel core immersed in a large "swimming pool" of ordinary water, which provided cooling for the fuel and shielding for the reactor. A swing mechanism, operated hydraulically by high-pressure water, oscillated a carrier containing two samples, such that each sample was located alternately at the centre and beyond the boundary of the core. Thus, a sample of unknown reactivity could be compared with one of known reactivity (e.g., a standard absorber such as boron). Unfortunately, because of the highly absorbing core and light-water moderator, the spectrum was not typical of a heavy-water reactor. Thus, measurements made in this facility usually required more interpretation than measurements in ZED-2.

In the sixties, an experimental program to measure the effective cross-section of fission products was undertaken. Samples of alloys of fissile material were irradiated in NRU. Reactivity measurements were made in PTR on the samples before and after irradiation. The samples were then dissolved and the uranium and plutonium contents determined from chemical and mass-spectrometric measurements. The change in reactivity due to the measured change in fissile content was subtracted from the measured reactivity differences of the samples, to obtain the reactivity of the fission products, from which their effective cross-section was inferred. Because the spectrum in PTR is not typical of a CANDU reactor, there was initially some uncertainty in the interpretation of these measurements. The samples were later sent to Studsvik, Sweden, for measurements in a very thermal spectrum, where good agreement with calculations using the program FISSPROD was obtained. The calculated effective thermal cross-sections were within 10 percent of those measured.

In 1968, a series of eight 19-element NPD uranium oxide fuel bundles, with irradiations ranging from 24 to 240 MWh/kgU (1,000 to 10,000 MWd/teU) were analyzed by the CEA (Centre d'Etudes Atomique) in France, using mass spectrometry. The bundles were selected to have an irradiation history typical of an unperturbed CANDU-type reactor with bi-directional on-power fueling. Agreement with predictions was good.

The results of all these programs provided a basis for accurately estimating the neutron balance for a typical CANDU reactor. Table 11.3 lists that neutron balance.

Table 11.3 also indicates the importance of fertile materials, even in a once-through cycle. There are almost as many neutrons produced from the plutonium nuclides as from uranium-235. The energy production is also distributed, with about 55 percent coming from uranium-235, 40 percent from the plutonium nuclides and about 5 percent from fast fissions in uranium-238. And there are significant amounts of fissile material remaining in the fuel. The composition of the discharged fuel relative to one initial uranium-235 atom (and 139 uranium-238 atoms) is approximately:

Uranium-238	137.0
Uranium-235	0.274
Uranium-236	0.100
Plutonium-239	0.350
Plutonium-240	0.155
Plutonium-241	0.035
Plutonium-242	0.010

Thus, operating on a once-through natural uranium cycle, CANDU extracts an energy that is approximately equivalent to that which would be generated by fissioning all the uranium-235 contained in the fuel. This is quite adequate to provide economical nuclear power at uranium prices of the last few decades. The objective initially defined in the development program had been clearly met in the CANDU reactor design with push-through, bi-directional, on-power bundle fueling.

The once-through uranium cycle provides less than 1 percent of the energy potentially available in uranium by fully utilizing the fertile component, uranium-238. Chapter eighteen discusses fuel cycles in which much more energy can be extracted from the natural uranium entering the cycle, by recovering and recycling the fissile and fertile components of the irradiated fuel. Since the early eighties the measurement program in ZED-2 has studied lattices containing typical recycle fuels, e.g., 37-element PuO_2/UO_2, 37-element PuO_2/ThO_2 and 37-element $(^{233}U)O_2/ThO_2$.

TABLE 11.3
Neutron Balance Sheet: Typical CANDU Reactor Operating Under Normal Conditions With an Equilibrium Core and an Exist Burnup of 180 MWh/kgU

Fates of 1000 fast neutrons produced			
Neutron Leakage			31
Non-fuel Absorptions			
Adjuster rods		9.8	
Control & misc.		0.4	
Moderator		15.7	
Calandria tubes		8.8	
Pressure tubes		23.1	
Coolant		0.3	
Sheaths		5.8	
Sub-total		63.9	64
Fuel Absorptions			
U235		245.2	
U238			
Resonance	93.3		
Non-resonance	242.0		
Sub-total	335.3	335.3	
Pu239		232.1	
Pu240		20.8	
Pu241		11.9	
Pu242		0.1	
Saturating fission products		32.1	
Other fission products		27.5	
Sub-total		905.0	905
TOTAL			1000
Sources of fast neutrons			
U235		510	
Pu239		440	
Pu241		24	
U238		26	
TOTAL		1000	1000

Reactor Control

H.K. RAE

INTRODUCTION

A reactor is controlled at a particular neutron flux or power (rate of fissioning) by maintaining the neutron consumption rate equal to the neutron generation rate; that is, maintaining the reactivity (k) at unity. This is usually discussed in terms of excess reactivity expressed in milli-k (mk) equal to 1,000(k-1). This control can be exercised by varying neutron leakage through changes in moderator level, or by varying neutron absorption through moving a control rod (a neutron absorber) in response to measured changes in neutron flux. The excess reactivity varies in response to many factors, such as coolant temperature, fuel temperature, coolant density, fission product concentration in the fuel, and so on.

It is necessary to control the reactivity for all power levels from the shutdown state to full power. It is also necessary to be able to reduce reactivity quickly to shut the reactor down in the event of any loss of control. Otherwise, a rise in reactor power could increase exponentially to the point of reactor destruction.

The complexities of describing the variation of neutron flux in both space and time in a large reactor core and its response to the host of parameters that may change, and to the control actions taken, require analysis by computer. Underlying CANDU control technology has been the evolution of ever more sophisticated codes for design and operational analysis. These are based on a large body of measurements in the zero-energy reactors, ZEEP and ZED-2, and have been verified at various stages in their development by experiments in NRX, NRU and several of the power reactors.

The principal thrusts of reactor-control-system development for CANDU have been to achieve highly reliable systems, to automate those systems and to control flux shape to maximize the total energy obtained from the fuel. Many inherent characteristics of heavy-water-moderated reactors have led to a number of unique features in CANDU reactor control systems.

Delayed Neutrons and Photoneutrons

Neutrons produced directly in the fission process are emitted promptly. However, most fission products are radioactive and decay, some of them emitting further neutrons. Because of the radioactive decay process, these neutrons are released some time after the fission process itself, with mean delay times ranging from 0.3 to 80 seconds. The fraction of emitted neutrons that are delayed depends on the nuclide that is fissioning. For the predominant types of fission in natural-uranium heavy-water reactors, the fraction is highest for fissions of uranium-238 caused by fast neutrons, intermediate for uranium-235 fission and lowest for plutonium-239 fission. In typical heavy-water research reactors, such as NRX and NRU, about 0.65 percent of all the neutrons produced in the reactor fall in the category of delayed neutrons, while for typical CANDU power reactors the fraction is closer to 0.5 percent, because of the plutonium grown into the fuel.

In heavy-water reactors, neutrons (called "photoneutrons") are also produced by interactions between gamma rays and the heavy-water moderator. As fission products build up in the core, the gamma-ray activity also builds up and provides a source of photoneutrons that have the same type of effect as the delayed neutrons discussed above. In a heavy-water reactor the fraction of neutrons produced that are photoneutrons is typically about 0.03 percent, but the mean delay times from the relevant fission event range from about 4 to 274,000 seconds.

The lifetime of a generation of neutrons in a heavy-water reactor is about one millisecond, and it might therefore be expected that controlling power would demand a very fast-acting control system. Fortunately, the existence of delayed neutrons and photoneutrons lengthens the effective neutron lifetime significantly, making the design of reactivity mechanisms to control the power completely practicable.

Figure 11.2 shows the effect of the delayed and photoneutrons on the time taken for reactor power, in a typical heavy-water research reactor that has been operating at steady power, to increase by 10 percent following a step change in reactivity. Shown are estimates of the times taken after the step change in reactivity to reach 110 percent of the previous steady power, in the actual case with delayed neutrons and photoneutrons, and in a hypothetical case with no delayed neutrons or photoneutrons, as a function of the size of the excess reactivity step.

From figure 11.2 it is apparent that a reactor power-control system having a response time of a few seconds will be adequate for maintaining steady operation and following changes in power demand. However, it is also clear that the shutdown systems, needed to counter any sudden large increase in reactivity, must respond in the order of a tenth of a second to a high-power trip signal. The thermal capacity of the fuel provides a delay of the order of a second before severe fuel damage can occur, and during this time the shutdown system must insert sufficient negative reactivity to reverse the power pulse. Because a sudden large reactivity insertion at low power could lead to an extremely rapid rate of rise of power before any high-power trip signal is activated, the shutdown system must also be actuated if the fractional rate of rise of power (or neutron flux) exceeds a safe value.

Although photoneutrons make a relatively small contribution to the effect shown in figure 11.2, they are important in permitting safe startups after prolonged shutdowns (of days or weeks), since they provide a source of neutrons that can be detected by the reactor instrumentation while the reactor is still significantly sub-critical.[3]

Reactivity Coefficients

The reactivity of the reactor core will change with changes in its physical properties; this is usually expressed in terms of an increase or decrease in reactivity per unit change in a particular property, and is referred to as a reactivity coefficient. The most important reactivity coefficients are those associated with

---

[3] A reactor cannot start up without some source of neutrons to initiate the chain reaction. There is always a small source present due to spontaneous fission and cosmic-ray effects. However, reliance on this source alone for initial startup is unsafe, since the normal instrumentation is not sensitive enough to detect its effects until the reactivity is very close to critical, and hence dependence on this natural source would lead to a "blind" startup.

FIGURE 11.2
Effect of Small Step Changes in Reactivity on Time Required to Reach 110% of Previous Steady Power

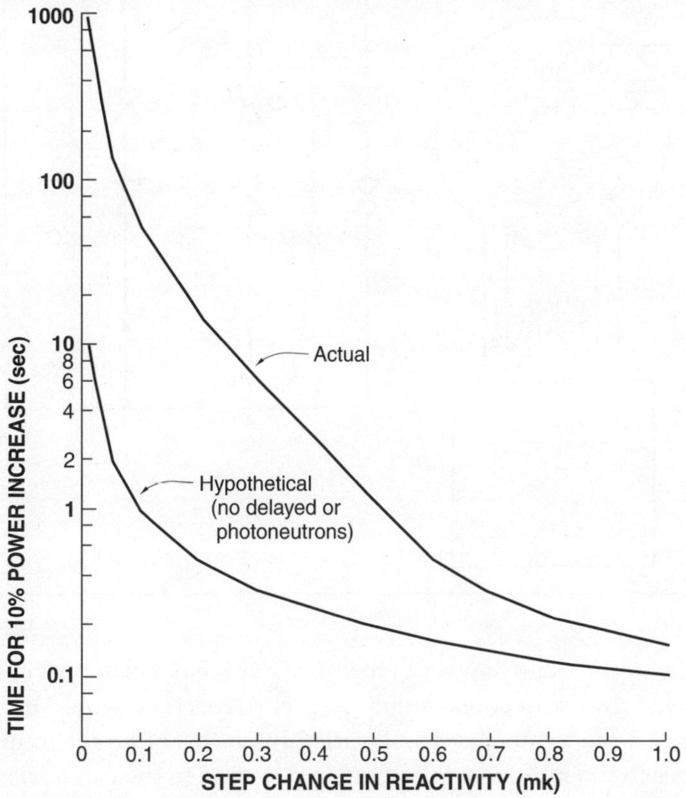

changes in temperature of the coolant and fuel, and changes in coolant density.

With fresh fuel, an increase in fuel temperature provides a significant negative contribution to reactivity; however, as irradiation proceeds, and plutonium isotopes accumulate in the fuel, this coefficient becomes less negative. An increase in coolant temperature, on the other hand, provides a positive contribution to reactivity through the reduction in coolant density. An increase in coolant temperature also causes a rise in average neutron energy which has an effect on reactivity that depends on the burnup of the fuel.

There have been many measurements of temperature coefficients. One example is a series of experiments in Pickering A to determine the overall temperature coefficient. The change in reactivity was measured as the temperature of the fuel and coolant were raised together from 50 to 270°C with the reactor at low power. Data for units 3 and 4 at initial startup showed an overall drop in reactivity of 9 mk and a temperature coefficient of reactivity at 270°C of -2.6×10^{-2} mk/°C. Data for unit 2 with fuel at near-equilibrium burnup showed an overall small rise in reactivity of 0.6 mk and a coefficient at 270°C of $+1.6 \times 10^{-2}$ mk/°C.

As reactor power is raised from a low value to full power, with the system already hot, the increase in fuel temperature provides further negative reactivity. This is principally due to increased absorption of neutrons by uranium-238. The power coefficient with the core at equilibrium burnup is small, about a -0.02 mk/percent change in power for a CANDU 6. This power coefficient is the principal in-reactor source of rapid negative reactivity change.

The important effect of coolant density on reactivity occurs when the coolant boils – effectively introducing voids within the channel. The resultant void coefficient is positive and of considerable significance in safety analyses, as well as in the control of CANDU units, where limited boiling is permitted during normal operation (Bruce, CANDU 6, Darlington). This positive void coefficient means that any accidental loss of coolant in the fuel channels with accompanying void formation will cause a reactivity increase and hence a power increase that will further increase the voiding. The shutdown systems must act quickly enough to check this developing power surge and bring the reactor to a sub-critical state (see chapter fifteen).

Multichannel Control Systems

Multichannel control systems were first used with NRU, added to NRX in the late fifties, and used at NPD and Douglas Point. The arguments in support of this decision are not too well documented, but the objective was clearly to develop a very safe system – that is, to avoid a failure that might cause a positive reactivity transient.

Aspiring to a more reliable system by increasing the number of parallel controlling channels requires a method of detecting a faulty channel and preventing it from taking part in the control. Several ways of doing this were developed. In NRU it was achieved by averaging four independent signals and also intercomparing them and arranging that any disagreement among the signals triggered the disconnection of the errant one from the average. With the moderator-level control system used at NPD, the parallel independent channels included the valves directly controlling the level, and the design ensured that any two channels could continue to provide adequate control whenever the third failed.

Moderator Level Control

The moderator level in NPD was controlled using a gas balance system (figure 11.3). Moderator was retained in the calandria by maintaining a differential helium-gas pressure between a dump tank below the calandria and the gas space above the moderator. The heavy-water moderator spilled through ports, which had a U-tube-like shape, at the bottom of the calandria into the dump tank, and was continuously returned to the calandria through pumps. The differential pressure was provided by several blowers, with the gas returning through three parallel lines. Valves (A, B, C in figure 11.3) in these lines controlled the differential pressure and thereby the heavy water level. Each pair of valves, AA, BB and CC, was controlled by one of the three parallel reactivity control channels. This arrangement ensured that adequate control continued even if one channel failed (either opening or closing its valves) or if a valve failed. The much larger valves D, E and F comprised the moderator dump system. They were normally closed; on opening they rapidly equalized the pressure in the dump tank with that above the moderator to drain the calandria in fifteen seconds. Each pair of valves, DD, EE and FF, was activated by one of the three trip channels, so that coincident signals from at least two channels were necessary for shutdown. This arrangement allowed the coincidence principle (see chapter four), and the ability to test a single channel without shutting down the reactor, to be extended

FIGURE 11.3
Schematic of Gas Balance System for Moderator Level Control

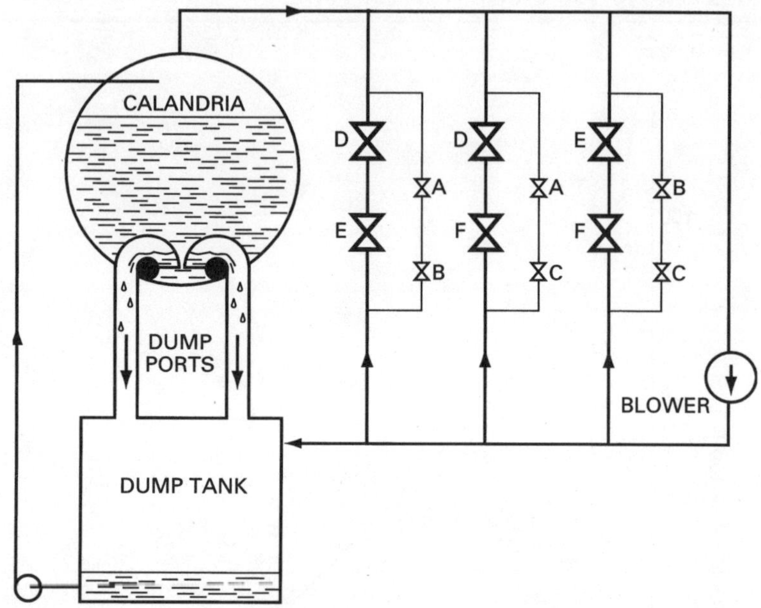

to the actual shutdown device itself. The gas balance system allowed the use of considerably smaller valves than would be required if moderator level was varied by means of liquid control valves in the moderator flow being returned to the calandria; it also provides a more practical fast-acting dump system.

Level control for NPD was chosen for its simplicity and reliability. It also allowed the fullest exploitation of the multichannel philosophy. The concept was first suggested during the studies by the nuclear power group in 1954 (see chapter ten), and was incorporated into the original NPD design – the lower hemispherical head of the reactor pressure vessel was to be the dump tank. An important advantage claimed for level control, at that time, over the use of control rods was that the latter displaced fuel channels in a vertical core and so reduced maximum power. By the time the change was made to a horizontal pressure tube reactor, which could have accommodated vertical control rods between the fuel channels without displacing fuel, level control was firmly entrenched as the control method of choice, and was retained.

In 1957, NRX was modified to permit the entire control of reactivity by varying moderator level. In this case, the multichannel system controlled parallel liquid control valves in drain lines from the calandria to the dump tank. The design and operation of this system provided valuable experience for some aspects of the NPD and Douglas Point systems.

While level control proved to be adequate for NPD, such a simple system alone could not meet the reactivity control demands of larger reactors. There were two principal shortcomings. Firstly, at various times, such as restarting after a shutdown of a couple of days or more, the fuel will have excess reactivity and the moderator level would have to be below some of the upper fuel channels. Not only do these calandria tubes then need to be cooled by a spray of heavy water (as was provided at NPD), but the reactor must operate at below full power for a time, while the short-lived saturating fission products regain their normal concentration. Douglas Point and later CANDUs are brought to full power under such circumstances by adding a soluble neutron absorber or poison (boron or gadolinium) to the moderator to counteract the excess reactivity, and then removing it at a controlled rate by ion exchange. Secondly, in order to get the maximum energy from the fuel throughout the core it is necessary in large reactors to have spatial control of reactivity, a task that moderator level cannot perform. Although Douglas Point, RAPP and Pickering A were provided with moderator level control systems, they were only used to perform secondary control functions.

Booster Rods and Adjuster Rods

The most troublesome saturating fission product is xenon-135 (half-life 9.2 hours), which has an extremely large neutron capture cross-section. It is formed by the radioactive decay of iodine-135 (half-life 6.7 hours) which is produced by fission. The xenon-135 concentration in the fuel during operation is set by its rate of formation and its rate of destruction by neutron capture. After a shutdown, the iodine-135 inventory continues to produce xenon-135, which nearly quadruples in concentration before decaying away to a low value about forty hours later. It is not economical in CANDU reactors to provide excess reactivity to counter the whole xenon-135 transient; the approach taken was to provide enough excess reactivity to allow the reactor to be restarted within thirty minutes of shutdown – requiring 10 to 20 mk. This allowed operation to resume if the cause of the shutdown could be eliminated quickly; otherwise, startup was delayed for a day and a half.

In NPD and Douglas Point this excess reactivity was provided by booster rods – enriched uranium fuel with its own low-temperature cooling system that can be inserted vertically into the moderator. A similar system was installed in the Bruce A reactors, but only used early in their operation on a few occasions: there were some safety concerns. With the greater reliability of the post-Douglas Point reactor-control systems, it was found that the causes of unanticipated shutdowns could either be eliminated in a few minutes, or required longer than the thirty-minute xenon-override period for corrective action.

In other CANDUs, neutron absorbers called adjuster rods are used. These reside within the core during normal operation and can be withdrawn to provide excess reactivity. In some reactors the absorber is cobalt, and the cobalt-60 produced is sold to be used in gamma-irradiation facilities at a price that offsets the value of the neutrons absorbed (see chapter five). While occasionally used for xenon override, the excess reactivity provided by adjuster rod withdrawal is more often required during periods of fueling machine unavailability. However, their main function is to produce a more uniform neutron flux throughout the reactor.

These adjuster rods are placed so as to depress the neutron flux in the central region of the core, and thus flatten the flux shapes both radially and axially. Normally, in the absence of flattening, the flux would peak in the centre of the core and fall off toward the edges. Fuel could operate at its design power rating at this peak position, but the average rating and thus the total reactor power would be considerably lower. Axial flattening therefore raises the average channel power and radial flattening increases the total power from a given number of channels, as well as allowing a higher-average channel burnup to be achieved.

Varying the burnup of the discharged fuel across the core is also used to achieve radial flux flattening. Fuel in the central channels is taken to a higher burnup relative to the outer channels, and the extra fission-product inventory of the fuel in these central channels depresses the flux in that region.

Fuel Management

The main long-term control of reactivity in CANDU is provided by on-power refueling. This normally proceeds at a rate sufficient to keep the zonal control elements in their mid-range.

The original concept was to refuel one bundle at a time, using a pair of fueling machines attached to the ends of a channel to insert a fresh bundle at one end and remove an irradiated bundle at the other end. This concept leads to uniform average burnup of all the fuel in a channel. A second aspect of the refueling concept was to fuel in opposite directions in adjacent channels, which results in the same average reactivity axially through the core.

This simple scheme worked well for NPD. To reduce the usage of the fueling machine at Douglas Point, four-bundle shifts were adopted when it was found that the penalty in burnup would be small. Since a channel contained twelve bundles, then each bundle would normally reside in three locations during its irradiation. Then, around 1970, it was found that fuel failures were associated with refueling (see chapter thirteen), and this led to a change to eight-bundle shifts. Similar difficulties at Pickering resulted in ten-bundle shifts being adopted as the optimum. For Bruce and later CANDUs, which have greater axial flux flattening, four- or eight-bundle shifts are used.

The refueling rate for a CANDU 6 station is seventeen bundles per day, on average, or about two channels per day with eight-bundle shifts. The selection of channels to be refueled is generally based on those calculated to have the highest burnups, but also must consider the effects of fresh fuel on channel outlet temperature and on local power distribution; it is also necessary to maintain a refueling pattern that has axial symmetry and equal rates in opposite directions.

On-power refueling avoids the need to shut down for refueling, and thus increases the annual capacity factor.

The equilibrium fueling rate observed at Pickering corresponds to a burnup of 170 MWh/kg, about 7 percent higher than predicted during design. This burnup value had been established by 1975, and in a sense represents the final point on Lewis's fever chart. Lewis's goal nearly twenty years earlier had been 10,000 MWd/tonne (240 MWh/kg). He had inspired developers and designers to practice strict neutron economy, and they had succeeded admirably.

Zonal Control and Flux Measurement

Refueling and the movement of reactivity control devices cause local changes in neutron flux that are unsymmetrical about the centre of the reactor. In a large reactor, such local perturbations can propagate because of the loose neutron coupling between widely separated regions of the core. In a CANDU reactor, the neutron migration length is 19 cm; this is the average distance travelled by a neutron from its appearance as a fission neutron to its absorption as a thermal neutron. The core radius in large CANDUs is seventeen to twenty times this migration length.

The propagation of local flux perturbations can lead to slow oscillations of the local flux, because of the positive feedback arising from the response of xenon-135 to a neutron flux change. If the flux in a region increases, the xenon-135 burnup will be increased initially, leading to a lower xenon concentration, hence a higher reactivity and a further increase in flux. The higher flux also leads to a higher iodine-135 production, and eventually an increase in xenon-135 concentration, thereby reducing the flux. If the reactor total power is constant, the flux can oscillate between different regions of the core, with a period of oscillation between fifteen and thirty hours, depending on which higher harmonic oscillations occur.

This instability phenomenon was first observed in American reactors, and was first explained in the mid-fifties by Arthur Ward at Chalk River who formulated the criterion for the threshold for these oscillations to propagate with time.

Such flux oscillations could result in fuel-bundle or fuel-channel power limits being exceeded, and therefore an on-line spatial flux control system is required. The problem did not arise in the small NPD core, and was of a limited magnitude for Douglas Point; in that case four absorber rods provided local control, one in each reactor quadrant. In Pickering and all later CANDUs, fourteen zones are established, each with its own reactivity controller.

A new control device was developed: the liquid control absorber. It is a small elongated tank within the moderator connected by small bore piping to a reservoir outside the reactor. The liquid is ordinary water, which in the CANDU core acts as a neutron absorber. A prototype tested in ZED-2 gave confidence in the device, which was installed in Pickering and all subsequent CANDUs. It has fulfilled the original expectations for high reliability.

The water level in each absorber is controlled by a valve, and controls the reactivity of its zone. All fourteen levels are also controlled in unison, to provide fine control of total power generation. The fourteen absorbers are arranged in six vertical assemblies as shown in figure 11.4.

To verify reactor kinetic and spatial control computer codes used to design the reactor control system, a series of transient experiments were performed in ZED-2. The core was divided into two halves, either separated by a row of neutron absorbers or an unfueled region. A step change was made in reactivity in one half of the core and the detailed neutron flux shape throughout the core was measured over time. This was compared to the code prediction.

In NRX, NRU, NPD and Douglas Point, the neutron flux was measured at several locations around the periphery of the core with ion chambers. Units of the required sensitivity and stability were developed in the early days at Chalk River by Hugh Carmichael and co-workers. The relationship of this peripheral flux to reactor power had to be included in the control system. However, for zonal control, the local flux or power must be measured, and ion chambers are not suitable for in-core use.

In NPD and Douglas Point, flow and temperature were measured at the outlet of each fuel channel, and these data gave the radial power distribution. At Douglas Point this was used to control the four zonal absorbers.

FIGURE 11.4
Reactivity Control Devices in CANDU

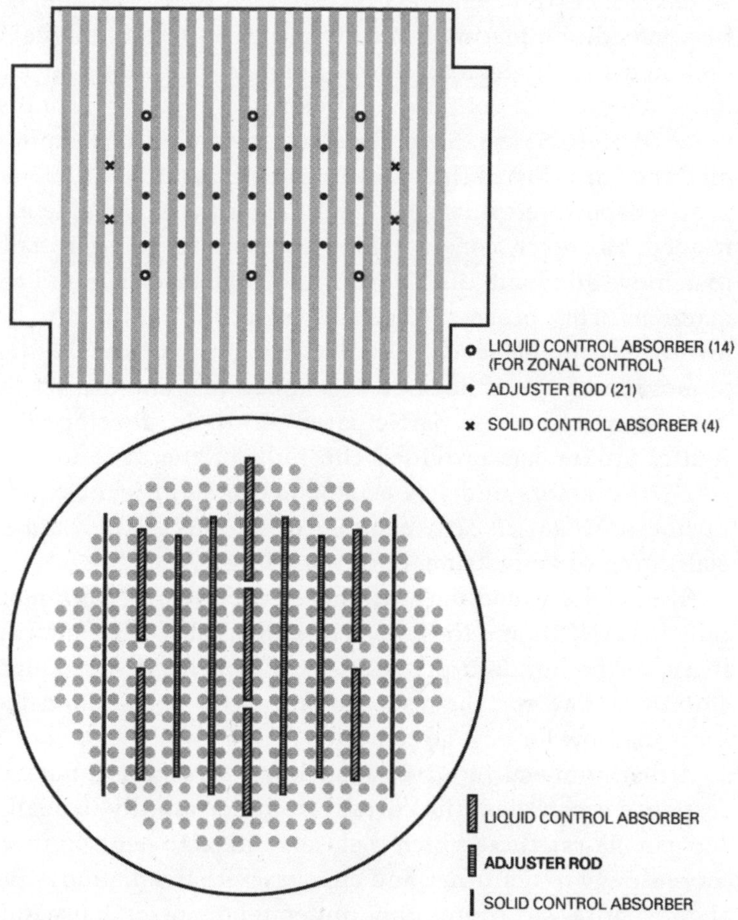

However, for Pickering, in-core flux measurements would be needed for zonal control. Fortunately, a new device, the self-powered neutron flux detector, had been invented by John Hilborn at Chalk River in the early sixties, based on earlier Russian work. It consists of a length of coaxial cable in which the inner conductor is a beta-particle emitter insulated from the Inconel sheath by a metal oxide. The beta particles must have sufficient energy to reach the sheath and produce an electric current. Emitters investigated were rhodium, cobalt, vanadium, platinum and zirconium. A suitable signal is obtained from a detector about 2 mm in diameter and 3 m long, wound as a coil around a central support over a distance of 30 to 85 cm. Several detectors on the same support give the variation of flux through a core sector.

A self-powered flux detector is, in principle, simple and rugged. However, considerable effort was required initially to achieve adequate quality control of the materials and the manufacturing process, to get the in-service failure rate below the acceptable level. A small Canadian company (Reuter-Stokes Limited) was licensed to manufacture and market the detectors, and they participated in their development. Reuter-Stokes has provided the bulk of the detectors for CANDU reactors and has established a significant export business. This is an early example of the successful commercialization of an instrument developed by AECL.

Ultimately, vanadium was chosen for accurate flux mapping in CANDUs and these detectors are used in large numbers (fifty to one hundred per reactor) in Bruce and subsequent CANDUs. However, the response time of the vanadium detector is too slow for on-line control.

A platinum emitter gives a much faster response, but this response is complex; the current responds rapidly to changing gamma-ray flux (which is closely linked to neutron flux), more slowly to neutrons, and changes with irradiation. Once the response was thoroughly understood, corrections could be made to extract the fast component. Then, the platinum detector (twenty-eight per reactor) became the basis for CANDU on-line control signals, the necessary corrections being made by on-line computer processing based on data from the vanadium flux mapping array available on a two-minute cycle.

In the case of Pickering, twenty platinum detectors are used, calibrated against power measurements from twenty-two instrumented fuel channels, for on-line control. The detectors originally installed had cobalt emitters, which gave a rapid response to changes in neutron flux, but had a rather limited useful life and needed to be replaced regularly.

Because of the possible variations in neutron flux about the desired flattened shape, each of the two independent shutdown systems has its own array of fast-response self-powered flux detectors to locate any unacceptably high local neutron flux. Detector locations are optimized, to ensure coverage of any flux shape that could arise. Each such array and its associated signal-processing electronics and two-out-of-three coincidence arrangements is called a regional overpower protection system. It can initiate a reactor shutdown, just as other parameters, such as high-rate-of-rise of neutron flux and high coolant pressure, can.

Figure 11.4 illustrates the diverse and rather complex array of control devices in a CANDU reactor of the eighties. Control requirements evolved in response to increases in size, power and fuel rating in CANDUs. This evolution has gone hand-in-hand with an ever-increasing degree of automation of reactor control, a trend found in power-reactor programs around the world.

Fast Automatic Startup

When fast automatic startup was incorporated into the design of NRU, it was a unique, pioneering step. It arose out of the need to overcome the rapid buildup of xenon-135 after a reactor shutdown. Since only limited excess reactivity could be provided, the time available to clear a fault and restart the reactor was limited, and it was argued that the startup time itself should be reduced to a minimum. It was then decided that the procedure should be automatic; otherwise, it would require the operator's full attention to the exclusion of monitoring other instruments (if, in fact, it could be done manually at all).

To enable the system to function safely, the neutron flux had to be measured over its entire range. In a heavy-water reactor, this range is only about six decades, since photoneutrons provide a lower limit. It was therefore possible to develop an ion chamber and amplifiers that could provide a signal throughout this range suitable for both automatic control and to warn of any excess rate of increase in reactivity.

Computer Control

Early in the design of Douglas Point, digital computers were being considered, not only for their data storage and display capabilities, but also for their ability to make rapid decisions.

A first step, in parallel with the Douglas Point design, was a computer-control experiment with NRU. Its tasks included scanning and digitizing reactor variables, alarm checking and recording, sensor recalibration, data storage and display, and closed loop control. It controlled the reactor power, a unique computer application at the time. More importantly, this experiment demonstrated the need for comprehensive self-checking methods to detect faults in either software or hardware.

At Douglas Point, a computer was installed to perform a wide range of functions; these included fuel-channel temperature monitoring, spatial control of neutron flux, moderator poison control, various process-control functions, and extensive data logging and processing. Operation of the zonal control system by the computer marked the first time a reactivity control element in a reactor had been positioned by a stored-program digital computer. Since the cost of a dual computer system for Douglas Point would have been excessive (the situation in 1962), the reactor could not be fully controlled by the computer, because of the latter's inadequate reliability. Therefore, the main control loop used an analogue system, and operator intervention was possible in those tasks the computer did control in the event that it broke down.

At Pickering, the decision to be made was to what extent the computer would become an integral part of the control and instrumentation system. To justify the cost of the computer, its integration should be as comprehensive as possible. Complete reliance on a computer system would impose stringent targets for its reliability. Alternative backup systems would add extensive extra equipment and design costs. These considerations led to the selection of a dual-computer configuration controlling all major reactor and power station processes. Only the simultaneous failure of both computers would require a shutdown. A major feature is that each computer contains all the control routines and a self-checking program that continually surveys its own performance. Only one computer is in control at any instant, but if it detects an internal malfunction that cannot be readily corrected, it transfers control to the other computer immediately; i.e., in ten milliseconds. This arrangement has proved to be extremely reliable.

The decision by Ontario Hydro to accept computer control coincident with introducing nuclear reactors into their power generation network was a remarkably bold one – not only on their part but also for the AECL designers and developers, who proposed it well before computer control was adopted elsewhere. Canada has been a pioneer in this field.

For CANDUs after Pickering, a similar dual-computer control system has been used, expanded to meet new requirements and modified to take advantage of evolving computer technology. These power-plant control centres embody in a very sophisticated form two of the concepts established early in the development of CANDU: a high degree of automation, freeing the operators for tasks of supervision and coordination to which they are much better suited, and designs permitting continual verification of reliability and safety.

At Darlington, all the logic functions in the shutdown systems are performed by digital computers. This replaces the conventional electronic relay logic used in all previous CANDUs. While this computer system is operating successfully, there were problems in guaranteeing the integrity of the software and in developing the appropriate software quality assurance.

ACKNOWLEDGMENTS

We would like to thank A. Okazaki for his helpful review of this chapter.

BIBLIOGRAPHY

Allan, C.J. *Response Characteristics of Self-Powered Flux Detectors in CANDU Reactors.* AECL-6171. 1978. (Presented at the International Symposium on Nuclear Power Plant Control and Instrumentation, 24-28 April 1978, Cannes, France.)

Critoph, E. "Fuel Cycle Physics." In *Canadian Engineering Centennial Conference: Proceedings of the Nuclear Power and Fusion Programs of the Canadian Nuclear Society, 18-22 May 1987, Montreal*, 111-18. Toronto: Canadian Nuclear Society, 1987.

_____. "Reactor Physics Aspects of CANDU Reactors." In *Proceedings of the Course on Reactor Theory and Power Reactors*, 215-61. Paper IAEA-SM4-44. Trieste, Italy: International Centre for Theoretical Physics, 1980. (Also AECL-6150.)

Duret, M.F., J. Griffiths, R. Bir, M. Chabrillac, and R. Vidal. *Plutonium Production in NPD: A Comparison between Experiment and Calculation.* AECL-3995. 1971.

Green, R.E., C.B. Bigham, I.H. Gibson, R.G. Jarvis, M.J. Halsall, C.H. Millar, and F.E. Driggers. "Lattice Studies at Chalk River and Their Interpretation." In *Proceedings of the Third International Conference on the Peaceful Uses of Atomic Energy, Geneva, 31 August-9 September 1964*, vol. 3, 89-99. Paper A/CONF.28/P/25. New York: United Nations, 1965. (Also AECL- 2025.)

Hilborn, J.W. "Self-Powered Neutron Detectors for Reactor Flux Measurements". *Nucleonics* 22 (February 1964): 69-74. (Also AECL-1896.)

Hinds, H.W., F.M. McDonnell, and D.H. Walker. *Kinetics Experiments on Coupled Cores in the ZED-2 Reactor.* AECL-4267. 1972.

McDonnell, F.N., and R.E. Green. "Spatial Power Control in Large CANDU Reactors: The Whys and Wherefores". In *Canadian Nuclear Association Eighteenth Annual International Conference, Ottawa, 11-14 June 1978. Full Papers*, Session 15, [first paper]. [Toronto]: Canadian Nuclear Association, 1978. (Also AECL-6212.)

Millar, C.H. "Fuel Management in CANDU Reactors." In *Nuclear Energy Maturity: Proceedings of the European Nuclear Conference, Paris, 21-25 April 1975*, vol. 4, 171-84. Edited by Pierre Zaleski, et al. Progress in Nuclear Energy, n.s. Oxford: Pergamon Press, 1976. (Also AECL-5090.)

Okazaki, A., and D.S. Craig. *Determination of Lattice Parameters Using a Few Rods.* AECL-2593. 1967.

Pearson, A. "The Control of Canadian Nuclear Reactors: An Account of Its Unique Features." In *Canadian Engineering Centennial Conference: Proceedings of the Nuclear Power and Fusion Programs of the Canadian Nuclear Society, 18-22 May 1987, Montreal*, 86-91. Toronto: Canadian Nuclear Society, 1987.

Pearson, A., E. Siddall, and P.R. Tunnicliffe. "The Control of Canadian Nuclear Reactors." In *Proceedings of the Second United Nations International Conference on the Peaceful Uses of Atomic Energy, Geneva, 1-13 September 1958*, vol. 11, 372-9. Paper A/CONF.15/P/213. Geneva: United Nations, 1958. (Also AECL-623.)

"Reactor ZED-2." In *Directory of Nuclear Reactors*, vol. 5, 223-4. Vienna: International Atomic Energy Agency, 1964. (Also AECL-2132.)

Roshd, M.H.M. *The Physics of CANDU Reactor Design.* AECL-5803. 1977.

Siddall. E. *Highly Reliable Control Systems for a 200-MW Power Reactor.* AECL-685. 1958.

Ward, A.G., and D.S. Craig. *Measurements of the Reactivity Change with Irradiation for Natural Uranium Samples Irradiated in NRX.* AECL-812. 1959.

Westcott, C.H., W.H. Walker, and T.K. Alexander. "Effective Cross-Sections and Cadmium Ratios for the Neutron Spectra of Thermal Reactors." In *Proceedings of the Second United Nations International Conference on the Peaceful Uses of Atomic Energy, Geneva, 1-13 September 1958*, vol. 16, 70-6. Paper A/CONF.15/P/202. Geneva: United Nations, 1958. (Also AECL-612.)

Chapter Twelve

Fuel Channels

C.E. ELLS

DESIGN

The safe operation of the fuel channel is of vital importance to reactor operation, and great effort has been expended in developing the CANDU fuel channel (figure 12.1) and studying its characteristics. In a fuel channel, the pressure tube contains the fuel and hot primary coolant, the spacer (garter spring) separates the pressure tube from the calandria tube, the calandria tube keeps the cool moderator from contacting the hot pressure tube, and the end fitting joins the pressure tube to the out-reactor circuit.

The evolution of fuel-channel design started in the mid-forties with the vertical channel in NRX, but NPD had the first of the CANDU channels. The NPD channel was designed in a cooperative effort between AECL and Canadian General Electric. Thereafter, channel design and procurement of channel components has been directed from the engineering section of AECL, now AECL CANDU. From NPD onwards, the channels maintained the schematic form of figure 12.1, but with differences at the ends of the channels based on fueling-machine operational needs.

There are four key requirements in fuel-channel design:

- the whole channel must be neutron-economic; to achieve this end, the components undergoing the high-neutron flux are made of zirconium-rich alloys and the amount of zirconium alloy is kept as low as possible;
- to meet the intent of accepted codes of practice for designing high pressure components, the channel is designed for adequate tensile strength;

---

[1] A metallurgist's recollection: Metallurgical problems did, of course, arise in the design, construction and operation of ZEEP, NRX and NRU, but space has not permitted a detailed description of that work here. Early metallurgical support for Chalk River was supplied by staff of Mines Branch in Ottawa. One of those metallurgists recalls being called to his supervisor's office in 1944, and being introduced to Sir John Cockcroft. The young man gulped and stammered "The John Cockcroft?" To which Sir John replied, "I am afraid so," then pulled a rod of an apparently black substance from his briefcase. He asked if the metallurgist would measure its thermal conductivity. To the obvious question, Sir John said that he was not allowed to say what the substance was. The young metallurgist, however, was quickly able to determine the density of the rod, and deduce that it was uranium. This was the start of nuclear metallurgy in Canada.

[2] An early metallurgical problem encountered with NRX involved the tendency of uranium metal to change shape during irradiation. Control of this shape change was perhaps the first purely metallurgical problem in which Dr. Lewis played an active role. One employee vividly remembers standing by Dr. Lewis in a fabrication facility in Ohio in June 1952, as the uranium rods passed through rolls.

FIGURE 12.1
Schematic Diagram of a Single CANDU Fuel Channel

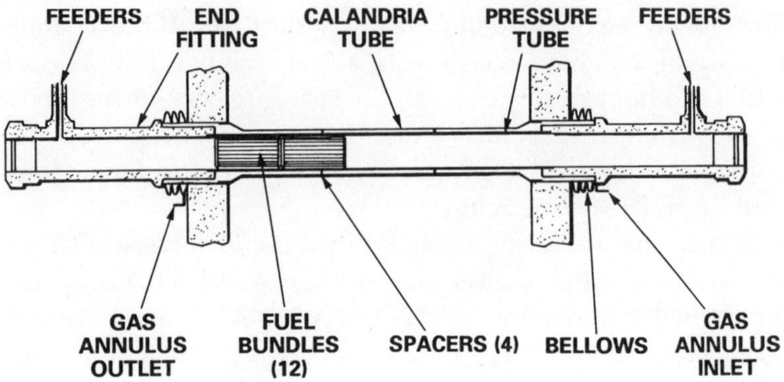

- a thirty-year operational life is targeted, and to this end the design must cover the diametral creep of the channel, its elongation from neutron-induced growth, and its sag from the weight of fuel and coolant;
- the effect of corrosion on metal loss and hydrogen ingress must be considered, as well as any effects of neutron irradiation on channel mechanical properties. Because of some inevitable degradation of channel components during reactor lifetime, the whole of the channel must be inspectable by available non-destructive testing techniques, and replaceable if necessary.

Since completion of the design of the NPD fuel channel, the design has been influenced both by knowledge gained in the laboratories and by operational feedback from the power stations. Some of the important changes explained and discussed later in this chapter are:

- whereas the pressure tubes in NPD, Douglas Point, Pickering A, Units 1 and 2 were made of cold-worked Zircaloy-2, all subsequent reactors used cold-worked Zr-2.5Nb (KANUPP and Gentilly-1 are exceptions);
- the garter springs have gone through several changes;
- details of the attachment of the pressure tube to the end fittings have changed significantly;
- in Bruce A, Unit 4, and later reactors, provision has been made to accommodate considerable elongation of the tubes, a requirement not recognized when the previous reactors were designed.

Again starting with NPD, AECL CANDU was heavily involved with procuring all components for the channels, inspecting those components and dealing with the many problems involved with their manufacture. It has also been involved with meeting the changing standards of the regulatory bodies, notably the Control Board. Latterly, staff of Ontario Hydro have been heavily involved with or taken over this work in Ontario. The story of this huge effort is outside the scope of this book, and in the remainder of this chapter we concentrate on the metallurgical development work that has led to the channel components in their current form.

PRESSURE TUBES

The resistance of pressure tubes to deformation is central to pressure-tube design and behaviour. The tensile strength of a pressure tube is derived from a test in which a specimen of the tube is forced to extend at a constant rate. The maximum load that the specimen can support before fracture occurs determines the "ultimate tensile strength" (UTS) of the material, via the expression:

$$UTS = \frac{\textit{maximum load}}{\textit{initial specimen area}}$$

The percentage elongation of the specimen at time of fracture is known as "total elongation," and is used as a measure of the ductility of the material. The creep strength of a material is derived from a test in which a specimen is extended under a constant load. Here the relevant property is the rate of extension of the test piece, known as creep rate. The test on any one specimen may continue for many days or indeed many years before specimen failure by creep rupture occurs. If the load is much higher, and rupture occurs in a much shorter period of time, the test is termed stress rupture and the derived property is the time to failure. Stress rupture, however, is not of concern to CANDU fuel-channel components.

The CANDU pressure tube is stressed both from the water pressure and weight of water and fuel. The water pressure stresses the tube in both circumferential and longitudinal directions, a principal effect being circumferential creep, resulting in diametral expansion. The weight of water and fuel causes the tube to creep downwards, termed sag. The strengths and deformations described here are all affected by the neutron environment in the reactor core, as discussed later in this section.

Early Work

Canadian work on pressure tubes was started in 1957 when the decision was made to change the pressure boundary of the NPD primary coolant from a single pressure vessel to the many smaller pressure tubes. The CANDU concept, based on natural uranium fuel with a heavy-water moderator, is practicable only because zirconium alloys can have the three essential properties required for the pressure tube:

- low wastage of precious neutrons;
- corrosion resistance against attack from the primary coolant;
- sufficient strength.

Various metals (e.g., ferrous or titanium alloys), would have the latter two properties, but the neutron capture cross-section would be unacceptably high. Zirconium satisfied the neutron-economy criterion, but required alloying additions to achieve adequate strength and corrosion resistance. In 1957, it was possible to envision the zirconium-alloy pressure tubes only because of the pioneering work undertaken for the United States' naval reactors.[3] There, Admiral Rickover had chosen to use for fuel sheathing the alloy known as Zircaloy-2, a zirconium-rich alloy containing some tin and oxygen and small amounts of iron, chromium and nickel. These alloy additions satisfied the requirements for corrosion resistance and strength.

Canada was not alone in developing zirconium-alloy pressure tubes. By 1957, both the Americans and Russians had started work on the concept and plans for the fabrication of tubes for the United States' Hanford N reactor were in progress. The first problem to be faced by the pressure-tube designers was the tensile strength of the tubes, a recurring theme in our pressure-tube design.

It was a long-established principle in pressure-vessel design that the maximum permissible stress should be UTS/4. We would have liked to use Zircaloy-2 in its fully annealed (most ductile) condition, where to a design engineer it is a material combining a number of desirable characteristics. For neutron economy in CANDU it is essential to keep the amount of zirconium alloy in the reactor core, or indeed any neutron absorbing material, as low as possible. Using enriched uranium, the Americans and Russians did not have this concern to the same extent. The obvious method of adding strength to the pressure tube was having the tube cold-worked before installation. The cold work increases the strength of the material, but decreases its ductility; we accepted 20 percent cold work as a compromise between these two factors. We did not know how much ductility was actually required, only that as much as possible was desirable. Because of the simple shape of the pressure tube, and the rigorous inspection we could use on it, we obtained permission from relevant authorities to design to a factor of 1/3 of the UTS rather than the conventional 1/4 UTS. This advantage, and the strength increment obtained from the cold work, enabled us to use pressure tubes sufficiently thin for acceptable neutron economy. With the decisions on material, its metallurgical condition and wall thickness fixed, we proceeded to procure the tubes for NPD, with Chase Brass of Waterbury, Connecticut the primary fabricator.

Even as the properties of the Zircaloy-2 tubes for NPD were being established, Dr. Lewis had already initiated work to

---

[3] Dr. Lewis was enthusiastic about information exchange on nuclear energy between countries, and vigorously encouraged it as a necessary part of supplying the world's energy needs. A tribute to the Canadian contribution appears in a 1989 publication from an American division of Sandvik, a Swedish corporation: "The work at AECL provided much of the technical data on which the commercial use of zirconium alloys was based." In the nineties, AECL has become much more jealously protective of its intellectual information and more interested in its commercial value.

develop a better tube.[4] We will however leave this search in abeyance and consider the factors that we now know affect the service life of the tubes.

Zirconium Hydride

It is not possible to appreciate the problems faced by the designers/operators of CANDU reactors without having some understanding of the properties and effects of zirconium hydride. Here we will use hydrogen to include all three of its isotopes – protium, deuterium and tritium – because, for our purposes, equal numbers of atoms of each have the same effect on the core components. Also, in quoting concentrations of hydrogen, we can consider the zirconium alloys to be pure zirconium; e.g., we can ignore the iron, chromium, nickel, tin in Zircaloy-2. Zirconium and hydrogen can interact to form several different compounds. However, it is sufficient to know that when the hydrogen concentration in the zirconium exceeds a critical amount, a compound termed zirconium hydride is formed. At the operating temperature of CANDU pressure tubes, the critical concentration is about 0.35 atom percent (at. percent); i.e., the percentage of hydrogen atoms in the alloy compound against the total of hydrogen + metal atoms. At room temperature, practically all hydrogen exists in the hydride form. A cross-section through a piece of zirconium containing hydrides is shown in figure 12.2. The black stringers represent edges of platelets of hydride. The hydride is normally a very brittle substance, particularly at room temperature (and below). It requires little imagination to appreciate that these hydrides can be points of weakness in the zirconium component, particularly if one of the platelets meets the surface of the component. Clearly, we should try to keep the components free of hydride. In practice it is not yet possible to produce a pressure tube that contains less than about 0.05 at. percent, and considerable attention is required to maintain this low value. More important, hydrogen (deuterium) enters the pressure tube from its corrosion by the primary coolant and interaction with the deuterium impurity annulus gas. We try, of course, to keep the corrosion low, but cannot eliminate it.

FIGURE 12.2

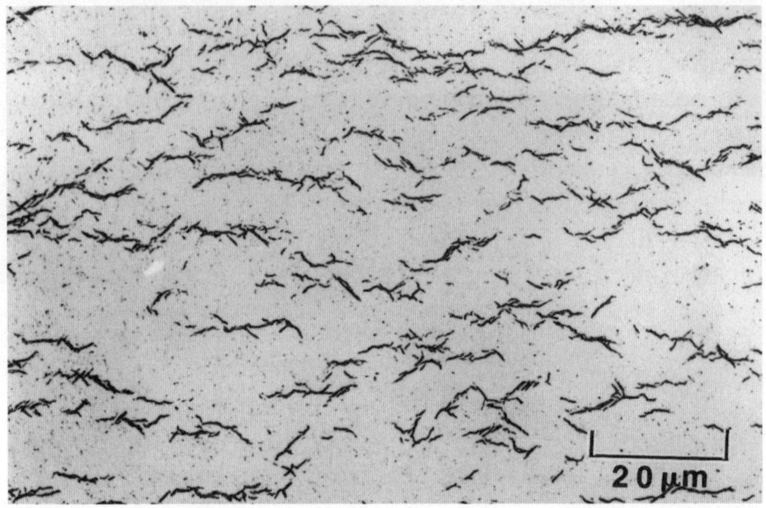

Photomicrograph showing zirconium hydride in Zircaloy-2. The metal surface is polished to be flat and smooth, and them lightly chemically treated. The longer black particles are the edges of hydride platelets. The black spots are from sections rich in Cr, Fe or Ni; in pure zirconium only the hydride would appear after this surface preparation. The hydrogen concentration in the specimen shown is 1.0 at.%. A scale marker is shown on the figure.

4 Developing an understanding of the metallurgical behaviour of the components in the CANDU core is an exceedingly complex task. The Canadian metallurgical community could not have done this efficiently in isolation from the world community. Extensive international contacts were required. Direct personal contacts took place in laboratories and meetings, and through the transfer of written documents. AECL documents had varied forms, e.g., internal memos, reports proprietary to AECL that might contain sensitive commercial information, reports that might be termed "official use only," containing information not completely verified or considered of little use to others, and journal publications submitted to the world for their use and comment. These last documents would receive intensive internal and external peer review before appearing in a journal. In turn, AECL metallurgists used journal publications, and documents of other types, depending on the particular exchange agreement we had with those organizations in which the document originated. The class of security varied from laboratory to laboratory: in 1958 at Savannah River, AECL visitors received a personal escort to the washroom.

By 1957 the Americans, principally at the Bettis Laboratory, had made preliminary studies of the effects of zirconium hydride on the behaviour of zirconium. The work continued there for many years, and also in the United Kingdom and other countries: the extent of the work in the USSR still is difficult to establish. Canada however had the larger stake in cold-worked material, and we have generated much of the world's information on hydride behaviour in this material. In the early-to-mid-sixties, we started studying the rate of migration of hydrogen in zirconium, and measured quantitatively the rate at which it migrated from a warm to a cold region of the component. We also examined the effects of hydrides on the tensile ductility of components, and on the effects when the components were shock loaded. From this work, we were able to show that, provided the hydride platelets were in planes parallel to the direction of stress, they caused little problem; if they were in perpendicular planes, the problems were serious indeed (figure 12.3). Even in this latter situation, the deleterious effect decreased with increasing temperature. We were fortunate that the initial fabrication route chosen for the pressure tubes resulted in hydride platelets forming in the least deleterious direction. From extensive laboratory studies, we were able to ensure that any small changes in fabrication procedure did not change the hydride orientation.

FIGURE 12.3
Schematic diagram of hydride platelets in a stressed zirconium alloy component

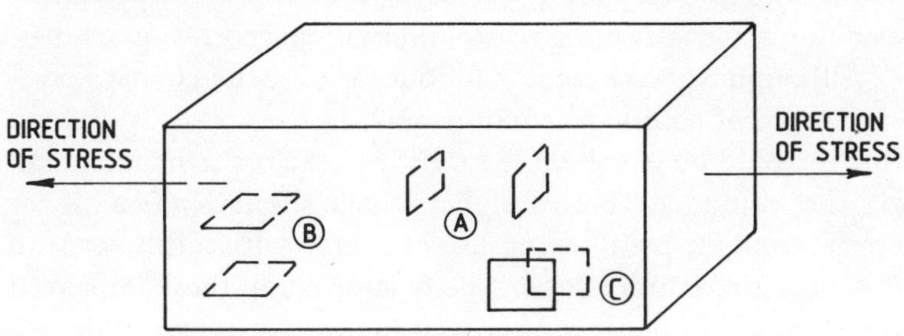

For the direction of stress shown the platelets at A could be of concern, those at B and C are comparatively harmless.

Creep and Growth

All stressed metals creep to some extent. In everyday use we can design components so that the creep is negligible for the desired lifetime (e.g., the frame of an automobile). In the reactor core the components must be kept as thin as possible, which results in stresses that are in the range that creep can be expected. Immediately following the decision to use cold-worked Zircaloy-2 pressure tubes, AECL initiated work to measure the creep properties of the material in tests in the laboratory (i.e., on material not being irradiated). We hoped to find that the diametral expansion of the tube might be no more than 3 percent (as required from expected coolant behaviour) in thirty years; careful measurement is required if tests of a year or few years are to give an accurate prediction over this period. Our early work, comprising tests in the laboratory, indicated that we might reach this goal. We knew from basic science that neutron irradiation could increase the creep rate, even though the UTS would be increased. Measurements of the creep rate in-reactor are extraordinarily difficult, but by the mid-sixties, we had found that a neutron flux as large as that in CANDU might increase the creep rate by a factor as high as ten in some situations. It was the in-service creep rate that was needed for design. The creep has two effects on tube behaviour; the tube diameter will increase and, under the weight of fuel and coolant, sag.

In 1957, no scientist or engineer associated with the Canadian nuclear industry realized that neutron irradiation in the absence of any external stress could cause a significant shape change in zirconium. However, work on this topic was underway or about to start at Harwell, United Kingdom, and in the early sixties we began to realize the importance of this factor. This shape change results in the elongation of the cold-worked pressure tubes, and for this reason it was decided to call it growth, a named used previously in the United Kingdom to describe a similar phenomenon in uranium. Design of the early CANDU reactors up to and including Pickering B proceeded without the knowledge of growth, and the subsequent tube elongation has involved costly modification to some reactors and has been the life-limiting factor requiring tube replacement in others.

Fracture

We noted previously that after 1957 it became possible to design pressure tubes so that the maximum stress was never more than UTS/3. If we had only to worry about failure in tension, this UTS/3 would indeed provide an excellent factor of safety. Since 1957 a new topic in the science of engineering, termed fracture mechanics, has evolved. Whereas the tensile UTS is based on tests of a crack-free near-homogeneous material subjected to an increasing stress (and normally loaded to failure in several minutes rather than days or years), fracture mechanics considers the size of an inhomogeneity or flaw that will grow without any increase in applied stress. Compared to the simple tensile failure, the crack may grow slowly over a longer period of time. In principle, a crack in a tube wall not yet through the wall can under stress suddenly extend through and along the wall, causing rapid failure. When the crack has reached a length at which this rapid growth occurs, the crack is said to have reached its critical crack length (CCL). Alternatively, the crack can grow through-wall, with subsequent leaking of the fluid contained in the tube, before the CCL is reached. This gives a comparatively easy means of detecting the crack in advance of CCL. Fear that tubes were approaching the set of conditions for which a CCL could be attained has thus far been the major life-limiting factor for pressure tubes.

Pressure-Tube Materials

The few years after 1957 were a period of intense activity on cold-worked Zircaloy-2. It was not certain that satisfactory tubes of cold-worked Zircaloy-2 could be fabricated. We were fortunate that organizations in the United States were also procuring tubes of Zircaloy-2, but we had to launch several development programs to determine whether the procedures would work for our particular requirements. It was necessary to show that experimental work being done in the laboratories, using rod and sheet, was relevant in predicting properties and behaviour of the actual pressure tubes. Much of the initial procurement of the tubes was undertaken in collaboration with Canadian General Electric, Peterborough, the main contractor for NPD.

The second project was to determine the effects of neutron irradiation on the properties of Zircaloy-2. From early work, we found that the strength from cold work did not anneal out during irradiation at reactor operating temperature; rather, the material strengthened further, but without a serious loss of ductility. We initiated the in-reactor creep tests; results from these experiments soon indicated the effect of irradiation to decrease the resistance to creep. A scientist at the CEGB in Berkeley, England, working with small specimens, warned us of the possibility of growth, but we had already started to examine this topic. We did not yet appreciate the concept of CCL, and were measuring the strength of sound (crack-free) tubes at failure with internal pressurizations. These tests showed that the tubes were even stronger than predicted from tests with the stress in only one direction (uniaxial) and coincident work at the Massachusetts Institute of Technology enabled us to understand the reason. It was in the mid-sixties, well after the commitment of tubes for Pickering 1 and 2, that we started to simulate the behaviour of cracked tubes in-reactor by testing tubes with artificial flaws.

A third project was to determine the corrosion behaviour of the cold-worked Zircaloy-2 in the environments of particular interest to us. Here we were able to draw on a lot of American and European experience, the Zircaloy-2 composition having originally been chosen for its corrosion resistance to steam at 400°C. By 1957, Zircaloy-2 was being used extensively as a fuel cladding material, giving rise to international corrosion studies.

Although we were content for our early reactors to have pressure tubes of cold-worked Zircaloy-2, in the sixties it was felt that it would be worthwhile for future reactors to have a material that could have both a higher tensile strength and a higher creep strength. By allowing use of a tube with a thinner wall (i.e., less zirconium alloy in the reactor core), these improved properties would allow improved neutron economy. Now we realize that these improvements in economy were small compared to those we could obtain from increased reliability of continuous reactor operation.

At an international conference on the peaceful uses of atomic energy in 1958, the USSR presented a great deal of information on their research into the properties and nuclear uses of zirconium alloys containing niobium and oxygen. The important features of their work were: a) the capture cross-section of niobium to thermal neutrons was sufficiently low that a zirconium alloy containing a few wt. percent niobium would be acceptable for use in CANDU; b) zirconium-niobium alloys appeared to have promise of acceptably low corrosion rates in water at the CANDU operating temperature; and c) additions of the niobium opened the possibility of strengthening via a quench-and-age heat treatment, a route not possible with the Zircaloys.

The addition of niobium to zirconium has a profound effect on the hardness and tensile strength available. Zircaloy-2 quenched from a high temperature, say 800°C, is somewhat harder than material slowly cooled from the same temperature. Zirconium-niobium alloys, on the other hand, are extraordinarily hardened by this quench; thus they are known as heat-treatable materials. From consideration of the USSR data, and an assessment of the effect on neutron economy in a CANDU reactor, Dr. Lewis and W.R. Thomas decided that Zr-2.5Nb had a good potential, and this alloy (containing 2.5 wt. percent niobium) has been used in Pickering 3 and all subsequent CANDU reactors.[5] Originally we had intended to have pressure tubes use Zr-2.5Nb in a heat-treated condition. There was a lot of experience on heat-treatable titanium alloys on which we could draw (titanium and zirconium have many characteristics in common), and early in the work we found that a water quench from about 880°C followed by holding from six to twenty-four hours at 500°C resulted in good strength. Tubes receiving this treatment are known as heat-treated tubes, and were installed in the Gentilly-1 and KANUPP reactors.

During development of the heat-treated Zr-2.5Nb tubes, the need was recognized for a test that would show that the desired properties were indeed obtained. The test would have to be done without damage to the tube, for which the term non-destructive testing (NDT) is used. Strength could be below that specified if tubes did not reach the correct solution temperature, or if the cooling rate during the quench was too slow. The correct cooling rate is the more difficult to achieve. Although there is no reason to assume that any of the tubes finally used in-reactor do have weak regions, the non-destructive testing problem was not really solved for heat-treated tubes. This led to the proposition that if the tubes were to be strengthened by cold-work rather than the quench-and-age heat treatment, confidence in the uniformity of strength in the finished tube would be significantly greater. With this realization, an exploratory program on cold-worked Zr-2.5Nb pressure tubes was initiated.

With the experience both in production of cold-drawn Zircaloy-2 pressure tubes and in extrusions of the Zr-2.5Nb tubes (for finishing by heat treatment), the development of cold-worked Zr-2.5Nb pressure tubes progressed rapidly to a successful conclusion. By cold-working the extrusions about 25 percent, and making use of controlled extrusion conditions, the tensile strengths of the cold-worked tubes were some 40 percent higher than the tubes of cold-worked Zircaloy-2, and still had acceptable ductility. With the essential properties of improved tensile strength and acceptable neutron-capture cross-section met, an extensive program was launched to determine the behaviour of the tubes in-reactor.

Although heat-treated Zr-2.5Nb tubes were chosen for FUGEN in Japan as well as for KANUPP and Gentilly-1, where the high tensile strength was nearly essential for neutron economy, there was considerable debate in Canada on which of the three candidate materials should be chosen for Pickering 3:

---

[5] Employee's recollection: Dr. Lewis maintained a great interest in all details of the metallurgy of the core components, right from the start of the development of CANDU. He pursued this interest with less concern for line management than might have been shown by heads of other sites. Staff members who worked at Chalk River have vivid recollections of answering their phone and hearing a secretary's voice say, "Dr. Lewis would like to speak to you." One sprang to mental attention, knowing that some incisive query would be forthcoming. It was incredibly difficult to be in advance of Dr. Lewis on any technical topic, once he had decided to master it. In later years W.R. Thomas would note that new employees did not have the inspiration of this personal contact.

- cold-worked Zircaloy-2;
- heat-treated Zr-2.5Nb; or
- cold-worked Zr-2.5Nb.

The choice had to be made with less than a desirable amount of evidence in hand, but in 1967 the cold-worked Zr-2.5Nb was chosen, and it has remained the choice for all subsequent CANDU reactors. The choice was based on tensile strength and creep resistance. Twenty-five years later we still feel that the choice was correct, even if made for the wrong reasons. We now know that the difference in in-reactor creep between cold-worked Zircaloy-2 and cold-worked Zr-2.5Nb is quite small, whereas the great advantage of Zr-2.5Nb is now recognized to be its very low deuterium pickup.

By the late sixties, with cold-worked Zr-2.5Nb fully established for reactor use, we turned our attention to finding an even better pressure tube. Again we were able to draw on the experience of the titanium industry, but for the pressure tubes we have the restrictive limitation that the neutron-capture cross-section of the alloy additions must be acceptably low. In the early seventies, we settled on a composition zirconium – 3.5 wt. percent tin, 0.8 wt. percent niobium, 0.8 wt. percent molybdenum – an alloy known in Canadian technical literature as "Excel". All of the commercial zirconium alloys, including Excel, contain about 0.1 wt. percent oxygen; this oxygen provides a significant increase in strength over otherwise oxygen-free alloy. Excel in an annealed condition has tensile strength as high as that of cold-worked Zr-2.5Nb, and equivalent or better in-reactor creep resistance. There are, however, doubts regarding its resistance to delayed hydride cracking, and the alloy has never been used commercially. Indeed, since the early seventies, there has been little activity on new zirconium alloy development directed towards pressure tubes.

In-Reactor Deformation of the Channel

We started measurement of the in-reactor creep of the alloys with tests on uniaxial specimens of Zircaloy-2. The word uniaxial is important here, because in-reactor the tubes are stressed biaxially. At AECL the design of a machine to test specimens uniaxially in reactor was initiated in 1960; the first reliable result we obtained was in 1963; it showed a marked increase in creep from neutron irradiation. These experiments, in order to be relevant to the stresses and temperatures of the pressure tubes in CANDU, are extraordinarily difficult. The specimen length changes are very small (e.g., 0.1 percent in one year), and the changes are very sensitive to test temperature, requiring temperature control to a few degrees Celsius (e.g., 3°C). To obtain this control on specimens in the middle of a reactor, with its attendant environment of changing temperatures and neutron flux, has required great skill and innovative ability. In the next twenty years the uniaxial tests were continued, and despite many failed tests provided us with a useful pool of information. These specimens were particularly useful for examining variation in temperature, stress, material and flux. The uniaxial tests have been supplemented with many experiments in which bent beams are irradiated, and the reduction in spring-back gives a measure of creep. These measurements are, in practice, much easier to perform than the direct measurement of creep rate. Small specimen tests on growth were not started at AECL till 1968, and these have continued to the present. Growth tests are inherently easier to perform than creep tests, and the information obtained from them has been more reliable and hence more useful than that obtained from the uniaxial creep tests.

In the final analysis, we were to require measurement on the tubes from the power reactors at Pickering and Bruce to be certain of the their behaviour. However, seventeen years prior to that time, in 1960, we had to attempt to study tubes by any available means. The work on tubes was initiated in 1961, starting with measurements of changes in diameter on an experimental tube installed in NRU. Just as for the uniaxial tests, gauging a tube to the accuracy required was an exceptionally difficult measurement, and Orenda Engines were involved in design of the experimental equipment from an early stage. The first diametral measurements were available during the summer of 1964, and by 1966 it was clear that the in-reactor creep of the tubes was significantly higher than out-reactor, the result predicted from the uniaxial tests. The diametral expansion of the pressure tubes has not been the life-limiting factor in any CANDU tube thus far, but may become more important when other problems are solved. The program was rapidly expanded, with both tubes of the diameter and wall thickness used in CANDU and tubes of much smaller size. The first gaugings of tubes in power reactors were done by AECL, starting of course with NPD tubes, in 1964-65. Later measurements were made on the other CANDU reactors, a program gradually taken over by Ontario Hydro for their reactors. The measurement of tube elongation was started later than the diametral gauging, and for the Ontario Hydro reactors has been done by the utility.

The objectives of the work on creep and growth are:

- to predict the tube diametral expansion and tube elongation as a function of time under service conditions, to allow operators to schedule tube changes, and to allow designers of future reactors to optimize the characteristics of new tubes, and
- to ensure that failure by creep rupture would not occur.

It was comparatively easy to achieve the second of these objectives, and show that creep rupture is not a credible occurrence. For the first objective we have produced a series of design equations starting in 1967 with a very simple form for diametral expansion:

$$\text{Creep rate} = \text{constant} \times \text{stress} \times (\text{T-160})$$

where T is the temperature in °C.
(valid in the temperature range 250 – 290°C)

As additional experimental information was accumulated, and guided by a large effort devoted to understanding the basic mechanisms of in-reactor deformation, the initial design equation went through successive modification. As the equations became more physically realistic, they became more complex in form.

Coincident with work on diametral expansion, we were concerned with the possibility of sag. The work was started in 1965; in the absence of any measurements this study was initially theoretical. The measurement of sag in-reactor required further equipment development, and by 1971 good sag measurements were in progress in NPD. The combined experimental and theoretical effort showed that about 90 percent of the channel sag resistance is provided by the relatively cool calandria tube. The type of sag data obtained is shown in figure 12.4. The agreement between the

FIGURE 12.4
Actual sag of a pressure tube in Pickering 1

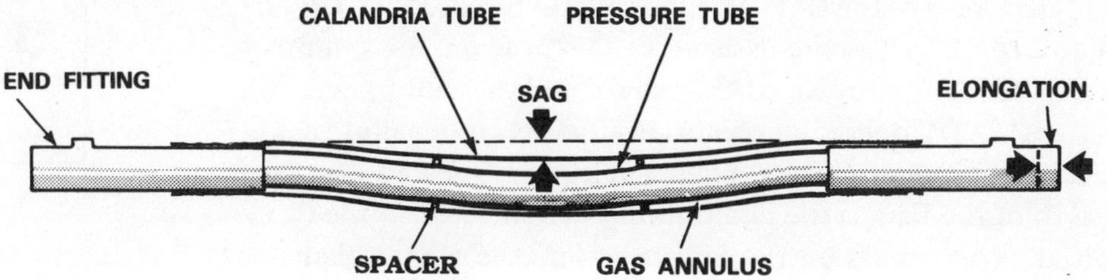

Sag and elongation of fuel channel

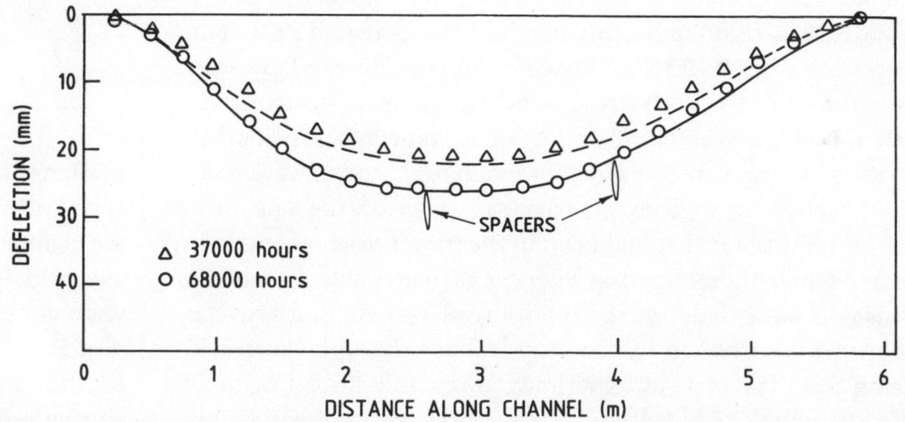

At the top, a not-to-scale drawing illustrates the sag and elongation of a fuel channel after some period of operation in the reactor. The bottom drawing shows the actual sag of a pressure tube in Pickering-1 for two periods of operation. An excellent agreement between the measured sag and the sag calculated from creep data, shown as solid lines, is apparent. Note that the spacers are not symmetrical along the length, having moved out of design position as described in the test.

theory and experimental results is excellent. The measured sag shown there is more than is desirable, and later reactors have used four rather than two garter springs to decrease the possibility of the pressure tube contacting the calandria tube.

Fracture of the Tubes

In predicting the behaviour of a large structure it is desirable to have the test specimen conform as much as possible to the geometry and size of the structure. Our development of pressure tubes started with the use of small tensile test pieces, but we realized that tests on lengths of tubes having the diameter and wall thickness of the CANDU tubes would be desirable. Orenda Engines Ltd. in Toronto assisted in this program for a number of years. We soon realized that sound tubes would never fail under CANDU operating conditions. Rather, tube failure could occur only if the tube contained a sufficiently severe flaw. The length of the flaw is the determining parameter (i.e., the CCL). Initial experiments on the failure of lengths of full diameter pressure tubes when internally pressurized were done with cold-worked Zircaloy-2 in 1959. From then till about 1965, many tests were done on this material, heat-treated Zr-2.5Nb and cold-worked Zr-2.5Nb. The work then was directed towards evaluating the CCL, with stress, temperature, hydrogen concentration, hydride orientation and irradiation being the principal variables. To measure the effect of irradiation, tubes were irradiated in NRU but we were not to be satisfied with the data until we could test tubes that had been in the power reactors and had reached much higher neutron fluences than available from NRU. It was necessary to define the type of realistic flaw, and a severe experimental problem was to stop leakage through the crack during test. The type of experimental result obtained is illustrated in figure 12.5. With all of the variables involved, many years were required to obtain a statistically satisfactory lower bound line. In an operating tube, a through wall crack, still shorter than CCL, will leak water into the annulus between pressure tube and calandria tube. With detection of the water before CCL is reached, leak-before-break (LBB) is achieved. We spent many years attempting to demonstrate that the leak-detection

FIGURE 12.5
Relationship between stress and crack length

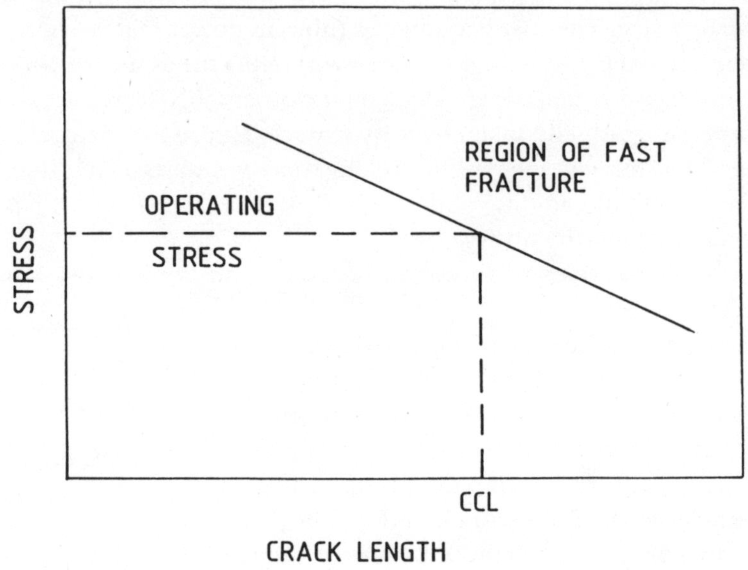

The solid line represents an idealized set of experimental data showing the relationship between stress and crack length which results in fast fracture. The dashed lines show how the critical crack length can be determined from the data when the operating stress is known.

systems in the reactor were sufficiently sensitive relative to speed of crack growth to assure that LBB would apply. In 1974, we claimed that we had succeeded and that we could use LBB as a principle of reactor safety. As seen later in this chapter, we were not correct in this claim. Work on this topic continues today. Having said that tests on specimens of full-diameter tubes are desirable, it remains that these tests are difficult and expensive, especially for irradiated material. Thus in 1962 we started work using small specimens designed to predict the fracture toughness and hence CCL of the tubes. It has been difficult to achieve the desired accuracy of correlation of results between the small specimens and full diameter tubes, but the prize is so important that we have continued trying and the work will certainly go on for several more years.

We have emphasized that any fracture of the tubes will, in the final analysis, occur because of some defect present in the tube. Mechanical defects in the tubes could occur during service, with a scratch from a wear pad on a fuel bundle being the most obvious source. In practice, these scratches have never caused problems; rather defects remaining after tube manufacture or installation have been the problem source. These defects are sought using various NDT techniques:

- visual examination,
- brushing with a dye penetrant, which seeps into cracks and remains apparent where the surface dye is removed; this is obviously useful only for surface defects,
- radiography, equivalent to a dental radiograph for cavities in teeth,
- ultrasonic inspection (UT), in which sound waves are sent through the metal and reflected differently from defects than from a sound material,
- eddy current (ET), in which the magnetic field resulting from an electrical current induced in the component is modified by a defect of some type.

The NDT procedures used for the pressure tubes have been dramatically improved since the procurement of tubes for NPD. These tubes were examined visually, then by ultrasonics of the quality available to the fabricators at that time. The NDT technique remained essentially unchanged for the whole of the sixties. NDT practice in Canada was to be improved remarkably by a policy of in-service inspection of American reactors, introduced in 1970. This policy caused the Control Board to request much more stringent inspection of CANDU components, but designers and operators would have, in any event, effected many improvements without the spur from the Control Board. To improve NDT practice, a quality control branch was established at CRL in 1971, and it immediately recognized that eddy-current technology must be added to the inspection procedure. At that time, there was very little knowledge of this technique in Canada, and the work was to result in a dramatic improvement of procedures both in Canada and the United States. Eddy-current inspection was not used on the pressure tubes of any of the first four Pickering reactors, but by the mid-eighties, became a standard addition for inspection by the tube fabricator.

Corrosion and Hydriding

Corrosion of the alloys raises two problems:

- wastage of material from formation of zirconium oxide on the exposed surfaces, and
- ingress of deuterium, from the water molecules supplying the oxygen.

A typical oxidation curve for the Zircaloys in water or gas is shown in figure 12.6. The post-transition corrosion is important for our purposes. Early in the sixties, we were accumulating relevant information on the way this post-transition corrosion was affected by temperature, metallurgical condition, water chemistry and irradiation. Generally, we found that for Zircaloy-2, oxidation *per se* would not be a problem, even though it could be markedly higher in a neutron flux than in the laboratory.

FIGURE 12.6
Oxidation of zircaloy

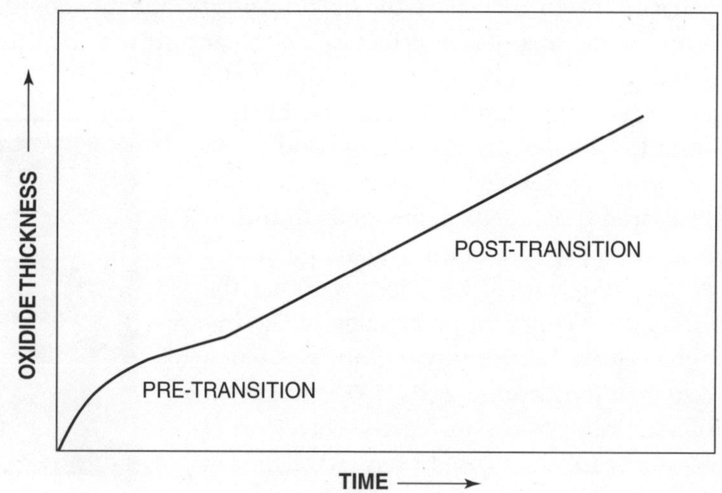

Typical form of the oxidation of Zircaloys in high temperature water as a function of time. At reactor coolant temperature, near 300°C, the transition typically occurs after 1000-2000 days.

The pickup of deuterium during the corrosion was recognized in 1957-58, and remains a major concern. The actual amount of deuterium entering the tube is less than 10 percent of that generated by metal oxidation, but is independent of the rate of oxidation for temperatures at which the CANDU reactors operate. It is this factor that gives the main impetus to keeping the corrosion low. By the mid-sixties, there was considerable conjecture that a twenty-year life and some hopes that a thirty-year life might be achieved. Doubts were all based on the still-open question of the magnitude of the increase in corrosion due to irradiation.

Very early in the sixties, we recognized that the corrosion behaviour of the Zr-2.5Nb alloy could be very different from that of Zircaloy-2. The oxidation rates were markedly dependent on metallurgical condition, and conditions were found for which the oxidation, irrespective of hydrogen pickup, was unacceptably high. Also, the rates in the laboratory were higher than for Zircaloy-2. Nevertheless, it was found that if the pressure tubes received some cold work (i.e., between quench and age or after extrusion), and if the aging was sufficient, the corrosion would be in the acceptable range. The nuclear industry had, by 1967, lived about ten years with the knowledge that irradiation often increased the corrosion rates for Zircaloy-2, although the size of the effect is not clear and is certainly dependent on the chemical condition of the water. It was startling therefore to find that, at least for some water-chemistry conditions, irradiation decreased the corrosion rate of cold-worked Zr-2.5Nb below that found in the laboratory. A second advantage found later for cold-worked Zr-2.5Nb was that the hydrogen pickup, as a percentage of the hydrogen released during oxidation, was much lower than for Zircaloy-2. By 1970, sufficient understanding of the in-reactor corrosion of cold-worked Zr-2.5Nb had been established to enable specification of its use for pressure tubes with increasing confidence. Because of their sensitivity to metallurgical conditions, we recognized that welds might be a particular source of concern and have avoided use of welded Zr-2.5Nb in any CANDU fuel channel. The later work on the corrosion of cold-worked Zr-2.5Nb, sorting out the synergistic effects of metallurgical condition, irradiation and water chemistry has occupied the full time of several people. The suppression of corrosion in-reactor has been shown to arise from the microstructural consequences of irradiation.

Delayed Hydride Cracking

The tenth of August 1974, was to be a traumatic day in CANDU technology; primary coolant appeared in the annulus gas in Pickering 3. It was a leaking pressure tube, the first of a number to occur in subsequent years. This event was completely unexpected, although at least LBB occurred. It was necessary first to find the leaking channel, then the leak itself, and then the cause of the leak. Once the tube was removed from the reactor the leak was found near the rolled joint. For a few days, the joint itself was thought to be the source. However, removal of the tube from the joint revealed a crack in the tube, and examination of the fractured surface of the tube revealed a pattern of varying oxide thickness in clearly defined steps (figure 12.7). Sixteen other tubes in Pickering 3 developed leaks, and eventually fifty-two

FIGURE 12.7

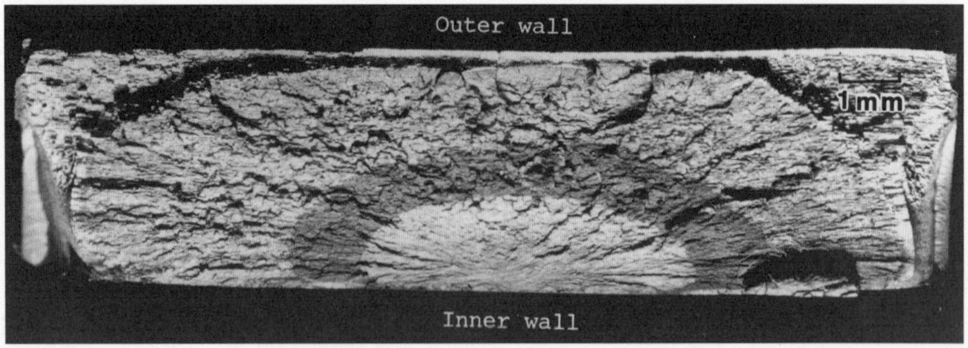

A through wall crack in a Zr-2.5Nb pressure tube from Pickering-3. The white section at the inner wall is the initial crack, and the other bands show how the crack grew by delayed hydride cracking, as in Figure 12.8. A scale marker is shown on the figure.

tubes were replaced in Pickering 4 as well. Both reactors had tubes of cold-worked Zr-2.5Nb, and it was significant that no failures had occurred in the longer running, but lower-tensile-strength Zircaloy-2 tubes in Pickering 1 and 2. On economic grounds alone, irrespective of safety of the reactor, a large development program was warranted to ensure that this form of failure could be eliminated.

We were soon able to demonstrate that the crack through the tube had occurred by a mechanism we now term delayed hydride cracking (DHC). The key to this mechanism is that individual hydrogen atoms will migrate to the region of enhanced tensile stress at the tip of a stressed crack (figure 12.8). As hydride forms there, it cracks, and a new hydride forms at the tip of the now-extended crack. Thus the defect can work its way through the tube wall, from applied stresses well below the UTS. Actually, in 1970, we had started work on the mechanism of delayed hydrogen cracking, but its occurrence in Pickering 3 still startled us. The key factor in these two reactors was a region of unexpectedly high tensile stress, due to the misapplication of an installation procedure at the rolled joints. All of the DHC cracks in these two reactors had been initiated before reactor startup, with the initial hydrogen concentration in the tube, although by the time of failure the hydrogen concentration near the rolled joints was higher than in the remainder of the tube. There were sufficient very small defects on the tube surfaces to serve as initiation points. The crack growth had occurred step-wise primarily at ambient temperatures during reactor shutdowns. It was easy to specify how the stress could be reduced. Study continues on how to reduce the deuterium concentration, because increased concentrations could make cracking possible at reactor operating temperatures. Nevertheless, all new CANDU reactors probably will have a sufficiently low stress from the rolling procedure to be immune to the failure at the rolled joint.

These cracks at the rolled joints resulted in renewed activity in various areas of pressure tube technology, apart from the work on DHC. The work on non-destructive testing required acceleration, both to inspect the radioactive tubes removed from the reactor and to inspect the tubes *in-situ* in reactors

FIGURE 12.8
Delayed hydride cracking

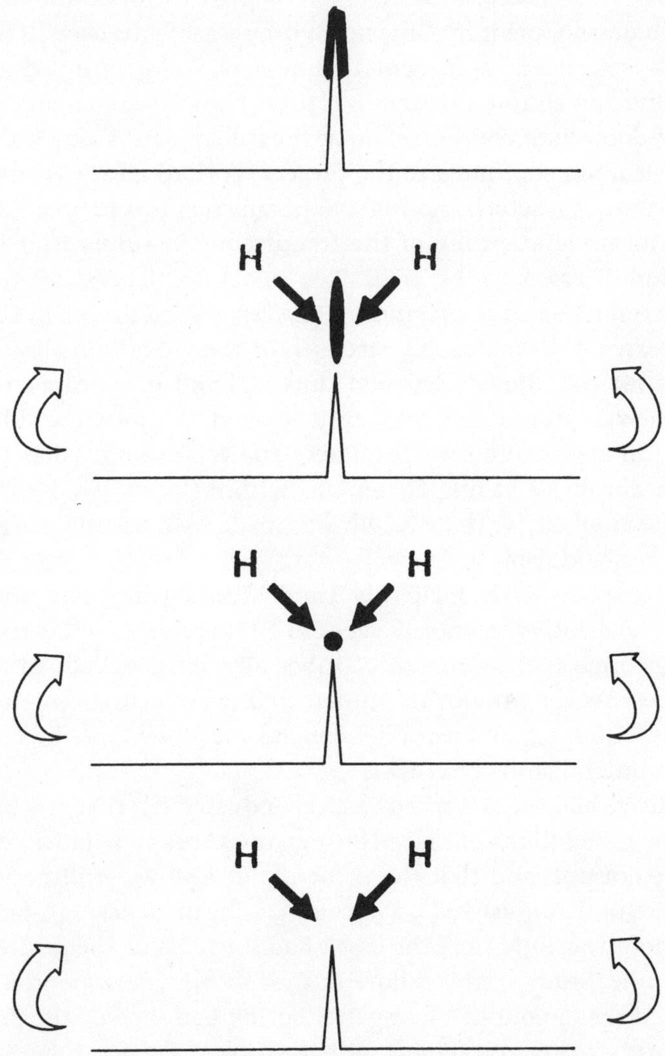

A schematic illustration of the process of delayed hydride cracking. Starting at the top figure, a sample of zirconium is stressed so that the material at the base of the crack is in tension. Hydrogen goes to the crack tip, forms a hydride, which fractures to extend the crack depth.

already constructed. Initially, the lead in this work came from the quality control branch at Chalk River, but at the Ontario Hydro sites, was gradually taken over by the operator, with much development by Ontario Hydro research division. Rapid advances in eddy-current technology were required and achieved. A second program was to determine the source of the high deuterium concentration at the rolled joint; study on the problem has continued to the present. A third area was an advance in characterization of the parameters involved in LBB, and an understanding of the fracture of the tubes after initiation of cracking by DHC. The selection of pressure-tube material was also re-examined. The crack growth rate in DHC increases with increasing strength of the zirconium alloy (at least for the alloys examined thus far) and it appeared that weaker pressure tubes (e.g., cold-worked Zircaloy-2) could be used in any future reactors. This issue was strongly debated, with Zircaloy-2 having champions within Ontario Hydro. The decision to stay with Zr-2.5Nb was made by a narrow margin, but has held firm.

The cracks in the Pickering 3 and 4 tubes, which had stimulated so greatly the improvement in NDT techniques, led naturally to more examination of tubes already installed. In this process, other previously undetected manufacturing defects were found; e.g., ferromagnetic inclusions observable by eddy current, and lap-type cracks.

Just when the Canadian nuclear industry felt that cracking at the rolled joints of Zr-2.5Nb pressure tubes was under complete control, and that the principle of LBB was fully established, on 1 August 1983, a Zircaloy-2 tube in Pickering 2 failed without warning. Here the crack had started near the centre of the tube. Study of this failure in the following few months revealed that an out-of-place garter spring had allowed the pressure tube to sag into contact with the cooler calandria tube. This contact, by cooling the pressure tube locally, resulted in a concentration of hydride at points of contact and the crack unzipped along these hydride blisters. The deuterium pickup in the Zircaloy-2 tubes was found to be far greater than in the Zr-2.5Nb tubes, and currently has eliminated Zircaloy-2 as a candidate material for pressure tubes. This event was the unexpected final justification for the choice of Zr-2.5Nb for Pickering 3, and 4.

The high deuterium pickup in these Zircaloy-2 tubes, associated with an unanticipated accelerating corrosion rate during the later years of tube life, raised the spectre that a similar acceleration might occur at some point in the life of the Zr-2.5Nb tubes. Also, for the first time, there was clear evidence of deuterium entry from the gas annulus between pressure and calandria tubes. Thus, work on understanding the corrosion mechanisms was increased. This high deuterium concentration in Zircaloy-2 tubes resulted in a decision to decommission NPD and Douglas Point (rather than retube), and to retube Pickering 1 and 2 with Zr-2.5Nb. (Pickering 3 and 4 have also been retubed, for a combination of reasons, including tube elongation and concern on blister growth.) To follow deuterium pickup during tube life, a method of taking thin scrapes of metal from the tube inner surface was developed; all other attempts to measure hydrogen concentration by true NDT methods to the accuracy required were unsuccessful. At present, the deuterium pickup in the Zr-2.5Nb tubes remains low, and study to determine the reasons for this behaviour continues. The Pickering 2 incident resulted in four other far-reaching actions:

- Development of non-destructive testing techniques to detect hydride blisters *in-situ* in the reactor;
- Development of non-destructive *in-situ* measurement of garter-spring locations;
- A huge development program to develop a method of repositioning garter springs in reactors already constructed; and
- A reduction in our perception of the amount of hydrogen in a pressure tube that was acceptable for continued service.

The Canadian nuclear industry is still learning more of the effects of hydrogen on pressure tube behaviour.

Pressure Tube Fabrication

The important task of procurement of the pressure tubes has been controlled jointly by AECL CANDU and the utilities, after the fabrication procedures had been qualified by Chalk River staff. Practically all tubes have used zirconium alloy made by

Wah Chang, in Albany, Oregon. Tubes for the early reactors were fabricated by an American Company, Chase Brass and Copper. Much of the fabrication (i.e., drawing, inspection, grinding, honing and autoclaving) has been moved to Canada under the direction of Nu-Tech Precision Metals of Arnprior, Ontario. The extrusion, however, is still done in the United States in the facilities originally owned by Chase Brass and Copper. The Canadian nuclear industry has always attempted to procure components in Canada, fulfilling a spin-off role in upgrading technology in Canadian industry. Thus, we attempted to have the alloys made by Eldorado in Port Hope, Ontario, but this company was unable to compete successfully with the American supplier. We persisted in attempting to have the latter stages of fabrication done in Canada. Initially, in 1967, the development was attempted at Wolverine (Canada) in London, Ontario. When this attempt failed, we attempted to continue the exercise at Noranda, in Montreal, Quebec. Extrusion billets were obtained from Wah Chang, extruded at Noranda who then tried finishing by both cold drawing and pilgering. When it became evident that the market for tubes would be smaller than initially hoped, Noranda withdrew before high-quality tubes were obtained. No further attempt to advance beyond the import of the extruded tubes is foreseen.

Pressure Tube Life

At the start of the fourth decade of the CANDU reactor many, perhaps all, of the problems in use of the Zr-2.5Nb pressure tubes have been identified and satisfactory solutions implemented. The initial concern on effect of irradiation on tensile strength was a non-issue, and the magnitude of the creep in-reactor is now well established and can be allowed for in design. For the first fifteen years, tube operation corrosion and deuterium pickup have been satisfactorily low. It is not possible to be certain that this situation will continue beyond twenty-five years. The low deuterium pickup has minimized the danger of hydride blister formation even in channels where the pressure tube contacts the calandria tube. Of the unanticipated problems, tube elongation can be covered by design and DHC will not be a problem unless there is a dramatic increase in deuterium pickup. Major shortcomings in the channel behaviour have not been metallurgical in nature, but from fabrication of the rolled joints and design of the spacers. Lifetimes of twenty-five to thirty years in new reactors now seem to be a realistic possibility. Development to improve the lifetime of the Zr-2.5Nb pressure tubes will continue, but very little work on radically different pressure tube materials will be done in Canada in the near future.[6]

SPACERS (GARTER SPRINGS)

With the decision to use a horizontal fuel channel it became essential to use a spacer between the pressure tube and calandria tube. Originally, the reason was to avoid unnecessary heat loss to the moderator when the sagging pressure tube came in contact with the calandria tube. Any heat loss to the moderator acts to lower reactor efficiency. Much later, in the eighties, we came to realize that avoiding hydride blister formation on the pressure tube was perhaps an even more important function of the spacer. Because of the temperature differences between the pressure tube and calandria tubes on reactor startup and shutdown, a primary requirement of the design was that the spacer had to be free to move on at least one of the surfaces. In addition, it was highly desirable that the spacer could move without scraping the protective oxide from the surface of either tube. Many concepts have been considered for spacers, including pads, rings, indented calandria tubes, continuous strips, and circumferential bellows. Eventually, an engineer at Chalk River suggested the use of a coiled spring (figure 12.9), free to

---

[6] As a material for use in nuclear-reactor cores, the metal beryllium has some exceedingly attractive nuclear and physical properties. In 1957 we started to investigate its possible use in the CANDU core, later particularly for WR-1. Laboratories in the United States, United Kingdom, France and Australia had also mounted similar and generally much larger programs. Although the metal was strong and corrosion-resistant, it has a fatal flaw of low ductility, and by 1965 the major attempts to use it in power reactors had ceased throughout the countries listed above.

move on both tubes. The spring needed to be strong enough not to be crushed by the weight of the pressure tube, to have sufficient corrosion resistance in the gaseous atmosphere of the annulus between the two tubes, and not to lose too much ductility from the neutron irradiation. A nickel alloy, Inconel-X 750, appeared to meet the requirements, and was adopted for

FIGURE 12.9

A garter spring in position on a section of pressure tubing. The outside diameter on the tube is about 110 mm. A scale marker is shown on the figure.

NPD. Once the decision to use a coiled spring was made, then it was necessary to fix the amount of tension in the spring on installation. The larger the tension, the more tightly the spring would be held against the pressure tube. With very little or no tension, the spring could be very loose on the pressure tube, or even tight against the calandria tube. The springs for NPD were tightly wound around the pressure tube, and gave excellent service for the twenty-five years of reactor operation. Similar springs are used in KANUPP.

In the early sixties, Dr. Lewis was maintaining his insistence on neutron economy, and kept pushing the designers to remove all possible neutron absorbers from the reactor core. A zirconium-alloy spacer would have less neutron absorption than Inconel, and Dr. Lewis demanded that one be developed. Zircaloy-2 was an obvious choice, but to have the required tensile strength it needed to be in a cold-worked condition, and the coiled spring would develop an undesirable hydride platelet orientation. It was known that small additions of copper added to zirconium alloys increased their corrosion resistance to some gases, and Dr. Lewis decided that an alloy of Zr-2.5Nb-0.5Cu should be satisfactory if used in a heat-treated condition. By coiling the springs before heat treatment much of the concern on hydride orientation could be eliminated (figure 12.10), and the heat-treated alloy would be as strong as or stronger than cold-worked Zircaloy-2. The development program on the Zr-2.5Nb-0.5Cu alloy at Chalk River included specifications of the heat treatment, which was to be significantly different from that for Zr-2.5Nb. The copper addition resulted in an alloy heat-treatable to much higher strengths than possible in Zr-2.5Nb, but accompanied by lower ductility. A full charge of Inconel-X 750 springs was fabricated for Douglas Point, but Dr. Lewis was able to insist that Zr-2.5Nb-0.5Cu springs be used.

The choice of material for the springs has remained a controversial topic to the present. The engineers undertaking the detailed design of the spring recognized several distinct disadvantages in using Zr-2.5Nb-0.5Cu compared with Inconel-X 750, and presented their viewpoint vigorously. One indication of the thought devoted to the topic is that one history of garter-spring development up to 1964 occupies ninety-two typed pages, and

FIGURE 12.10

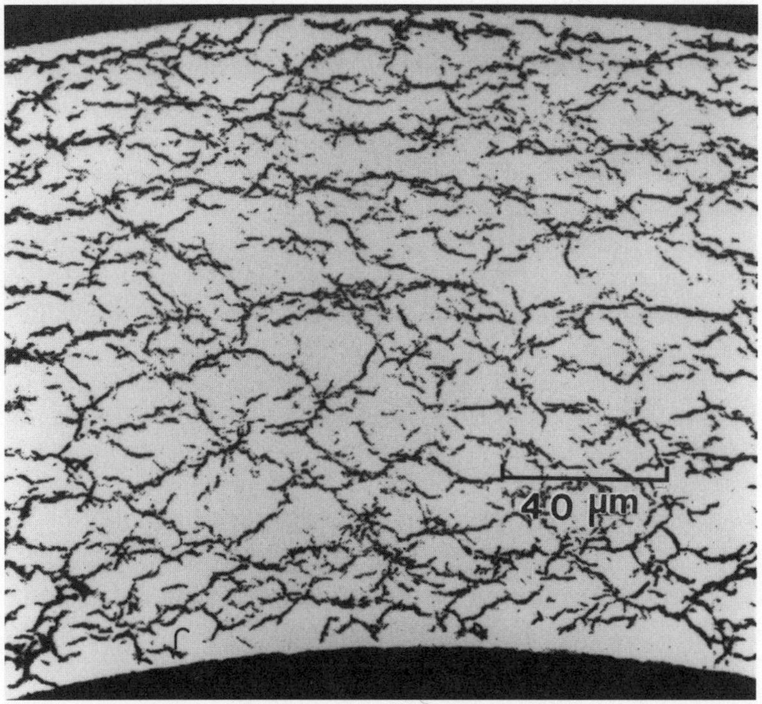

This photomicrograph shows the edges of hydride platelets in a section of garter spring made from Zr-2.5Nb-0.5Cu. The hydride orientation is seen to be random. Although an ideal orientation would have all platelets in the circumferential plane, the orientation shown here is the best compromise available. The specimen has about 2.5 at.% hydrogen. A scale marker is shown on the figure.

a second history covering up to the present time occupies fifty-two additional typed pages. The engineers were concerned that a spring of Zr-2.5Nb-0.5Cu would fail at some isolated section if tightly wound against the pressure tube, and that the joint between the ends of the spring would be a particular point of weakness. It is not clear that Dr. Lewis was ever exposed to the detailed arguments, and he insisted on the use of Zr-2.5Nb-0.5Cu for the spring material. Metallurgists working at Chalk River at that time recall discussions on the spring materials as being the most emotionally charged of any in their careers, and Dr. Lewis seems to have taken an uncharacteristically dogmatic attitude on the topic. The spring eventually designed was to fit loosely on the pressure tube, and tightly against the calandria tube. This spring design was used for Douglas Point, and all CANDU reactors up to Bruce 8. For Bruce 8 and all subsequent CANDU reactors, we reverted to Inconel-X 750 springs, fitted tightly around the pressure tube.

By 1977 evidence from sag measurements indicated that some springs were not in the design position, and the important failure in 1983 of the Zircaloy-2 tube in position G-16 of Pickering 2 was directly attributable to incorrectly positioned garter springs. Subsequent work has shown that much if not all of the movement of garter springs from the location immediately after channel installation occurred during construction and commissioning of the reactor. Once the overall problem was appreciated, equipment was designed to locate the position of the garter springs in an already assembled reactor, and thence move the springs back to approximately the design location. This was done for Bruce B, Units 5, 6, and 7 and Pickering B, Units 7 and 8, before these reactors went critical. Some channels in reactors that have been in operation have received the same treatment. Although the laboratories gave some assistance to this work it was largely done by Ontario Hydro and AECL CANDU. The problems arising from the garter-spring movement have been very costly to Ontario Hydro and AECL, perhaps the most costly of any single factor of design. It is not known whether garter springs of Zr-2.5Nb-0.5Cu tightly fitted around the pressure tube would be satisfactory, but there is no indication that this concept will be empirically tested in the foreseeable future.

CALANDRIA TUBES

The initial requirement for the calandria tube arose from the need to insulate moderator water from the pressure tube. It was not until much more information on in-reactor creep was obtained that we realized the calandria tube was providing the major part of the resistance to sag of the channel. Aluminum was chosen for the NPD calandria tubes, probably because the technology of fabricating thin-walled aluminum tubing was better established than for any other material of acceptable neutron capture cross-section. These tubes were quite satisfactory throughout that reactor's life. The reactor designers realized, however, that there would be advantages of both strength and corrosion resistance if the tubes could be made of Zircaloy. The calandria tubes have a diameter of about 127 mm (5 in.) and wall thickness of about 1.37 mm (0.054 in.); there was little industrial experience in fabricating Zircaloy tubes of this diameter and wall thinness, and none in Canada. The desirable tolerance on dimensions is difficult to achieve by the extrusion and drawing process used for pressure tubes. The fabrication process selected for the calandria tubes (with the exception of tubes for KANUPP) involves starting with sheet (which can be rolled to the required thickness tolerance), then bending and welding the sheet into a tube. The tube is annealed in the final stage of fabrication and thus has a lower tensile strength than any of the pressure tubes. Although skill is required to obtain the desired straightness and (lack of) ovality, the key to obtaining a successful tube is to have a sufficiently small grain size at the weld. From their experience with titanium, Carpenter Technology of San Diego, California, were able to fabricate the calandria tubes from Zircaloy, and they supplied the tubes for Douglas Point, Gentilly-1 and Pickering 1 and 2. Grain-size control at the weld was greatly improved from Pickering 1 onwards. For KANUPP, Canadian General Electric obtained seamless calandria tubes from a fabricator in France.

For obvious reasons, both the federal and provincial governments have encouraged the production of CANDU components in Canada, and in the late sixties it was decided to attempt to make the calandria tubes in Canada. Canadian Westinghouse was chosen to do this, with technical assistance from AECL. The fabrication procedure was developed in time to produce the tubes for Pickering 3. The tubes made in Canada, as well as those obtained offshore, have operated successfully in-reactor. In addition to the made-in-Canada policy, Ontario Hydro has a policy of trying to obtain bids from more than one supplier. Although Canadian Westinghouse, now Zircatec Precision Industries, was firmly established as a supplier of acceptable tubes, Ontario Hydro and AECL in the late seventies encouraged Bristol Aerospace, Winnipeg, to become a supplier. This effort was successful and Bristol supplied tubes for both the Wolsong-1 and Gentilly-2 reactors.

By 1977, we had developed sufficient information on growth to indicate that the elongation of the calandria tubes could result in undesirable loading of the end shields. To evaluate this problem we undertook a major irradiation project in a high flux reactor in Idaho Falls, United States. This enabled us to obtain in a few years a neutron fluence equivalent to thirty years or more in CANDU. The results demonstrated that, despite our early concerns, calandria-tube growth would not cause problems.

Throughout the history of CANDU reactors, no problems have arisen from any unexpected behaviour of the calandria tubes.

END FITTING AND CHANNEL CLOSURE

The zirconium-alloy pressure tube must be joined by some means to the ferrous alloy out-of-core circuit. It would be far too expensive to have the complete circuit fabricated from zirconium alloy. Welding or brazing of the pressure tube to a ferrous alloy might be considered, but this type of joint would have severe limitations. Rather, we adopted a rolled joint, in which the pressure tube is forced against the inner surface of a stainless-steel hub and thereafter the circuit is completed with standard methods of chemical/mechanical engineering. Before "rolling-in", the hub has grooves, three or four, into which the pressure tube is forced to form ridges on the tube (figure 12.11). This ridge-groove system makes a watertight seal and mechanically attaches the tube and end fitting; the latter is required because the pressure of the coolant puts a longitudinal stress on the whole of the fuel channel. Rolled joints of this type were a standard industrial item before NPD was designed, and we merely adopted and modified it for our particular purpose. A rather different end fitting was used for Gentilly-1; a description of it is outside of the scope here. Except for KANUPP (and two channels in NPD), all end fittings have been made of the 403 grade of stainless steel.

FIGURE 12.11

The photograph shows the ridge and groove scheme in a disassembled rolled joint. The section of rolled joint is at the left, the grooves in it machined before assembly. The section of pressure tube at the right would have been quite smooth before joint fabrication. The outside diameter of the pressure tube is about 110 mm.

The 403-stainless was chosen for the end fitting because it had adequate corrosion resistance to its environment, is strong enough to make the leak-tight joint needed, and because its thermal expansion coefficient is acceptably close to that of zirconium alloys such that the joint would not degrade from the thermal cycles experienced during years of service. The end fittings for NPD were cast. Thereafter, they have been forged or extruded. It has been possible to have the major part of the fabrication done in Canada, principally by Donlee, of Weston, Ontario.

The principal problem faced in the use of the 403 stainless end fittings arose from licensing their use. Over much of the world, standards set by the American Society of Mechanical Engineers (ASME) are accepted as the gospel to be followed in the design of components for nuclear reactors. In 1972, the relevant ASME committee made some radical changes in the requirements for toughness of materials used in constructing nuclear reactor vessels. The CANDU end fittings, because of the strength required for a successful rolled joint, did not meet the letter of the ASME code. Engineers at Chalk River were able to demonstrate both theoretically and experimentally that the end fittings were indeed very safe in our application. In a long series of discussions with the Control Board, ending in 1978, Canadian designers, reactor operators and the Control Board were convinced of the safety of the end fittings.

Like the calandria tubes, the end fittings per se have never been a source of trouble in CANDU reactors, and have fulfilled successfully all design functions. Problems at the rolled joint did cause the pressure-tube failures in Pickering A, Units 3 and 4. The combination of design of the joint and design and operation of the tooling used in making the joint resulted in unplanned deformation of the pressure tube immediately at the inner end of the end fitting. This deformation in turn caused a high residual tensile stress in the pressure tube at that region. When this procedural flaw was recognized, the joints for Bruce A, Units 1 and 2, were already fabricated. The joints in the latter two reactors were repaired by heating them to relieve the stress, and for subsequent reactors the design of the joint and the installation tools were modified.

The necessity to refuel the channels while the reactor is at full power means each end of the fuel channel must be sealed by a removable channel closure. Sealing the pressurized hot water in the channels such that there is a capability of resealing is a very demanding requirement and is accomplished by means of a self-energizing seal disc (figure 12.12 shows a typical closure and seal disc). The seal disc is made from high strength 17-4PH material, onto which a soft nickel gasket is plated to create a face seal with an insert shrunk into the 403-stainless-steel-channel end fitting. The seal disc is held in the end-fitting by a closure mechanism that is operated by the fueling machine. During installation the fueling machine applies an initial pre-load deflection to the seal disc that is locked-in by the closure mechanism. Hydraulic pressure force from the channel coolant heat-transport system deflects and loads the seal disc further during normal operation, to increase

FIGURE 12.12

Photograph of two faces of a Douglas Point seal disc. The soft nickel gasket is evident on the face on the right. The diameter of the flat surface is about 96 mm.

the sealing force and produce a leak-tight seal for the channel closure. Some unacceptable leakage occurred with early seal discs at Douglas Point. The leakage was associated with closure installation forces and initial pre-load deflection tolerances. In 1968-72, AECL CANDU, in collaboration with Chalk River, performed redesign on the seal disc. The disc's contour shape was modified and the fueling machine installation and removal force tolerance bands were modified to provide a more repeatable installation setup. The fabrication process involving the soft nickel gasket was improved by providing heat treatment of the disc after the nickel plating. During the redesign period of the seal disc, consideration was given to changing the disc material from 17-4PH to a titanium alloy. One solution evaluated was to use very pure, and hence soft, titanium for the seal, rather than plating nickel onto titanium, which presented a metallurgical problem. Several seal discs made from pure titanium were fabricated by Canadian Westinghouse. This program was not completed, as the problem was solved by changing the seal-disc shape and installation procedure. There have been no further problems with seal-disc leakage since the Douglas Point episode.

BIBLIOGRAPHY

Biefer, G.J., L.M. Howe, A. Sawatzky, and F.H. Krenz. *Hydrogen Pick-up in Zirconium Alloys: A Review of Data up to June 1,1959.* AECL-919. 1959.

Clendening, W.R. *CANDU Fuel Channels.* Report TDAI-337. Mississauga, Ont.: Atomic Energy of Canada Limited, CANDU Operations, 1985. (Describes the rolled joint, end fitting and channel closure.)

Ells, C.E., C.E. Coleman, V. Fidleris, E.T.C. Ho, and A.R. Causey. *The Behaviour of the CANDU Calandria Tubes.* AECL-9514. 1987.

Ells, C.E., and A. Sawatzky. *The Effect of Neutron Irradiation and Hydrogen Concentration on Tensile Properties of the CANDU Coolant Tube Spacer Alloy.* AECL-2272. 1965.

Langford, W.J. *Applicability and Interpretation of Fracture Test Methods for Metals.* AECL-6199. 1978. (Presented at a Seminar on Failure Analysis and Fracture Control, Queen's University, Kingston, Ont., 8-9 May 1978.)

Perryman, E.C.W. *"Pickering Pressure Tube Cracking Experience."* Nuclear Energy (British Nuclear Energy Society) 17 (1978): 95-105. (Also AECL-6059.)

Price, E.G. *Highlights of the Metallurgical Behaviour of CANDU Pressure Tubes.* AECL-8338. 1984.

Ross-Ross, P.A. *Fuel Channel Development for Canada's Power Reactors.* AECL-3126. 1968. (Presented at the Annual General Meeting of the Engineering Institute of Canada, Halifax, N.S., 29-31 May 1968.)

Ross-Ross, P.A., W. Evans, and W.J. Langford. *Experience with Zirconium-Alloy Pressure Tubes.* AECL-4262. 1972. (Presented at the Third Inter-American Conference on Materials Technology, Rio de Janeiro, 14-17 August 1972.)

Wyatt, A. *"NPD-2, Canada's Prototype Power Reactor."* Nuclear Energy (Institution of Nuclear Engineers) 1 (1962): 192-201. (Also AECL-1568.)

Chapter Thirteen

Fuel

A.S. BAIN

INTRODUCTION

Fuel for the CANDU nuclear power reactor has to meet a very demanding set of criteria which evolved and were refined over the course of the fuel development program:

- Operation must be safe, with a very low chance of release of radioactive products to the reactor systems;
- The fuel assembly must have a low parasitic absorption of neutrons;
- It must operate to the required performance specifications of power output, residence time, dimensional stability and mechanical integrity under all postulated normal and abnormal operating conditions of the reactor;
- It must have predictable performance under all potential accident conditions;
- It must be capable of being easily and safely handled before and after irradiation, during temporary storage and, eventually, during more permanent storage or disposal;
- It must be economical to fabricate.

The challenges were to identify suitable materials, establish a design, develop the fabrication techniques, and do the laboratory and in-reactor experiments to prove the proposed product would meet the criteria. A complication to the challenge was that initially there was very little, and at times zero, information in the worldwide literature on specific aspects of fuel for nuclear-power reactors. There was considerable information on fuel for research reactors; this provided useful background information as a starting point. However, it soon became evident that a totally new approach had to be followed to develop the fuel for the CANDU reactor.

INFLUENCE OF REACTOR DESIGN

The original design for the NPD reactor called for a vertical pressure vessel with full-length fuel elements. Later the design was changed to a horizontal orientation using pressure tubes instead of a pressure vessel. Also, it was decided to have on-power refueling, where fresh fuel would be inserted and used fuel (that which had reached its target burnup) would be discharged while the reactor was in operation. These reactor design changes had a profound effect on the design of the fuel. Fuel assemblies had to fit inside the pressure tubes, be short and fully compatible with the fueling machines, be free to move along the pressure tube, and be designed to allow sufficient coolant to pass through the assembly to remove the fission heat.

As the CANDU reactor developed there was a demand for more power per channel by any or all possible methods, such as increased bundle[1] diameter, more elements per bundle, and increased power per unit length of element. As a consequence, the fuel design had to evolve to meet the new performance specifications.

PROGRAM ORGANIZATION

The development of the CANDU bundle is a tremendous success story that utilized all aspects of good engineering. At first, efforts concentrated on developing a thorough scientific understanding of the potential components that could meet the rigid operating and performance specifications of a CANDU reactor. Then they focused on engineering aspects, with progressive emphasis on individual fuel elements, clusters of elements, full-sized fuel bundles, and the behaviour of the fuel under all postulated operating and accident conditions. The challenge was to produce and remove up to 1 MW of heat from a 50-cm-long fuel bundle that had to remain intact and undistorted in hot pressurized water for one to four years, with fission, transmutation and massive radiation damage, using materials that have a low parasitic absorption of neutrons.

The program was led by AECL, with the enthusiastic support and participation of government departments (especially Energy, Mines and Resources), Eldorado Mining and Refining, the provincial utilities that owned and operated the reactors (not only Ontario Hydro, the main user, but also Hydro-Quebec and New Brunswick Electric Power Corporation) and private industry, particularly Canadian General Electric (CGE) and Westinghouse Canada Limited, who invested millions of dollars to set up fuel-fabrication departments in their organizations. As part of the incentive for this investment, CGE and Westinghouse (and later Combustion Engineering) were awarded the royalty-free right to produce the fuel for the CANDU reactors.

AECL awarded contracts to these fuel fabricators to do specific research or development work on their own premises. Also CGE, Westinghouse, Combustion Engineering, the utilities, clients or potential clients, and a few other interested organizations sent people to Chalk River and Whiteshell to work with the AECL research personnel on AECL-funded programs. These contracts and the hands-on attachment program proved to be very efficient methods of developing information and transferring technology – both AECL technology to the manufacturers and their feedback on fabrication know-how that led to optimized design and fabrication methods.

There were frequent progress meetings, and every few months an overall program meeting. At first, people from competing private organizations were reluctant to participate in meetings where all the AECL-funded fuel programs were discussed in detail. They were concerned that their company secrets could be inadvertently revealed. However, AECL insisted on full participation, and the fuel information meetings became an example of how people from different organizations can co-operate to achieve the objectives of a program. At times over one hundred people attended these meetings. Lively discussions were common. This direct interaction between scientists, fuel engineers, fabricators and product users provided the synergism necessary for success.

Although CGE and Westinghouse Canada Limited (and later Combustion Engineering Canada) were branch plants of American-based multinational organizations, the Canadian organizations were universally loyal to the Canadian program. There was no evidence or suspicion of any transfer of proprietary information to the parent company – in fact it appeared that the Canadian groups proved themselves to be equal to, or often ahead of, the larger American organizations.

Dr. W.B. Lewis became involved very early in the overall strategy and progress of the fuel program, and whenever possible would drop in to the laboratories to see some new results, or to discuss his thoughts on problems or directions for the program directly with the staff. Dr. Lewis chaired special fuel meetings and was prepared to comment on all aspects of the program. As in all of his endeavours, Dr. Lewis was rarely wrong, and his quiet guidance combined with his clearly

[1] At first the fuel assemblies were called slugs; this was later changed to bundles.

stated objective to develop a simple, efficient, safe and neutron economical fuel for CANDU stimulated, and at times pressured the staff to produce a magnificent engineering achievement. Very early in the design one requirement above all was emphasized: the fuel-cycle costs should contribute no more than 1 mil/kWh[2] to the total unit energy cost of power from the CANDU nuclear-power reactors. This economic target had a significant impact on the design and fabrication of the fuel. And, as explained later in this chapter, this target was achieved.

The apparent simplicity and ultimate reliability of the final product is a result of a twenty-five year R&D program, starting in the fifties, by several hundred people, and costing several hundred million dollars. The program was in two main areas: development of the individual components, such as the fuel and sheath, and development of the assembly of elements into a bundle that could produce the required power in a safe and economical manner.

INDIVIDUAL COMPONENTS

Fuel

The fuel material must be able to produce the required power, and perform safely under the required operating conditions. Originally, the CANDU fuel was to be uranium metal, based on experience of the fuel elements from the Chalk River research reactors. Uranium metal is attractive as a fuel material because of its high density, and its relatively good thermal conductivity. However, it was found that uranium metal has two disadvantages: it reacts with hot water very rapidly if a defect in the sheath should occur; such oxidation results in an excessive increase in volume, sheath distortion and high radioactivity releases. Secondly, it swells and distorts under the irradiation exposure demanded for power reactor fuel, giving unacceptable dimensional changes. To make matters worse, the distortion is not uniform, due to the anisotropic crystal structure of the uranium. The corrosion and swelling could be controlled by alloying, and many uranium alloys were studied. However, the absorption of neutrons by the alloying ingredients was a serious concern and this option was not pursued.

The development of an alternative fuel material was given high priority. Fortunately, the Americans and the British had been working on fuel materials for their nuclear programs, and had concentrated on the development of uranium dioxide as a prime candidate. Through the cooperative program with the United States, AECL became aware of the work on uranium dioxide. Was uranium dioxide suitable as a fuel for CANDU reactors? The answer was not obvious; concerns existed about the low thermal conductivity that would lead to high fuel temperatures, particularly at the centre of the pellets. It was expected that these high temperatures would cause extensive thermal expansion and the release of volatile fission products from the material, which in turn might cause a pressure buildup inside the element, and consequently an expansion of the fuel element. However, after careful consideration of available results, particularly American tests in NRX, uranium dioxide was, in 1955, chosen as the fuel material for the CANDU reactor, and AECL launched a full-scale program to develop uranium dioxide as the fuel for CANDU.

Multidisciplinary teams worked together on fundamental properties, specifications, fabrication methods and irradiation performance. Experimental facilities had to be built, and inspection technologies developed. As the research progressed the work at Chalk River Laboratories became a worldwide focal point, with many international scientists/engineers attached to work with AECL experts. Cooperative programs were encouraged, particularly with the United States and the United Kingdom. Other smaller, but still significant, cooperative programs were carried out with France, Germany, Holland, Norway, Sweden, Italy, Argentina, Mexico, Romania, Korea, Japan, the USSR, India and Pakistan.

Parameters studied included:

Composition – Uranium dioxide can contain more or less than the stoichiometric amount of oxygen, to form uranium dioxide$_x$.

---

2 1 mil = 0.1 cent.

Fabrication and performance over a wide range of stoichiometry was studied, with the final decision that the optimum composition was $UO_{2.00}$ to uranium $UO_{2.05}$. Material above $UO_{2.05}$ was not acceptable due to poorer irradiation performance, especially with regard to the release of fission products.

Density – The goal was to produce a high-density uranium dioxide in order to incorporate as much fissile component as possible using natural uranium.

Fabrication – Many routes were possible to convert the uranium dioxide powder into the dense ceramic material needed for the fuel elements. Candidates included sintered pellets; vibratory-compacted powder; oversized sheathing swaged to the final size, with the swaging serving to densify the contained powder; extrusion of uranium dioxide cylinders; and microspheres produced by a sol-gel process. The sintered pellet option proved to be the most promising to achieve high density, and work on the others was dropped.

The development of a process to produce high-quality sintered uranium dioxide pellets started at a laboratory scale, where each part of the process was investigated. The uranium dioxide powder was poured into a die, then pressed to form a pellet. This stage results in a pellet of about 60 percent of the required final density – called a green pellet. The green pellets are then sintered at high temperature to achieve the required density. The uranium dioxide powder originally supplied by Eldorado was not suitable for sintering. It lumped together and would not flow freely into the pellet die. The green pellets would not remain intact, so an organic binder had to be added. This organic binder had to be totally removed by the sintering procedures. When this powder plus binder was used, a pellet reject rate occasionally up to 50 percent occurred, mainly due to chipping at the edges.

Experimental work at AECL and at the Mines Branch in Ottawa showed that uranium dioxide powder produced from ammonium diuranate (ADU) could be compacted, then granulated to form a powder that flowed easily to fill the pellet die, and had the distinct advantage that the green pellets would hold together without the addition of an organic binder. Thus, the process was simplified and the potential for adding an impurity to the uranium dioxide pellets was removed. This technology became the standard powder production route.

There were several options for sintering, with the three main variables being the temperature, the covering gas and the duration. The original process called for a two-stage sintering: the first in an inert atmosphere to produce the required density and the second in a hydrogen atmosphere to remove excess oxygen. A single-stage process was needed to reduce costs to help meet the 1 mil/kWh target, and to provide a process that could be strictly controlled. The chosen route was sintering in a hydrogen atmosphere at 1,700°C for about two hours to produce pellets of a minimum of 96 percent of theoretical density.

Other single-stage processes included using carbon dioxide or steam as the covering gas during sintering. These had the advantage of sintering at much lower temperatures. However, the product had excess oxygen, and thus was eliminated from consideration, due to unacceptable irradiation performance. A reduction cycle could have been added to bring the product to the stoichiometric composition, but this was too expensive to put into practice.

Experimental evidence showed there was a need for distributed voidage inside the element to accommodate the thermal expansion at the centre of the pellet. As this requirement became better defined, the uranium dioxide pellet fabrication process was refined. Dishes were incorporated into the ends of the green pellets, which, on sintering, resulted in the proper voidage (see figure 13.1).

A production route was established so that the pellets were deliberately oversize, and ground to the correct diameter on a centreless grinder to provide the pellet-to-sheath diametral clearance (about 0.1 mm) required for good operation. After grinding, it was found to be essential to dry the pellets thoroughly, since any retained moisture could cause problems during irradiation.

The fuel fabricators at CGE and Westinghouse worked with AECL to refine the fuel fabrication process. This team approach was successful in that the pellet rejection rate dropped to about

FIGURE 13.1
UO_2 Pellet showing End Dishing

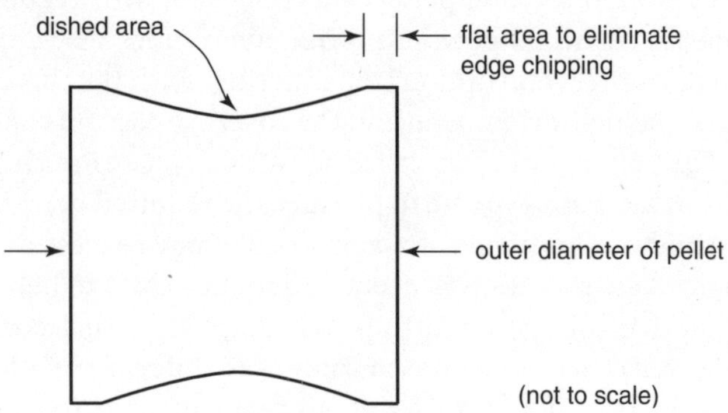

5 percent. The scrap was recycled, with the added bonus that because the amount was so small it could be added to the dry powder and did not need total reprocessing.

Irradiation performance – The initial irradiation of test samples was an exciting experience. AECL had the NRX reactor, and later the NRU reactor, both of which had extraordinarily good irradiation facilities. Separate systems within the reactor, called loops,[3] provided the operating pressure, temperature and coolant conditions pertinent to a CANDU power reactor. There were small-diameter loops to take strings of single elements, three-element clusters, or seven-element bundles. Eventually the larger-diameter NRU loops took a string of up to seven full-size CANDU bundles, where coolant conditions could be varied to give single- or two-phase flow, with the amount of boiling up to 10 wt. percent steam.

In NRX a hydraulic facility called the hydraulic rabbit allowed rapid insertion and removal of a single test specimen with the reactor at power, and consequently a precisely controlled duration of irradiation ranging from a few seconds to several hours, at a constant power. Longer irradiations were possible, but the resultant fission-product activity became too high for the facility shielding and the transport flasks. Four specimens were usually run in sequence – while the operating conditions of the reactor were constant – to show the difference in behaviour by varying one parameter, such as power output, duration of irradiation, fuel-to-sheath clearances, stoichiometry, density, etc. Other essential facilities, such as shielded cells with remote-handling capabilities (hot cells), chemical analyses, X-ray spectrometry and neutron radiography were also available to assess the performance.

Initially, the determination of fuel burnup was done by wet chemistry methods to analyze for fission products. This was a long process. Results were not very accurate, and often were quite different from burnups estimated from calorimetry in the loops and measured values of the neutron flux in the reactor. This meant that calculations on power output, and particularly fission-product production and percentage release, were suspect. The research chemistry branch had built a mass spectrometer. Funds for this were partly justified by the need for more accurate burnup analyses, by using the spectrometer to measure the uranium-235/uranium-238 ratio of samples removed from irradiated fuel, from which data the burnup could be calculated to a high degree of accuracy. The first few samples were done with enthusiasm by the research chemists, to prove that their system worked. However, when samples began to be submitted on a frequent basis there was a minor rebellion: "The mass spectrometer was a research tool, not a service machine for the fuel people".[4] Heated discussions followed, but managerial diplomacy prevailed; essential samples were processed, with the promise that another setup would be assembled to

[3] The first loop for NRX was built by Westinghouse Atomic Power Department (WAPD), of the United States. They also supplied the special pumps required for the safe operation of the system. This loop was used extensively by WAPD to develop the fuel for the American nuclear-powered submarines and its first civilian nuclear-power station at Shippingport.

[4] A memorable statement from an agitated research scientist whose program was being constantly interrupted. His objections led to the establishment of an analysis group that was essential for the success of CANDU fuel development.

provide the burnup analyses essential for a full understanding of fuel performance, and this was done.

One important aspect in the early days of the experimental irradiation program was the quickness with which an experiment could be conceived, authorized, designed, fabricated, irradiated, examined and results made available. Red tape was at a minimum. Experiments were proposed in writing, then reviewed by a reactor irradiation committee. Questions were answered and the irradiations approved, or, rarely, rejected. Many times the main discussions centered on reactor utilization: what had priority, the fuel experiments or production of radioisotopes? (This question was never fully answered; situations were handled on a case-by-case basis.) In simple tests in the hydraulic rabbit, a test could go from conception to examination in a couple of months. In loop tests it took a bit longer, because of scheduling in the loop. This rapid procedure helped greatly to investigate many parameters and to identify suitable and unsuitable paths to follow. The joy and excitement of working in this program was that everyone felt part of the team. People came in at all hours of the day or night to be there when experiments were being done. Participants in the discussions before, during and after an irradiation included fuel researchers, the reactor and loop operators, and the fuel fabricators. The first-hand knowledge and feedback led to a better appreciation of the total program, and helped to develop the cooperative environment essential for success.

Power levels varied from very low to those that produced central melting of the uranium dioxide (uranium dioxide melts at 2,800°C; central temperatures sometimes exceeded 3,000°C). In many instances it was the first time such an irradiation had been done, and often the hot-cell areas were crowded with people eager to see first-hand what had happened during the irradiation.

In an early test, the irradiated specimen was cut open and examined with the hot cell microscope. The uranium dioxide had a mysterious hole in the centre (see figure 13.2), indicating it must have melted. But it couldn't have melted because the power level was too low. So what happened?

FIGURE 13.2
Fuel with Central Void

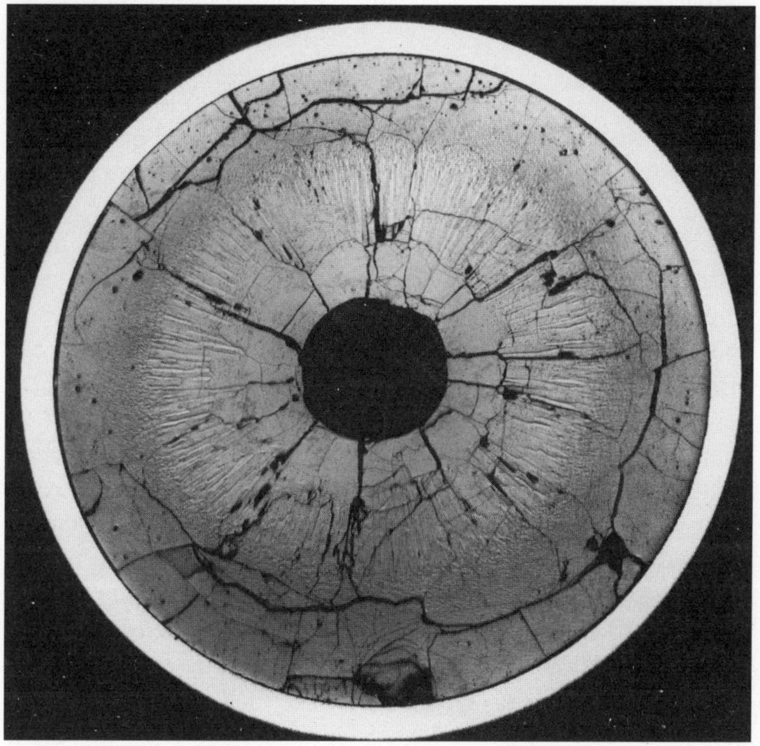

Solving that mystery evolved into a large program, involving AECL and personnel from laboratories around the world. Arguments arose particularly between AECL and American laboratories and fuel companies. AECL was convinced that melting had not occurred, whereas the American organizations were certain that the centre of the pellet had melted. The difficulty was the lack of a verified expression for the thermal conductivity of the uranium dioxide under irradiation over the temperature range of interest. At AECL the important aspect was to correlate power output with fuel behaviour. (Whether fuel melting had occurred was not of prime importance.) To do this, AECL correlated the observed fuel behaviour with the integrated thermal conductivity of the uranium dioxide (between the surface and central temperature of the uranium dioxide pellets).[5] This approach eliminated the need to know the central temperature. However, outside AECL, particularly in the United States and particularly for safety analyses, the use of the integrated thermal conductivity was not widely accepted. A precise value of the thermal conductivity was considered to be essential, as was an unambiguous statement on whether the central part of the pellet had melted.

In many laboratories, including AECL, studies were intensified on thermal conductivity, the effect of pellet density, shape and distribution of porosity, effects of cracks in the uranium dioxide pellets, diametral clearance, pellet-to-sheath gap conductance, including the effect of filling gases and surface roughness of the pellets and sheath on heat transfer, and the effect of impurities, including moisture and excess oxygen.

The results showed that at temperatures above about 1,400°C, uranium dioxide is not a static object during irradiation; rather, it becomes a dynamic system. Above 1,400°C, uranium dioxide grains will increase in size (grain growth). AECL laboratory tests showed that the appearance of a structure resembling one in which central melting has occurred can result from migration of lens-shaped (lenticular) voids in the uranium dioxide (see figure 13.3) to the central zone via a vaporization/condensation process at temperatures well below the melting point. Normal pores become part of the lenticular voids, and with sufficient time porosity can move to the centre, giving a central hole similar to that characteristic of melting. This process was demonstrated clearly in laboratory tests at temperatures well below the melting point of uranium dioxide, as well as in irradiation experiments. Tests also showed that cracks in the uranium dioxide will heal and that the grain structure in the hot central regions will vary significantly with very minor changes in pellet density.

These tests proved that the structure seen in post-irradiation examinations of the uranium dioxide and the formation of a central void could occur well below the melting point of the uranium dioxide. This interpretation was supported by other evidence, such as irradiation with inert (tungsten) markers, and a demonstration of a discontinuity in radial sheath strain at central melting.

FIGURE 13.3
Fuel with Lenticular Voids

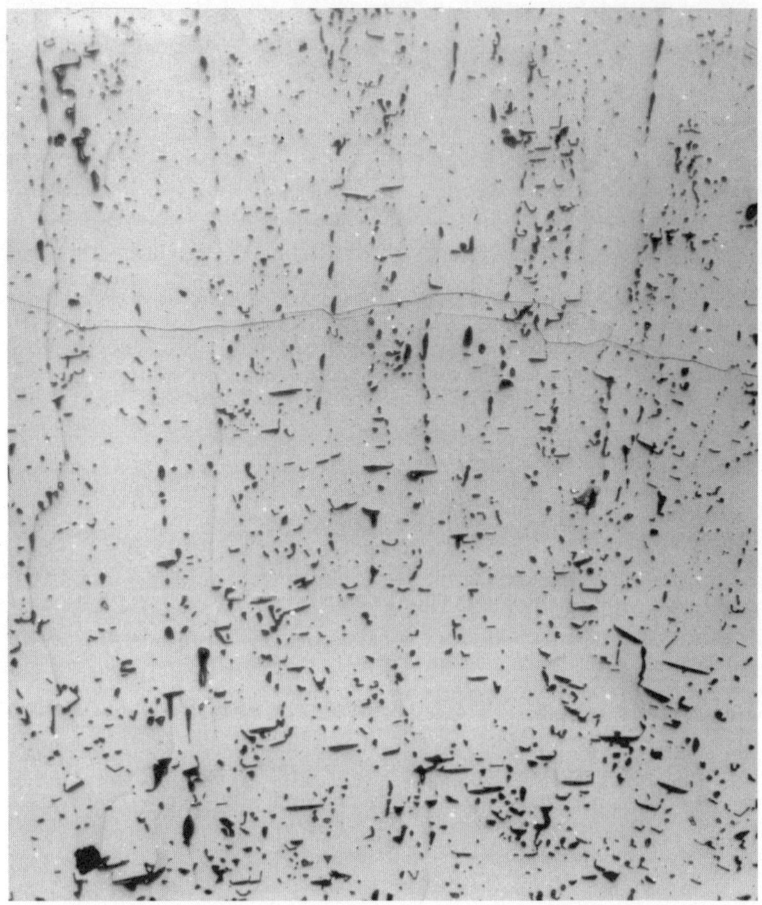

The extent of grain growth and pore movement was shown to influence greatly the release of volatile fission products from the uranium dioxide, and is now an important part in computer models developed to predict the performance of uranium dioxide fuel under normal and abnormal (accident) operating conditions.

[5] The integrated thermal conductivity can be readily calculated from measured values of the diameter of the element, power per unit length, and enrichment of the uranium dioxide.

Early in the uranium dioxide experimental program potentially devastating news came from General Electric's Vallecitos, California Labs. Several of their Zircaloy-sheathed elements had disintegrated during irradiation, and General Electric was considering abandoning Zircaloy in favour of stainless-steel sheaths for the boiling-water reactors. If no market existed for zirconium-based sheathing in the United States, it would be doubtful if there could be a viable industry to supply only Canadian demands; and stainless steel absorbed too many neutrons to be used with natural-uranium fuel in CANDU reactors. AECL had not experienced such disintegration of Zircaloy-sheathed fuel and questions were raised on material impurities. American analyses showed a high concentration of fluoride in the uranium dioxide, probably from residual fluorine due to incomplete conversion of the uranium hexafluoride used in the uranium-enrichment process, which is required for fuel in American reactors. AECL postulated that such fluorine concentrations could lead to sheath disintegration. AECL did a definitive test with two levels of fluoride. Within Canadian specifications there was no problem; but with high fluoride concentrations sheath disintegration occurred. This information was quickly dispatched to the United States; Zircaloy was kept as the preferred sheathing. Without this prompt action and successful elimination of the problem, the entire CANDU system could have been in jeopardy.

A major initial concern with the use of uranium dioxide was what would happen if a highly powered fuel element defected in-reactor, and there was contact between the very hot (possibly molten) uranium dioxide and the coolant. Several times during the early tests, fuel elements with very high power outputs per unit length (linear output) failed while under irradiation in the reactor. A gaping hole formed in the sheath, and the outer parts of the pellet were lost; this resulted in direct contact between the coolant and central uranium dioxide. Polished cross-sections of the specimens at the defect area gave the general impression that the coolant had cooled the uranium dioxide and there was no large flow of molten material into the coolant. Further, there was never any pressure surge indicated in the instrument read-out of any loop test, and there was no evidence of damage to the pressure tubes of the loops or the flow tube of the hydraulic-rabbit facility. At the time, the importance of these observations was not fully realized. Later, however, the data showing that even a large defect in highly powered fuel did not cause a pressure surge and did not damage adjacent components were very valuable in assessing the consequences of an accident. This information became an important part of the presentations for licensing of not only CANDU reactors, but also of light-water reactors throughout the world.

AECL scientists were the first in the world to report irradiation-induced densification of the UO_2, which was later to plague the original low-density light-water-reactor fuel. AECL also mounted in-reactor tests to show how initial grain size of the sintered pellets affected fission-gas release. In the mid-to-late seventies there was a concerted program to better understand uranium dioxide properties. This came largely from a detailed examination of fuel from the Pickering reactors. In particular, factors affecting dimensional changes in the uranium dioxide and the elements (swelling, densification, fission-gas release and grain growth) were studied, and results were incorporated into fuel design.

After the early tests had defined the short-term behaviour of uranium dioxide fuel elements, the program concentrated on long-term irradiations in the loops to determine the behaviour of fuel elements and bundles at burnups appropriate to a nuclear power reactor. Topics of importance were dimensional stability, mechanical integrity, temperature distribution, fission-product release and sheath corrosion. Of these, the release of the fission-product gases was of particular importance, because of the very small free volume at power, and therefore the potential for fuel-element expansion.

Performance models – In the early days (B.C. – before computers), the data from the laboratory work and irradiation experiments were accumulated and calculations were done with slide rules, simple calculators or even by hand. AECL generated a great deal of data from the extensive experimental program, particularly on deformation and gas release, and how these relate to linear power outputs and hence temperature of the fuel.

Correlations between individual parameters were assessed and put into empirical models. With the introduction of computers in the mid-sixties, the models were transferred into computer codes. Figure 13.4 shows a schematic of the relationships between important parameters; this served as a basis for early computer codes to model the irradiation performance of uranium dioxide fuel elements.

Initially success was spotty; but as the amount of experimental evidence and programming expertise increased, and computers became more advanced and more available (e.g., personal computers operating alone or linked to a mainframe unit), the fuel performance models became more sophisticated. Codes such as ELESIM and ELESTRES were developed and used within AECL, and also used by the Control Board and the utilities, as standard programs to assess the performance of existing fuel, and for analyses of new concepts. These codes were used by other organizations (national and international) under a licensing arrangement.

The AECL codes were compared to, and correlated with, codes of other organizations to produce current versions that accurately reflect the world knowledge on nuclear fuel.

FIGURE 13.4
Important Parameters Affecting Fuel Performance

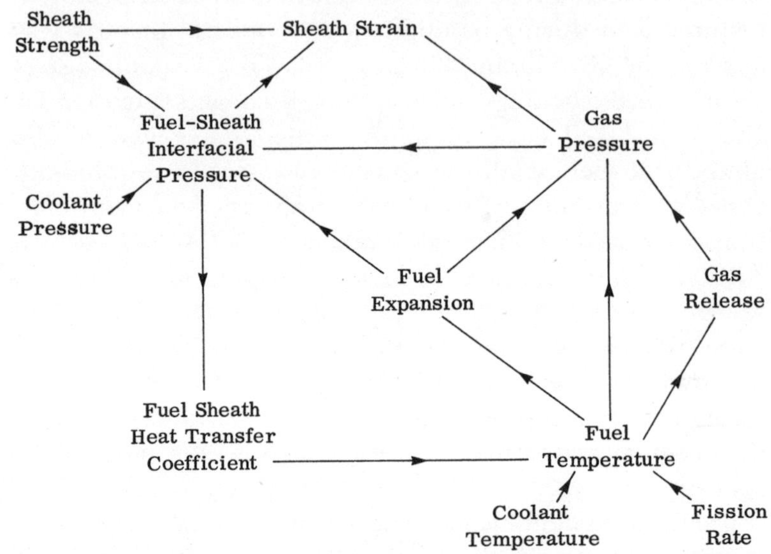

In-reactor instrumentation – In addition to the measurements of heat output (calorimetry) and thermal neutron fluxes, the development of the CANDU nuclear fuel called for the establishment of a totally new technology: in-reactor instrumentation of experimental fuel elements. The first attempts at instrumentation were to measure uranium dioxide temperatures up to about 2,800°C using thermocouples. Thermocouple manufacturers across North America and Europe were contacted. Some were unable to help. More adventurous companies worked with AECL and other nuclear research organizations around the world to develop suitable instruments; gradually, the expertise was accumulated to get reliable temperatures of operating fuel elements.

An example of the innovative approach was the fabrication of a tungsten sheath for the first thermocouple. Personnel at Chalk River vapour-deposited tungsten on a thin copper tube. When the deposit was thick enough to stand alone, the copper was dissolved away, leaving the required tungsten tube. Beryllium oxide beads were used to insulate the thermocouple wires from each other and from the sheath. Many thermocouple wires were tested, with the final choice being tungsten – 5 wt. percent rhenium vs. tungsten – 26 wt. percent rhenium, because they would work at high temperatures and because there was no change in response due to effects of the high exposure to neutrons.

Other demands arose and instrumentation was developed to measure long-term gas pressure, instantaneous pressure changes during power changes, sheath diametral strain, element elongation and bundle distortion. A unique piece of apparatus was a null balance pressure switch to measure the internal pressure in a fuel element. This simple system allowed accurate measurements of the fission-gas pressures, thus providing extremely important data for the understanding of gas release. Another important device was an in-reactor diameter-measuring rig, which used cantilever beams and motion transducers to accurately record very small diameter changes.

To give rapid changes in power the loop pressure tube and the assembled elements, plus specialized measuring devices, were placed inside a coiled tube containing helium-3. (Helium-3 has a high thermal neutron capture cross-section. With helium-3 in the coil the neutron flux was reduced; when the helium-3 was rapidly removed, the neutron flux, and hence the power, increased, producing a power ramp.) Thus the changes in diameter, internal pressure or other parameters could be measured during power ramps and power cycling.

A fuel-moving device provided single-fuel-element and fuel bundle movements from low- to high-power positions in the loop, to simulate load-following operations and power ramps, typical of CANDU fueling operations, and to measure the force needed to move bundles at power.

A totally different field of instrumentation was also essential – the detection of radioactivity in the loop coolant that would signify the presence of a defect. A rapid indication of a sheath defect was required, to determine when a defect occurred, so later analyses could correlate all the results in establishing a full understanding of the irradiation behaviour. When success was achieved in rapid measurement of radioactivity, the requirements were made more demanding. The fuel specialists wanted to know which isotopes were released. This information was important for several reasons:

- To determine which fission products existed at the fuel/sheath interface (e.g., whether iodine was present for possible reaction with the sheath to form stress-corrosion cracks);
- To gain more information on fission-product mobility in the uranium dioxide; and
- To obtain data for studies on fission-product release to the heat-transport system and containment building in the event of a major accident.

This added complication caused some consternation, but the challenge was met and the needed data were obtained. The data showed that the major isotopes detected at the counting station were xenon and krypton (noble gases), cesium and iodine. Later examination of the tubing leading to the counting station showed deposits of tellurium and molybdenum.

This part of the program is another example of two separate disciplines of AECL working together to develop what proved to be one of the best research-oriented, in-reactor fission-product detection systems in the world.

An outcome of this R&D work was a cooperative program with the CANDU designers and with Ontario Hydro, which culminated in the development of systems for CANDU power reactors to quickly detect and locate a fuel bundle with a defective element. The R&D results also indicated that some fission products would probably plate out very quickly in the plant systems, and not be released to the reactor containment hall or to the environment.

Sheath

The requirements for a suitable sheathing material were many: formability to the proper diameter and wall thickness; adequate corrosion resistance at the 300°C operating temperature of the coolant; sufficient strength and durability to withstand the stresses due to coolant pressure and fuel expansion; capability to securely attach end caps and spacers; low cost; and most importantly, low neutron capture. Stainless steel would meet all but the last requirement. Aluminum/nickel was considered but it was eliminated by 1960 because the thickness required to achieve adequate strength cost too many neutrons. Efforts concentrated on zirconium alloys, although initially they were expensive. Much of AECL's initial information came through its cooperation with the Americans. The Bettis Laboratories in the United States had developed a zirconium alloy containing tin and smaller amounts of iron, chromium and nickel. This alloy – Zircaloy – became the standard. Later, other zirconium alloys were considered, but none offered any significant improvement over Zircaloy. Zircaloy-4, containing no nickel, was selected because of slightly better corrosion resistance and significantly lower uptake of hydrogen during normal reactor operation.

Because of the emphasis on neutron economy, the sheath had to be as thin as possible, consistent with safe and reliable operation. Originally, as in the seven-element NPD bundles, the sheaths were 0.6 mm thick, but this caused too much absorption of neutrons. Eventually the concept that evolved was to have a thin sheath that would collapse uniformly due to the coolant pressure, and be supported by the uranium dioxide pellets. For the nineteen-element bundles to be used in the central channels of NPD a thickness of 0.4 mm was chosen. Laboratory and irradiation tests were done to determine its acceptability. With rigorous control of the diametral clearance, no problems occurred. But if the clearance were too large, longitudinal ridges could form. On repeated starts and shutdowns, the ridges could crack at the crests, due to fatigue failure. Other tests were done to determine whether the sheath thickness could be reduced while maintaining the specified pellet/sheath clearance. Too thin a sheath again resulted in the formation of crack-prone longitudinal ridges. The originally chosen thickness-to-diameter ratio was confirmed; this has resulted in trouble-free operation in the power reactors. Fine-tuning of this ratio might lead to some economies, but at an undetermined risk of fuel failures that would be difficult to justify.

The strength of the Zircaloy also received considerable attention. For NPD, AECL chose stress-relieved cold-worked material, unlike the American specifications, which called for annealed material. The initial fuel for the Douglas Point reactor had sheaths with higher strength, but lower ductility, than the NPD sheathing. A few defects occurred (less than 0.1 percent of the fuel). The defects seemed to occur mainly in the sheaths with the highest strength/ductility ratio. The sheathing specification was changed to a lower strength/higher ductility ratio, which was obtained by a stress-relief treatment for the Zircaloy that was well short of the full anneal used for the light-water-reactor fuel sheaths. However, subsequent, more detailed analysis failed to substantiate any relationship between the initial mechanical properties and the propensity to defect.

Initial flaws in the tubes could have a significant effect on fuel-element integrity; therefore a low-cost method of completely inspecting incoming tubing had to be developed. Any flaw greater than 10 percent of the wall thickness (i.e., 0.04 mm) was cause for rejection, and this required a considerable improvement of ultrasonic testing sensitivity over that available in 1960. Using water immersion for sound transmission, tubes were rotated past a transducer head assembly to obtain a complete helical scan. The inspection system was a major achievement, developed mainly by the fuel fabricators – no fuel elements are known to have defected due to initial flaws in the fuel sheaths.

End Closures

The uranium dioxide must be contained within the fuel sheath and end closures. The end closures posed three problems: design, supply of components and method of sealing. The design had to consider neutron economy, displacement of uranium dioxide, neutron flux peaking,[6] effect on temperature distribution, sheath integrity and ability to withstand the stresses that could be applied during irradiation. The components had to be free of leaks, dimensionally consistent to the specification, and low cost. The designs were established, but the cost target was more difficult. Mass-production techniques had to be introduced to take economic advantage of high-volume throughput.

Of the various methods available for fixing the end caps, the fuel fabricators concentrated on welding – initially tungsten-inert gas (TIG) welding, then by resistance welding. Welding parameters had to be defined to ensure a consistent, 100 percent leak-free closure that would cope with any stresses that could be applied during irradiation. Areas of metallurgical interest included sheath thinning, projection of weld material, impurity migration to concentrate in the weld line, formation of fine cracks that could act as stress raisers, surface contamination that could result in unacceptable localized corrosion, and changes in mechanical strength of the adjacent sheathing to give a non-uniform behaviour along the length of the element. Fabrication procedures were developed, including gas evacuation, back-filling with helium or a helium/argon mixture, and specifications of the electric current required for the resistance welds. Extensive development effort led to a process suitable for

[6] Flux peaking occurs at the ends of the elements due to lower absorption of neutrons by the Zircaloy end caps and end plates, rather than by the uranium dioxide pellets in the body of the element. Such neutron-flux peaking could result in a higher localized temperature.

high-rate, low-cost production. Initially, the welds were done in an inert atmosphere, but procedures were developed to allow the weld to be done in air, thus saving time and money. The helium filling allowed inspection by helium-leak detectors to ensure 100 percent weld integrity.

Fuel Elements

From an external aspect, the finished fuel element must be within dimensional specifications to allow assembly into fuel bundles. Internally, there are also severe specifications that must be met. The fuel/sheath clearance is critical: if it is too small, the sheath will be stretched by the expansion of the hot uranium dioxide; if it is too large, the sheath may deform locally (ridges), and there will not be a good sheath-to-uranium-dioxide contact to promote efficient heat transfer. The extensive test programs defined the acceptable clearance range and, equally important, the consequences of being off specification (defects, higher uranium dioxide temperatures, more fission-gas release and dimensional instability).

Early in the fuel development program a potential problem – internal hydrogenous contaminants such as water or oil leading to sheath hydriding and failure – was recognized and dealt with: tests with swaged fuel elements showed high concentrations of hydrogen in the sheath. The hydrogen must have come from hydrogenous contaminants (probably water) inside the element, since the amount of excess hydrogen far exceeded the amount produced from the external oxidation of the sheath. Also, through its cooperative programs with American laboratories, AECL became aware of the failures of some American fuel elements. The failures were attributed to hydriding of the sheath, with the hydrogen coming from internal contaminants; in this case, oil. AECL assessed the potential for defects in CANDU fuel. Although the CANDU fuel pellets had a higher density than light-water-reactor pellets, and hence a lower probability of water being absorbed into the pellets, AECL considered it prudent to call for drying of the pellets before they were inserted into the fuel sheaths. A test program in the NRX and NRU loops showed that a hydrogen limit of 1 mg per element should be specified for CANDU fuel. This control achieved its purpose, in that no CANDU power-reactor fuel element has failed due to excess hydrogen introduced during fabrication.

The heat produced during the fissioning process must be removed by the primary coolant. With the thin collapsible sheathing there is good contact between the fuel pellet and sheath. Such contact may be widespread, or may be only at discrete points where points of roughness contact. The greater the physical contact, the more efficient the heat transfer. In the early days, to assist in the heat transfer, the fuel element was evacuated, then filled with a high-conductivity gas, helium. Later, a sizable test program elucidated the factors controlling fuel-to-sheath heat transfer, which emphasized the importance of uranium dioxide/Zircaloy contact, and the relatively minor role of the filling gas. This helped to define the demanding specifications for pellet and sheath smoothness. AECL's work alleviated concerns that fuel-to-sheath heat transfer would deteriorate during irradiation, due to degradation of the heat-transfer characteristics of the filling gas by released fission-product gases. CANDU fuel still contains helium filling gas, to allow leak detection by mass spectrometry.

FUEL BUNDLES

Requirements

In the early days, it was decided that CANDU fuel would be purchased on a performance specification. The detailed method of fabrication would be up to the suppliers of the fuel bundles. Therefore, for any specific reactor, the external appearance of the fuel bundle would be constant, but interior details of the elements and method of fabrication could be different from each of the manufacturers.

The performance specifications required the fuel bundles to:

- Be capable of producing the maximum specified power and burnup, with adequate margins to accommodate unusual conditions;
- Remain dimensionally stable, so that they can be moved along the pressure tube during on-power fueling and remain

mechanically intact during post-irradiation handling, at-reactor storage in water bays, and transportation to, and handling at, storage and disposal vaults;
- Fit the fueling machine, so that they can be loaded and unloaded without damage to the bundles or to the fueling machines;
- Maintain inter-element and element-to-pressure-tube spacings, to allow the specified flow of the heavy-water coolant;
- Have bearing pads designed to ensure that loading and movement of fuel will not wear grooves in the pressure tubes; and
- Eliminate stagnant areas of coolant that could cause local temperature variations and accelerated corrosion.

The target was to have a fuel bundle that would meet the specifications, yet not contribute more than 1 mil/kWh to the total unit energy cost. From an operational viewpoint, Ontario Hydro wanted as high a uranium content as possible, and was willing to pay a bonus for such.

Each of these demands was a separate area of study, yet they had to be considered as a whole. The development of the current bundle design took years of testing to set the specifications, with many potential designs being dropped as tests and analyses showed they were inappropriate. Some of the early irradiations were done in a test reactor at Savannah River, Georgia, before the AECL facilities at Chalk River were available.

As the design of the CANDU reactor evolved, the fuel bundle had to be modified to provide increased power. In NPD the original fuel bundle for the outer channels had seven elements. For the inner channels, where the power output was higher, a 19-element bundle was designed and developed with the same overall bundle diameter, but with smaller-diameter elements, to ensure that the central temperature and surface heat fluxes of the uranium dioxide did not reach unacceptably high values. This bundle was also used for Douglas Point. For Pickering, with its larger-diameter pressure tubes (10 cm compared with the previous 8 cm),[7] a 28-element bundle was designed, using the same size elements as for the 19-element Douglas Point bundle. For the Bruce, Darlington, and CANDU 600 reactors, with their increased power densities, the fuel bundle consisted of 37 elements, each element being slightly smaller in diameter than the Pickering fuel, but with the overall bundle diameter unchanged (see figure 13.5).

End Plates

In the bundle the end plates, and their attachment to the elements, have to give the structural integrity demanded by the loading and discharge systems, plus the stresses applied from the flowing coolant. The end plate can not be excessively thick, because of neutron economy and displacement of fuel from the channel, and it must not seriously hinder the coolant flow. The fuel bundle also has to remain intact during spent-fuel storage and handling in the storage bays. The process finally chosen[8] to join the elements to the end plates is a resistance-projection weld made under the protection of a localized argon purge. The process was designed to control the amount of molten material during the welding, with no sharp cracks or other stress raisers being tolerated. No bundles have failed in-reactor due to breakage of an end cap/end plate weld.

Spacers

The initial NPD bundles had wire wrap spot-welded to the sheath, in a helix, to give the spacing and, it was thought, to promote mixing of the coolant (see figure 13.6). However, thermalhydraulic tests using dye injections and a transparent flow tube showed that the wire wrap did not enhance the good mixing already existing in the bundle.

Also, there were indications that fretting could occur where the wire wrap touched the sheath of an adjacent element, and that accelerated corrosion could occur in the crevice between

---

7 The first in-reactor loop in NRU (the E-20 loop) had an 8 cm inner diameter pressure tube. This size, as well as the fuel developed for irradiation in E-20, was carried through to NPD and Douglas Point.

8 The original NPD 7-element bundles were riveted together, because the process for welding was not developed.

FIGURE 13.5
Fuel Bundle for Pickering Reactor

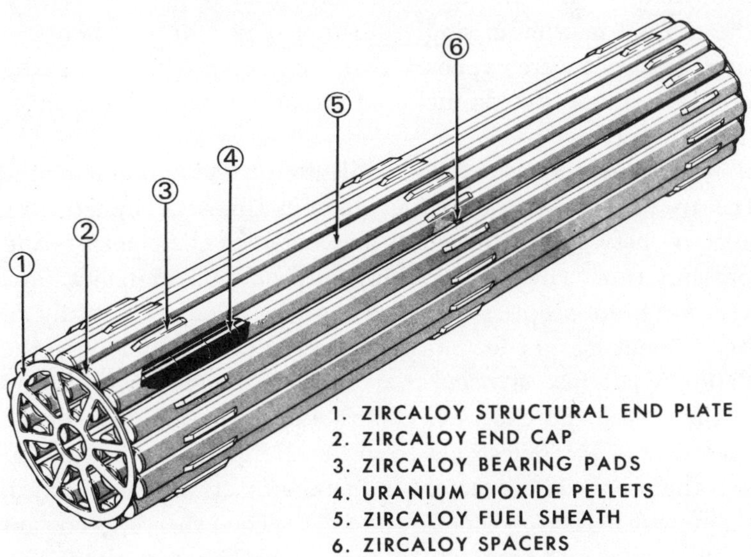

FIGURE 13.6
Example of Wire Wrap

the wire wrap and the sheath; laboratory tests also showed that a side load such as that applied by the fueling machine latch could tear off the wrap, leaving a cavity in the sheath and loose wire in the machine. This was potentially more dangerous in the thinner sheathing of the 19-element bundles, since the hole might penetrate the sheath. Therefore, there was an incentive to change the design for four reasons: to reduce the amount of Zircaloy, and to eliminate the potential for fretting, crevice corrosion and tear-off of wire wrap. (In fact, tear-off of wire wrap was never observed on the wire-wrapped bundles of the NPD power reactor.)

The decision was made to change from wire wrap to small pads as spacers. After extensive testing it was decided to use Zircaloy appendages brazed to the sheath at precise locations, and at carefully designed angles to the axis of the sheathing (see figure 13.7).

The brazing introduced problems that had to be fully assessed. The braze alloy (Zr+5 weight percent beryllium) was chosen because of its melting temperature and corrosion resistance. Beryllium is a toxic material, so special handling systems had to be installed. A second possible problem was the local change in metallurgical structure of the fuel sheath caused by the high temperature of the brazing process, which could lead to different mechanical properties and corrosion behaviour. To get away from brazing, many alternative designs and attachment methods were studied. None gave as good a performance as the brazed-on spacers. Therefore a process-control standard for the brazing procedure had to be established, since an out-of-control sequence could result in excessive erosion and alloy penetration into the sheath. A criterion was set that 90 percent of the minimum design-wall thickness must be maintained for the life of the fuel, taking into account the corrosion rate of the braze alloy. To date, no fuel is known to have defected because of problems with the appendages.

FIGURE 13.7
Bearing Pads and Spacers

Bundle Testing

The specified bundle, power and burnup depended on the reactor. In the early reactors – NPD and Douglas Point – the power output per unit length per element was low, but in later reactors it was raised. The individual elements could sustain such power demands, but the bundle configuration imposed additional requirements, particularly in the heat input to the flow channels within the bundle. When more power was demanded from the fuel (for the Pickering and Bruce reactors), extensive thermalhydraulic studies were done to ensure that the critical sub-channels in the bundles had sufficient flow to remove the heat under all expected conditions, without the sheath drying out.

AECL did not have thermalhydraulic rigs to do the required tests, so while the experimental facilities were being built the first tests were contracted out to Columbia University in New York. When built, the AECL out-reactor facilities had up to 12 MW of electric power input to simulate full-scale in-reactor conditions for a single channel in the Bruce and the CANDU 600 reactors, with sufficient electric power to go well beyond the normal maximum single-channel power. Experiments in these facilities were expensive, with costs per test series being up to $5 million, which included the fully instrumented simulated fuel string.

A major factor in the fuel-bundle design was maximizing the fuel-to-coolant ratio in the fuel channel; this necessitated close spacing between the elements and between the elements and pressure tube. The target spacing was set at 0.050 inches, (i.e., 1.27 mm), but sceptics expected many problems, especially an insufficient margin to the critical heat flux (CHF). Above the CHF, dry patches can occur that could cause severe overheating of the sheath and excessive corrosion, leading to failure of the fuel. One of the milestones from the thermalhydraulic testing was the proof via in-reactor[9] and laboratory tests (early sixties) that under normal coolant flow there was good sheath-to-coolant heat transfer and an adequate margin to CHF with the 0.050-inch spacing.

Two other areas of concern in the thermalhydraulics assessments were investigated. The first concern was whether the results from the in-reactor vertical loops were valid for horizontal pressure tubes. Out-reactor tests showed that in the horizontal channels, stratification of the coolant into hot and cold layers was not important at the flow rates of the power reactors; hence the vertical test results could be applied to CANDU. This was fully verified when the 12 MW horizontal tests mentioned above were completed. The second concern was that when the pressure tube increased in diameter due to creep, there could be preferential flow in the excess clearance, and reduced flow within the bundle, leading to a significant decrease in CHF in the critical sub-channels of the bundle. Analyses indicated that a decrease in CHF would occur, but it was small in relation to the size of the margin to CHF, and therefore there was no cause for worry.

---

[9] The AECL instrumented in-reactor tests were the first such tests in the world, and were widely praised by experts in the field.

The reactor designers predicted that boiling would occur in many channels of the Bruce reactors. Laboratory tests, followed by in-reactor loop experiments, were done under simulated reactor conditions to produce coolant boiling. Also, the NPD reactor was converted to boiling for a few months. Results showed that CANDU fuel could withstand some boiling of the coolant without deleterious effects, and that boiling in the Bruce reactors could be tolerated.

The fuel bundles would be subjected to a high-velocity, high-mass flow of coolant. Experience in such systems showed that vibration and fretting damage (due to constant impact or rubbing) could occur. AECL, jointly with Ontario Hydro, set up experiments in horizontal rigs at the Sheridan Park Laboratories of AECL to study the effect of coolant flow on the CANDU bundles, and on the bundle/pressure tube contact points. Flow rates and duration of exposure were varied, with some tests lasting several thousand hours.

Wear of spacers and bearing pads was measured by mechanical gauges and by optical methods that projected the internal spacing of the bundle onto a viewing screen for measurement. The susceptibility of the bundles to wear was determined under the various test conditions. In normal conditions the tests showed that there should not be any problems associated with vibration and fretting of the fuel bundles. However, abnormal situations can occur. For example, the coolant enters a fuel channel radially through the perforated wall of a liner tube. Usually there is no fuel at this position, except during fueling operations, when bundles pass the coolant inlet and are subjected to a cross flow. The testing program showed that a fuel bundle can withstand the short periods of exposure to cross flow during fueling; but if a fueling procedure is delayed by any of several reasons a bundle could be stuck in a cross flow for considerable time. Severe damage would occur. This was confirmed at the Pickering station where, during fueling operations, a bundle was stuck for up to twenty-eight days under cross-flow condition. Damage was indicated by an increase in radioactive iodine in the coolant. The bundle had vibrated apart and a fuel element had ruptured; special procedures had to be set up to remove the individual elements.

The flow tests and in-reactor experience emphasized the need for close interactions between fuel and system designers, to ensure that flow conditions, equipment design and bundle design were compatible, and, equally important, that the potential consequences of unusual situations were well known to reactor operators. The physical integrity of the CANDU fuel bundle has been very good in the Pickering and Bruce reactors, and for those reactors has not posed a significant problem for routine operation.

SYSTEM CHEMISTRY AND CORROSION

The chemistry of the primary heat-transport coolant was shown to have a significant impact on the behaviour of the fuel in the reactor. Parameters studied included the concentrations of oxygen and dissolved deuterium, and the pH (strictly pD) of the heavy water. Other areas of investigation included the corrosion of the out-reactor part of the primary heat-transport system, the tendency to deposit corrosion products – "crud" – on the fuel, and if deposits should occur, their effect on the behaviour of the fuel.

It was fairly obvious that excess oxygen should be suppressed to keep down the corrosion of the sheath. This was done by additions of deuterium gas to the coolant. However, the amount added had to be strictly controlled. With high concentrations of deuterium in the coolant, the extent of deuterium absorbed by the Zircaloy sheathing was found to be more than expected, and certainly much more than wanted.[10] A balance had to be struck; the compromise was to have between 0.5 to 2 ppm of dissolved deuterium gas per kilogram of coolant. At this level the oxygen concentration was maintained at less than 5 ppb; corrosion was minimal and at the end of irradiation (average of two years) the hydride concentration was in the acceptable range of zero to 20 ppm deuterium by weight in the sheath (equivalent to 10 ppm hydrogen). With much more deuterium gas in the coolant the deuterium concentrations reached an unacceptable level of up to 120 ppm.

[10] The deleterious effects of hydrogen (deuterium) in zirconium alloys is discussed in chapter twelve.

The coolant pH also had to be kept under strict control. The out-reactor piping of the primary heat-transport system was mild steel. At a neutral pH this material would corrode, but at pH-10 the corrosion rate was acceptably low. Tests were done to determine what additions could be added to maintain a pH of 10, yet have no unacceptable effect on the fuel. The demand was for a strongly basic system; candidates were lithium, sodium and potassium hydroxides. Activation products from sodium and potassium were not acceptable; hence lithium hydroxide (LiOH) was chosen at a concentration of 0.5 to 2 ppm Li by weight in the coolant.

Other tests showed that care had to be taken in the fuel and channel design to eliminate the possibility of LiOH concentrating in an area to produce a pH of greater than 12. At this pH Zircaloy corrosion will increase rapidly. This circumstance occasionally occurred in the power reactors, where there were sufficiently high concentrations of LiOH between the bearing pads and the pressure tubes to cause crevice corrosion. Pits were observed on the inside of NPD pressure tubes and on the surfaces of bearing pads. Alternative designs of the bearing pads were tested, and modifications were made to minimize such corrosion.

In the early stages of the fuel development, the fuel bundles were autoclaved to give a shiny black protective oxide coating, and as a proof test to show that the material had the expected corrosion resistance. Many bundles showed a layer of white zirconium oxide, which is a thicker, less protective oxide layer. These bundles were rejected for power-reactor use because of the worry that accelerated corrosion could jeopardize the integrity of the sheath. This became a significant economic penalty, so a program was launched to determine the consequences of such a layer on fuel-bundle performance. Some of the bundles were irradiated in loops; when no problems arose others were inserted into the power reactors. The bundles operated properly, with no adverse effects. Hence the quarantined bundles were declared suitable for irradiation. As a consequence of these results it was decided that the autoclaving proof test was no longer needed and autoclaving was dropped from the fuel-bundle fabrication. This cost-saving action proved to be a good decision; there have been no defects from corrosion of fuel bundles in any CANDU power reactor.

A minor corrosion concern was noted in bundles from NRU loop experiments, where rings of white corrosion products were observed on the transition zone between the regular sheathing and the material in the appendage area affected by the heat from brazing. These rings were caused by the temperature-instigated migration of impurities that concentrated on the surface of the sheathing, and the grain structure of the Zircaloy within the transition zone. Considerable effort was put into braze sequences and other remedies. However, experience from power-reactor fuel showed that the corrosion remained as a surface effect, and did not affect the performance of the fuel; therefore, no remedial action was necessary.

FUEL PERFORMANCE

The experimental program established allowable parameters for successful operation of the CANDU fuel bundle under normal operating conditions. The low power-level-per-unit length of NPD fuel elements was raised in later reactors as confidence and understanding was gained, and ambitious power-level specifications were set, particularly in comparison with light-water-reactor fuels.

A few defects occurred in Douglas Point, and in the early seventies in the Pickering reactors the bundle defect rate during operation rose from about 0.1 percent to almost 1 percent (for elements, the defect rate was a factor of 10 lower). The reactors did not have to be shut down due to radioactivity levels, but it was close. Examinations showed that the defects were small cracks through the sheath. Discussions with nuclear organizations in other countries revealed that similar defects had been found. International discussions occurred, with a relatively free flow of information on the defects, but some results on promising remedies were withheld because of proprietary interests.

AECL launched an extensive program to determine the cause of the failures, and to put in place a remedy. Metallurgists, chemists, physicists, engineers, reactor operators, fuel fabricators and anyone else who could contribute were assembled into a team. Analyses proceeded in four areas:

1 Immediate cause: the cause of failure was shown to be a significant increase in power after a prolonged period at lower power; e.g., power increases resulting from bundle shifts into the central regions of the core, or insertion of fresh fuel into adjacent channels. In the United States, the defects were attributed to a rapid power increase, but AECL tests showed that for CANDU the defects could occur with power increases taking several days to complete (see figure 13.8). Also, AECL analyses showed that a short duration at the higher power level and a return to a lower level did not result in defects; for example, during on-power fueling some irradiated bundles were moved temporarily into high flux positions and then moved to lower power positions, without defecting.

2 Remedial action: Ontario Hydro was able to restrict the magnitude of power changes as a result of refueling for those bundles that were to remain in high-power positions. This caused some loss of flexibility in reactor operations, but was accommodated without undue concern.

3 Defect mechanisms: Considerations focused on four mechanisms: failure of the sheath due to inappropriate starting mechanical properties, localized stress concentrations over cracks to produce tensile failures, embrittlement due to irradiation damage, and stress-corrosion cracking (SCC) from the inside of the sheath (see figure 13.9). Certainly, all the requirements existed for SCC: a material susceptible to SCC, firm contact between the fuel and sheath, stress concentrations over cracks in the uranium dioxide pellets, and the presence of iodine which is known to cause SCC in zirconium alloys.

4 Potential solutions: AECL brainstormed the defect problem for potential solutions. Suggestions included modifications to the pellet length and shape, changing the pellet/sheath clearance, hollow pellets, different sheath materials (Zr-2.5Nb or a zirconium-chromium-iron-tin alloy called Ozhennite), elements internally pressurized with helium, annealed or beta heat-treated Zircaloy sheathing, low-density pellets, installation of graphite discs between the pellets, and the deposition of a graphite – or later a siloxane – interlayer on the inside of the sheath.

FIGURE 13.8
Graph of Power Ramp Defects

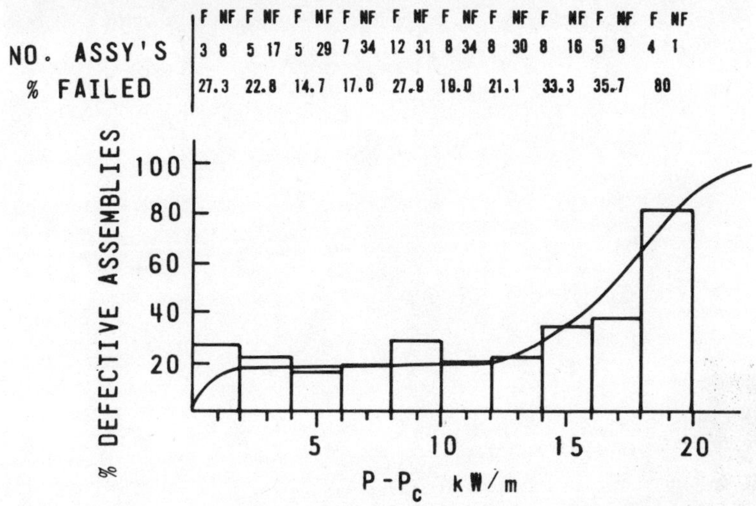

The deposition of a graphite layer was suggested because extensive laboratory tests had shown that a graphite lubricant between the pellets and sheath greatly reduced the localized strain and hence the stress concentrations over pellet cracks, thereby reducing the potential for SCC. Also, graphite was known to trap iodine, as evidenced by the use of charcoal filters in gas masks and for ventilation systems.

Seventeen different design modifications were tested. The results showed that the mechanical properties of the sheath and embrittlement were not a factor, since even fuel annealed in a hot cell between the low-power soak and the high-power test defected. And, elements with alternative sheathing defected. Four types of modified elements, namely those with a graphite interlayer on the inside of the sheath, a siloxane interlayer, graphite discs between the pellets, and low-density pellets, survived the power ramps. Fuel incorporating the graphite

FIGURE 13.9
Stress Corrosion Cracking in Fuel Element Sheath

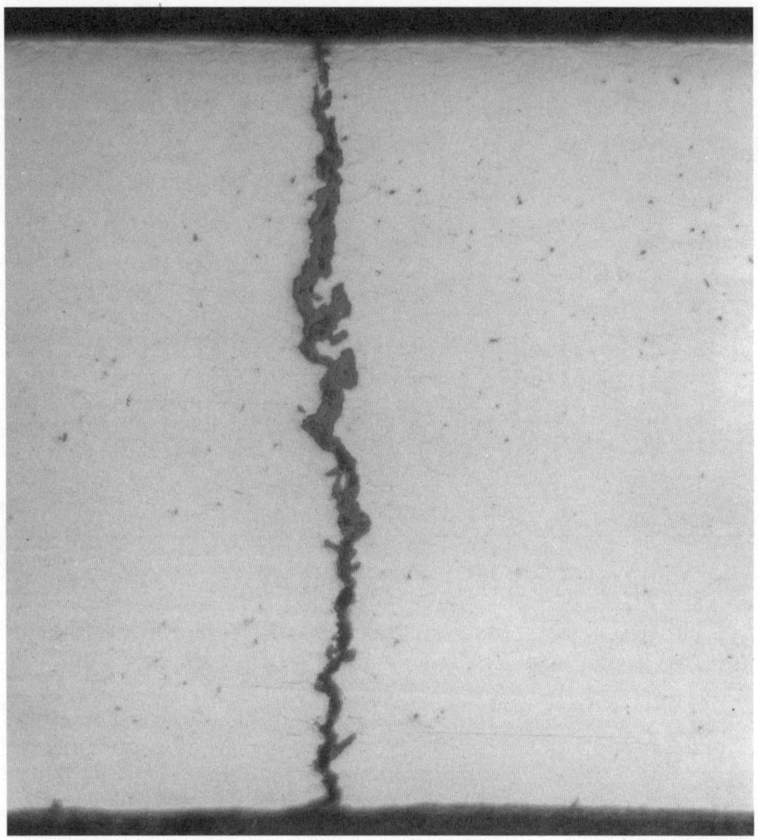

interlayer (referred to as CANLUB fuel) was selected for production, because:

1 it had been given the most severe power ramp,
2 it had a very low adverse effect on the neutron economy of the fuel, and
3 it would increase fabrication costs very little.

The success of the graphite interlayer to reduce the stress concentrations and/or to absorb iodine indicated strongly that the defects were due to SCC from the inside of the sheath. Support for SCC being the defect mechanisms was obtained from an irradiation in which fresh uranium dioxide pellets had been inserted into irradiated sheaths. No defect occurred under irradiation conditions that caused failures in regular fuel elements.

The graphite interlayer – CANLUB – is now a standard component of CANDU fuel.

Restrictions on magnitude of power changes and the CANLUB layer have essentially eliminated power ramp defects from current CANDU nuclear reactors. The success of this investigation is a fine example of inter-organization cooperation within the Canadian nuclear industry. From the appearance of the problem in 1971, it took the incredibly short time of two years to introduce modified fuel into the Pickering power reactors.

In the late sixties, a Zircaloy-sheathed uranium-dioxide fuel element that had defected in one of the loops of the NRU reactor was being routinely handled in the hot cells. The operator picked it up by one end, and the element broke in two. This caused great concern, and detailed examinations were started to determine what had happened to cause the sheath to be so drastically weakened.

The cause was found to be localized hydriding of the sheath. Water had entered the element via the defect and reacted with the uranium dioxide to produce a higher oxide of uranium, and hydrogen. Some of the hydrogen reacted with the sheath to form concentrated areas of zirconium hydride, which is brittle at room temperature. The cross-sectional appearance was like rays emanating from the sun – hence they were called "sunburst" defects. The location of the sunburst defects could be recognized from the exterior of the sheath because of a blister-like appearance – hence the descriptive phrase "blister" defects (see figure 13.10).

Hydrogen diffusion experts within AECL were immediately consulted. Explanations were developed that described the absorption of hydrogen, the solubility and diffusion of hydrogen in the Zircaloy, the precipitation of zirconium-hydride platelets in stress fields, and the formation of a solid zirconium-hydride zone. Zirconium hydride has a much lower density than Zircaloy; hence the increase in volume and the blister-like appearance.

As a result of such defects, an extensive program was directed at determining the behaviour of defected fuel in the reactor, and

FIGURE 13.10
Sunburst and Blister Defects on the Inside Surface of a CANDU Fuel Sheath

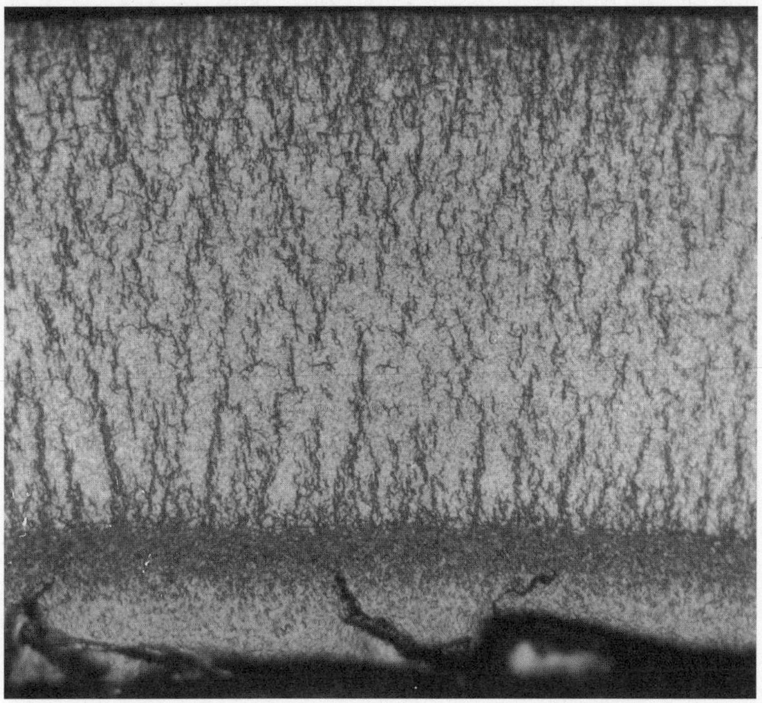

during storage in the water-filled bays. The objective of the program was to provide the reactor operators with information on how long they could leave defected fuel in the reactor without causing unacceptable problems of fission-product release to the coolant, sheath or bundle disintegration in the reactor, and potential problems during discharge and storage. The test program showed a direct correlation between sheath deterioration, power output per unit length of element, and the duration of exposure after the defect occurred. At low-power outputs, there was very little release of radioactivity to the coolant, and no noticeable deterioration. At higher powers, the activity release increased significantly, and sheath deterioration could occur. The results of the program showed that even for a CANDU bundle operating at its maximum power there is sufficient time (several days) for the operators to detect and identify the defective bundle and discharge it before there is any deterioration that could cause a large release of fission products or threaten bundle integrity.

A small but significant point about the power-reactor fuel is that the first-core fuel bundles were always developed and delivered on time to the original reactor schedule; availability of fuel never caused a delay, even with strikes at the fuel manufacturers.

URANIUM SILICIDE FUEL

In the late sixties and early seventies, to obtain higher uranium content and lower fueling costs, much effort was expended examining uranium silicide as a potential fuel for CANDU reactors. As the result of a large development program, it was determined that the resistance to corrosion and swelling under normal operating conditions was acceptable, but post-irradiation tests showed that the expected swelling under some accident conditions would be unacceptable. Much of this early work was used as the basis for study in the early eighties to perfect a uranium silicide-based fuel to satisfy low-enrichment uranium-fuel requirements in research reactors, because of non-proliferation demands: namely, research reactors must be fueled with material with less than 20 wt. percent uranium-235 in the uranium content. The current concept is uranium silicide or uranium silicide dispersed in an aluminum matrix, with co-extruded aluminum sheath. There has been a successful demonstration in NRU (see chapter two).

SUMMARY, AND FUTURE TARGETS

CANDU fuel is the result of a cooperative program of many groups within the Canadian nuclear industry, combined with knowledge gained from international sources. The product was new, and had to be developed essentially from scratch, starting in the mid-fifties. Scientific knowledge, combined with engineering skill and good experimental facilities, led to a set of specifications for the fuel bundle. Fabrication expertise was accumulated by private industry, along with extensive power-reactor operating experience by the utilities.

Interactions were two-way and quick. Many problems were anticipated and eliminated during the development program; when problems did occur in a CANDU reactor, members of the fuel team worked together to determine why the problem occurred and how to solve it for the short and long term.

By the mid-sixties the reference CANDU fuel bundle design was firmly established and a high uranium/zirconium ratio of >0.9 was achieved. Fabrication specifications had been written, efficient methods of manufacture were established at the fuel suppliers, and the cost target was met. The irradiation performance has been good. The introduction of the CANLUB coating combined with modified fueling procedures has ensured that the number of defects in the CANDU power reactors remains at a low, acceptable level. As of mid-1991, about 750,000 bundles had been inserted into the Pickering, Bruce, Gentilly-2, and Point Lepreau reactors; of these, about 600 have defected, for a defect rate of less than 0.1 percent.

Development work continues, particularly for higher burnup and for the use of alternative fuels, such as those enriched with plutonium, or based on thorium.

ACKNOWLEDGMENTS

Acknowledgment is made to J.R. MacEwan, R.D. Page, A.M. Aikin, R.D. MacDonald, A.J. Mooradian and I.J. Hastings: fuel experts who reviewed progressive drafts of this chapter, and whose comments and additions are incorporated into this final version.

BIBLIOGRAPHY

Bain, A.S., and J.E. LeSurf. *Oxidation and Hydriding of Zircaloy in the NPD Reactor.* AECL-3065. 1969.

Bain, A.S., and J.A.L. Robertson. *UO_2 Irradiations of Short Duration.* AECL-806. 1959.

Bain, A.S., J.A.L. Robertson, and A. Ridal. *UO_2 Irradiations of Short Duration. Part II.* AECL-1192. 1961.

Bain, A.S., J.C. Wood, and C.E. Coleman. "Fuel Designs to Eliminate Defects on Power Increases." In *Nuclear Fuel Performance: Proceedings of the International Conference, held on 15-19 October 1973 at the Institution of Civil Engineers, London,* 56.1-.5. London: British Nuclear Energy Society, 1973. (Also AECL-4518.)

Belle, J., and B. Lustman. *Properties of UO_2.* Report WAPD-184. Pittsburgh, Pa.: Westinghouse Electric Corporation, Bettis Plant, 1957.

Chalder, G.H., N.F.H. Bright, D.L. Paterson, and L.C. Watson. "The Fabrication and Properties of Uranium Dioxide Fuel." In *Proceedings of the Second United Nations International Conference on the Peaceful Uses of Atomic Energy, Geneva, 1-13 September 1958,* vol. 6, 590-604. Paper A/CONF.15/P/192. Geneva: United Nations, 1958. (Also AECL-602.)

Fanjoy, G.R., and A.S. Bain. "CANDU Fuel: Fifteen Years of Power Reactor Experience." In *Nuclear Power and Its Fuel Cycle: Proceedings of an International Conference on Nuclear Power and Its Fuel Cycle, Salzburg, 2-13 May 1977,* vol. 2, 595-606. Paper IAEA-CN-36/184. Vienna: International Atomic Energy Agency, 1977. (Also AECL-5711.)

Hastings, I.J., T.J. Carter, R. daSilva, P.J. Fehrenbach, D.G. Hardy, and J.C. Wood. *CANDU Fuel Performance: Influence of Fabrication Variables*, 60-131. AECL-MISC-250 (rev.1). 1984. (Presented at IAEA/CNEA International Seminar on Heavy Water Fuel Technology, Session 3, San Carlos de Bariloche, Argentina, 27 June – 1 July 1983.)

Hastings, I.J., T.J. Carter, G.M. Macgillivray, R.D. MacDonald, and J. Judah. "Canadian High Burnup Fuel Experience." In *Proceedings of the Canadian Nuclear Society 6th Annual Conference,* 6.26-.32. P.M. French, G.J. Phillips, eds. Toronto: Canadian Nuclear Society, 1985. (Also AECL-8772.)

Haywood, L.R., J.A.L. Robertson, J. Pawliw, J. Howieson, and L.L. Bodie. "Fuel for Canadian Power Reactors." In *Peaceful Uses of Atomic Energy: Proceedings of the Fourth International Conference on the Peaceful Uses of Atomic Energy, Geneva, 6-16 September 1971,* vol. 8, 183-99. Paper A/CONF.49/P/156. New York: United Nations; Vienna: International Atomic Energy Agency, 1972. (Also AECL-3979.)

Lewis, W.B., M.F. Duret, D.S. Craig, J.I. Veeder, and A.S. Bain. "Large scale Nuclear Energy from the Thorium Cycle." In *Peaceful Uses of Atomic Energy: Proceedings of the Fourth International Conference on the Peaceful Uses of Atomic Energy, Geneva, 6-16*

September 1971, vol. 9, 239-53. Paper A/CONF.49/P/157. New York: United Nations; Vienna: International Atomic Energy Agency, 1972. (Also AECL-3980.)

MacDonald, R.D., and I.J. Hastings. *Graphite Disc-Oxide Fuel Elements for High Powers and Extended Burnup*. AECL-8627. 1984. (Presented at IAEA Specialists' Meeting on Improved Utilization of Water Reactor Fuel with Special Emphasis on Extended Burnups and Plutonium Recycling, CEN/SCK, Mol, Belgium, 7-11 May 1984.)

MacEwan, J.R., A.S. Bain, M.J.F. Notley, and R.W. Jones. "Irradiation Experience with Fuel for Power Reactors." In *Peaceful Uses of Atomic Energy: Proceedings of the Fourth International Conference on the Peaceful Uses of Atomic Energy, Geneva 6-16 September 1971,* vol. 10, 239-56. Paper A/CONF.49/P/158. New York: United Nations; Vienna: International Atomic Energy Agency, 1972. (Also AECL-3981.)

Notley, M.J.F., and I.J. Hastings. *A Microstructure-Dependent Model for Fission Product Gas Release and Swelling in UO_2 Fuel*. AECL-5838. 1979. (Presented at the IAEA Specialists' Meeting on Fuel Element Performance Computer Modelling, Blackpool, UK, 13-17 March 1978.)

Notley, M.J.F., and J.A.L. Robertson. "Zircaloy-UO_2 Failures Tied to Fluorides." *Nucleonics* 19 (March 1961): 77-9. (Also AECL-1218.)

Page, R.D. *Canadian Power Reactor Fuel: Nuclear Power Symposium, May 1974.* [Toronto: Canadian Nuclear Association; Atomic Energy of Canada Limited, 1974.] (An updated version was issued by AECL in 1976 as AECL-5609.)

Page, R.D., D.G. Hardy, A.J. Mooradian, J. Howieson, G.R. Fanjoy, and D.B. Nazzer. "Engineering and Performance of UO_2 Fuel Assemblies." In *Proceedings of the Third International Conference on the Peaceful Uses of Atomic Energy, Geneva, 31 August-9 September 1964,* vol. 10, 65-74. Paper A/CONF.28/P/18. New York: United Nations, 1965. (Also AECL-2018.)

Robertson, J.A.L. $kd\theta$ in Fuel Irradiations. AECL-807. 1959.

Robertson, J.A.L., A.S. Bain, A.H. Booth, J. Howieson, R.F.S. Robertson, and W.G. Morison. "Behaviour of Uranium Oxide as a Reactor Fuel." In *Progress in Nuclear Energy,* ser. 5, *Metallurgy and Fuels,* vol. 3, *Basic Materials and Phenomena,* 278-320. New York: Pergamon Press, 1961. Originally published in *Proceedings of the Second United Nations International Conference on the Peaceful Uses of Atomic Energy, Geneva, 1-13 September 1958,* vol. 6, 655-76. Paper A/CONF.15/P/193. Geneva: United Nations, 1958. (Also AECL-603.)

Wood, J.C., D.G. Hardy, and A.S. Bain. *Power Ramping Performance of Water Reactor Fuel*. AECL-6676. 1979.

Chapter Fourteen

Heat Transport System

H.K. RAE

KEY CONSIDERATIONS

From a chemical and metallurgical viewpoint, the materials comprising the heat transport system (HTS) and their interaction with the heavy water circulating through it are important. From a safety viewpoint, the integrity of the pressure boundary and the maintenance of an adequate flow of heavy water through the fuel channels are paramount. From an economic viewpoint, the heavy-water inventory, heavy-water leakage and component reliability are the important factors. From a thermodynamic viewpoint, the heat removal from the fuel and the efficient operation of the steam generators are the essential purposes of the HTS. The designer must integrate all these and many other considerations into the overall specification of the HTS.

The main materials that contact the heavy water in the HTS are carbon steel (piping and vessels, such as the pressurizer), stainless steel (end fittings, pumps, and many of the valves), zirconium alloys (fuel sheaths and pressure tubes) and nickel alloys (steam generator tubing). The last is the dominant contact area. A variety of nickel alloys have been used for steam generator tubing in successive CANDU designs, the selection of each representing an evolution in improving overall system performance. The corrosion behaviour of these materials depends on the chemistry of the water – its purity, acidity or alkalinity (that is, its deuterium ion (D_30^+) concentration or pD[1]) and oxygen content. The latter is influenced by the extent of radiolytic decomposition of the water when it passes through the reactor core. An important part of the HTS is a purification circuit for chemistry control and to remove unwanted materials from the water. Two aspects of corrosion are important: adequate allowance in design must be made for the amount of metal (wall thickness) consumed by corrosion during the expected reactor life (e.g., 30 years), and the release of corrosion products to the water should be small, since corrosion products (metal oxides or ions) in the water may deposit on the fuel to impede heat transfer or coolant flow and to produce radioactive species (^{59}Fe, ^{58}Co, ^{60}Co, ^{51}Cr, etc.), which can then move to other parts of the HTS. During reactor shutdowns the radiation fields from these radioactive corrosion products may hamper maintenance work and be the dominant source of radiation exposure for station staff.

The HTS operates at high pressure, typically 10 MPa. The pressure boundary of the HTS is one of the principal barriers to any release of fission products from

[1] pD is analogous to pH and equal to -log C where C is the concentration of deuterium ion in moles per litre. pD will be used when discussing heavy-water chemistry, while pH will be used in discussing the chemistry of loop experiments where light water is used.

the reactor core to the public, and its integrity must be as high as practicable. This integrity is achieved through design, component selection, care in fabrication and construction, and painstaking analysis and quality assurance at all stages.

The reactor control system, fuel design and heavy-water flow rate determine the heat removal rate from the fuel. The efficient transfer of this heat to raise steam in the steam generators is dependent on their design and fabrication, and on the maintenance of adequate chemistry control of both the heavy-water coolant and the steam generator water. The potential for reduced heat transfer in the steam generator due to deposits of corrosion products, and for tube failure by corrosion or vibration, are concerns.

Reactor operation is dependent on a reliable HTS pumping system. The main area of concern in pump performance is the frequency of shutdowns due to pump shaft seal failure.

Economic power from CANDU demands that heavy-water losses be extremely low. The target for a large reactor is an unrecovered loss rate of less than 0.5 kg/h. At the 1985 value of $300/kg this target would contribute no more than 0.03 cents/kWh to the cost of power. Heavy-water leakage depends on system design, component selection, fabrication, construction, quality assurance and maintenance. The loss rate is dependent on the leakage rate and on the provision of a variety of heavy-water recovery systems and their operating efficiencies. In the recovery operation, mixing of ordinary water and heavy water will occur, and the recovered water must be reconcentrated to reactor concentration, 99.8 percent heavy water; this is called upgrading. Thus, the total cost of managing heavy-water containment, referred to as the heavy-water upkeep cost, is the sum of the costs of the recovery and upgrading operations plus the cost of the heavy-water make-up to replace losses. The target in 1985 dollars is an upkeep cost of less than 0.06 cents/kWh.

Although the HTS contains only about a third of the reactor heavy-water inventory, it is by far the major source of heavy-water leakage. The moderator system, containing the rest of the heavy water, is at low pressure and is only an important source of heavy-water loss when an occasional major acute leak occurs.

A final key consideration is the presence of tritium in the heavy water. Although the rate of neutron capture by deuterium is very low, it nonetheless results in a steady production of tritium, which accumulates in the heavy water as TDO molecules. Tritium disintegrates with a twelve-year half-life to helium-3 and emits a very low energy electron (beta particle). It is not a source of external radiation dose. However, since it occurs as part of a water molecule, it is easily absorbed into the human body either through contact with the skin or the lung tissue (see chapter three), and then it is a source of internal radiation dose. Thus, reactor operators and maintainers must avoid contact with heavy water containing significant amounts of tritium, and exposure to air containing such heavy-water vapour; the design of reactor building ventilation systems should prevent air-borne tritium being dispersed beyond those areas where heavy-water equipment is located.

Many aspects of the HTS have had important R&D input. Those discussed in this chapter are heavy-water upkeep, tritium control, corrosion and corrosion-product movement, pump seal reliability, and steam generator performance. The thermalhydraulic limits to fuel bundle power are discussed in chapter thirteen. Those safety aspects of the HTS that have required a major R&D program are discussed in chapter fifteen.

EARLY EXPERIENCE WITH WATER CHEMISTRY AND CRUD

In the early fifties, the operation of the loops in NRX for fuel testing gave the first opportunity to study HTS chemistry. Indeed, problems with high concentrations of iron oxides in the water and the formation of thick iron oxide deposits on the fuel-element surfaces in these early tests made it obvious that water chemistry and control of corrosion needed attention. A loop chemistry program began.

Tests with neutral water in these stainless-steel loops resulted in oxide deposits on the fuel that were typically 0.1 mm thick at the end of a month. This raised concerns about possible overheating of the fuel sheath in a long irradiation because of the low thermal conductivity of the oxide deposit, and about possible blockage of the narrow passages between fuel elements in a bundle (1.3 mm) resulting in local overheating. These deposits were chiefly composed of magnetite, Fe_3O_4, with some hematite (Fe_2O_3), nickel oxide (NiO) and chromia (Cr_2O_3). Magnetite is the stable oxide of iron in water in the absence of oxygen, so that the presence of hematite indicated the presence of oxygen and this was confirmed by the water analysis.

The radiolytic decomposition of water had already been defined reasonably well (see chapter seven). It produces four main species: molecular hydrogen, hydrogen peroxide, and two free radicals: atomic hydrogen and hydroxyl. At high temperature the peroxide immediately decomposes to oxygen and water, especially at high pH. The free radicals rapidly react with molecular hydrogen and oxygen (which are in solution in the water) in such a way that water is reformed and the free radicals are regenerated. Once a low concentration of free radicals accumulates, the rate of recombination of molecular oxygen and hydrogen equals the rate of water decomposition. The result is a small steady-state concentration of molecular hydrogen and oxygen, as well as a much lower steady-state concentration of free radicals, and no net decomposition of water.

Thus, the oxygen concentration can be controlled by adding excess hydrogen. Recombination reactions will produce an excess of hydrogen radicals, which will react with the oxygen as soon as it forms to keep its concentration at a very low steady-state level. This procedure was instituted in the loops, with the excess hydrogen concentration being controlled in the range from 0.5 to 2 mg/kg of water. The oxygen concentration was then found to be below the limit of detection and no hematite was formed. Hydrogen was added periodically as a compressed gas. If the loop leakage rate was low enough, much of the excess hydrogen would be supplied by the normal corrosion rates of the surfaces, where metal reacting with water produces oxide plus hydrogen gas that dissolves in the water.

However, oxygen control alone did not solve the "crud" problem. The term crud has been used in the nuclear industry to denote the oxide deposits on the fuel sheaths and the oxide particles in the water. It is said, apocryphally perhaps, that the origin of the use of the word crud to describe these deposits was in the initial informal reports of the examination of the first fuel test at Chalk River for the United States navy that referred to a "Chalk River unidentified deposit". Despite the navy's undoubted predilection for acronyms, it is equally plausible that the originator of the use of crud in this context was simply aware that one of the accepted definitions of the word is "a deposit or coating of an impure or alien substance".

It was clear that although the corrosion rate of stainless steel in the loops was quite acceptable from the viewpoint of metal wastage, the release of corrosion products to the water was too large. It was known that alkaline conditions would give less corrosion, and that it should reduce the solubility of magnetite and reduce the stability of colloidal particles that were an important component of crud under neutral conditions. Tests at a potassium hydroxide concentration of 4 mg K/kg, or 10^{-4} moles per litre (equivalent to a pH of 10), resulted in fuel elements virtually free of any deposit. Later, the potassium hydroxide was replaced by lithium hydroxide to avoid producing radioactive potassium-40 in the water.

In another early test, potassium sulfate at a concentration of 2×10^{-4} moles per litre was used. While it was not as effective as the hydroxide, the sulfate did reduce deposition substantially. Although it was tested as an alternative reagent that would reduce colloid stability, it now seems likely that its effectiveness was due to increasing the pH, since hydrolysis of sulfate ions to bisulfate ions would produce alkaline conditions.

The loops were provided with sidestream purification systems where the water could be degassed and filtered, and soluble impurities (ammonia, carbonate, silicate, etc.) could be removed by ion exchange. Using a mixed-bed ion exchanger with the cation resin saturated with lithium ions and the anion resin in the hydroxyl form helped to control the pH, so that additions of lithium hydroxide were needed less frequently. The experience gained in developing the loop purification systems formed the basis for specifying the systems needed for CANDU reactor HTS chemistry control.

An important experiment in 1955 to investigate the mechanism of crud deposition involved switching water chemistry from alkaline to neutral to alkaline for different periods of the same fuel irradiation. After a period at pH 10, the fuel was examined and found to be clean. After a month at neutral (pH 7) conditions, a thick crud deposit had formed. A further month of irradiation at pH 10 resulted in the fuel returning to its clean condition. Thus, crud deposition was shown to be reversible. At the time, the main mechanism for deposition was believed to be the radiation-induced flocculation of colloidal particles, and the benefit of increased pH was thought to be mainly a much lower colloidal particle concentration. Reversibility was attributed to hydraulic forces removing particles from the deposit.

A decade later, when detailed studies of oxide solubility as a function of temperature and pH began to be done at various laboratories, it seemed more likely that the main mechanisms for crud release and deposition were dissolution and precipitation. Not only was magnetite much less soluble at pH 10 than at pH 7, but at pH 10 the solubility increased with increasing temperature, while at pH 7 solubility exhibited the opposite change with temperature and a much larger change. So at pH 7 the fuel surfaces, the hottest area in the loop, corresponded to the lowest iron solubility, and magnetite precipitated here while tending to dissolve at all other surfaces. At pH 10 the lowest solubility occurs at the coldest point, the out-reactor cooler, and very little iron deposits in-core. When fuel covered with iron oxide was placed in a loop at pH 10, this crud dissolved and moved to the cooler surfaces.

For CANDU, the choice had been made during the design of NPD to use carbon steel piping. For the original NPD this was justified on the basis of the relatively low area of piping involved, the good corrosion performance of carbon steel if water chemistry was adequately controlled, and the cost savings relative to stainless steel. When NPD was re-designed as a pressure tube reactor, the area of pipe walls became more important – 20 percent of the total heat transport system area. However, the choice of carbon steel was retained. At pH 10 in the absence of oxygen, its long-term corrosion rate was not much higher than that of stainless steel, and it was free of any possibility of intergranular corrosion; with more extensive piping the potential cost saving was even more attractive.

For the startup of NPD there was concern that crud concentrations might be very high initially in a new system, especially if rust had formed during construction. What was needed was a means to condition the HTS to produce a stable magnetite film before the fuel was loaded. Tests were made in an out-reactor loop at Chalk River, to which large areas of fresh carbon steel and rusty carbon steel were added. It was found that with oxygen-free pH 10 water, the conditioning process could be completed in a few days at 150°C, and crud concentrations in the water fell to less than 0.01 mg/kg.

Before installation, the carbon steel components at NPD were either chemically or mechanically cleaned. The conditioning process outlined above was successfully completed using the main circulating pumps to raise the water temperature to 150°C. Then the fuel was loaded into the reactor. Crud concentrations were always low in the NPD HTS, and deposits on the fuel extremely small. Variants of this conditioning procedure have been used for all CANDUs, with similar excellent results.

Around 1960, experience with the initial operation of the NRU heavy-water system (see chapter two) gave the first indication of the importance of oxide solubility in controlling the movement of corrosion products in a non-isothermal recirculating water system. In this case, the oxide of interest was alumina formed on the surface of the aluminum fuel sheaths in NRU. A large concentration (60 mg/kg) of colloidal alumina accumulated in the heavy water and ultimately was controlled when a small evaporator was installed to remove the colloidal material on a continuous basis. Laboratory studies and analysis of the NRU data suggested that the oxide on the fuel sheath was dissolving and the aluminate ions so formed were precipitating from the heavy water when it was cooled before returning to the reactor. The aluminate ions precipitated as gibbsite, a form of alumina of lower solubility than the corrosion product on the sheaths. To confirm this mechanism, NRU was operated for two months at a pD of 5.7 by adding nitric acid. This deuterium ion concentration ($2x10^{-6}$ moles per litre) corresponded to the minimum solubility of alumina, and the gibbsite formation rate was observed to decrease substantially.

We normally think of metal oxides like alumina or magnetite as being insoluble. Indeed. their solubilities are low – of the order of 0.01 mg/kg. But the flows in a large reactor are very high, so that even a small difference in solubility between the reactor inlet and outlet temperatures offers the potential to transport hundreds of kilograms of iron per year. Once this was recognized, the case for stringent chemistry control was compelling.

ACTIVITY TRANSPORT

The selection of a lithium deuteroxide (LiOD) concentration of 2.5 mg per litre (0.7 mg Li/L)[2] and a negligible oxygen concentration (1 to 5 mg/kg added deuterium) as the appropriate water chemistry conditions for the CANDU HTS was based largely on NRX loop experience. However, there were also extensive corrosion tests of the HTS materials, which confirmed both low corrosion rates and low rates of release of corrosion products from the surfaces under these conditions. Carbon steel, stainless steel and Zircaloy are common to all CANDUs; for Pickering Unit 3 and all later reactors the pressure tubes are a zirconium-niobium alloy. The steam generator tube material has varied from Inconel[3] at NPD, Monel at Douglas Point and Pickering, Inconel-600 at Bruce to Incoloy-800 at the CANDU 6 stations and Darlington. Up to the mid-eighties, all these materials have shown satisfactory corrosion performance in CANDUs, with one exception. After about ten years of service, Zircaloy pressure tubes experienced a period of greatly accelerated corrosion[4] accompanied by absorption of much of the deuterium thus generated. As described in chapter twelve, this led to a tube failure by hydride cracking and the removal of all Zircaloy pressure tubes from service by the mid-eighties. In recent years, steam generator tube failures due to corrosion (pitting and stress-corrosion cracking) have occurred.

Although corrosion product concentrations in the heavy-water coolant have nearly always been very low in CANDUs, typically 0.01 mg/kg or less, and deposition on the fuel has been low, there has been nonetheless a constant movement of corrosion product particles and ions from surface to surface throughout the HTS. Although small in mass, this transport of corrosion products can cause a significant problem, due to radioactive nuclides being produced by neutron capture while the corrosion products are resident on the fuel sheath surface. Typical nuclides are iron-59 from iron-58, cobalt-58 from nickel-58, chromium-51 from chromium-50 and cobalt-60 from cobalt-59. They are transported to and accumulated on the inner surfaces of the HTS piping, steam generator tubing and other components. This process is referred to as "activity transport".

Because these nuclides are gamma-ray emitters with half-lives varying from a month to five years, they are the major contributors to radiation fields around HTS equipment during reactor shutdowns. Thus, they may impede maintenance and are a major source of potential radiation exposures of station staff.

The predominant contributor to these radiation fields after the first year or so of operation is cobalt-60. It has a 5.3-year half-life and emits penetrating gamma rays (1.2 and 1.3 MeV). Cobalt in the HTS arises from two sources: high-cobalt alloys used for hard facings on valve seats, seals and bearings, and cobalt impurities in steels and nickel alloys. In the later CANDUs, new hard-facing materials have replaced most of the high-cobalt alloys.

Radiation fields due to activity transport were moderate at NPD. However, at Douglas Point a very severe problem arose when activity transport created extremely high radiation fields in

---

2 This is 10^{-4} moles of lithium deuteroxide per litre of heavy water, or a pD of 10.9 when measured at 25°C, equivalent to a pH of 10 for this concentration of lithium hydroxide in ordinary water at 25°C. Prior to 1988, using the old value for the ionization constant of heavy water, this pD would be reported as 10.7.

3 Monel, Inconel, and Incoloy are trademarks of International Nickel. The use of these trade names does not imply that all the steam generator tubes were supplied by International Nickel; they are used to designate alloys of the following compositions by weight percent:

	Ni	Cr	Fe	Cu
Inconel-600	72+	14-17	16-10	
Monel	63-70		2.5	balance
Incoloy-800	30-35	19-23	balance	

4 While not fully understood, it appears that as the oxide thickens on Zircaloy in a reactor, it eventually cracks and becomes porous, allowing oxygen and/or lithia to concentrate within the oxide and increase the rate of oxidation of the underlying metal; this has not been observed with zirconium-niobium.

the boiler room by the third year of operation. This was the result of periods of poor chemistry control and inadequate water purification during the first two years, exacerbated by some equipment failures. These periods had been characterized by high concentrations of corrosion products in the water; although this was recognized as a danger signal, there were so many other problems demanding attention in the early operation of Douglas Point that establishing adequate chemical control in the HTS was delayed. A major cause of these high radiation fields was low deuterium concentrations in the coolant, resulting in oxygen levels sufficiently high (greater than 10 mg/kg) that the corrosion-product release rate from the Monel steam generator tubing, which contained 1,500 mg Co/kg, became excessive.

These high radiation fields led to the establishment of a major development program, which had two objectives: reduction of the fields at Douglas Point, and a detailed understanding of the activity transport process, leading to ways to control it at Pickering and later reactors.

The first approach to field reduction, which was the easiest to implement, was to add substantially more HTS purification capacity to remove particulate material. Then, a series of thermal shocks and chemical shocks in combination with reactor shutdowns were administered over a period of several months by varying the coolant temperature, and by stopping hydrogen addition or adding oxygen, and then restoring reducing conditions by adding hydrogen again. This released large amounts of corrosion products, which were removed by the filters, and reduced the steam generator radiation fields by a factor of about four.

The longer-term approach to radiation field reduction was to develop a decontamination process – that is, a method to dissolve the corrosion products and remove them from the system. Existing decontamination processes required concentrated solutions of chemicals and were deemed to be impracticable for a heavy-water system. If the chemicals were added to heavy water, it would be difficult and expensive to recover all the heavy water after the decontamination. If ordinary water solutions were to be used, then the heavy water would have to be displaced by ordinary water, since the fuel had to remain immersed to remove the heat of radioactive decay, and this displacement operation would result in excessive heavy-water upgrading costs.

Over a period of five years, a suitable process was developed, tested in an NRX loop and then in NPD, and finally applied to Douglas Point in 1975. Known as the CANDECON® process, it uses a dilute solution of organic acids to dissolve the metal oxides, and chelating agents to complex the metal ions and promote the dissolution. The reagents are added to the heavy water and the solution is recirculated through a temporary purification system containing cation exchange resin, which removes the metallic ions and regenerates the acid. Finally, the reagents are removed by mixed-bed ion exchange resins. Thus, the corrosion products and their associated radioactivity are transferred from the HTS system surfaces to beds of ion exchange resin. Equipment for handling and storing this resin is already part of each CANDU station. The decontamination of Douglas Point reduced fields associated with the carbon steel piping by a factor of six.

The CANDECON process was also successfully applied to the CANDU-BLW at Gentilly. The process has been licensed to industry and used to decontaminate parts of various light-water reactor circuits. Decontamination of the whole HTS of any other CANDU reactor after Douglas Point has not been necessary, because of the much better control of activity transport, as described below. The feeder pipes of the Pickering units were decontaminated using a modified CANDECON process, to reduce radiation exposure during retubing.

Activity transport occurs in all water-cooled reactors. By the mid-seventies there was a substantial international R&D effort devoted to elucidating the controlling mechanisms. The work in Canada, led by AECL, was well in the forefront of this effort.

An obvious requirement in material selection for the HTS became its cobalt content. A limit was set for cobalt in carbon steel of 60 mg/kg. The evolution in steam generator tubing from Monel to Inconel to Incoloy was largely to reduce the release of cobalt to the coolant; the allowable cobalt concentration in these alloys was set at 150 mg/kg.

Fundamental studies of equilibria at metal surfaces in high-temperature water (see chapter seven) and loop studies of mass transfer rates led to the development of an overall model of the activity transport process that encompassed the roles of both corrosion-product particles and species in solution. This model allowed prediction of the growth of radiation fields, and gave a means to assess the effects of changes on the rate of increase of these fields. For example, an optimum pD was defined that minimized transport, and the need for close control limits on that pD (10.9 to 11.4) was demonstrated.

Corrosion-product particles were found to play an important role in activity transport even when their concentration in the coolant was very small. Therefore, filtration was studied extensively as a means of reducing the rate of radiation field growth. To be effective, such a filter must handle a high flow rate, much higher than the flow in the HTS purification system, where the heavy water is cooled down so that ion exchange resins can be used. Therefore, these new filters must operate hot. Two types were developed to the stage of successful pilot tests at NPD: a deep bed of graphite particles, and a bed of steel balls placed in an electromagnetic field to trap magnetite, which is weakly magnetic. However, both were considered to be too expensive to include in a large CANDU reactor system, and with the low rates of activity transport achieved in Bruce A and later CANDUs they were deemed to be unnecessary.

All this effort on activity transport involved AECL Power Projects and Ontario Hydro as well as the research groups at Chalk River and Whiteshell. The good communication that this fostered meant results were quickly applied in design and operation, and occupational radiation doses at CANDU stations after Douglas Point have been low. A convenient measure is the total dose integrated over the station staff for a year, divided by the total electricity produced, expressed as millisieverts per electrical megawatt-year (mSv/MW.a). Up to 1985, the occupational dose has been typically 5 to 6 mSv/MW.a at Pickering A, and about 2 mSv/MW.a at Bruce A and the later CANDU stations[5]. Compared with light-water reactors over the period 1980-85, the average total occupational dose due to corrosion products at CANDUs was a third of that at pressurized heavy-water reactors (PWR) in Japan or Germany, and a sixth of that at PWRs in the United States. However, using similar philosophies for material selection and chemistry control as those just described for CANDUs, many of the latest PWRs are also experiencing very low radiation fields and therefore low total doses. The average individual dose to workers at Canadian nuclear power plants has tended to decrease over time, from around 15 mSv/a in 1970 to about 5 mSv/a in the early eighties.

HEAVY-WATER UPKEEP

Essential objectives of the CANDU HTS design are low heavy-water leakage and efficient recovery of this leakage, so that heavy-water losses are very low. Heavy-water upkeep comprises this recovery, restoring the recovered heavy water to reactor grade, and the replacement of losses. Early in the CANDU program a target was set for a heavy-water upkeep cost of no more than 5 percent of the cost of electricity. This translated to a loss rate of less than 0.5 kg/h, and a recovery and upgrading cost roughly equivalent to the cost of that make-up rate.

Joints, seals, valves, pumps, heat exchangers, steam generators and fueling equipment are all potential sources of leakage. Much of the leakage is heavy-water vapour which can be recovered by drying the air in the rooms containing the heavy-water equipment. However, the secondary side of steam generators and heat exchangers contains ordinary water, and its leakage into the same air can downgrade the recovered heavy water.

Such downgrading is inevitable, although its extent can be controlled by good design and adequate maintenance. The recovered heavy water is restored to reactor grade (99.8 percent heavy water) by distillation, and packed distillation columns for this purpose, supplied by Sulzer Bros. (Switzerland), have been installed at each CANDU station (except NPD). In addition, Ontario Hydro had a large, multicolumn unit (using sieve trays rather than packing) installed by Lummus at Pickering, to provide additional upgrading capacity for their nuclear system. Heavy-water distillation is discussed further in chapter seventeen. At Chalk River, a large electrolytic unit (see chapter two) has been used to upgrade water recovered at the research reactors and

[5] These doses include the contribution from tritium, which is about one quarter of the total.

NPD, and to supplement upgrading capacity at Douglas Point and Gentilly-1; it also has provided a contract upgrading service for non-AECL reactors when required.

Heavy-water upkeep cost was high initially at NPD. Modification or replacement of leaking components, improvements to sealing of the reactor vault and heavy-water equipment rooms, and improvements to the dryers brought the cost down substantially. By 1970, losses were only 0.3 kg/h and the efficiency of recovering heavy-water leakage was 96 percent.

In the design of Douglas Point the assumption was made that the HTS could be made leak-tight and kept in that condition. This proved to be much too optimistic, at least with the components selected for its construction. There was little segregation of heavy- and ordinary-water systems, so that heavy water recovered from the boiler room was severely downgraded. During the early years of operation, the systems were modified to provide some segregation. Rooms were sealed and dryers were added. Severe leakage was experienced from packed-stem valves and flanged pipe joints. Even when this was reduced, the total upkeep cost was about six times higher than the target.

This experience at NPD and Douglas Point led to various measures to reduce heavy-water upkeep costs that were applied in the design of Pickering and later reactors. Heavy-water and ordinary-water systems were segregated as far as possible, and light-water valves, joints, pumps and other leak-prone equipment were excluded from heavy-water areas. Those parts of buildings containing heavy-water equipment were sealed and ventilation flow was minimized; the air within these areas was continually circulated through dryers and all air exhausted from here was also dried before discharge. The number of mechanical joints and valves was drastically reduced – e.g., there were 2,000 packed valves in heavy-water service at Douglas Point and only 170 per reactor at Pickering – and those selected were of high quality. Welded joints were used extensively, even if it meant cutting pipes in future to replace or repair components. Pump shaft seals, heat exchanger flanges and large valve stems were fitted with two seals, so that heavy-water leakage into the interspace between the seals could be recovered in a closed heavy-water collection system for direct return to the HTS.

The early NPD and Douglas Point experience also led to an extensive R&D program on many aspects of heavy-water upkeep. The most significant efforts were in dryer performance, improved packed valves, steam generator tube integrity, leak detection and leak location.

To provide design data for the air-drying systems, several desiccants were tested, and it was quickly shown that molecular sieve (a porous alumino-silicate) appeared best for this heavy-water recovery operation. It absorbed water efficiently and reduced the dew point of the treated air to about -30°C. The drying system adopted contained two parallel beds of granular desiccant; when one approached saturation, the flow of air being dried would be switched to the second bed, while the first would be regenerated. In regeneration, the water would be desorbed by a reverse flow of heated air recirculated through the bed and a condenser to recover the water. Optimum conditions for regeneration and recovery were determined experimentally. Probably the most important insight to come from the analysis of dryer system design was the advantage of using deep beds of desiccant, rather than the shallow beds normal in conventional air-drying practice. The cost of the extra desiccant required in a deep bed was more than offset by the greater flexibility in drying cycle length and the insurance excess capacity provided for handling periods of high heavy-water leakage.

Although the use of double packing for valve stems (referred to above) substantially reduced the leakage of heavy water into the air of rooms containing the HTS, this in itself was not a solution to the problem of leakage past valve packings. In the initial years of Pickering's operation, leakage past the primary packing on a valve stem increased each time the valve was actuated. Eventually, the heavy-water collection system would be overloaded and some leakage past the secondary packing might occur. To counter deteriorating valve performance, compression of the packing had to be increased by tightening the packing gland nuts, or the valve stem had to be repacked. The result was excessive valve maintenance requirements.

An R&D program at Chalk River examined the factors that control leakage past packings. Major contributors to leakage were shown to be packing porosity and insufficient packing compression. Low-porosity packing materials were identified, such as graphite-impregnated woven asbestos. The compressive force on the packing was found to decrease with valve actuation, due to packing wear and consolidation. If the compressive force was not restored, leakage led to erosion of the packing and further deterioration.

The solution developed in the laboratory was to apply the compressive force to the packing through a spring-loading arrangement that provided sufficient stored energy to maintain a constant force, despite the wear and consolidation of the packing. In effect, these "live-loaded" valves behaved like conventional valves that were having their gland nuts continually adjusted (see figure 14.1).

FIGURE 14.1
Live-loaded Valve Packing Arrangement

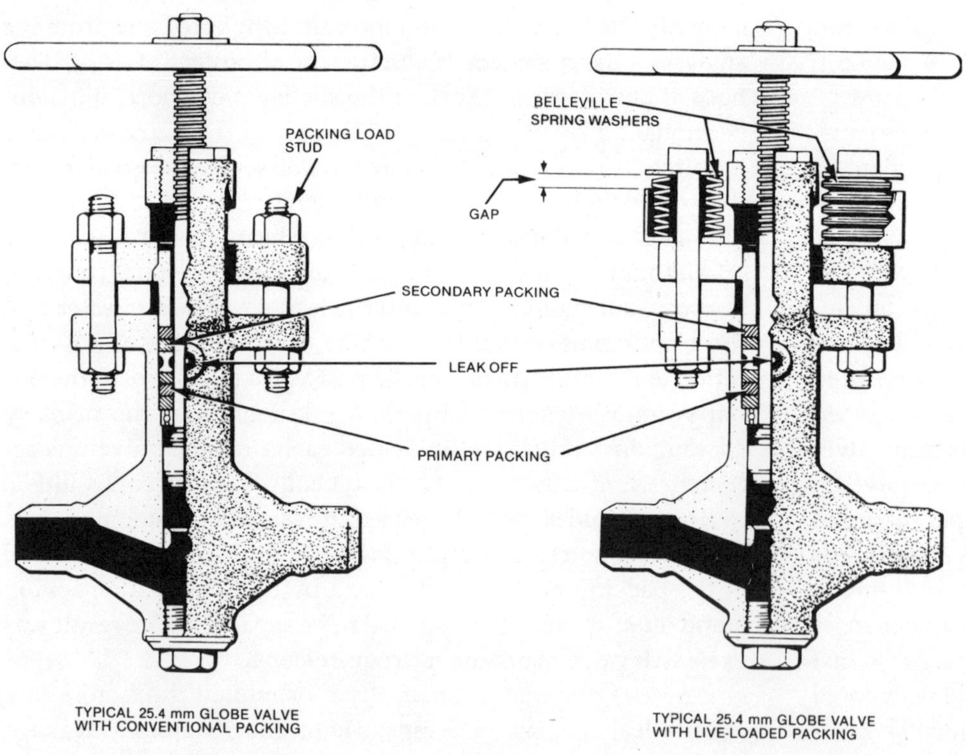

When live-loading was added to the HTS valves in Pickering in 1973, leakage past packing was greatly reduced, but more importantly, valve maintenance time was cut by a factor of more than one hundred. An annual savings in maintenance costs at Pickering of $1 million was achieved. This technology was transferred to industry and used at all subsequent CANDU stations.

The steam generators contain a very large area of thin-walled tubing, which separates the heavy water of the HTS from the ordinary water of the steam cycle. In a Pickering reactor the total surface area of this tubing is 24,000 square metres. The heavy water is at a greater pressure than the steam cycle water, so there is no danger of dilution of the HTS should there develop any defects in the tube walls or leaks at the welds that seal the tubes to the tube sheet. However, out-leakage of heavy water through such defects could be a major source of heavy-water loss. We will discuss tube integrity later; here we are concerned with the detection of such leaks before a major loss can occur.

To detect the leakage of heavy water into the steam cycle water, or into the cooling water of heat exchangers in the HTS or the moderator system, on-line monitors were developed.

Techniques for measuring heavy-water concentration by infrared absorption spectrometry were developed at Chalk River around 1960 (see chapter seventeen). A laboratory version of an on-line infrared monitor was developed to determine concentrations in the 100 to 200 parts per million range, suitable for detection of heavy-water leaks into ordinary water (which contains about 150 ppm heavy water). The sensitivity was one part per million which, in most situations, would be equivalent to a leak rate of the order of 0.3 kg/h. The infrared technique does not give an absolute concentration, so the instrument had to be calibrated frequently against a reference water sample. This technology was developed into an engineered process instrument by Barringer Research Limited.

Two of these instruments were installed at each of the Pickering A reactors – one for sampling the twelve boilers and one to handle the various cooling water samples. They were soon abandoned, because the sample systems proved to be prone to flow blockage, and inadequate sample purification made any measurements unreliable. Instead, Pickering operators relied on manual sampling of the various boiler water and cooling water streams on a weekly basis. This was a costly lesson that an instrument of this complexity cannot simply be installed and hooked up to a standard sampling system; instead, an overall integrated engineered system is required.

An intensive development program to improve the reliability of these leak-detection systems was undertaken at Chalk River. Much-improved sample purification methods were defined, and the design of the purification units was tailored to each application. The operating and environmental parameters for good instrument performance were defined and a sound maintenance strategy was developed. Prolonged testing showed that the detection sensitivity was limited to a few ppm by instrument stability and noise.

This improved leak detection system was demonstrated at NPD and eight units were installed in the four Bruce A reactors. These have operated satisfactorily. In the first two years of operation, four major leaks were detected, ranging from 3 to 14 kg/h. Similar systems have been installed in the other CANDU reactors, and those at Pickering A have been refurbished and recommissioned. AECL and Barringer developed an improved infra-red monitor that is simpler and more reliable, requiring much less maintenance.

An alternative to the infrared monitor was also developed at Chalk River. It determined heavy-water leakage by measuring the tritium concentration in steam generator water or cooling water using a plastic scintillator to detect the low-energy beta particles emitted. Under laboratory conditions a sensitivity equivalent to an increase in heavy-water concentration of 0.1 ppm seemed possible (except during the first few months of reactor operation, when the tritium content of the heavy water would be low). However, in field use, this sensitivity was found to deteriorate rapidly, because of fouling of the scintillator. A unit installed at NPD was compared with the improved Barringer infrared monitor during a steam generator tube leak. The tritium-in-water monitor had a considerably lower sensitivity and its calibration was less stable.

Two types of area monitors have been used to alert operators to significant increases in heavy-water leakage into the various parts of the reactor building: tritium-in-air detectors using ionization chambers developed at Chalk River and conventional hygrometers.

A variety of leak-location instruments were developed that have had varying degrees of success at CANDU stations. Amongst those that proved to be practical were ultrasonic detectors attached to the fueling machine, to locate leaking end closures on fuel-channel end fittings, and portable monitors to measure tritium or moisture.

The measures taken in the design of Pickering to achieve low heavy-water upkeep costs were generally successful. In the period 1971-75, the overall average upkeep cost was only 40 percent above the target figure, that is about 7 percent of the cost of power. However, as further improvements were made, the upkeep cost was brought down to the target of 5 percent of the cost of electricity. By 1983, the average heavy-water upkeep cost for the three Ontario Hydro stations then operating was less than 4 percent of the cost of electricity, well within the target set in 1964.

The severe requirement of containing heavy water imposed a discipline on attaining excellence in design, fabrication, installation and operation that not only achieved a satisfactory degree of leak-tightness, but has contributed to the overall high capacity factor that has characterized CANDU performance.

TRITIUM CONTROL

Tritium accumulates in the heavy water until it reaches a steady state, where the production rate equals the loss rate by decay, leakage and removal. In the coolant its concentration is generally limited to less than 1 mg/kg (<1.4 Ci/kg) by the low production rate, because of the low residence time in the neutron flux, and the replacement of losses with fresh heavy water. On the other hand, in the moderator the concentration could approach 50 mg/kg (70 Ci/kg) after many years of operation. This water would be extremely hazardous to handle. This possibility led Ontario Hydro to decide in the late seventies to build a tritium removal plant at Darlington, to control tritium concentrations in all their heavy water. The plant began operating in 1987. Meanwhile, a program of moving moderator water from the oldest reactors to a number of newer reactors kept tritium concentrations below 25 mg/kg.

Low heavy-water leakage and the dry conditions in the reactor building achieved by efficient dryer performance limit exposure of staff to tritium ingestion. In situations where tritium concentrations in air are high, or risk of contact with liquid water exists, protective clothing and hoods with external air supply are used. The radiological hazards, the effectiveness of protective clothing, and tritium monitoring are discussed in chapter three.

By the mid-seventies, at Pickering, the occupational dose from tritium had risen to a third of the total dose. Concern that the tritium dose would continue to rise also contributed to the decision to build the tritium removal plant; it spurred improvements to the control of heavy-water leakage, and changes to the air-flow patterns within the reactor building, to move tritium away from work areas. Improved work procedures were also implemented, to reduce the potential for exposure. The result was a reduction in occupational dose due to tritium from about 2.5 mSv/MW.a in 1976 to 1 mSv/MW.a in 1981 at Pickering. At the newer CANDUs this dose has generally been about 0.5 mSv/MW.a.

The Ontario Hydro tritium removal plant is based on Swiss and French technology. The tritium is transferred from the heavy water to a stream of deuterium gas using a vapour-phase catalytic exchange process,[6] and then pure tritium is separated from the deuterium by cryogenic distillation. At the time this plant was committed, AECL was developing a much simpler and cheaper liquid-phase catalytic exchange process for the first step, but it was still at the laboratory stage – see water-hydrogen exchange in chapter seventeen. A tritium-extraction pilot plant was built at Chalk River using this new process and cryogenic distillation, to demonstrate the process and to remove tritium from AECL heavy water. However, with the shutdown of NPD and Douglas Point, the need for AECL detritiation capacity has decreased substantially, and this, combined with additional expenditures required to prepare detailed licensing documents, resulted in the pilot plant being mothballed.

Once it was decided to extract tritium from reactor heavy water, it became necessary to develop techniques for its safe storage. Chalk River developed the expertise to handle pure tritium and the technology for its storage as titanium tritide; this technology was put into practice at the Darlington tritium-extraction plant. The tritium handling expertise at Chalk River has subsequently been applied to developing technology for fusion reactors. This program, together with Ontario Hydro's extensive experience with tritium, forms a major part of the activities of the Canadian Fusion Fuel Technology Project – see the national fusion program in chapter nine.

STEAM GENERATORS

In a CANDU 6 station there are four steam generators, each with a capacity to transfer 525 MW of heat. Figure 14.2 illustrates the main features of one of these units. Twenty metres in overall height, the unit consists of three main sections: a steam-water separator, the two-metre diameter shell containing the bundle of 3,550 tubes, and the hemispherical bottom head for heavy-water entry and exit. Each U-bend Incoloy-800 tube is twenty metres in length and 16 millimetres in diameter, giving a total area of about four-thousand square metres (about one acre) to transfer heat from the hot heavy water within the tubes to the boiling water within the shell. Since even a pin-hole size leak would lead to an unacceptable heavy-water loss, a major concern

---

[6] Tritium occurs in heavy water as the DTO molecule. The exchange reaction, $DTO + D_2 \longrightarrow D_2O + DT$, provides a low concentration of DT in D_2 for cryogenic distillation. A further exchange reaction, $2DT \longrightarrow D_2 + T_2$, allows pure T_2 to be recovered. It can be marketed for fusion research and for tritium-luminescent devices, such as exit signs.

FIGURE 14.2
600 MW Steam Generator

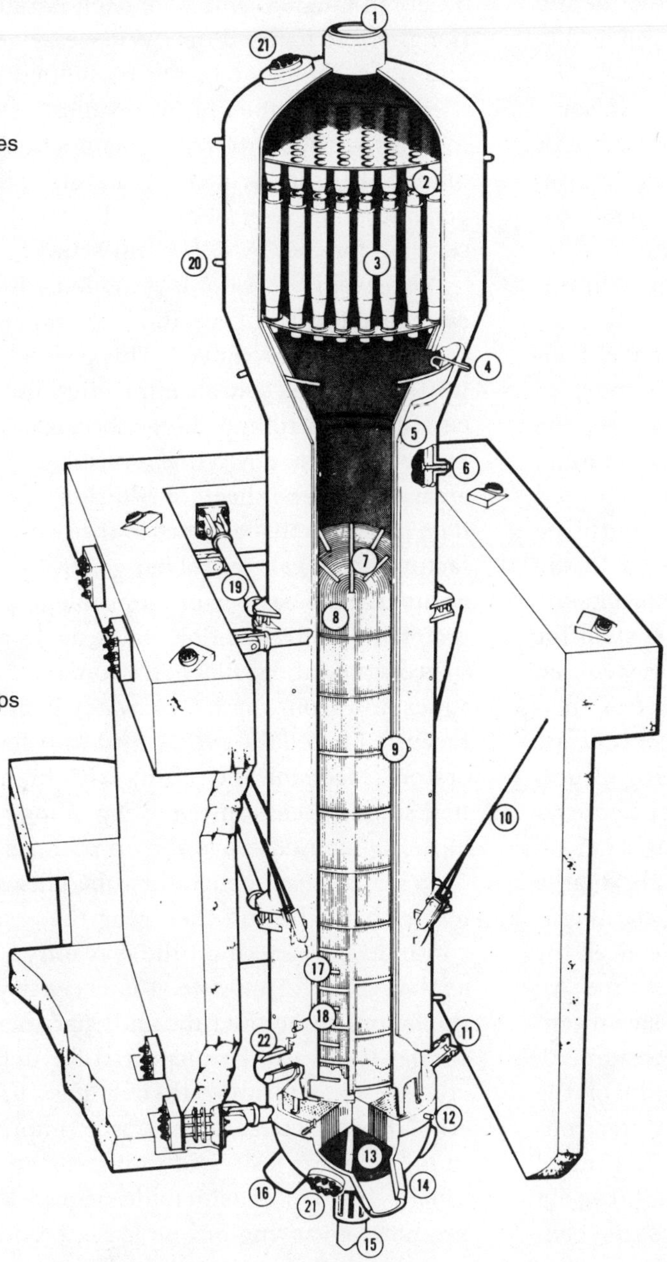

1. Steam Outlet Nozzle
2. Secondary Steam Cyclones
3. Primary Steam Cyclones
4. Chemical Feed Nozzle
5. Downcomer Annulus
6. Emergency Water Supply Nozzle and Header
7. U-Bend Supports
8. Tube Bundle
9. Tube Support Plate
10. Back-up Supports
11. Observation Port
12. Blowdown Nozzle
13. Divider Plate
14. D_2O Inlet Nozzles
15. Base Support
16. D_2O Outlet Nozzle
17. Baffle Plate
18. Preheater
19. Lateral Supports
20. Water Level Control Taps
21. Manway
22. Feedwater Nozzle

is to achieve and maintain a very high degree of tube integrity. The tubes are rolled into the thick tube sheet at the bottom of the shell and welded to give 7,100 seals separating ordinary and heavy water. The spacing between the tubes in the bundle is maintained by a number of tube-support structures, which contain openings for each tube to pass through.

The design of the CANDU steam generator has evolved from the rather small units at Pickering – twelve per reactor, each about 160 thermal MW – to the largest at Darlington – four per reactor, each 730 thermal MW.

Steam generators in fossil-fueled (coal, oil or gas) power stations have a long history of development and increasing sophistication in design. However, the steam generator for a nuclear station is quite different, since the heat source is hot water within the tubes, rather than a high-temperature flame external to the tubes. Boiling in the nuclear unit occurs outside the tubes, resulting in precipitation and deposition of solids within the tube bundle, especially in the region immediately above the tube sheet. The chemical conditions within such deposits and sludge determine the environment within which the tubes and their rolled joints must survive.

Corrosion was clearly a major concern in the design and operation of steam generators. Initially, studies focused on pitting and cracking due to chemicals that might concentrate in the boiling water. The condensed steam is returned to the steam generator through a series of feedwater heaters, where corrosion products may be picked up and carried forward to accumulate in the steam generator. On its way, the feedwater is deaerated, to remove the bulk of oxygen picked up in the condenser (from air in-leakage), and hydrazine is added to destroy the remaining traces of oxygen. Any leakage of condenser cooling water into the condensate will introduce impurities such as sulfate, chloride and organics. Chloride in-leakage from sea water cooling can lead to severe problems and must be monitored carefully. For CANDU, this situation arises only at KANUPP, Point Lepreau and Wolsong.

Within the steam generator, impurities such as sulfate or chloride can produce aggressive acidic conditions as they are concentrated. However, from fossil-fueled steam generator technology, it was expected that this could be controlled by adding sodium hydroxide and sodium phosphate, and controlling the degree of alkalinity.

Studies were performed at Chalk River in small model boilers that could duplicate the effects of concentrating impurities on tube corrosion, especially in the crevice formed between the tube and the tubesheet. Monel was found to be tolerant of a wide range of chemistry conditions, and could be expected to give satisfactory service at Pickering. However, with the preferred alloys, Inconel-600 and Incoloy-800, used in the later reactors, there was some cracking in concentrated alkaline solutions. Therefore, an "all-volatile" treatment system for steam generator water was recommended. In this system, an organic amine is added, to provide the desired degree of alkalinity, but it is sufficiently volatile that it does not concentrate in the steam generator. This system is used at Pickering, Bruce, Darlington and the CANDU 6 stations, except Point Lepreau, which uses alkaline phosphate. Since there is no buffering action, which the alkaline phosphate provides, the all-volatile treatment requires stringent monitoring of cooling-water impurity ingress. When this is detected above the tolerable level, the station must shut down to plug the leaking condenser tube. An alternative is to provide for full-flow purification of steam generator feedwater, and this was adopted for Point Lepreau.

AECL has made an extensive fundamental study of the chemical species produced when steam generator water containing impurities from cooling water in-leakage is concentrated in crevices or deposits of corrosion products. The results can aid CANDU station staff to assess the level of risk to the steam generator posed by a given in-leakage occurrence.

These corrosion and chemistry studies, centred at Chalk River, have also involved Ontario Hydro research division and others in Canada, and have been paralleled by a large international effort focused on PWR steam generators. Collaboration and open exchange of results helped to define problems and their solutions quickly. For example, when unexpected cracking on the tube interior (coolant side) was reported in regions of high stress in Inconel tubes in some light-water reactor steam generators, Incoloy (shown to be immune from such attack) was selected for the CANDU 6 units and Darlington.

The steam generator is provided with a drain line, so that the concentrated solution and accumulated sludge of corrosion products can be removed. This process is known as blowdown. It has controlled soluble impurities, but has been much less successful in controlling sludge, because the small size and limited number of blowdown lines constrain their effectiveness to local areas of the tube sheet; in addition, the sludge tends to consolidate in a mass that adheres to the tube surfaces. Although satisfactory chemical control has generally been maintained, sludge accumulation has become a problem, promoting corrosion and distorting hydraulic flow patterns. Deposits of corrosion products on the tube walls, both from the boiling water outside the tubes and from the heavy water inside (iron transport), have recently limited the overall heat transfer rate in some steam generators. Mechanical cleaning with high-pressure water jets has had some success in removing sludge, and recently chemical cleaning methods have shown promise.

When a steam generator tube fails, it becomes necessary to locate that tube and then plug the ends so that operation can continue. The leaking tube is readily located using a dye such as fluorescein. However, it is necessary to find the position of the failure along the tube and characterize it. Then adjacent or similar tubes can be inspected for defects (pits and incipient cracks), to determine their fitness for continued operation. AECL has devoted a large effort to improving eddy current techniques for this task, and to increasing the eddy current probe sensitivity to a considerable degree. The eddy current probe produces a varying magnetic field, which induces electric currents in the tube; changes in the shape of the magnetic field are detected, which can be correlated with defect size and shape. These techniques have also proved useful for inspecting heat exchanger tubes, and they have been applied on a commercial basis both in the nuclear industry and elsewhere (e.g., the chemical industry).

Although steam generator tube failures have for the most part been due to corrosion, a number of mechanical failures due to vibration and fretting have also occurred. The potential for such mechanical damage was recognized from the beginning and a program to study these phenomena was begun about 1970.

The objectives of this work were to avoid flow-induced vibration, to understand vibration failure mechanisms (mainly fretting between a tube and the tube-support structure), and to provide guidance for steam generator design, modification and repair. This required the development of technology in the areas of thermalhydraulic analysis, vibration excitation mechanisms, multi-span tube dynamics and fretting-wear damage behaviour. The ultimate goal was a fretting-wear prediction capability.

A computer code was developed to perform a comprehensive three-dimensional thermalhydraulic analysis, to predict flow velocities and steam fractions at all points in the steam generator. This code has been used to suggest design changes to eliminate regions of high velocity that could lead to excessive tube vibration, and to assess stagnant regions prone to concentrate impurities. It has been important for modelling the deleterious effects of deposition on flow distribution and velocities.

There are three mechanisms of flow-induced vibration excitation: periodic wake shedding, random turbulence, and fluid-elastic instability. The latter is the most important, setting a maximum velocity for stable operation. Extensive laboratory tests for many geometries and special situations typical of CANDU tube bundles have been done using liquid and two-phase cross-flow and parallel flow. The resulting data allow calculation of tube vibration amplitude and frequency, and reaction forces at points of support, for any steam generator design. Damping of the vibration by liquid films between the tube and the support structure are also important in this assessment. When deposits impede this damping, vibration may be increased; this appears to have been the case recently in some Bruce A steam generators, leading to fatigue cracks in the U-bend region, which caused tube failures.

The reaction force times the sliding distance of a tube rubbing on a support defines a work function that can be correlated with fretting damage. A long and laborious series of tests were conducted to measure the very low fretting rates (10^{-8} m/h) that are consistent with a thirty-year tube life. A range of forces, clearances, vibration frequencies and tube arrangements were studied at steam generator operating conditions in special high-pressure autoclaves. Reproducible results were elusive for many years; however, with improved experimental techniques it has now been possible to obtain satisfactory data.

Thus, reliable predictions of fretting wear can be made for operating steam generators, or for new designs. This technology has also had many useful applications to heat exchangers through commercial contracts.

In 1986 the performance of CANDU steam generators was reported to be excellent. Only 35 tubes had failed out of 506,000 then in service. This failure rate was a factor of one hundred lower than the average failure rate for all PWRs to that date. However, by 1992 several thousand tubes had failed and been plugged in CANDU steam generators, mainly in Bruce A, unit 2 and Pickering B, unit 5; only half of the operating CANDU reactors had fewer than ten steam generator tubes plugged, including five with none. More recently, further CANDU steam generator tube failures have occurred. This deteriorating performance is the result of accumulated sludge and tube deposits where corrosion can occur, coupled with the generation of an aggressive environment during sporadic periods of inadequate water chemistry control.

As we have seen, a complex interaction of mechanical design, materials selection and operating chemistry determines steam generator performance. Early recognition of this interaction, extensive chemical and corrosion studies, and good communication amongst the R&D groups, designers, manufacturers and operators were intended to achieve highly reliable CANDU steam generators. Key factors were proper material selection, good design, good quality assurance, full control of fabrication procedures, and rigorous control of water chemistry during operation. Unfortunately, the latter three factors were not fully achieved with every steam generator.

PUMP SEALS

The HTS pumps are vertical, electric-motor-driven centrifugal units (see figure 14.3). The seal between the shaft and the pump body is provided by two or three mechanical seals in series. These are end face seals, in which the upper flat surface of a thick ring attached to the shaft and the lower surface of a similar ring attached to the pump cover form a thin radial gap. A small flow of cool heavy water radially through the gap across the seal faces, which are almost in contact, provides lubrication; this water goes to the collection system. Pump speed is thirty revolutions per second. Pump rating has increased from 0.4 MW (500 hp) at NPD to 8 MW at Bruce, with a corresponding increase in physical size.

The mechanical seals form a package that can be replaced by removing the space coupling (see figure 14.3) between the motor and pump shafts. One of the seal rings has a flexible mounting allowing axial movement, but controlling tilt and eccentricity to very small tolerances. Hydrostatic forces then hold the faces close together, but also maintain a thin lubricating film of heavy water flowing to the collection system. The rotating face is a hard material, tungsten or titanium carbide, and the stationary face is a soft carbon-graphite composite. The essence of good seal design is to achieve a balance of forces acting on the seal that provides a very thin (a few micrometers), stable lubricating film under all operating conditions. Work at AECL in the mid-sixties showed that a slightly tapered face on one seal component, giving a gap that converges in the direction of flow, is essential for stable lubrication. When the seal is unpressurized, springs act on the flexible seal component to keep the gap closed.

Reliable pump performance is essential to achieving high reactor capacity factor. This objective was achieved through redundancy in the number of pumps in early stations: two operating and one standby at NPD, eight operating and two standby at Douglas Point, and twelve operating and four standby at Pickering. For Bruce and later CANDUs, four operating pumps were provided with no standby, reflecting increased confidence in pump and pump seal reliability. Using several seals in series within the seal package not only increases seal

FIGURE 14.3
Heat transport system pump for CANDU reactor service

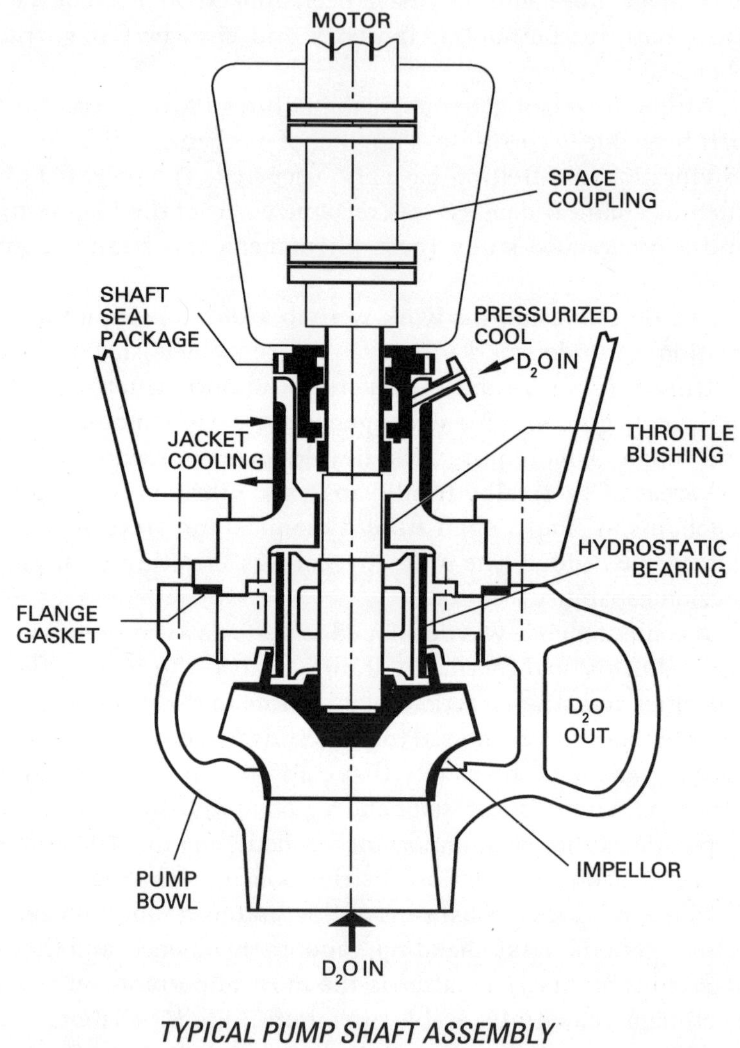

FIGURE 14.4
Bruce Pump and Motor on Test Stand at Byron-Jackson Canada, Scarborough, Ontario

lifetime by lowering the pressure drop across each one, but allows satisfactory operation with one failed seal, so that its replacement can usually be delayed until a scheduled reactor shutdown.

Early experience at NPD and Douglas Point made it clear that pump seals would require significant engineering development. Initially, seal life at NPD was only a few hundred hours before heavy-water leakage to the collection system became too high. Rubbing wear was followed by substantial erosion-corrosion due to the high leakage velocities. The main cause of seal failure was found to be adverse deflections and warpage of the seal faces. Seal performance was further impaired by gas coming out of solution at the seal faces, but this was readily corrected by venting and improved degassing of the heavy-water coolant. At Douglas Point, a close-clearance bushing was the primary seal, with an end-face seal as backup. However, there were still many problems during the first year of operation. In both reactors, seal wear resistance was achieved using a cobalt-alloy facing, and this led to cobalt contamination of the HTS.

An aggressive program to develop improved designs and materials was begun. AECL Power Projects identified sintered titanium carbide as their preferred material for erosion and wear resistance, using this as the rotating seal face in the development of a novel elliptical seal design. Effort at Chalk River focused on controlling face deflections and enhancing lubrication by various hydrodynamic and hydrostatic features. Through contracts with Dilworth, Secord, Meagher and Associates, another new seal design was developed. This and the elliptical seal in a two-stage arrangement were retrofitted to the NPD pumps; by 1971, seal lifetime had been extended to about three years. An elliptical seal was also substituted for the backup seal in the Douglas Point pumps, similarly improving their reliability.

Testing of Pickering pump seals began at Chalk River in 1971. By this time, the pump manufacturer, Byron Jackson (BJ), had adopted and standardized on titanium carbide and carbon-graphite as the two materials for their seals, with much-improved results in American boiling-water reactors. This design was chosen for installation in Pickering A in equal numbers to NPD-type seals developed by AECL. Testing of both helped Chalk River to identify the critical dimensions, degree of flatness and other factors in seal installation and maintenance that station staff must achieve for good seal performance. The BJ seals eventually performed better, attaining an adequate average life of about three years, and the NPD-type seals were phased out. Although there have been several hundred seal failures at Pickering A and B over the period 1970-90, they have had little impact on reactor reliability, because four of the sixteen pumps in each unit are spares.

In preparation for developing the large Bruce A pumps, BJ, their manufacturer, installed a facility (see figure 14.4) to conduct the full-scale pump proof tests required by their contract. Seal problems soon arose because of the size increase over Pickering: 187 mm in diameter instead of 130. This was potentially disastrous for the reliability of Bruce, because of the bold commitment to only four pumps per unit, with no installed spares. Pump delivery was stalled in 1972 until proof tests could be completed successfully. By now, the Chalk River development work had reached the stage where detailed analysis of seal component deformation due to pressure and thermal effects (using finite-element techniques) could be done accurately. This allowed the design of a seal that remained dimensionally stable and was compatible with the Bruce seal cartridge. Once tested at Chalk River to AECL's satisfaction, the seal and its design and manufacturing information were supplied to BJ.[7] Its success in allowing the pump proof tests to be completed was outstanding, and well appreciated by BJ. The decision was made for BJ to supply both the AECL design and an improved version of their own for equal installation in Bruce A. The former proved to be superior, and have been used (with further improvements) in all subsequent CANDUs. This AECL seal design has successfully met the target of a five-year life.

With Bruce B and Darlington planned, Ontario Hydro decided they needed additional full-scale pump test capability. A facility was built at their research centre for acceptance testing and for further investigation of pump and seal performance.

At Chalk River, dynamic studies of the effects of misalignment and eccentricity have defined the tolerable levels of these inaccuracies, and led to improved design criteria. Detailed computer models for predicting seal performance have enabled further evolution in the AECL seal design, so that it is almost insensitive to temperature and pressure transients.

The development program has been essential in establishing stringent quality control in seal manufacture, installation and maintenance. It has demonstrated the importance of highly trained pump seal maintainers, to ensure that each replacement seal is correctly installed so that it performs reliably.

The knowledge about pump seals developed by the AECL researchers exceeds any possessed by the pump manufacturers. This has led to a number of commercial contracts to resolve pump seal problems in LWRs. A part of the seal technology involves the use of O-rings to seal the components of the mechanical seal to the shaft or the pump cover. This expertise in O-ring materials and design enabled Chalk River to contribute to the solution of the problem that destroyed the space shuttle Challenger (see chapter twenty).

[7] The first Bruce seal cartridge with AECL seals was assembled and spirited out of the laboratories at night, headed for Toronto in the back of a pickup driven by the designer. Otherwise it was going nowhere, because of the only extended strike and picket line in the history of Chalk River Laboratories.

ACKNOWLEDGMENTS

I would like to thank a number of colleagues who have reviewed this chapter and given advice and assistance: H.A. Allsop, W.J. Holtslander, D.H. Lister, R. Metcalfe, M.J. Pettigrew, N.E. Pothier, P.A. Ross-Ross, R.L. Tapping and G.F. Taylor.

BIBLIOGRAPHY

Barber, D., and D.H. Lister. "Chemistry of the Water Circuits of CANDU Reactors." In *Water Chemistry and Corrosion Problems in Nuclear Power Plants: Proceedings of an International Symposium on Water Chemistry and Corrosion Problems of Nuclear Reactor Systems and Components,* 149-61. Paper IAEA-SM-264/15. Vienna: International Atomic Energy Agency, 1983. (Also AECL-8217.)

Dombra, A.H. "Location of Heavy Water Leaks in Canadian Power Reactors." *Transactions of the American Nuclear Society* 10 Suppl. (1967): 10-11. (Also AECL-2755.)

Doubt, G.L. *Reference Book for the Design of Valve Packings, Sealing High Temperature Water.* AECL-5120. 1976.

Hatcher, S.R., and H.K. Rae. "Formation and Control of Turbidity in Aluminum-Water Reactor Systems." *Nuclear Science and Engineering* 10 (1961): 316-30. (Also AECL-1210.)

Hodge, R.I., J.E. LeSurf, and J.W. Hilborn. "Steam Generator Reliability: Canadian Practice." In *Reliability Problems of Reactor Pressure Components: Proceedings of a Symposium on Application of Reliability Technology to Nuclear Power Plants,* vol. 2, 177-97.

Paper IAEA-SM-218/25. Vienna: International Atomic Energy Agency, 1978. (Also AECL-6253.)

Keenleyside, G., and W. Kalechstein. *Advanced Instrument Development: A Case Study. Heavy Water Monitors: The Second Generation.* Chalk River R&D, no. 1. AECL-8632. 1985.

LeSurf, J.E. "Control of Radiation Exposures at CANDU Nuclear Power Stations." *Journal of the British Nuclear Energy Society* 16 (1977): 53-61. (Also AECL-5643.)

LeSurf, J.E., G.F. Taylor, D. Barber, and L.J. Sennema. "Occupational Radiation Exposures at Canadian CANDU Nuclear Power Stations." In *Nuclear Power Experience: Proceedings of an International Conference on Nuclear Power Experience, Vienna, 13-17 September 1982*, vol.4, 543-56. Paper IAEA-CN-42/47. Vienna: International Atomic Energy Agency, 1983. (Also AECL-7769.)

Lynch, G.F. "On-Line Infrared Heavy-Water Instruments: Status, Economics and Prospects." In *Canadian Nuclear Association Eighteenth Annual International Conference, Ottawa, 11-14 June 1978, Session 10, Instrumentation and Control,* Paper 78-CNA-523. Toronto: Canadian Nuclear Association, 1978. (Also AECL-6205.)

Metcalfe, R., and R.D. Watson. "Pump Seals for Nuclear Service: Understanding and Improving Their Performance." In *Proceedings of the Fifth Symposium on Engineering Applications of Mechanics, University of Ottawa, 16th and 17th June 1980*, 209-16. Report NRCC-18370. Ottawa: National Research Council of Canada, 1980. (Also AECL-7052.)

Montford, B. "Techniques to Reduce Radiation Fields." In *Canadian Nuclear Association Sixteenth Annual Conference, Toronto, 13-14 June 1976,* vol. 8, 41-58. Toronto: Canadian Nuclear Association, 1976. (Also AECL-5523.)

Rae, H.K., G.M. Allison, A.R. Bancroft, W.D. Mackintosh, J.F. Palmer, E.E. Winter, J.E. LeSurf, and S.R. Hatcher. "Experience with the Chemistry of Water in Moderator and Coolant Systems." In *Proceedings of the Third International Conference on the Peaceful Uses of Atomic Energy, Geneva, 31 August-9 September 1964*, vol. 9, 318-25. Paper A/CONF.28/P/20. New York: United Nations, 1965. (Also AECL-2020.)

Robertson, R.F.S. *Chalk River Experience in 'Crud' Deposition Problems.* AECL-1328. 1961. (Written for presentation at Tripartite Conference on Transport of Materials in Pressurized Nuclear Systems, 28 February and 1 March 1961.)

Ross-Ross, P.A., E.J. Adams, D.F. Dixon, and R. Metcalfe. "Performance and Reliability of Primary Circuit Components in CANDU Reactors." In *Nuclear Energy Maturity: Proceedings of the European Nuclear Conference, Paris, 21-25 April 1975*, vol 6, 85-109. Ed. by Pierrer Zaleski, et al. Progress in Nuclear Energy, n.s. Oxford: Pergamon Press, 1976. (Also AECL-5092.)

Tatone, O.S., and R.L. Tapping. *Steam Generator Tube Performance: Experience with Water-Cooled Nuclear Power Reactors During 1985.* AECL-9724. 1988.

Tomlinson, M., K.A. Burrill, G.F. Taylor, H.E. Thexton, B. Montford. *Materials Research in AECL.* [J.E. LeSurf, ed.] AECL-5227. Summer 1975.

Chapter Fifteen

Safety Research and Development

R.E. GREEN

INTRODUCTION

The safety of nuclear reactors has been a matter of major interest from the beginning of Canada's nuclear program. Reactor-safety philosophy, and its application to the research reactors, are discussed in chapter four. The safety-related research and development (R&D) activities relevant to the CANDU power reactor system are described in this chapter.

CANDU power plant design and the related safety philosophy have evolved in parallel over a period of about forty years, and continue to evolve today. For the first power plants, NPD and Douglas Point, the designs were based on well-established physical principles and engineering judgement, and included large factors of safety, consistent with the newness of the technology and the lack of operating experience. However, in many areas, particularly those related to safety, the engineering judgement was backed up by laboratory experiments and larger-scale engineering tests. The successful operation of these first plants encouraged the designers to propose the construction of much larger units (to benefit from economy of scale), having greater fuel-channel powers, thereby improving the overall competitiveness of the CANDU concept vis-à-vis other nuclear-reactor concepts, and other electricity-generating sources.

The CANDU designers and the researchers both realized that for the larger, more complex commercial plants more information would be needed on how the various plant systems would behave, both at normal operating levels and during various postulated accidents, to ensure that the plants were safe and posed no significant threat to the public and the environment. They also realized that the regulatory agency, the Atomic Energy Control Board (AECB), would require more accurate information on the safety margins and on the possible consequences of various accident scenarios, since these new plants represented a greater potential hazard because of their increased size with larger fission-product inventory, and their higher fuel operating temperature associated with higher channel power. Another important factor was that one of the large commercial units (Pickering) was to be located near a large city (Toronto).

In the early days (the late fifties and early sixties) of the nuclear power program, safety analysis dealt mainly with the consequences of the failure of individual critical process systems and/or pieces of equipment, as identified during the design process. However, as the program developed and the plants became larger, more attention was paid to dual failures, situations that could occur much less frequently but could have much greater consequences. One impetus for increased interest in these low-frequency events was a study[1] carried out for the United States Nuclear Regulatory Commission (USNRC) in the early seventies, which showed, using very pessimistic assumptions, that the consequences of some very low-probability accidents were unacceptable. This led nuclear regulatory agencies around the world to request more analysis of low-frequency accident scenarios. These factors led to a large increase in the amount of safety analysis required to support the application for a plant operating licence. This trend to more detailed safety analysis also coincided with the increased availability of powerful computing methods, which permitted extensive mathematical modelling of the behaviour of plant components and systems, both for normal operating conditions and for hypothetical accident situations. This increase in systems analysis led to the need for a larger, more comprehensive R&D program to provide component performance data and system properties for a wide range of conditions, required to define the parameters used in the mathematical models.

Thus, reactor safety R&D programs all over the world expanded greatly in the seventies, and concentrated more and more on investigating the consequences of low-frequency, worst-case accident scenarios. This thrust for more detailed safety analysis tends to be self-ratcheting; i.e., the more detailed the model, the more questions that are raised, requiring more R&D to answer them.

Before describing the evolution of the CANDU safety R&D program, it is worthwhile to examine how safety is incorporated into the design of CANDU power plants and how safety analysis has attempted to address the issues raised by the designers and the AECB, since these factors have determined the scope and depth of the safety R&D activities over the years.

ENGINEERING FOR SAFETY

From the outset, the CANDU designers had safety very much in mind, and CANDU power plants have always been designed to limit the release and transport of radionuclides from the plant to the environment and, hence, to the public. The sources of radionuclides in a CANDU plant are neutron-activated impurities (e.g., corrosion products and other particulates) in the primary coolant and the moderator, tritium (produced by neutron capture in deuterium) in the coolant and moderator, and fission products and actinides (heavy elements created by successive neutron capture in the isotopes of uranium) in the fuel. The designers made use of the inherent safety features of the CANDU concept, and used multiple barriers to minimize the release of such radioactive material to the environment.

The CANDU reactor has several inherent safety features, one of which is the use of natural uranium as fuel. The inherently low reactivity of natural uranium requires a reactor lattice geometry near the reactivity maximum (to maximize fuel burnup); consequently, lattice distortion or dispersion cannot cause a significant increase in reactivity. The use of natural uranium also means that there is only a small amount of excess reactivity in the reactor core at any time, thereby reducing the likelihood of a severe loss-of-regulation accident (LORA). Another safety feature is the use of many small-diameter, high-pressure, high-temperature fuel channels surrounded by a large pool of relatively cool (70°C) heavy-water moderator at atmospheric pressure. The cool moderator could serve as a heat sink during an accident in which the emergency cooling system is impaired.

There are several barriers to the release of radioactive material from a CANDU plant to the environment. The first barrier is the fuel itself. Since the fission products generated in the fuel represent the bulk of the radioactive material in the plant, they represent the greatest hazard. However, the uranium oxide fuel is a ceramic material, with a high melting point (about 2,850°C), and is relatively chemically inert in water or steam. Radionuclides are created within the fuel during fission and most remain trapped there. Virtually all of the solid and more than 90 percent of the gaseous fission products are retained by the uranium oxide at normal operating temperatures.

The second barrier is the Zircaloy sheathing in which the fuel is contained. The sheathing is designed to withstand the stresses resulting from uranium oxide expansion, fission-gas pressure and external cooling-water pressure during normal

[1] Described in the WASH-1400 report.

operation, and the mechanical loads imposed by fuel handling. Even if the sheathing were breached, there would be only a minor release of fission products, unless the fuel were grossly overheated. Thus, keeping the fuel from becoming overheated prevents a potential danger from becoming a real one.

The third barrier is the heat transfer system, comprising the pumps, boilers, fuel channels, and associated piping. This sealed and monitored system is designed to a high standard of quality, and can be counted on to perform reliably, to keep the fuel cool, and to retain most of the fission products, for all of the expected modes of operation and all of the accident scenarios, except a loss-of-coolant accident (LOCA).

The fourth barrier is the containment system. In all Canadian single-unit stations, the reactor and piping systems are enclosed in a concrete structure designed to contain any radionuclides released into it, while the multi-unit stations have a vacuum building that serves essentially the same function for all of the station's reactors. Dousing systems are also provided to rapidly condense any escaping steam and to ensure that the containment building pressure is kept within safe limits.

There is also an exclusion area (with a radius of about 1,000 m) surrounding each power plant. Public access to this area is controlled, and the atmospheric dilution between the reactor building and the boundary of this zone would reduce the concentration of any fission-product release by a factor of at least 100. There is also emergency response planning and preparation to protect the public, the operations staff and the facilities, in the unlikely event of a release of radionuclides from the plant.

However, the CANDU reactor has a positive "void coefficient", whereby a reduction in coolant density, or a loss of coolant that results in voids in the fuel channel, results in an increase in core reactivity, hence greater reactor power, which tends to drive off more coolant. If unchecked, this positive feedback process would result in a large increase in reactor power. Safety shutdown systems are provided to ensure that the nuclear reaction, hence power generation, can be stopped when required, even during accident situations. In all the CANDU plants built in Canada after Pickering A, there are two completely independent shutdown systems having short reaction times, rapid insertion rates, and sufficient reactivity worths to overcome the reactivity effect of any failure in any of the process systems. The Pickering A reactors also have two shutdown systems, but one of these (moderator dump) is not as fast as the systems in the more recent plants.

An emergency core-cooling system is also provided to refill the primary coolant system and to provide cooling of the fuel for an indefinite period after any loss-of-coolant accident. The emergency cooling water is supplied from three sources: high-pressure water from storage tanks, medium-pressure water from the dousing tank and at a later stage, if necessary, water recycled from the reactor building floor where it would have collected after leaving the reactor and dousing system.

SAFETY ANALYSIS

Before a CANDU power plant receives a licence to operate in Canada, the designers, operators and the AECB must all be convinced that releases of radioactive material from the plant during various postulated accidents will not exceed levels prescribed by the AECB. To convince the AECB requires a detailed assessment of the behaviour of the plant and its many safety systems for various hypothetical accident scenarios.

It is likely that more safety analysis is done for a nuclear power plant prior to it receiving its operating licence than for any other industrial activity. As in any industrial plant, a large number of hypothetical accidents can be postulated to occur. However, some judgement must be exercised with respect to accident severity. For the early plants, the safety analysis was based on assessing the consequences of the failure of critical components in the various process systems. More recently, safety design matrices have been used to identify the critical accident sequences (see chapter four). Accidents that require the intervention of one or more of the safety systems to prevent fuel failure and/or the release of radioactive material to the environment are considered most important. These are defined as serious process failures and the Canadian regulatory approach requires that the consequences of such events, both when considered alone and in combination with the unavailability of one of the safety systems, must be assessed. These two situations are the so-called single-failure and dual-failure accident scenarios, respectively. Together they cover a wide spectrum of accidents, whose analysis enables an overall risk assessment of the plant to be made.

Postulated single-failure accidents are analyzed using pessimistic assumptions regarding equipment performance and process behaviour. This approach ensures a conservative safety system design that would be fully capable of handling real accidents. The single-failure accidents are categorized according to the process system where such failures are postulated to occur, mainly the primary coolant system, the steam and feedwater system, the fueling machine, and the control system. The last could result in an uncontrolled increase in reactor power (a LORA) until the increase activated the shutdown systems.

Analyses are also performed for serious process failures combined with unavailability of either the emergency cooling system or the containment system. Analysis of loss of reactor shutdown capability is not required where two completely independent, each fully capable, shutdown systems of high reliability are provided.

Since it is not feasible to determine experimentally the consequences of the various accident scenarios using the power plants themselves, this must be accomplished using mathematical simulations based on extensive experimental data describing both system and component behaviour. These simulations include computer models of the various plant components and subsystems, and follow the behaviour of these after the onset of the postulated accidents. These models are based on extensive, well-established performance data for the various system components, such as pumps, valves, etc. Some of the important questions that must be addressed by the safety analyses are:

- Will any fuel fail during a LORA?
- How much fuel will fail during a LOCA, both with and without action by the emergency core cooling system?
- If fuel fails, what fraction of the fission products will escape, and where will they go?
- Will any pressure tubes fail during a LOCA, with and without emergency cooling?
- How does the horizontal orientation of the fuel channels affect their integrity during a LOCA?
- Will a large quantity of hydrogen be produced during a dual LOC/LOEC[2] accident, and how could the resulting potential for a hydrogen/oxygen explosion affect the performance of the containment system?
- Will the reactor containment building withstand the pressure rise associated with a LOCA, and possible combustion of hydrogen?

The ultimate question is, of course, how much radioactive material could be released to the environment and what radiation doses could members of the public receive? For the plant to receive its operating licence, these calculated releases and consequent radiation doses must be below the limits set by the AECB.

The role of safety analysis is to provide credible answers to questions such as the above. The credibility of these answers depends on the validity of the mathematical models that comprise the plant simulations, and this is where reactor safety R&D plays an important role.

SAFETY R&D

The first CANDU-specific safety R&D activities began in the late fifties and early sixties in response to concerns raised during the design of the NPD and Douglas Point reactors. Experiments to address these concerns were done in the laboratories and research reactors at Chalk River and at various engineering laboratories, principally at Canadian General Electric (the designer of NPD), AECL's Nuclear Power Plant Division (NPPD) in Toronto, and Ontario Hydro. The rapid response of these groups and the results obtained enabled those first plants to obtain their operating licences. For the larger commercial plants, as noted earlier, much more comprehensive safety analysis was required and a more comprehensive safety R&D program was established to support the analysis.

From the outset, the objectives of the CANDU reactor safety R&D program have been to develop a thorough understanding

[2] LOEC stands for Loss of Emergency Coolant

of the phenomena that might occur during reactor accidents, and to develop and verify the mathematical simulations used in the plant safety analyses. This would ensure that the latter would describe the events that might occur during postulated accidents with a credibility satisfactory to the designers, the operators and the AECB. Since the models could not be verified by comparison with real accidents, the approach taken was to check the individual submodels that comprise the complete plant model against laboratory experiments that examined a particular property or subsystem of the power plant (called "separate-effects" tests), and by using special test facilities that included many of the features of the actual plant, but sometimes at less than full size (called "integrated" tests). The overall objective was to establish a sound theoretical basis for the plant model by verification with experiments of increasing scale and complexity. In this way, the model should be able to describe the events that might occur during a postulated accident in a full-size plant.

The safety R&D program divides naturally into several areas:

- reactor physics, mainly concerned with the reactivity effects associated with loss-of-coolant accidents;
- thermalhydraulics, dealing primarily with the behaviour of the primary coolant system and the emergency core-cooling system during loss-of-coolant accidents;
- fuel behaviour, dealing with how the fuel responds during various accidents;
- fission-product release and transport, which deals with the fraction of the fission products released from the fuel during an accident and how these migrate after they are released, the latter strongly dependent on fission-product chemistry;
- fuel-channel behaviour, which examines how the pressure and calandria tubes behave, particularly during a combined LOC/LOEC accident;
- containment behaviour, dealing with how this system responds to the pressure rise during a severe LOCA, how well it retains fission products and how it copes with any hydrogen that might be produced.

Each of these areas will now be discussed in more detail.

Reactor Physics

Reactor physics is involved in all reactor core behaviour, both during normal operation and during accident situations. It is particularly important for analyzing LOCAs in CANDU due to the reactor's positive void coefficient. From the beginning, the designers of NPD and Douglas Point wished to know as accurately as possible how the core reactivity would vary with time if coolant were to be lost from the reactor. One important reason this information was needed was to help decide whether dumping the heavy-water moderator was going to be sufficient to overcome the reactivity increase on coolant loss. Since there was concern about the ability of the reactor physics codes to calculate the coolant-loss reactivity effect with sufficient accuracy, appropriate experiments were done in the ZEEP and ZED-2 lattice-testing reactors at Chalk River in the late fifties and early sixties. These experiments confirmed that coolant voiding would lead to an increase in core reactivity and showed that for the smaller CANDU reactors, such as NPD and Douglas Point, moderator dump was sufficiently fast to shut the reactor down safely during a LOCA. However, this was not the case for the larger reactors and this led to the introduction of fast-acting shutdown systems for all CANDU reactors after Douglas Point.

Another experiment done in ZED-2 showed that removing moderator from around a single fuel channel reduces the core reactivity, as predicted. This is because the lack of moderator increases the likelihood that neutrons will escape from the reactor before they can be slowed down to the point where the fission reaction is more likely to occur. This result confirmed the hypothesis that for a worst-case accident, where fuel channels might fail, the steam entering the moderator would displace the latter, resulting in reactor shutdown.

Thermalhydraulics

One of the most important safety-related issues in the thermalhydraulics area concerns the rate of coolant loss from the primary circuit in the event of a major pipe break. This issue was first raised in the late fifties in connection with NPD, and later for Douglas Point. The rate of coolant loss was a critical parameter, since it would affect the rate of pressure rise in the reactor building and the amount of positive reactivity inserted into the reactor core before the shutdown system could take effect. Experiments were done at both Chalk River and NPPD in Toronto to measure coolant loss rates from pressurized vessels through pipes of different diameters and different lengths. The results were used in the safety analysis for the early reactors.

Safety analysis involves many assumptions and approximations, since many of the situations analyzed are too complex to be solved exactly. For example, one of the LOCA scenarios that must be analyzed assumes that a large break occurs in the primary coolant circuit. Such an event would cause a sudden drop in pressure in the circuit and the conversion of much of the pressurized coolant to steam, so that the escaping coolant is a two-phase mixture of steam and water. The first computer models used by the designers assumed that the steam and water moved at the same velocity, were of equal temperature, and were uniformly distributed in the fuel channels and piping, and that all of the individual fuel channels behaved in the same "average" way. It was also assumed that the emergency cooling water entered all of the fuel channels simultaneously and that the water was distributed uniformly across the fuel channel, so that all of the fuel elements were cooled at the same rate. However, because of the complex piping arrangement, and the horizontal orientation of the CANDU fuel channels, there was some concern that not all of the fuel channels would receive emergency cooling at the same time and also that when the emergency coolant reached the fuel channels there would be stratification of water and steam, with the water flowing along the bottom of the channel, and the steam along the top. It was felt that this could result in a large temperature difference between the top and bottom fuel elements.

To investigate these phenomena, experiments were done at a special facility located at Westinghouse Canada Limited (now the Stern Laboratories) in Hamilton, Ontario. This facility, called the Cold Water Injection Test (CWIT) facility, was built in 1972-73 and consisted of two horizontal Zircaloy flow tubes the same size as CANDU pressure tubes, each containing a 37-element bundle of electrically heated elements to simulate CANDU fuel (see figure 15.1). The two channels were separated in elevation by 5 m and were connected to inlet and outlet headers (located 10 m above the bottom channel) via CANDU-type feeder pipes. In this way some of the important geometrical features of the CANDU plant could be modelled. Superheated steam was used to heat the pressure tubes up to about 300°C (the CANDU operating temperature) and the electric heaters were used to raise

FIGURE 15.1
Cold Water Injection Facility

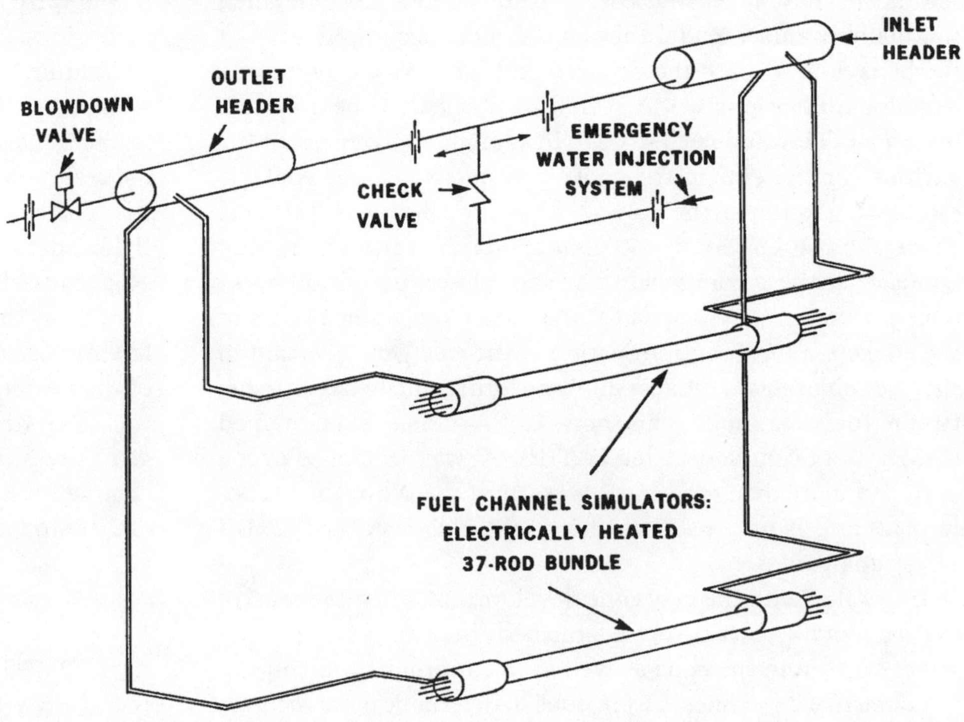

the fuel-element temperatures to the desired level (350°C to 500°C). The loop was then depressurized using a quick-acting valve (to simulate a LOCA), and cold water was automatically injected once the loop pressure fell below the injection pressure.

One of the first experiments performed in the CWIT facility in 1973 on cold-water injection into preheated horizontal rod bundles indicated that the injected water ran along the bottom of the fuel channel. The heat rejected by the elements in contact with the water produced steam, which cooled the upper elements. However, because the water and steam cooling rates were different, the temperature difference between the upper and lower elements could be quite large (up to several hundred degrees). As cold-water injection proceeded, the water rose in the channel and more fuel elements were progressively wetted, until eventually the whole rod bundle was wetted and the channel was filled with water.

Because the existing codes could not treat this type of situation accurately, it was necessary to develop a more detailed thermalhydraulics model that would more accurately predict the behaviour of both the primary and secondary coolant systems for the various LOCA scenarios that had to be assessed. Such a model would need to contain submodels to represent the various components of the coolant systems; i.e., the coolant-circulating pumps, the flow-distribution headers, the fuel channels, the boilers, the secondary circuit and the piping connecting these components. It should also permit different temperatures and velocities for the steam and water phases of the coolant, allow for stratification of the emergency coolant in the fuel channels, and also allow for different behaviour between fuel channels. This new LOCA model (first named RAMA, but eventually called CATHENA) was developed over a period of more than ten years by a team at the Whiteshell Laboratories and is now used by the AECL designers for CANDU safety analyses.

In parallel with the new code development, a comprehensive experimental program was established, both to verify the parameters used in the code as well as to confirm its predictions. Fundamental experiments were done to determine more accurate values for heat-transfer coefficients for flow-boiling[3] situations. These included two series of coolant blowdown[4] experiments: one using hot, pressurized water, and another employing Freon. (Freon was used because it behaves like water, but at much lower temperature and pressure, thus making the experiments much easier and less costly.) In the latter experiments, high-speed films were used to visualize the flow development at various locations along the transparent pipe used.

Early LOCA analyses indicated that fuel-sheath temperatures depend significantly on coolant pump behaviour in two-phase flow, since the pumps affect the pressure and flow distributions in the primary circuit, hence heat removal from the fuel. The existing pump models were based on empirical data and assumptions that led to quite conservative predictions (i.e., high values) of fuel temperature. Experiments were undertaken to develop more accurate pump models for use in the two-phase flow regimes common to all LOCA analyses.

Over a fifteen-year period, many different types of experiments were conducted in the CWIT facility at Westinghouse, to provide a better understanding of fuel-channel thermal-hydraulics for various reactor upset conditions. These included symmetric and asymmetric header-break/coolant-injection simulations, tests with stagnant liquid-filled initial conditions, natural-circulation flow-stability tests, and tests to determine the time between the injection of emergency cooling water into the headers and its arrival at the fuel channel, in the absence of a pressure difference between the inlet and outlet headers, a condition that could occur for some large-break LOCAs. The latter experiments showed that water entered preferentially one of the feeders of each fuel channel, and that channel rewetting was delayed until the filling of that feeder established a sufficient pressure differential to drive the water through the fuel channel. A model for this time delay was subsequently incorporated into the LOCA simulation.

[3] Flow-boiling is the term used to describe how water flowing in a pipe is converted to steam via heating and/or depressurization.

[4] Blowdown is the term used for the escape of fluid from a hot, pressurized system after a break occurs.

A unique feature of the CANDU design is the flow-distribution headers. These headers (or manifolds) are long (about 6 m), large-diameter (about 0.3 m) pipes that connect the inlet feeder pipes to the coolant pumps (called the inlet headers), or the outlet feeder pipes to the steam generators (the outlet headers). In a postulated LOCA, emergency coolant is assumed to be injected into these horizontal headers, which then distribute the flow, via the feeder pipes, to the fuel channels. Because of the geometrical arrangement, flow stratification in the headers could affect the distribution of flow to the feeders during emergency coolant injection. To investigate this effect, a full-scale header facility representative of the Pickering plant was constructed at Whiteshell in 1974. In this facility, which had 110 feeder pipes, the header temperature could be brought to the desired value using electrical heaters, and water injection was initiated using a quick-acting valve. During coolant injection the liquid level and pressure were measured at several positions along the header, and feeder flow rates were monitored, for a range of conditions. This facility was dismantled in 1975 to make room for the RD-12 facility, described below.

FIGURE 15.2
Schematic of Full-Scale Header Test Facility

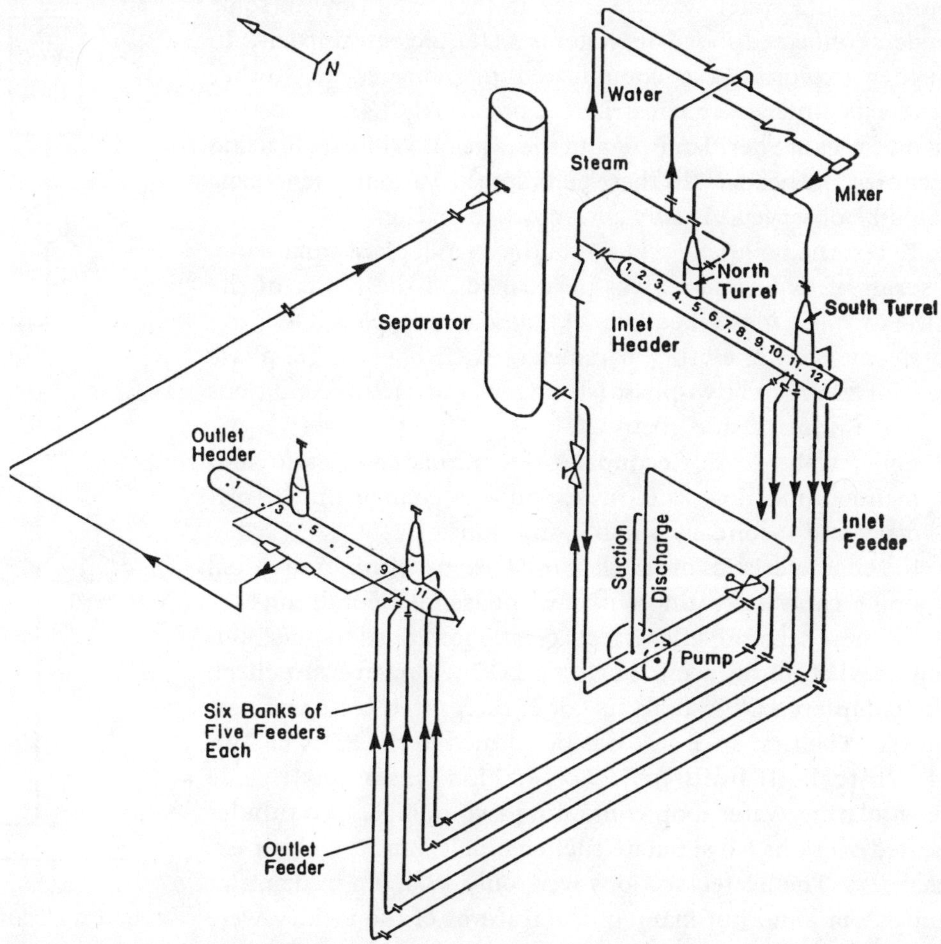

In the late seventies, a small-scale model header facility was constructed at Whiteshell. This facility had 6" diameter headers and fifty feeders of 3/4" diameter. A key feature of this facility was the use of two transparent windows that permitted direct observation of phenomena that could occur in the two-phase (steam/water) mixture during emergency coolant injection. However, because this small-scale facility did not permit investigation of all the phenomena of interest, another full-scale header facility was constructed in the early eighties at Westinghouse (see figure 15.2). This facility comprised two full-diameter, half-length CANDU headers with feeder takeoffs geometrically similar to those in the current CANDU plants. Blowdown from, and emergency coolant injection into, one or both of the headers was possible in this facility, which was used to study the behaviour of two-phase flow in headers under LOCA conditions. The tests showed that stratification of the steam and water phases occurred and that steam flowed preferentially to the feeders connected to the header above the water level. In some feeders, steam was detected even though the connection to the header was below the steam-water interface (called vapour pull-through) and water was observed in the steam flow in some feeders connected above the water level (liquid entrainment). To further explore these complicated phenomena, two other header facilities were constructed, one at AECL's engineering laboratories at Sheridan Park and the other at Whiteshell, using lucite for the headers. In these facilities the various phenomena could be observed visually and recorded on film.

To test the boiler model used in the overall plant simulation, a series of experiments was performed on the boiler of the RD-12 facility (described below). These tests explored:

- steady-state operating parameters with both single-phase (i.e., water) and two-phase (steam plus water) flow conditions at the primary side inlet;
- boiler water level, secondary-side steam/water ratio, and recirculation flow rate, in response to changes in the primary and secondary circuit conditions;
- the characteristics of the flow oscillations within individual boiler tubes operating with two-phase inlet conditions.

All of these separate-effects tests greatly improved understanding of what might occur during a LOCA. However, to check the complete LOCA analysis code, integrated tests were required. The first such test was performed in the RD-4 facility at Whiteshell, built in 1972-74. This was a small-scale recirculating water loop containing two pumps, two tubular heated sections (to simulate fuel channels), and two heat exchangers. The heated sections were only 13.4 mm in diameter and 1.5 m long, but many of the features of the facility were relevant for the verification of the analysis methods for pressure-tube reactors. Coolant blowdown experiments were done to study the effect of pipe-break size on depressurization rate, flow distribution and heated-section temperature. Experiments were also done with cold water injected simultaneously into each header at the end of depressurization.

The RD-4 facility was succeeded in 1976-77 by the much larger RD-12 facility, which had heated sections four metres long with bundles of seven 15 mm diameter electrically heated elements to simulate the fuel (see figure 15.3). This facility also

FIGURE 15.3
RD-12 Recirculating Pressurized Water Loop

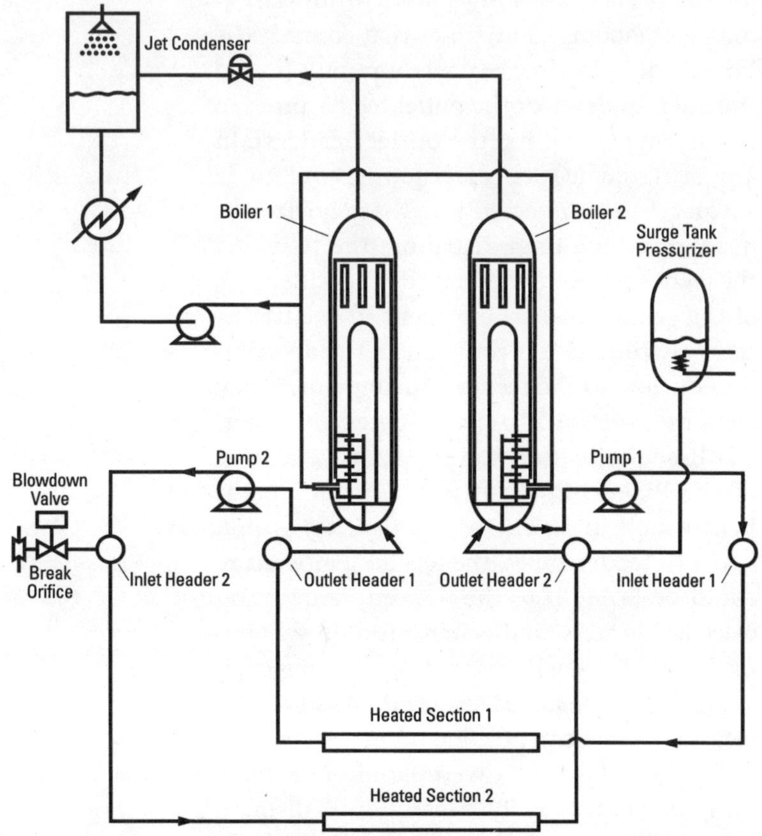

had recirculating U-tube type boilers, with an interacting secondary circuit and a much wider range of operation than RD-4. A pressurized cold-water injection system was also provided, to supply water simultaneously to the four headers when the loop pressure fell below a preset value, thereby simulating a LOCA with emergency coolant injection. The experimental program included tests with various pipe-break sizes, located at different points in the loop circuit, various cold-water injection pressures, and several modes of boiler cooling. In all of these experiments detailed measurements were made to determine coolant flow rate, pressure, temperature and density distribution throughout the loop, as well as fuel-element surface temperatures, and differential pressures across various loop components.

The RD-12 facility was followed in 1982-83 by an even larger facility, called RD-14, which was a model of a primary coolant loop with the various components arranged geometrically to reproduce the gravitational effects in a CANDU plant (see figure 15.4). It consisted of two full-scale, full-power (6 MW each) fuel channels, each containing a full-length, 37-element, electrically-heated bundle to simulate the fuel, plus full-size feeders and two full-height steam generators, all arranged in the CANDU figure-of-eight configuration. The steam generators had full-size U-tubes, but the number of tubes was reduced in proportion to the number of heated channels, to give the correct heat transfer area per fuel channel. The loop was designed so that fluid mass-flow rate, transit times and pressure/enthalpy terms in the primary system of the loop were the same as those in a typical CANDU under both forced- and natural-circulation conditions. The emergency coolant injection (ECI) system was also designed to permit simulation of the various ECI systems found in CANDU reactors.

The RD-14 facility, which cost more than $10 million, was one of the largest of its kind in the world, with an overall height of about 34 m. It was used to investigate many LOCA-related phenomena, such as:

- blowdown tests for many different pipe-break sizes, including the critical break size, where the pressure difference across the reactor core becomes close to zero, resulting in a stagnant flow condition in the fuel channels for an extended period (tens of seconds);
- two-phase thermosyphoning tests, to study the situation that would arise from a loss of the coolant pumps, or during small-break LOCAs, where the fuel channels are cooled mainly by natural circulation.

This facility was also used to perform several secondary-side depressurization tests.

The large size and height of RD-14 required the construction of a special building to house it at the Whiteshell Laboratories. Its operation was quite spectacular, since each blowdown test resulted in the release of a large quantity of steam (through a pipe on the roof of the building), which could be seen for miles around on the flat Manitoba landscape.

In the mid-eighties, the RD-14 facility was converted from a two-channel to a ten-channel facility, called RD-14M, with the channels at different elevations (see figure 15.5). The total loop power is the same as for RD-14, so each of the RD-14M channels operates at one-fifth the power of the original RD-14 channels. This new arrangement has allowed the study of the interaction between parallel fuel channels for both natural-circulation cooling and LOCA situations.

Many useful results were obtained from the tests done with the RD-14 and RD-14M facilities and these were used to

FIGURE 15.4
Schematic Diagram of RD-14

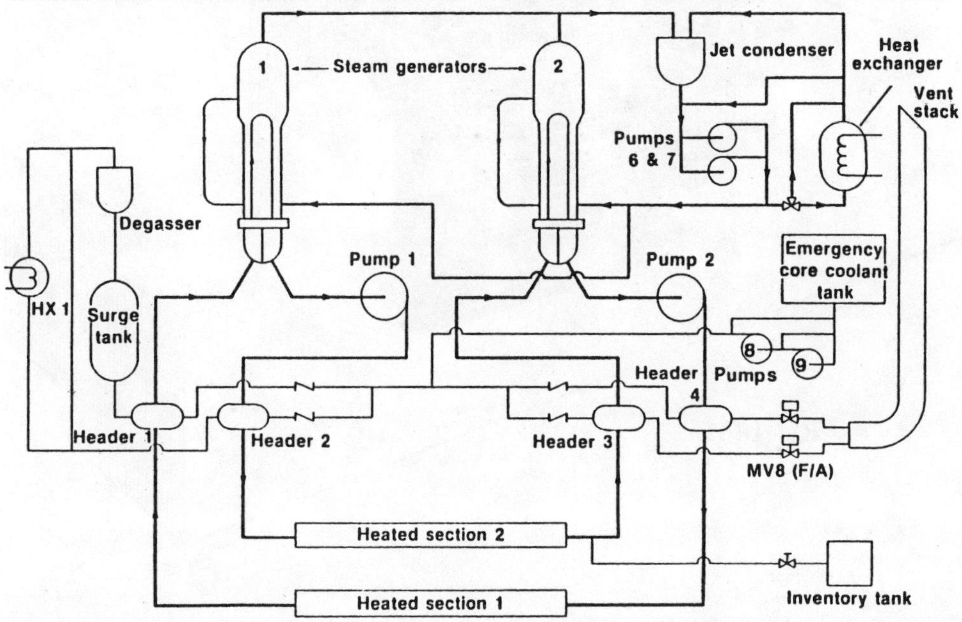

FIGURE 15.5
Schematic of RD-14 Multiple-Channel Facility

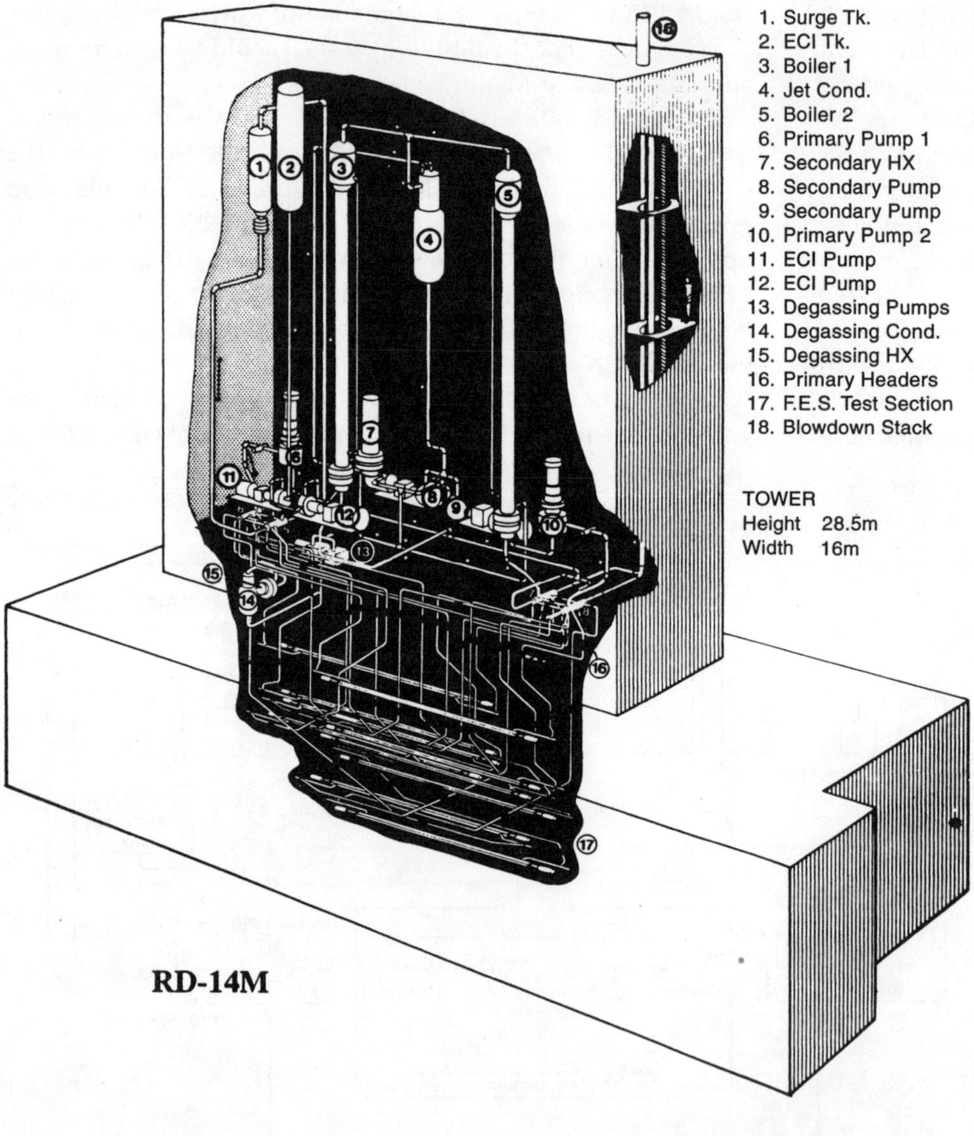

refine and verify the CATHENA code. The results of these tests show that for all LOCAs, of any break size, with or without power to the coolant pumps, fuel sheath temperature will not exceed about 600°C, hence fuel should not fail, as long as the emergency coolant system is available.

The thermalhydraulics experiments required the measurement of many parameters, such as absolute and differential pressures, temperature, cross-section-averaged flow rates, cross-section-averaged coolant densities, and the distribution of the vapour and liquid phases within the fuel channel. Some of these measurements required that existing commercial instruments be modified to operate under high-temperature, high-pressure conditions, and in some cases new instruments had to be developed. One important development was a gamma-ray densitometer which employed three collimated beams from a single gamma-ray source to determine the spatial distribution of steam and water within the fuel channel. The three beams traversed different paths across the channel and from the difference in attenuation of the radiation by the steam and the water the distribution of the two phases could be determined. The method is similar to X-ray tomography, used in CAT scans to examine the internal features of various objects, as well as of humans.

Fuel Behaviour

As explained earlier, the greatest hazard associated with the operation of a nuclear power plant is the harm that might result from a release of the radioactive fission products and actinides contained in the fuel elements – and the hazard is greater the longer the fuel has been irradiated. In the early years of nuclear power development (the mid-to-late fifties), little was known about the behaviour of power reactor fuel of the CANDU design at any level of irradiation. At that time, the research programs were directed at obtaining a basic understanding of fuel behaviour under normal operating conditions and establishing limits for safe operation of the fuel. However, many exploratory experiments were done with operating parameters well beyond what were later specified as safe operating limits. For example, fuel was operated at power ratings high enough to produce central melting of the uranium oxide. Some defects occurred, and in some cases the hot central part of the fuel came in direct contact with the coolant. Operating conditions were varied to permit extensive boiling of the coolant, with the vapour content approaching 100 percent. In one test, conditions were arranged to produce dryout[5] of some parts of the fuel sheathing. The heat transfer coefficient between the fuel and the coolant is reduced when dryout occurs, which can result in high fuel and sheath temperatures, increased corrosion, and eventually sheath defects.

During the design phase for NPD and Douglas Point there was some concern as to how the fuel for these reactors might behave during a major LOCA; in particular, how the fuel would withstand the thermal shock associated with the injection of the emergency coolant, and whether serious fuel sheath rupture would occur due to a large increase in internal fission-gas pressure. In 1961-62, several experiments were done with unirradiated fuel elements at Canadian General Electric to address these issues.

To investigate the thermal shock phenomenon, a 19-element fuel assembly mounted in a Zircaloy pressure tube was heated to about 1,100°C in a salt bath, after which room-temperature water was injected into the fuel channel. Some of the fuel elements bowed, causing the fuel bundle to jam in the tube, but there was no significant fuel-element damage. To investigate the effect of internal gas pressure, several fuel assemblies were pre-pressurized internally and heated to about 1,100°C. Several fuel sheaths failed but there was no major damage. In other experiments, where one fuel element was heated to about 1300°C, all elements withstood the initial internal pressures. The fuel sheaths expanded by about 30 percent, and the sheath damage consisted only of cracks 6 to 12 mm long.

These early experiments gave the designers confidence that serious fuel failure was not likely to occur even for the worst accidents envisaged at that time. However, they did indicate that there were many areas that required further study, particularly as it was planned to operate fuel in the commercial plants at significantly higher power ratings.[6] Thus, over a period of several years, starting in the sixties, a comprehensive R&D program was established to determine CANDU fuel behaviour for the various postulated accident scenarios. This work was mainly done at Chalk River and included both analysis and experiments to investigate:

- the rate of heat transfer between the fuel and the sheath, and between the sheath and the coolant;
- the extent of fuel melting and the reaction between molten fuel and the sheath;
- the amount of fuel swelling due to both thermal expansion and the expansion of the contained fission-product gases;
- the temperature history of the Zircaloy sheath and its subsequent mechanical behaviour;
- reactions between the sheath and the coolant that could lead to oxidation and embrittlement;
- the potential for coolant flow blockage due to mechanical deformation of the fuel bundle.

The performance of these experiments required the development of special techniques and equipment. For example, an electric heater positioned inside a string of annular fuel pellets was used to simulate a CANDU fuel element. With this simulated fuel element, power ratings up to three times the normal operating levels for the outer elements of a CANDU fuel bundle were possible in the laboratory.

---

[5] Dryout occurs when the surface heat flux becomes so great that a liquid film cannot be maintained on the outer surface of the fuel sheath.

[6] Power rating here is defined as power output per unit length of a fuel element.

Since laboratory tests could not supply all the data needed, some in-reactor tests were necessary. Originally, the NRX and NRU loops were not designed to accommodate continued operation with defective fuel. If a serious fuel defect did occur, the reactor would have to be shut down to minimize the release of radioactive material to the loop, the reactor hall, and eventually the environment. From the late sixties to the late seventies one loop in NRX (the X-4 loop) was used for testing fuel elements with coolant conditions ranging from hot water to superheated steam, at sheath temperatures up to 500°C. These experiments helped to define the time that fuel could operate with elevated sheath temperatures without defecting. However, the loop was not designed to cope with large amounts of radioactive material. So additional facilities had to be built to investigate fuel/fuel-channel behaviour during postulated accidents.

In the early seventies, the X-2 loop in NRX was modified to allow operation with defected fuel, either as single elements or clusters of three elements (called trefoils) (see figure 15.6). The fuel assemblies were instrumented to measure fuel-centre and surface temperatures, sheath temperatures, fuel-element internal gas pressure, coolant temperature and coolant pressure. Six experiments were done in the X-2 loop and many useful results were obtained. However, the modified loop could not provide all the information needed. Thus, a new loop (the X-9 loop) was designed for NRX, but uncertainty over the long-term future of NRX resulted in cancellation of the project.

The experiments in the X-2 loop facility gave many valuable results. For example, they

FIGURE 15.6
Schematic Diagram of X-2 Loop

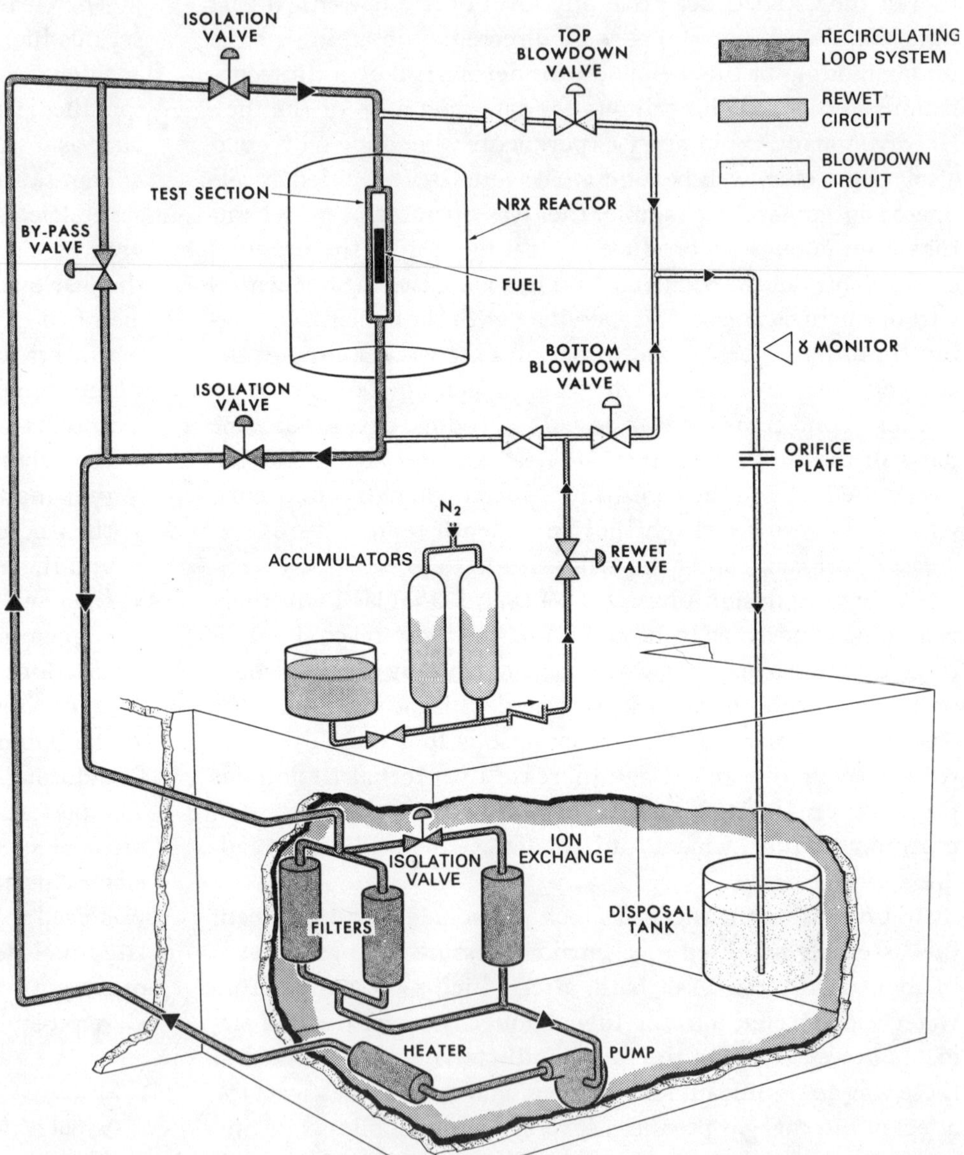

showed that central melting of the fuel did not cause sheath failure. The molten fuel did not flow out between the pellets; it remained at the centre of the fuel-pellet stack even when sheath failure occurred by accelerated corrosion at power, at temperatures greater than 1,000°C. No fuel/sheath interactions were observed. The deterioration of defective fuel operating in dryout conditions was slow for sheath temperatures below about 800°C, and the extent of deterioration was not dependent on whether the fuel was initially defected. It was also confirmed that a small diametral expansion of the sheath was effective in reducing the internal gas pressure in the fuel element. This is due to the relatively small initial internal void volume in CANDU fuel elements. A small sheath expansion causes a significant percentage increase in internal void volume, hence a decrease in internal pressure, the driving force for sheath expansion. This is very different from the behaviour in LWR fuel, which is much longer than CANDU fuel, has a large plenum void volume, has free-standing sheaths, and is pressurized to prevent sheath collapse by creep. During a LOCA in a light-water reactor, the sheath heatup is localized to the high-power region of the reactor core. The entire filling-gas volume inside the element is available to strain the sheath at the hottest region, and strains to failure are possible.

Experiments were also done in the U-1 loop at NRU, where fuel elements were operated in and out of dryout on many occasions during a five-week period. These tests showed that CANDU fuel could operate safely for short periods of time (tens of minutes) in dryout at relatively low temperatures, such as might occur during a loss of spatial power control, when the fuel in a region of the reactor might operate above the normal operating level.

The irradiations in NRX and NRU also showed that when a fuel element failed at a power level high enough to cause central fuel melting, there was no violent reaction between the fuel and the coolant. No large pressure peaks were recorded and there was no damage to the pressure tube. The initial release of fission products was significant, however, and would serve as a signal to the reactor operators that a fuel defect had occurred, as discussed further in chapter thirteen.

In another series of tests at Whiteshell, CANDU fuel bundles were heated in vacuum to 1,400°C and 1,600°C and held there for thirty minutes to study their mechanical behaviour at elevated temperatures. In these tests the bundles slumped into contact with the pressure tube and bundle elements slumped into contact with their neighbours. However, there were no sheath failures and the bundle end-plate/end-cap welds remained intact.

After the cancellation of the X-9 loop facility, plans were made for a new facility to be installed in the U-1 loop position in the NRU reactor. This new facility, called the Blowdown Test Facility (BTF), was to be capable of creating the conditions that CANDU fuel might experience during LOCAs (see figure 15.7). The overall objective of the project was to perform integrated tests to provide a database of information on fuel behaviour and the release of fission products from severely damaged fuel to the primary heat transport system, for a range of sheath temperatures. It was clear that the BTF facility would take quite some time to build and commission and, furthermore, it would not be possible to reduce the coolant flow and simultaneously provide a power pulse to simulate the power surge that would occur during a large-break LOCA in a CANDU reactor. However, a facility to perform such tests, called the Power Burst Facility (PBF), existed in the United States, and arrangements were made to test CANDU fuel there while BTF was being built.

Four CANDU fuel elements were tested in the PBF in the early eighties. Three of these had been previously irradiated in the NRX reactor to produce the fission-gas inventory and fuel conditions typical of high-power fuel elements in a CANDU reactor, while the fourth was unirradiated. During the test the fuel-element power was adjusted to match the temperature vs. time profile expected in a LOCA, and reflooding of the test section after coolant blowdown was used to simulate the emergency coolant. Only one of the four elements failed during the test, that being the irradiated element that had been back-filled to a pressure 2.5 times the maximum predicted for a CANDU fuel element. This element expanded about 26 percent before failing, and the failure was a small

FIGURE 15.7
Blowdown Test Facility (BTF)

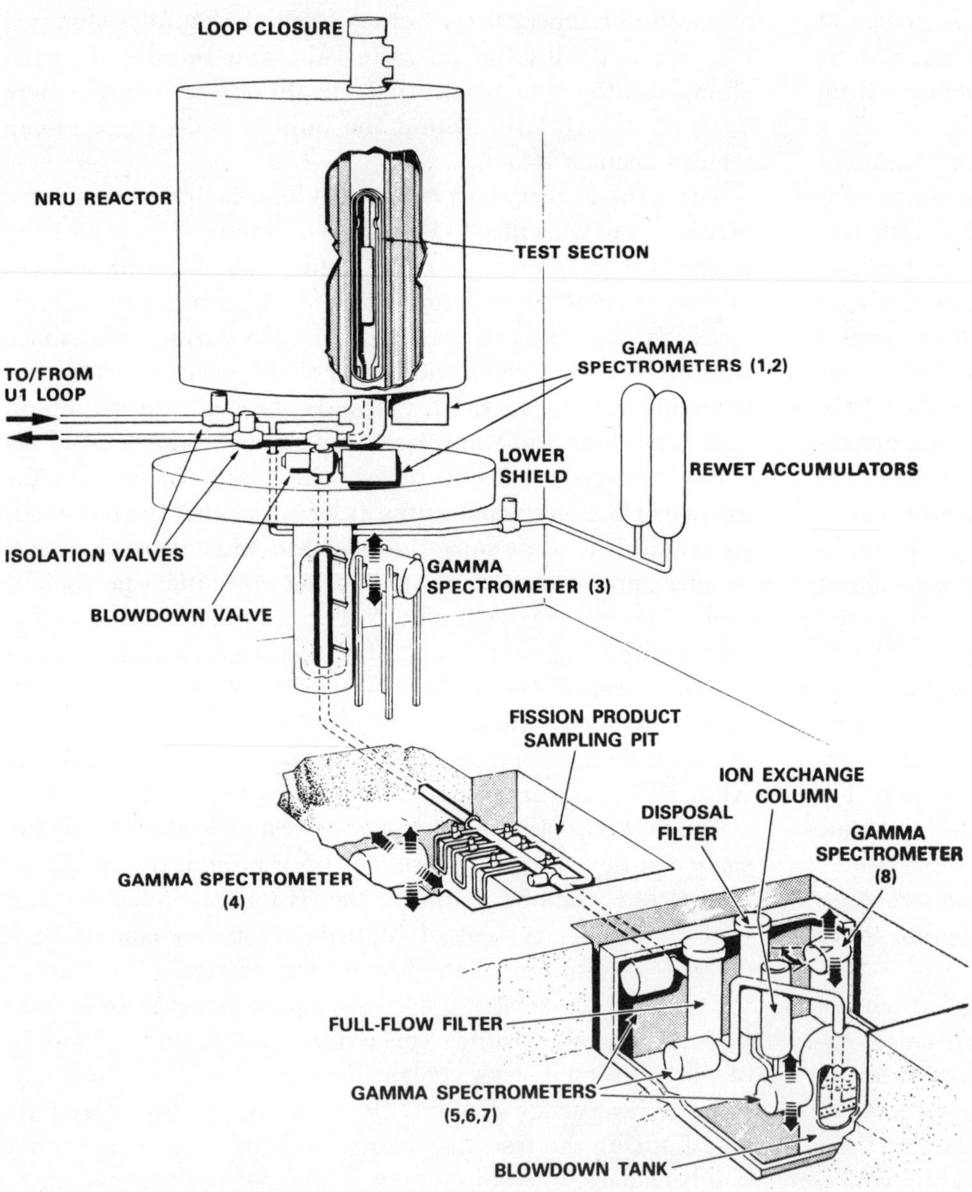

crack. The other three elements showed strains (diametral expansions) of 3 to 4 percent, but did not fail. The important conclusion from the PBF test was that for CANDU fuel elements, even with the maximum design burnup, there was no indication of failure due to the power pulse expected during a large-break LOCA.

In the early seventies, in parallel with the experimental program, theoreticians were developing codes to describe the behaviour of CANDU fuel under accident conditions. This was a challenging task, since the behaviours of the various materials comprising the fuel bundle were not simple. For example, the behaviour of the Zircaloy sheathing was quite complex for the range of temperatures that could occur during a LOCA. Deformation of the material depended on the initial strength, the grain size and the crystal structure. The grain orientation was also important since the alloy deforms differently in different directions (i.e., it is anisotropic). Also, at 820°C it begins to undergo a phase transformation which is complete by 950°C. The deformation is the result of "creep" processes (see chapter thirteen), which behave differently at different temperatures. The fuel sheathing also oxidizes, and at increasing rates at higher temperatures. The oxidation reaction is also exothermic (i.e., it produces heat). At the rapid oxidation rates expected during a LOCA with impaired emergency cooling, the heat produced would cause a further significant increase in sheath temperature. Zircaloy also absorbs oxygen and if the concentration were to exceed a critical value the sheath could become embrittled, so that it might fail during the injection of emergency cooling water. This initially caused considerable concern since work in the United States for light-water reactors suggested that sheath temperatures should not exceed 1,200°C. However, further work showed that failure of the fuel sheath due to oxygen embrittlement is very unlikely during a LOCA transient to 1,350°C, because the time at temperature is too short. All of this information was incorporated into a microstructural model for creep deformation that would be used to predict with high credibility the behaviour of a CANDU fuel sheath during a LOCA.

Two different approaches were taken initially to develop models of fuel bundle behaviour during accidents. One led to the development of the ELOCA code which treated the problem in only one dimension (radial) but was firmly based on the data obtained from the many fuel experiments. The other approach used a code called FAXMOD and was an attempt to treat the fuel behaviour problem in three dimensions, starting from the basic physical properties. The latter approach was found to be extremely difficult and was eventually dropped, partly because of the inordinate amount of effort that would have been required, but also because ELOCA, which was later upgraded to two dimensions, was providing results with the required credibility. With ELOCA it is possible to predict how CANDU fuel bundles would behave during the various accidents postulated for the CANDU reactor. Later, ELOCA was incorporated into the CANSIM code suite which is able to treat the interactions between the fuel bundles, the fuel channels, and the fuel-channel thermalhydraulics, thereby giving a complete picture of what happens in the reactor core during an accident.

Fission-Product Release and Transport

Perhaps the most important concerns related to safety in any reactor are the fraction of the fission products that will escape from the fuel during an accident, and where they will go. Thus, from the beginning, an important part of the safety R&D program has been to investigate how fission products escape from the uranium oxide fuel matrix and how they migrate throughout the plant. As early as 1956, experiments were done to measure fission-product release from irradiated UO_2. These early tests showed considerable fission-gas release from UO_2 pellets heated to 1,400°C, and that gas release was less for higher UO_2 densities. These early tests were followed by more extensive experiments in the early sixties using the X-2 loop in the NRX reactor, where a sweep-gas system was employed to carry the volatile fission products to a location where they could be measured using gamma-ray spectrometry.

Other experiments were performed in heavily shielded cells (called hot cells), where the test parameters could be more carefully controlled than in the reactor. Such experiments were used to measure the release of fission products from uranium oxide fuel previously irradiated to burnups typical of those for CANDU reactors. The behaviour of the uranium dioxide fuel at elevated temperatures is important here. Oxidation of the uranium increases as the temperature increases and oxides with higher oxygen content are formed. The result is a decrease in fuel density, a collapse of the grain structure and release of the fission products held in the grain boundaries. However, some fission products (20 percent of the iodine and cesium) remain in the fuel grains until they are released via a slower mechanism such as diffusion.

These studies provided excellent data on the diffusional release of iodine and cesium, as well as the oxidation-enhanced release of several other fission products. The deposition of fission products on surfaces, as functions of material and surface temperature, and the movement of iodine, cesium and xenon along piping were also measured.

In the early days of safety analysis it was assumed that a large fraction of the volatile[7] fission products would escape from the reactor, and the protective systems were designed accordingly. This assumption was strongly influenced by the fact that during the Windscale research reactor accident in the United Kingdom in 1958, most of the volatile fission products escaped from the plant. Clearly, if no more than a

[7] Volatile here means those fission products that occur in the gaseous form at the temperatures expected during an accident, normally the fission gases and a few other elements, notably iodine and cesium.

small fraction of this inventory were to escape the risk to the public would be considerably reduced. (The Windscale reactor had no effective containment and the accident involved the burning of the graphite moderator and the uranium metal fuel.)

Two of the most hazardous fission products are iodine and cesium, and since iodine is volatile, the worst-case scenario assumed that a significant fraction of the iodine would be released to the environment. Of all the fission products, iodine poses the greatest risk to the public because of its ability to enter the food chain (e.g., via milk products). Thus, during the design of NPD and Douglas Point, there was considerable concern as to how much iodine might escape from the reactors and how much would escape to the environment via the stack during a major accident. W.B. Lewis and G.C. Laurence believed that most of the iodine would probably be trapped inside the plant by the large quantities of water and steam that would be present in the reactor containment building due to operation of the emergency cooling and dousing systems. Some experiments were done at Chalk River in 1960 to investigate this. The experiments showed that most of the iodine was indeed trapped by the water and steam, with the ratio of iodine in the air to that in the water being certainly less than 0.1 percent and probably less than 0.01 percent. It was assumed that any airborne iodine would be removed by filters in the plant stack and work was done at the same time (1960) to find the best material for the stack filter. Several materials were tested, e.g., charcoal, and gauzes of copper and silver-coated copper. Charcoal was found to be the best, with a removal efficiency of better than 99 percent.

Despite this early work at Chalk River, some safety analysts still assumed that a significant fraction of the iodine would be released to the environment. Indeed, USNRC Nuclear Regulatory Guides issued in 1974 (and still in effect at the time of writing) stated that, for design purposes, it must be assumed that 50 percent of the iodine inventory in the reactor core would be released to the primary containment and that 25 percent would be available for leakage from containment. Those values were presumably influenced by the Windscale accident.

However, many believed that the Windscale results did not necessarily apply to water-cooled reactors, such as CANDUs and light-water reactors, and research programs were continued in various laboratories to investigate the forms in which iodine, and cesium, could exist during a LOCA in a water-cooled reactor. One such program was started at Whiteshell in the mid-seventies, before the accident at Three Mile Island in 1979 where, despite earlier dire predictions of the consequences of such an accident, only 15 curies out of a total reactor inventory of 67 million curies of iodine-131 (the critical iodine isotope) were released; the remainder was retained within the reactor and the plant buildings. This incident resulted in an intensification of the work at Whiteshell and elsewhere. These studies concentrated on the tendency for iodine and cesium to form compounds and the stability of these compounds under LOCA conditions. The kinetics of the relevant thermodynamic relationships were derived and incorporated into a computer code. The subsequent analysis, supported by laboratory experiments, showed that under LOCA conditions in water-cooled reactors iodine and cesium would combine to form cesium iodide, and that cesium would also form cesium hydroxide. Since cesium iodide and cesium hydroxide are salts that readily dissolve in water, and since in water-cooled reactors (such as CANDUs and light-water reactors) there would be plenty of water present during a LOCA, only a small fraction of these fission products should be released. The results of this work explained the fission-product behaviour observed during the Three Mile Island accident.

What happened at Three Mile Island was in sharp contrast to the situation that occurred in 1986 at Chernobyl, where the reactor's graphite moderator caught fire and there was little opportunity for fission-product compounds to dissolve in the cooling water. Consequently, there was a large release of fission products which were carried for thousands of kilometres by the wind. The large release to the environment resulted from the poor containment system at Chernobyl. This was similar to the situation at Windscale in 1958. In that case, the reactor was air-cooled and a large release of fission products also occurred, due to lack of a proper containment system.

The CANDU power plant has several large liquid systems: the primary cooling system, the moderator, the emergency coolant system and the dousing system. It is now believed that even in a worst-case accident the majority of the fission products that would escape from the fuel would be dissolved in water, and that any release to the environment would be extremely small. In addition, all CANDU plants have well-engineered, highly reliable containment systems.

Fuel Channel Behaviour

One of the key features of the CANDU reactor is its use of several hundred horizontal pressure tubes to contain the fuel and coolant, rather than a single pressure vessel surrounding the complete reactor core, as in a light-water reactor. While the use of small pressure tubes immersed in a cool moderator, rather than a large vessel operating at high pressure, is believed to have certain safety advantages, questions were raised from the beginning about the integrity of the CANDU pressure tubes and about the possible consequences of a pressure-tube rupture with the reactor at full power. A particular concern was whether a bursting pressure tube might cause the rupture of surrounding tubes, leading to a chain reaction that could result in a catastrophic failure of the reactor core.

To address these concerns, a series of pressure-tube burst tests was conducted between 1958 and 1962. In the first of these, a full-size mockup of a segment of the NPD core was constructed, consisting of a 3×3 array of unirradiated fuel channels located in a large tank of water (see figure 15.8). The central pressure tube was artificially flawed and then ruptured by overpressure. Depending on the test conditions, the calandria tube did or did not fail. It was found that for NPD conditions the calandria tube only failed if the initiating crack was longer than 13 cm. If the calandria tube failed, neighbouring calandria tubes were damaged, but no pressure tubes failed. In a later test series, a Douglas Point fuel channel was used as the central channel in the test array. Because the Douglas Point fuel channel had an open annulus between the calandria and pressure tubes, no calandria tubes failed during these tests. The NPD annulus was more restrictive, with four vent holes at each end, so that the annulus pressure could

FIGURE 15.8
Pressure Tube Burst Test Facility

build up after pressure-tube rupture, increasing the probability of calandria-tube failure.

Parallel experiments were conducted at NPPD in Toronto in 1962 on the explosive rupture of a pressure tube, as might result from a runaway accident, in which the fuel would rapidly overheat. These experiments also involved an array of nine channels located in a large, water-filled tank. Starting with water in the pressure tube at supercritical conditions (24 MPa,

380°C), a thermite bomb was used to simulate the rapidly heating fuel. Despite the extreme test conditions and the almost explosive failure of the test channel, none of the remaining pressure tubes was ruptured, except in one case where a pressure tube failed near a weld remote from the rupture site; this failure was attributed to poor welding. These tests indicated that the propagation of pressure-tube failures in a CANDU core was highly unlikely.

After Douglas Point, the CANDU designs have a restricted outlet from the annulus, more like NPD, and concerns were again raised that failure of a pressure tube might lead to consequential failure of the calandria tube. One fear was that the resulting pressure pulse might damage the shutoff-rod guide tubes that penetrate the reactor lattice between the fuel channels, thereby preventing these rods from dropping into the core to shut the reactor down. Detailed analysis showed that this was extremely unlikely so no further full-scale pressure-tube burst tests were conducted. This decision was confirmed when a major failure occurred in a pressure tube in the Pickering Unit 2 reactor in 1983 with the reactor operating at full power. The bellows at the ends of the fuel-channel annulus ruptured, as expected, but the calandria tube did not fail and the reactor shut itself down safely.

While the early experiments were concerned with the consequences of a pressure-tube rupture, there was still the question of the likelihood of such a rupture. This was of particular concern for a postulated combined LOC/LOEC accident where there would be no emergency coolant available to take away the heat generated in the fuel. In this situation it was calculated that the energy from the fuel would be transferred to the cool moderator by a combination of thermal radiation and conduction. The heat would flow from the fuel, through the pressure tube to the calandria tube, and finally to the moderator, where it would be removed by convection and boiling. In this case the moderator, not the primary cooling system, provides the heat sink.

There was concern, however, that as the fuel temperature rises, the fission-product decay heat would be supplemented by heat generated by the exothermic reaction between steam and the zirconium of the fuel sheath. The high temperatures expected might also lead to fuel-bundle and pressure-tube deformation and, depending on the pressure in the channel, pressure tubes could sag or balloon into contact with their calandria tubes. Depending on the pressure-tube temperature, the contact area and the contact thermal resistance, the heat flux at the outside surface of the calandria tube might be large enough to produce dryout of the calandria tube surface, thereby reducing the heat transfer coefficient between the calandria tube and moderator, possibly leading to failure of the fuel channel.

To address these questions, a comprehensive program on fuel-channel behaviour was established. To obtain a sound understanding of the phenomena involved, separate-effects studies were undertaken in many areas, including:

- pressure-tube sag and ballooning at high temperature (greater than 700°C);
- the production of hydrogen via the steam-zirconium reaction;
- effect of contact conductance on the heat transfer between the pressure tube and calandria tube if the former were to deform into contact with the latter;
- emissivity of Zircaloy at high temperature;
- embrittlement of pressure tubes by hot hydrogen/steam mixtures, and by oxygen.

The results of these studies were used in the development of the fuel-channel code CHAN which describes fuel channel behaviour during accidents.

A series of integrated tests was also performed to simulate some of the fuel-channel situations that might arise during a severe accident (see figure 15.9). In some of these tests the pressure tube was allowed to balloon out into contact with the calandria tube. It was found that as long as the outside surface of the calandria tube did not experience dryout, it remained cool and did not deform. Such ballooning would be expected if the pressure in the fuel channel remained high for a time after the start of a LOC/LOEC accident.

If the pressure in the channel were low, however, the pressure tube would initially deform by sagging into contact with the calandria tube, and the latter would only remain cool as long as dryout did not occur on its outer surface. The temperature of the moderator is an important factor in this scenario and many experiments were done to investigate the relationships between pressure-tube temperature, thermal contact conductance and moderator temperature. The results of this work help to define the maximum value of moderator temperature if the latter is to serve as an effective heat sink for such events and calandria tube failures are to be avoided.

Another experiment involved measurement of the temperature distribution around the circumference of the pressure tube under coolant stagnation. In such a case, coolant stratification could occur, with liquid coolant lying at the bottom of the pressure tube, and steam at the top. Thus, the top of the pressure tube would be much hotter than the bottom, thereby raising a concern as to pressure-tube failure before contact with the calandria tube had been established. The internal pressure could cause the pressure tube to deform at the hot spot, causing it to balloon or rupture into local contact with the moderator-cooled calandria tube. Experiments were conducted to determine the circumferential temperature variation as a function of channel power, coolant pressure and temperature. In fifteen such tests, only one resulted in failure of the pressure tube, due to localized heating at the top of the tube. (This type of accident scenario in a CANDU reactor should not result in failure of the calandria tube since the fuel-channel pressure is too low. If the pressure were high, the pressure tube would rapidly balloon into contact with the calandria tube and would be cooled, via heat transfer to the moderator.)

CANDU fuel bundles are supported in the pressure tube by bearing pads. During a LOCA, some heat from the fuel bundle would be transferred to the pressure tube via conduction through these pads. Concern was expressed as to the integrity of the pressure tube due to local deformation at the fuel-bundle bearing-pad/pressure-tube contact point. Experiments were done to explore this effect and the results showed that there was very little strain in the region of the hot bearing pads. Thus, the probability of pressure-tube failure during ballooning should not be affected by the bearing pads of overheated fuel bundles.

There was also some concern that during a rapid rise in temperature in a fuel channel molten material may drop onto the pressure tube, causing a hot spot and possibly tube failure. An experiment was set up to study this scenario, in which a pressure tube was heated internally and pressurized to make it balloon into contact with its calandria tube. Further heating melted a piece of Zircaloy which then fell onto the ballooned pressure tube. The heat was transferred through the pressure/calandria tubes to a surrounding tank of water. These tests showed that while localized dryout of the outer surface of the calandria tube can occur, as well as a small amount of deformation, the integrity of the pressure tube is not threatened as long as the calandria tube is not in film boiling prior to the molten material contacting the pressure tube.

The approach taken in this program was to perform several series of experiments, progressing from a single fuel-element simulator to a cluster of seven elements, and then to a full-size fuel channel. The results of the tests were used to refine and validate both the CHAN fuel-channel code and the CATHENA system code.

FIGURE 15.9
Fuel Channel Integrity Test Rig

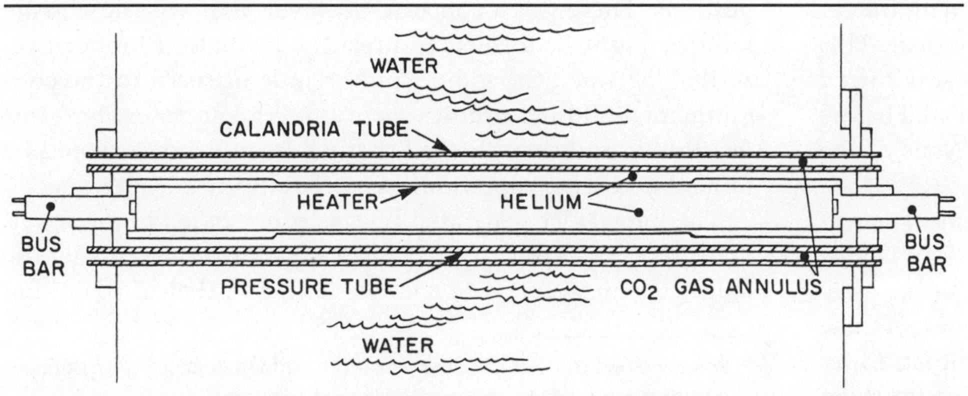

Containment System

In current CANDU power plants, the containment system provides the final engineered barrier between the reactor, with its fission products, and the environment. In single-unit stations it is a concrete building surrounding the reactor and the steam-generating equipment, while in multi-unit stations it comprises the reactor buildings, the vacuum building, and the ductwork connecting these. In both cases the containment systems are engineered to withstand the pressure rise that would occur due to a break in the primary cooling system followed by the ejection of a mixture of hot water and steam into the reactor building. They are also designed to retain any fission products that might be released from the fuel. A major potential threat to containment integrity is posed by the possible combustion of hydrogen that may be generated during the course of an accident. Research programs have been developed over the years to address all of these issues.

During the design of NPD, several containment system concepts were evaluated. One novel one was the use of a large balloon, which would be inflated by the pressure rise that would occur immediately after a large pipe break, and before the dousing system could take effect. This idea was rejected because of uncertainty as to the uniformity of the balloon material; i.e., the possibility of rupture at thin spots.[8] The solution adopted at NPD was a pressure-relief duct with rupture discs that would fail during the initial pressure transient, thus relieving the pressure to the outside. Once the dousing system operated, the relief duct would be sealed by a gravity-driven trap door. This approach was based on the fact that the pressure pulse occurred immediately after the pipe break and would be over well before any fuel would fail and release fission products. None of the reactors after NPD has such a vented containment.

To reduce the size of the pressure rise in containment associated with the release of steam following a large-break LOCA, dousing systems are used in CANDU plants. Some of the early safety experiments involved measuring the effectiveness of these dousing systems. Tests were performed at NPPD in the late fifties and early sixties to determine optimum parameters for these systems, such as droplet size, flow rate, and temperature. These tests proved the viability of the concept and provided basic data for the design of all of the subsequent CANDU dousing systems.

A computer code (called PRESCON) was also developed to analyze the behaviour of the containment system during a LOCA. This code calculates the rise in pressure in the containment buildings resulting from the expulsion of steam from the reactor core, with allowance for the condensation of steam on cold surfaces and heat transfer to the dousing sprays. The code also includes models to describe the dynamics of the dousing system and the pressure-relief valves between the reactor buildings and the vacuum building in multi-unit stations.

The behaviour of the fission products in the containment system is of great importance from a safety perspective, and of special interest is the behaviour of iodine. Its behaviour in containment will depend on the chemical forms released into the coolant during an accident. The studies on fission-product release mentioned earlier showed that for the reducing steam conditions expected in the primary cooling system during a LOCA, the iodine would be released in the form of cesium iodide, which on contact with water dissolves to form a nonvolatile aqueous ion. The subsequent behaviour in the containment building would depend on a number of factors; e.g., the pH of the water, the temperature, the radiation field and impurities in the sump water in the building. An analysis of containment chemistry showed that for the conditions expected during a LOCA, the dominant chemical forms of iodine would be aqueous ions. There was a concern, however, that volatile organic iodides might be formed as a result of radiation-induced reactions between the iodine and organic material in the containment building. Studies were undertaken to explore this possibility, and they showed that no organic iodide would be formed under the conditions expected during a LOCA.

The knowledge generated by this program led to the development of a computer code (called LIRIC) that can predict the iodine chemical forms in containment, and their distribution

[8] Someone remarked at the time that if the balloon concept were pursued it would probably be the world's largest condom!

between the aqueous and gas phases, as a function of time following a severe reactor accident. The LIRIC code is a chemical kinetics model containing about 100 differential equations that represent the rates of formation and decay of all the species involved in the thermal and radiation chemistry of aqueous solutions containing iodine, including the effects of additives (e.g., thiosulphate and hydrazine) that might be used to suppress the formation of volatile iodine.

To investigate the behaviour of iodine in containment in an integrated fashion, a special facility, called the Radioiodine Test Facility (RTF), was built at Whiteshell in the mid-eighties to simulate the conditions that would exist in a containment building after an accident (see figure 15.10). The facility included a 400-litre cylindrical reaction vessel with shielding adequate for a cobalt-60 radiation source designed to reproduce the radiation fields that would be encountered during postulated accidents. The inner surface of the vessel could be changed by using liners (painted or unpainted) made of stainless steel, carbon steel, or concrete, to simulate different containment building surfaces. The temperature of the vessel could also be varied from room temperature to about 80°C. Extensive instrumentation was provided to monitor the various iodine species.

Tests were done using surfaces such as zinc primer, vinyl, polyurethane and epoxy, with and without radiation present. They showed that zinc primer can provide an important absorbent surface for iodine. This was a surprise, since iodine was not expected to react with this surface. The presence of radiation also produced a large increase in the amount of iodine deposited on surfaces exposed to the gas phase. The information gleaned from this program is used to improve the performance of CANDU containment buildings, and to provide more accurate assessments of the release of iodine from these buildings.

Since the rapid combustion of hydrogen can produce very high temperatures and very high pressures in confined volumes (which could threaten the survival of equipment and the surrounding building), hydrogen generation and behaviour is an important safety issue in nuclear power plants. Consequently, research on this topic has been performed for many years. There are two main sources of hydrogen to consider in a CANDU power plant: the exothermic reaction between the Zircaloy fuel sheathing and steam (which might occur during a combined LOC/LOEC accident), and radiolysis of the heavy-water coolant and moderator, the latter leading to a buildup of hydrogen and oxygen in the helium cover gas in the calandria vessel.

During the design of NPD and Douglas Point, the formation of explosive mixtures of deuterium and oxygen in the helium cover gas above the moderator was a major concern.

FIGURE 15.10
Schematic of Radioiodine Test Facility (RTF)

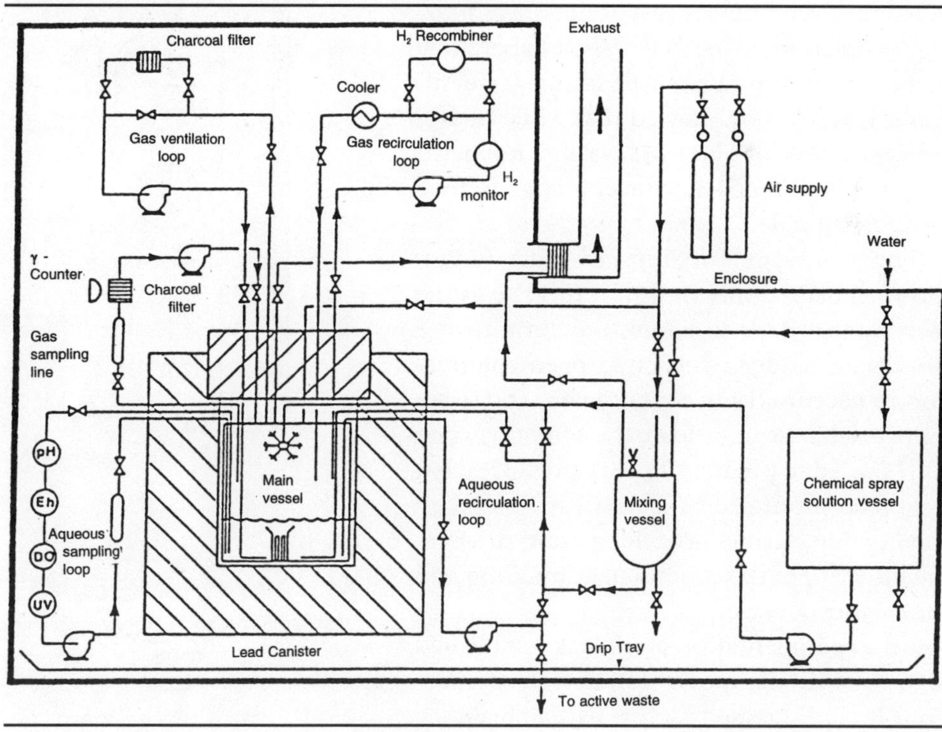

At one point, consideration was given to using a different cover gas than helium, one that would combine with the hydrogen (or oxygen) to prevent the buildup of explosive mixtures. However, the alternatives had other undesirable features (e.g., they would eventually become corrosive in the radiation environment), so helium was retained. Experiments were done at Chalk River to determine the cover-gas concentrations of deuterium and oxygen that might be expected during reactor operation. This work led to the use of deuterium-oxygen recombination units on all CANDU plants, to maintain safe levels of these gases in the calandria.

The suspected formation of a so-called "hydrogen bubble" in the Three Mile Island reactor during the accident there in 1979 caused a great deal of concern at the time. However, well before that accident, AECL scientists and engineers were cognizant of the potential hazard posed by hydrogen combustion, and a special facility to study hydrogen behaviour in containment, called the Containment Test Facility, was nearing completion at the Whiteshell Laboratory at the time of the Three Mile Island accident. (Later investigation showed that while hydrogen was formed at Three Mile Island, it burned off safely and there was never a large bubble ready to explode.)

The Containment Test Facility (CTF), shown schematically in figure 15.11, was the focus for an R&D program to determine the conditions under which hydrogen combustion can occur within containment, and to acquire a database to validate the computer code VENT, used to predict the peak pressures that might be produced as a result of hydrogen combustion during an accident, and to assess the integrity of the containment building and the calandria vessel.

An experimental program was established to investigate all aspects of hydrogen combustion, including such factors as flammability limits, laminar burning velocities and the effects of gas diluents, including steam, on these. Also studied were flame acceleration due to venting, the effectiveness of ignitors that might be used to produce controlled burning of hydrogen, and the effect of turbulence on flame propagation. These experiments were conducted in the CTF and the Hydrogen Laboratory at Whiteshell.

The CTF comprises two large-scale vessels (a 2.3-metre-diameter sphere and a 5.3-metre-long × 1.5-metre-diameter cylinder) that simulate containment structures and have pressure ratings of 10 MPa (100 times atmospheric pressure). These vessels were used for large-scale experiments to determine pressure transients and integrated combustion effects. Tests with peak pressures of 1.5 MPa (fifteen times atmospheric pressure) have been performed in the CTF.

FIGURE 15.11
Containment Test Facility

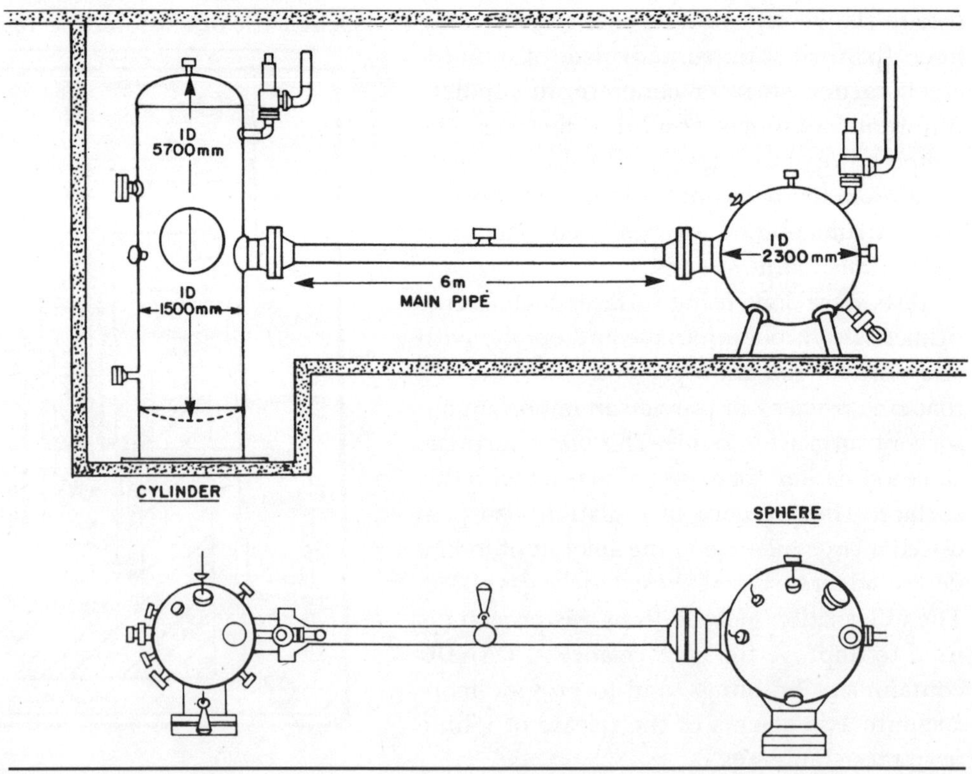

More fundamental studies of the detailed physics and chemistry of hydrogen combustion were also performed using bench-scale apparatus supported by special analytical tools for measurement and data collection, including microsecond Schlieren photography, laser Doppler anemometry, fast-acting pressure transducers and high-speed data acquisition systems.

More recent studies (in the mid-eighties) have investigated turbulent burning velocities, detonation limits and the transition from deflagration to detonation. A hydrogen detonation would exert a much greater pressure on containment equipment than deflagration. While a full-scale detonation is not possible in a CANDU containment building during a LOCA, because there would be insufficient hydrogen available, the possibility of local detonations was a concern. In localized areas of high hydrogen concentration, detonation may be externally induced (e.g., by a local high-energy discharge) or self-induced by the transition of a deflagration to a detonation. The latter normally results from acceleration of a deflagration to a high velocity followed by the development of a detonation kernel around some obstacle in the flame path. These experiments have led to the derivation of mechanisms and criteria for the transition to detonation in the presence of typical obstacles.

In addition to the research on combustion phenomena, work was also done on the development of mitigation technologies. One example is ignitors which, when properly deployed, can be used to burn off excess hydrogen in an innocuous manner. Another example is the demonstration of the effectiveness of an alternative hydrogen combustor for use in vault cooler ducts where deuterium can accumulate.

These studies have shown that it is extremely unlikely that conditions could arise where hydrogen combustion could threaten the integrity of a CANDU containment system, if proper precautions are taken.

CONCLUSION

The CANDU safety R&D program within AECL has evolved over the years to address the many questions that have been raised both by CANDU designers and the AECB as the various CANDU plant designs have evolved and proceeded through the licensing process. From the beginning, the CANDU developers appreciated the critical safety issues and initiated appropriate R&D programs to address them. Because it was realized that the phenomena associated with postulated reactor accidents are complex and difficult to calculate, the approach taken was to rely heavily on experiments to elucidate the phenomena involved, and to develop the plant simulations based on this firm experimental foundation.

The safety program and other related R&D programs have resulted in changes in the design of the safety systems for CANDU plants over the years:

- measurements of the coolant void reactivity effect led to the use of fast-acting shutdown systems, rather than moderator dump, for the larger CANDU reactors;
- work on the radiolysis of heavy water led to the use of hydrogen-oxygen recombiners in the moderator cover-gas circuit on all CANDU reactors;
- thermalhydraulics studies showed that high-pressure emergency cooling systems were required for some of the accident scenarios, rather than gravity-driven systems;
- the work on iodine and cesium behaviour in containment showed that a dousing system is not required for washout of the iodine and cesium during LOCAs;
- the results of the tests on pressure-tube integrity during combined LOC/LOEC accidents led to the determination of the optimum temperature for the CANDU moderator;
- blackening the inside of the calandria tubes, to enhance the heat transfer from the fuel to the moderator during combined LOC/LOEC accidents;
- the use of computers in the Darlington safety systems to enhance their reliability, based on the successful application of computers in the control of CANDU plants, as discussed in chapter eleven.

The safety R&D programs have also confirmed the soundness of the original CANDU concept and of the multiple-barrier, defence-in-depth approach used by the CANDU designers. The results obtained from these programs have increased our confidence in the safety of the CANDU reactor and have shown that even for low-probability accidents the regulatory standards can be met. They have provided a sound basis for the safety analysis of CANDU plants, without which these plants would not be able to obtain their operating licences. Indeed, some licences have been granted contingent upon the successful completion of safety R&D studies.

As the safety R&D program developed, close ties were also established with groups performing safety research in other countries, and Canada has had over the years co-operative arrangements with major agencies sponsoring reactor safety research, such as the USNRC, the OECD and the IAEA. In this way, the Canadian program has benefited from the work done by others, both where it is applicable to CANDU reactors, and where it is not reactor specific (e.g., the atmospheric dispersion of radioactive materials).

While many of the CANDU safety issues have been addressed by the R&D program, there will probably be a need for an ongoing program to deal with issues related to new reactor design features and any new questions that the regulatory agency may have.

ACKNOWLEDGMENTS

Significant contributions to the content of this chapter by W.T. Hancox, H. Tamm, H.E. Rosinger, R.D. MacDonald, P.A. Ross-Ross, N.H. Sagert, K.N. Tennankore, D.J. Wren, P.J. Ingham, V.G. Snell and G.L. Brooks are gratefully acknowledged.

BIBLIOGRAPHY

Bancroft, A.R., L.C. Watson, and R. Hewitt. *Tests on Collectors to Remove Iodine from Ventilation Air.* AECL-1131. 1960.

Beahm, E.C., C.F. Weber, T.S. Kress, and G.W. Parker. *Iodine Chemical Forms in LWR Severe Accidents: Final Report.* Report NUREG/CR-5732. Washington, D.C.: U.S. Nuclear Regulatory Commission, Office of Nuclear Regulatory Research, 1992.

Hancox, W.T. *Safety Research for CANDU Reactors.* AECL-6835. 1982. (Presented to the IAEA Technical Committee Meeting on Thermal Reactor Safety Research in Moscow, 1-4 December 1981)

Hancox, W.T., W.C. Harrison, and H. Tamm. "CANDU Reactor Safety Research Program." In *Thermal Reactor Safety: Thermal Reactor Safety Meeting, Sun Valley, Idaho, 31 July-5 August 1977,* vol. 1, 183-202. La Grange Park, Ill.: American Nuclear Society, 1977.

Lemire, R.J., J. Paquette, D.F. Torgerson, D.J. Wren, and J.W. Fletcher. *Assessment of Iodine Behaviour in Reactor Containment Buildings from a Chemical Perspective.* AECL-6812. 1981.

Morison, W.G. "Engineering for Safety." In *Canadian Engineering Centennial Conference: Proceedings of the Nuclear Power and Fusion Programs of the Canadian Nuclear Society, 18-22 May 1987, Montreal,* 58-64. Toronto: Canadian Nuclear Society, 1987.

Ross-Ross, P.A. "Experiments on the Consequences of Bursting Pressure Tubes in a Simulated Power Reactor Arrangement." In *Preprints of the 2nd International Conference on Structural Mechanics in Reactor Technology, Berlin, 10-14 September 1973,* Comp. by Thomas A. Jaeger., vol. 2, pt. F, paper F2/2. Brussels: Commission of the European Communities, 1973. (Also AECL-4877.)

Tamm, H., V.S. Krishnan, H.E. Rosinger, A.C. Vikis, J.C. Wood, and D.J. Wren. *Overview of the Technical Basis for the Safety of CANDU Reactors.* AECL-9559. 1988.

United States. Nuclear Regulatory Commission. *Reactor Safety Study: An Assessment of Accident Risks in U. S. Commercial Nuclear Power Plants: Main Report and Executive Summary.* Report WASH-1400. Washington, D.C.: U.S. Nuclear Regulatory Commission, 1975.

Watson, L.C., A.R. Bancroft, and C.W. Hoelke. *Iodine Containment by Dousing in NPD-II.* AECL-1130. 1960.

Chapter Sixteen

Boiling-light-Water and Organic-liquid-cooled CANDU Reactors (CANDU-BLW and CANDU-OCR)

R.G. HART

INTRODUCTION

When the Canadian nuclear power program was beginning to take shape the choice of the moderator, heavy water, was clear but the choice of the coolant to go with it was not. Consideration was given to pressurized heavy water, boiling light water, steam, fog (steam with entrained water droplets), organic liquid, carbon dioxide and helium. Even after pressurized heavy water was chosen there were reservations about its high cost and the ability to control leakage to a degree where heavy-water makeup costs would not be prohibitive. Accordingly, it was considered prudent to initiate work on alternative coolants as backup to heavy water. The alternative coolants chosen for backup study were boiling light water and organic liquid; reactors based on those coolants were designated CANDU-BLW and CANDU-OCR. Serious work on the concepts started in the early sixties and late fifties respectively.

CANDU-BLW

The conceptual natural-uranium-fueled CANDU-BLW power reactor (see figures 16.1 and 16.2) was similar to the CANDU-PHW (see chapter ten) in that the heavy water moderator was contained in a large, low-pressure calandria and the fuel and coolant were contained in pressure tubes. However, in the CANDU-BLW, the pressure tubes were oriented vertically in the core and the fuel bundles, which were strung on a central support tube, could be inserted and removed from the bottom only. As the light water coolant passed upward through the core it boiled and exited the core at about 290°C, 7.7 MPa[1] and 20 percent quality (i.e., with 20 percent of the water converted to steam). The coolant was then directed to a steam separator from which the steam passed directly to the turbine generator. The water from the separator was mixed with the steam condensate from the turbine generator and recirculated. To obtain a satisfactory energy output (burnup) from each fuel bundle, the fuel strings had to be removed periodically to the fuel bays, where the bundles would be rearranged on the central support tube and returned to the reactor for further irradiation.[2] The potential attractions of the CANDU-BLW (relative to the CANDU-PHW) were that it would have a lower capital cost because of less heavy water, no steam generators and a slightly higher heat-to-electricity conversion efficiency, and it would have lower operating costs because

[1] This is to be compared with 310°C, 10 MPa and close to 0 quality in the CANDU-PHW (see chapter ten).

[2] This is accomplished in the CANDU-PHW by bidirectional fueling.

it eliminated the problem of heavy-water leakage from the high-pressure coolant system. The main potential difficulties were that the coolant absorbed more neutrons than the pressurized heavy-water coolant, making it more difficult to achieve an economical burnup in natural uranium fuel, and that boiling led to variable neutron absorption in the coolant, making the reactor difficult to control.

Accordingly, development work focused on designs of fuel and fuel channels that would maximize neutron economy, on definition of the limits of steam quality that could be safely used and on methods of controlling the reactivity variations that would result from inevitable differences in boiling rates in the fuel channels. Methods also had to be developed to control coolant chemistry in the boiling regime, so that excessive corrosion and solids deposition on heat transfer surfaces could be avoided.

The fuel design chosen (see figure 16.3) was similar to the successful CANDU-PHW design (see chapter thirteen), except that the fuel pins were fatter, to improve the uranium dioxide to zirconium ratio, and the fuel pin spacing was less, to reduce the amount of coolant in the channel. Because of the fatter fuel pins, the maximum channel power had to be reduced by about 20 percent relative to the CANDU-PHW, to keep the uranium dioxide temperature near the centre of the pellets at acceptable levels. This meant that the CANDU-BLW had to have about 20 percent more channels than a CANDU-PHW with the same power output. Further, because of the close element-to-element spacing, the sheath strain generated by the volumetric expansion of the uranium dioxide and the pressure from released fission gases had to be severely limited. This was aided by using hollow end plugs to provide extra space for the fission gases. Irradiation testing

FIGURE 16.1
Schematic Illustration of a CANDU-BLW Power Station

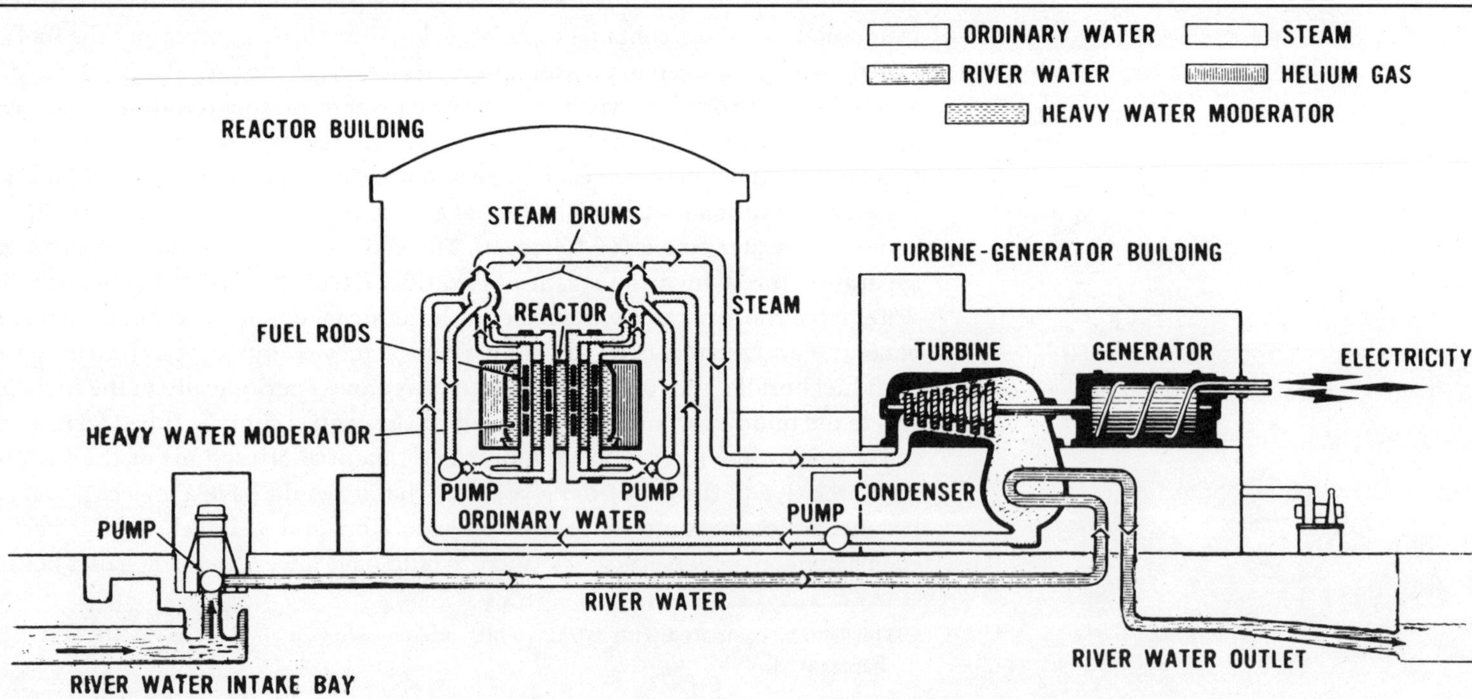

FIGURE 16.2
Cutaway of a CANDU-BLW Reactor Building

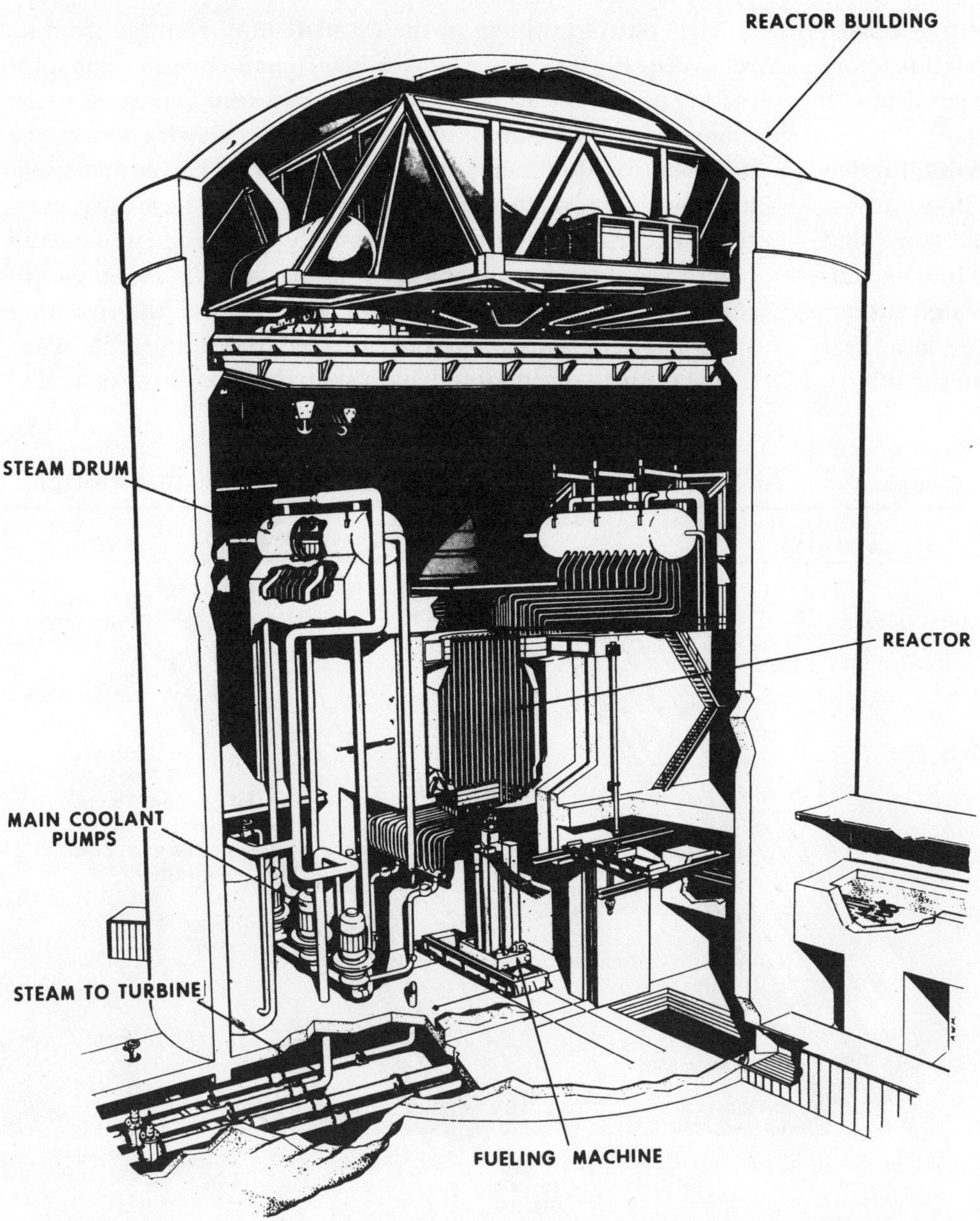

of fuel pins of this general design started in boiling light water loops in NRX in 1961, and by 1964 full bundles were being irradiated in a large boiling-light-water loop in NRU. In general, the fuel performed as expected and the tests verified that fuel of this design could give acceptable performance in a boiling-light-water-cooled reactor. Further, the fabrication of the test sections by the CANDU-PHW fuel suppliers confirmed that fuel for CANDU-BLWs could be manufactured on the same fuel fabrication lines as the CANDU-PHW fuels, with only minor alterations. This gave confidence that the fuel for CANDU-BLWs could be fabricated at the low costs that were already being achieved for the CANDU-PHW fuel. However, even with all the precautions taken to save neutrons, the average discharge burnup from a natural-uranium-fueled CANDU-BLW power reactor was expected to be only about 180 MWh/kgU, compared with 240 MWh/kgU for a CANDU-PHW. Taking into account slight differences in fabrication cost, the overall fueling cost was expected to be about 25 percent higher. This, at least partially, offset the expected lower capital and operating costs.

As for the fuel, the fuel channels for the CANDU-BLW (see figure 16.4) were similar to those already developed for the CANDU-PHW (see chapter twelve). The pressure tube material was Zr-2.5Nb, as chosen for the CANDU-PHW, but because of the difficulty in achieving acceptable neutron economy in the CANDU-BLW, it was used in the heat-treated rather than the cold-worked state. The increased tensile strength provided by the heat treatment allowed the thinnest possible tube to be used (chapter twelve). The end fitting design was of course different, to accommodate the different design of fuel assembly and fueling from one end only. The single-ended fueling allowed the channel to be designed for easy replacement. This was a very attractive feature because it was expected that the fuel channels in all CANDU reactors would have to be replaced at least once during each reactor's projected life.

To define the steam quality operating limits, AECL funded the construction of out-reactor test loops, both in-house and at Westinghouse Canada, to determine under what fuel power and coolant flow conditions the fuel cladding would go into dryout. Dryout is a condition where the water film on a heated surface breaks down, causing a precipitous decrease in the local heat transfer coefficient and a resultant sharp increase in the surface temperature. It is a condition to be avoided in reactors, as the high temperature leads to an accelerated deterioration of the fuel cladding. In the out-reactor tests, the fuel pins were simulated by electrically heated tubes, and thermocouples were buried in the tube walls at strategic locations, to detect when the surfaces went into dryout. The results allowed the scientists to develop a generalized correlation that could be used to predict the conditions that would lead to dryout in a power reactor channel. The correlation was then confirmed in thermocoupled fuel experiments in the in-reactor loops. The predictions were expressed as a critical power ratio (CPR), which is the ratio of the fuel power at which dryout will occur to the design operating power in the highest powered channel. It was found that the CPR was 1.5 for operation at 20 percent outlet quality but only 1.2 for operation at 40 percent. Thus, at 40 percent with typical in-core power variations, periodic excursions into dryout on some fuel sheaths could be expected. Although experiments indicated that a fuel could be developed to withstand these periodic excursions, it was considered prudent to avoid them, so the design outlet quality for CANDU-BLW power reactors was set at 20 percent.

The control problem in the CANDU-BLW resulted from its relatively high positive power coefficient. Even at constant reactor power, it was impossible to always keep a constant steam quality in each channel; when the local quality increased, the local neutron flux increased and the local fuel power increased, driving the local steam quality even higher. These local reactivity excursions would not necessarily lead to a loss of overall reactor control, as the changes in spatial flux shape would increase overall neutron leakage. Nevertheless, the resulting spatial power oscillations were expected to be intolerable. Thus it was deemed essential to have a spatial flux control system to

FIGURE 16.3
Comparison of Fuels for CANDU-PHW, CANDU-BLW and CANDU-OCR Reactors

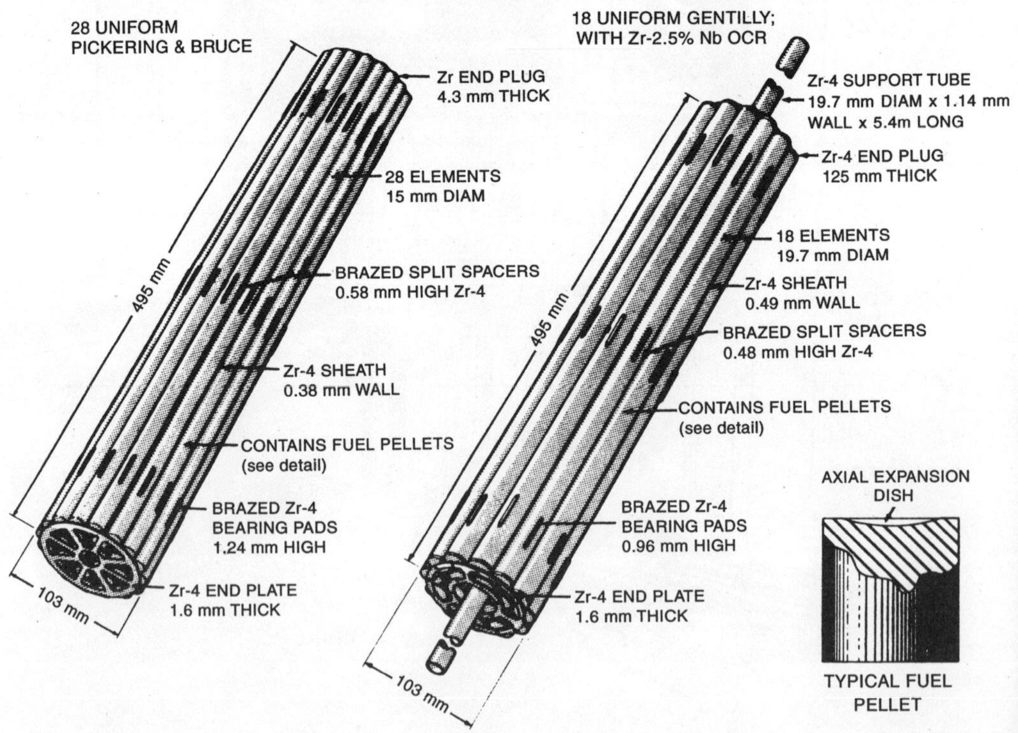

limit them to acceptable levels. To design such a system, it was necessary to be able to calculate the magnitude of the flux perturbations to be expected, their location in the reactor and the rate of reactivity change. To do such calculations, the scientists at Chalk River developed a computer code that they were able to check against experiments in the ZED-2 reactor. Calculations showed that the reactivity excursions could be controlled by absorber rods, strategically located in the core, which would automatically move to counteract the flux perturbations when they occurred.

As in the CANDU-PHW, the coolant in the CANDU-BLW had to be kept alkaline, to prevent excessive corrosion of both the fuel cladding and the out-reactor circuit, and excessive deposition of corrosion products on the fuel. In the CANDU-PHW this is done by adding lithium hydroxide, but lithium hydroxide could not be used in the CANDU-BLW because it would concentrate in the water phase, making the water more alkaline and the steam and steam condensate less alkaline than desired. Thus in the CANDU-BLW the coolant was kept alkaline by the addition of ammonia. Ammonia is volatile and would circulate with the steam phase, but it would also decompose with time, so a continual makeup was necessary. A series of tests in the in-reactor loops showed that if the concentration of ammonia in the exit water was maintained at seven parts per million (ppm), both fuel-cladding corrosion and deposition of out-reactor corrosion-product iron oxide on the fuel cladding remained at acceptable levels. The tests further showed that the growth of the iron oxide film on the cladding was reversible: if the film grew thicker because of a temporary loss of control of the ammonia concentration, it would return to its former equilibrium thickness when the ammonia concentration was returned to its specified level. This gave added confidence that corrosion-product deposition on the fuel could be adequately controlled, but led to concerns that radioactive corrosion products would be transported to, and deposited in, the out-reactor circuit.

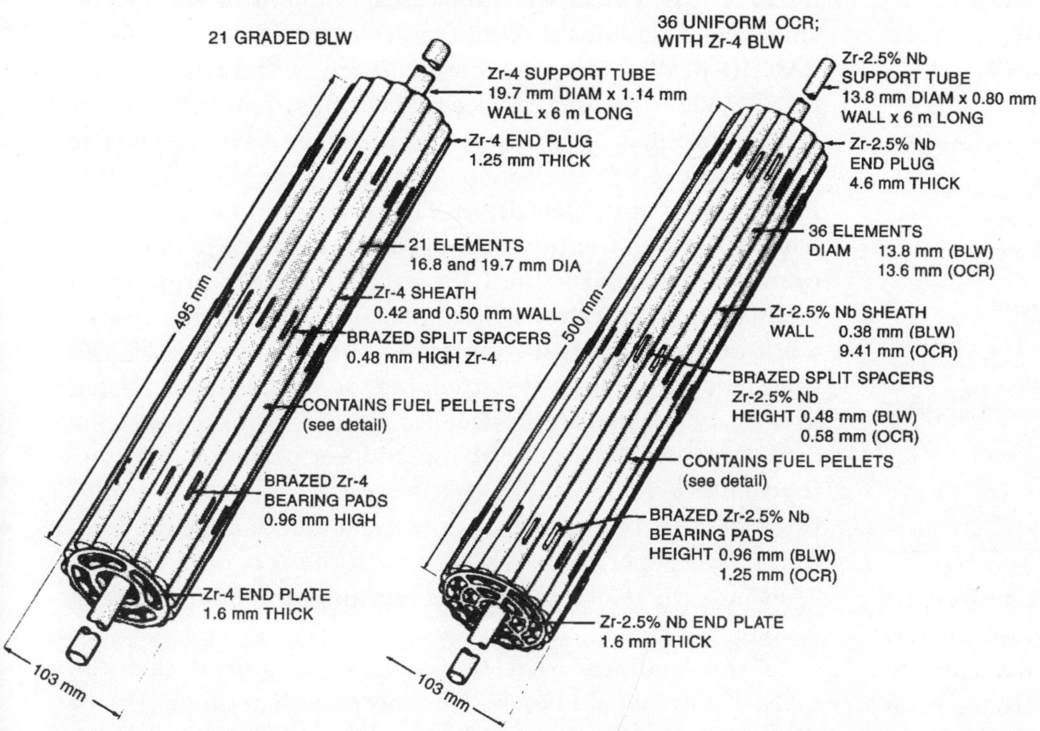

By 1965, Hydro-Quebec had become interested in gaining experience in nuclear power. Although Hydro-Quebec still had potential water power sources available, these sources were getting further and further from the load centres. In 1966 the federal and provincial governments reached an agreement whereby AECL would build a prototype 250 MWe natural-uranium-fueled CANDU-BLW power station to be known as Gentilly, on the south bank of the St. Lawrence River, near Trois Rivières. The arrangements were similar to those made with Ontario Hydro for Douglas Point, in that AECL would pay for the station while Hydro-Quebec would operate it and

FIGURE 16.4
Typical Fuel Channel for the CANDU-BLW

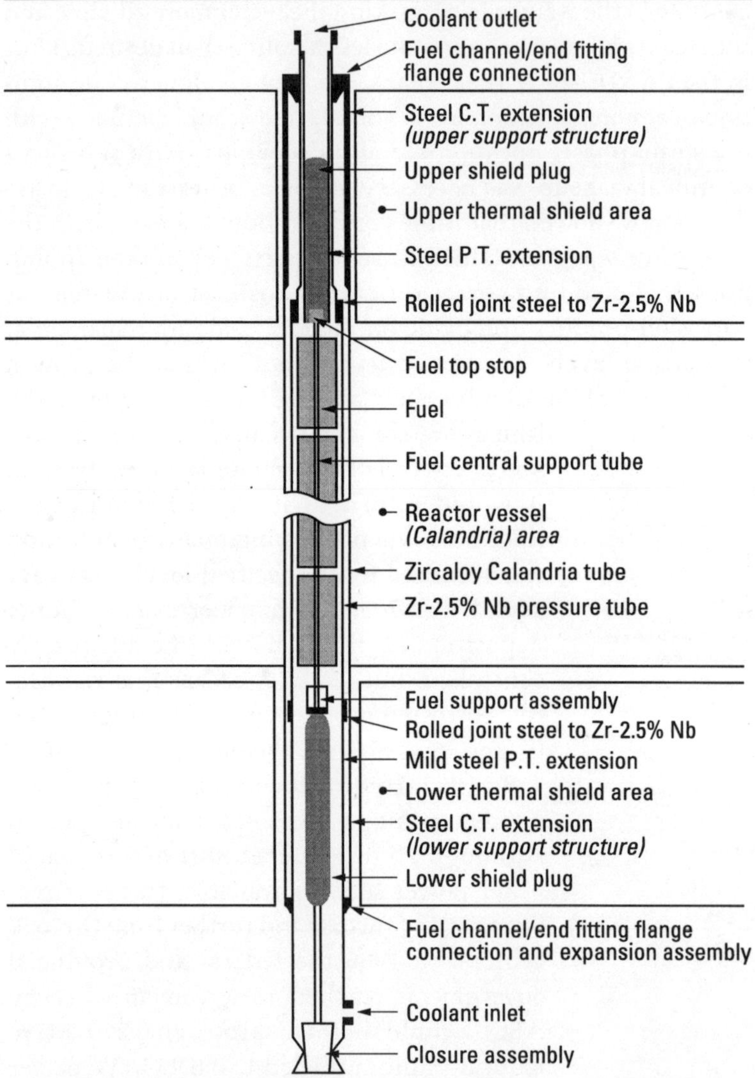

pay for the power at an agreed rate consistent with the formula used in the Ontario Hydro/AECL agreement for Douglas Point. A number of engineers transferred from Chalk River to the AECL design offices in Toronto, to form the nucleus of a team to do the detailed design; by 1970, design and construction had been completed. First power was produced in April 1971 and full power was reached in May 1972, but the station did not perform as well as expected. The absorber rods were located too close to the centre of the reactor, resulting in insufficient control in the outer regions, with consequent excessive flux tilts, particularly during booster rod insertions. In November 1972 there was insufficient heavy water available in Canada to start up Pickering 3, so it was decided that the heavy water from Gentilly-1 would be loaned to Pickering 3 and, while it was away, the absorber rods would be moved to correct the control problem. The absorber rods were moved but the heavy water was not returned until late 1974. When the reactor restarted it was plagued with corrosion problems in the condensers, apparently related to the start up, in the interim, of new industrial plants on the St. Lawrence river upstream of the cooling water intake. Further, even after the repositioning of the absorber rods, the reactor remained difficult to control. By 1977 it was clear that concern about control would preclude the commitment of any larger natural-uranium-fueled CANDU-BLWs. Also, the performance of Ontario Hydro's CANDU-PHW stations had been excellent, serious heavy-water leakage problems had not materialized, and Hydro-Quebec had committed to the construction of a 600 MWe CANDU-PHW power unit, Gentilly-2. The question now arose as to what to do with Gentilly-1. A decision was made to modify it to produce steam for the LaPrade heavy-water production facility then under construction at the Gentilly site. Design work on these modifications proceeded through 1977 and some modifications were started, but the work was terminated in early 1978 when Hydro-Quebec agreed to supply the steam from Gentilly-2 at a price that could not be matched by using Gentilly-1. Gentilly-1 continued to generate electricity through 1978, but its operation was sporadic and the information it was generating was not worth the cost of operation. Thus in April 1979 it was shut down and by 1984 it had been decommissioned.

In the laboratories it was always recognized that the CANDU-BLW would be a better concept with uranium-235- or plutonium-enriched fuel, rather than natural fuel. With enrichment there wasn't such a high premium on neutron economy, so the fuel pins could be thinner, the channel power higher and the lattice pitch tighter, resulting in a considerably lower inventory of heavy water. Further, with enrichment, the fuel could be taken to about 580 MWh/kgU, vs 180 MWh/kgU for natural uranium fuel, mitigating any fueling cost penalty associated with enriched fuel fabrication. But most importantly, with enrichment, the positive power coefficient would be substantially reduced, making it much easier for the control rods to correct any reactivity excursions resulting from boiling imbalances. Accordingly, even after problems materialized with Gentilly-1, work continued on the CANDU-BLW concept, with emphasis on its use as a plutonium burner (CANDU-BLW-PB). The development program had two basic components: 1) to demonstrate the improved controllability, and 2) to demonstrate the fabricability and irradiation performance of plutonium-bearing fuels. Similar designs were being developed in the United Kingdom and Japan. To demonstrate improved controllability, AECL built a hybrid computer system that would simulate an operating CANDU-BLW-PB. This system had two state-of-the-art analogue computers that could be programmed to represent the reactor system behaviour in real time. This behaviour could be controlled by a digital computer in exactly the same way that the computer would control the reactor. By 1975, this system had confirmed that a large plutonium-burning CANDU-BLW power reactor could be controlled without difficulty. To demonstrate the fabricability and irradiation performance of the plutonium-bearing fuel, AECL planned to fabricate plutonium-bearing CANDU-BLW fuel in a recycle fuel fabrication facility (see chapter eighteen) and irradiate it to the target burnup in a boiling-light-water loop in the NRU reactor. The fuel was to be essentially identical to the Bruce 37-pin bundle (see chapter thirteen), with the central pin replaced by a central support tube. For various reasons, completion of the recycle fuel fabrication facility was seriously delayed, and by the time it was completed the time scale for recycle of plutonium in Canada had receded to the distant future. So work on the CANDU-BLW was stopped. Thirty-six-pin plutonium-burning fuel assemblies were later successfully irradiated to the target burnup in pressurized water coolant, but not in boiling light water. The United Kingdom's enriched-uranium-fueled equivalent of the CANDU-BLW, the Steam-Generating Heavy-Water Reactor, operated successfully until 1990, and the Japanese plutonium-burning equivalent of the CANDU-BLW, FUGEN, is still operating.

CANDU-OCR

In the mid-fifties the United States Atomic Energy Commission (USAEC) was funding a program on enriched-uranium-fueled, organic-cooled-and-moderated reactors managed by Atomics International, a division of General Atomics in Los Angeles. Engineers at Canadian General Electric who were following the progress of this program concluded that a reactor cooled with organic liquid and moderated with heavy water could be used with natural uranium, and would have some attractive features for power production. They suggested the concept to AECL and, in 1959, AECL agreed to help fund it; the CANDU-OCR program was born.

The conceptual natural-uranium-fueled CANDU-OCR (see figures 16.5 and 16.6) was similar to the CANDU-BLW, in that the pressure tubes (see figure 16.7) were oriented vertically in the calandria and the fuel bundles were strung on a central support tube (see figure 16.8). Like the BLW, fueling was from one end only and the fuel strings had to be periodically removed for bundle shuffling, but in the OCR fueling was from the top rather than the bottom. The steam generating system was, however, similar to the CANDU-PHW in that the steam to drive the turbine was generated on the secondary side of a heat exchanger. The potential attractions of the CANDU-OCR relative to the CANDU-PHW, were its lower capital cost, resulting from a lower heavy-water inventory, a lower coolant pressure and a higher coolant temperature (higher conversion efficiency), and its lower operating costs resulting from the elimination of heavy-water leakage from the primary circuit and the minimal transport of radioactivity by the coolant (ease of maintenance). As in the CANDU-BLW, these potential advantages were partially offset by higher fueling costs, but it

FIGURE 16.5
Schematic Diagram of a CANDU-OCR Power Station

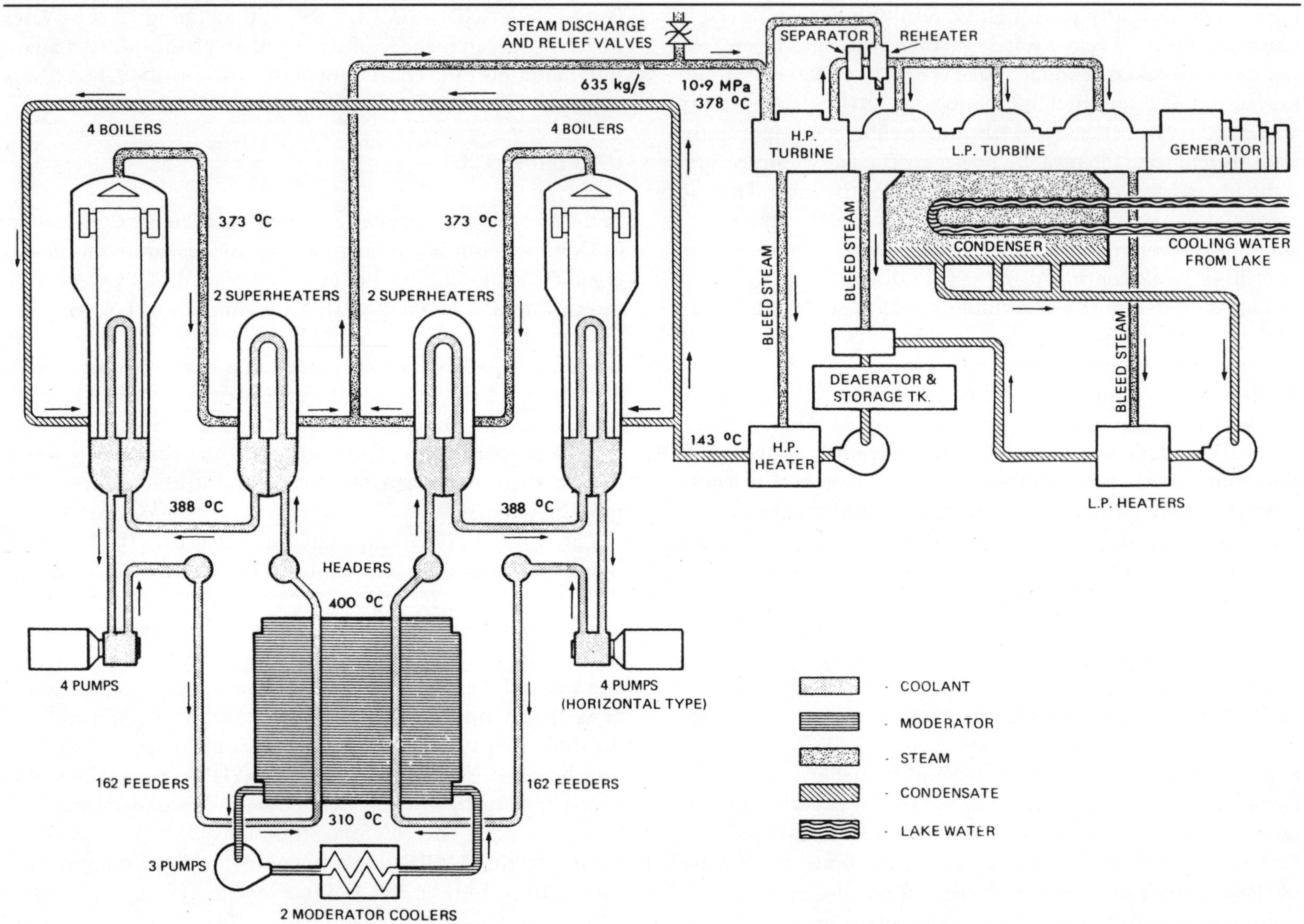

was felt that in the OCR this fueling cost disadvantage might disappear if uranium carbide or uranium metal, with their higher uranium densities and their better thermal conductivities, could be used as fuel. This seemed possible in an OCR, because the corrosion problems that prohibited the use of uranium carbide or uranium metal with water coolants did not exist with organic coolants. The main potential difficulties of the OCR were that the coolants gradually decomposed and polymerized under heat and radiation, that fuel surfaces could foul if the coolant chemistry was not right, that the fuel cladding and pressure tubes had to withstand higher coolant temperatures, and that the coolant was flammable. Development work was required to identify the most stable practical coolants, to determine the coolant treatment necessary to avoid fouling, to determine the heat transfer properties of the selected coolants, to develop suitable fuels and pressure tubes, and to determine the best methods of preventing and handling fires. The development work was focused on Whiteshell Reactor-1 (WR-1), a 40 MWt, heavy-water moderated, organic-cooled research reactor designed and built by CGE under contract to AECL.

As a result of screening tests done in the United States, the initial reference coolant for WR-1 was a commercially available mixture of terphenyls known as Santowax-OM. It was produced by the Monsanto Chemical Co. as a by-product of the manufacture of biphenyl from benzene. The screening tests showed that, of organic compounds available at low cost, the terphenyl family offered the best radiation and thermal stability. Santowax-OM was chosen over other available terphenyl mixtures because it had the lowest melting range. However, even Santowax-OM was highly viscous at room temperature, and with it some trace heating of reactor circuits was deemed necessary. Thus one of the earliest tasks of AECL scientists assigned to the OCR program was to find a coolant that would not require trace heating. After some study they concluded that HB-40, a partially hydrogenated mixture of terphenyls marketed by Monsanto as a plasticizer and heat transfer

FIGURE 16.6
Cutaway of a CANDU-OCR Reactor Building

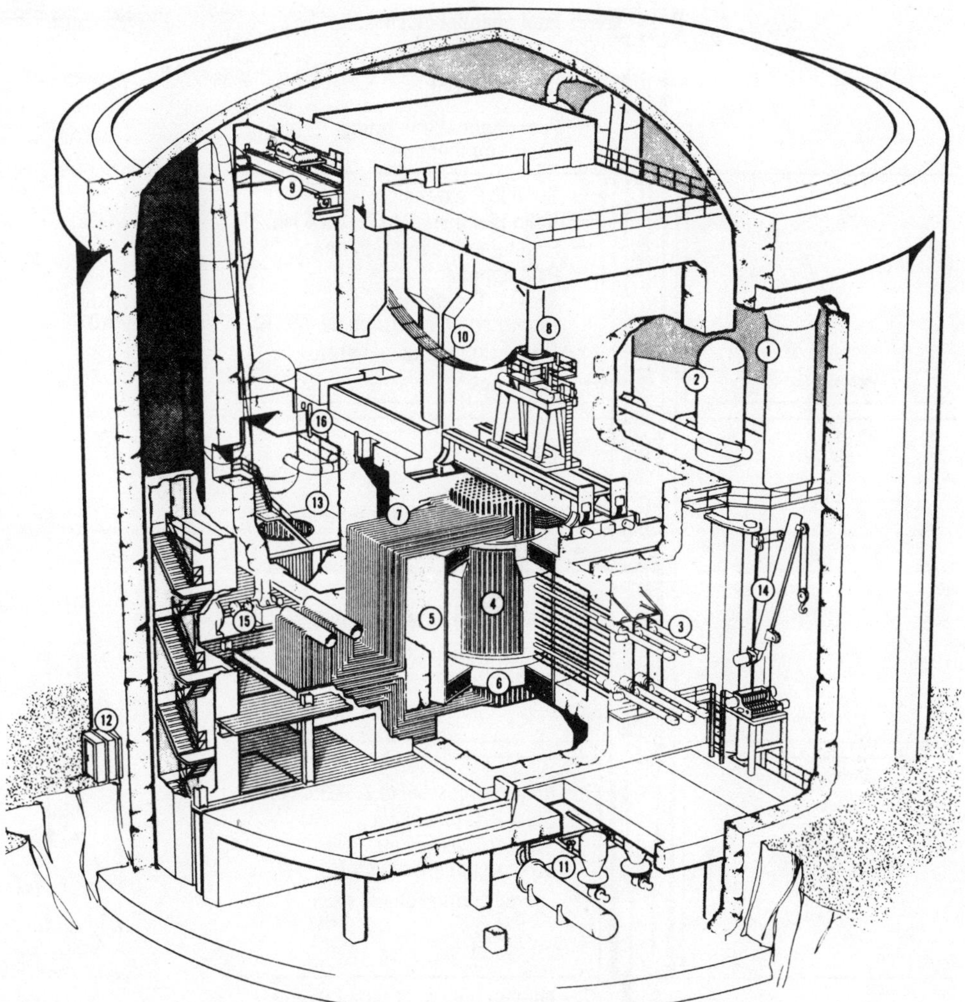

1. Boilers (8)
2. Superheaters (4)
3. Booster Rods
4. Calandria Assembly
5. Shield Tank
6. End Shield
7. Feeders
8. Fueling Machine
9. F/M Service Crane
10. F/M Vault Door
11. Moderator System
12. Emergency Airlock
13. Fuel Transfer Bay
14. Booster Flask Crane
15. Primary Pumps (4)
16. Fueling Machine Ports

FIGURE 16.7
Typical Fuel Channel for the CANDU-OCR

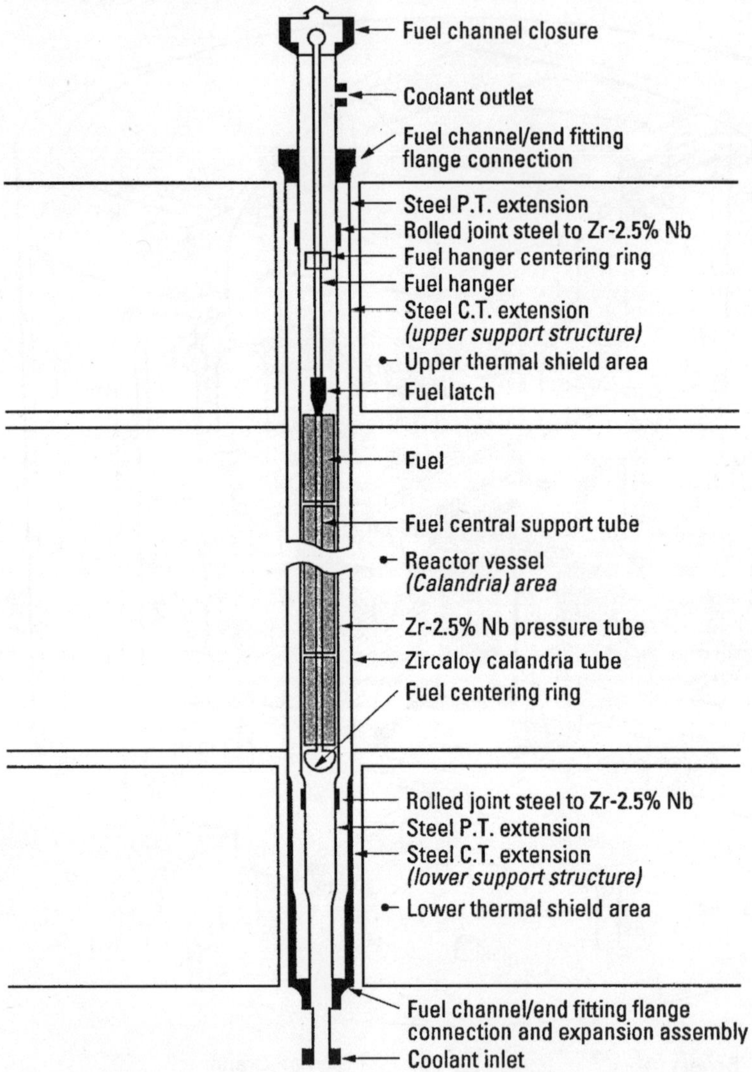

FIGURE 16.8
Early Design of a WR-1 Fuel Bundle Assembly

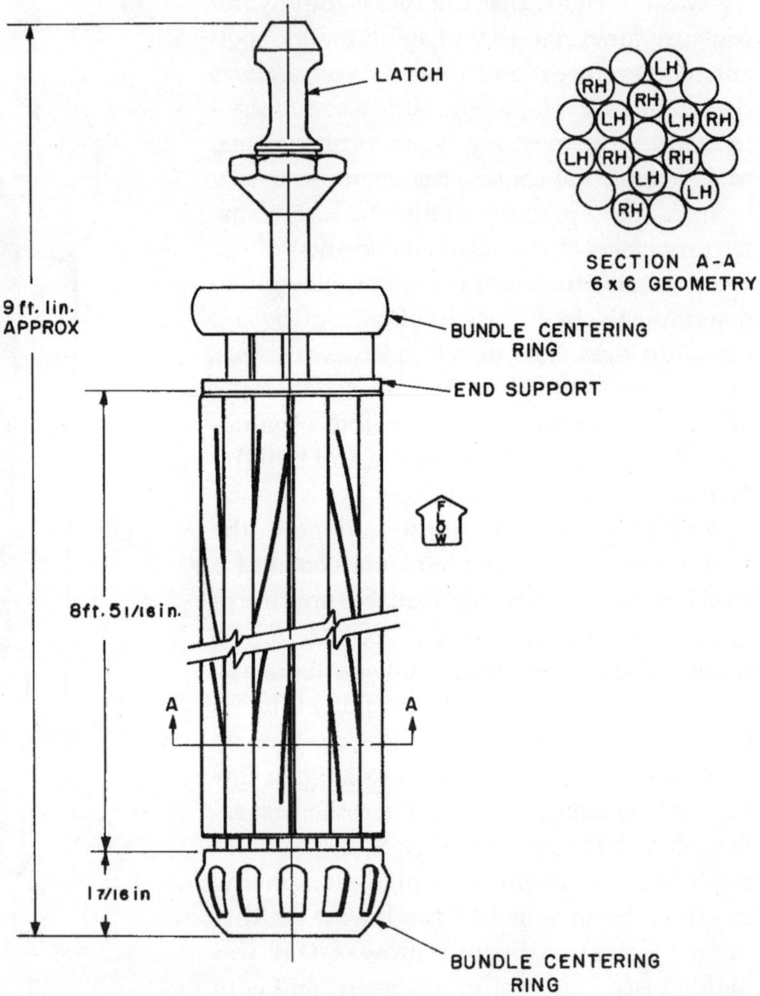

agent, showed enough promise to be used at least as the startup coolant for WR-1. Initially it was thought that, because of the higher decomposition rates of HB-40, the outlet temperature of the reactor would have to be limited to 350°C. However, under heat and radiation the composition of the coolant changed and the decomposition rate decreased to a point where, at equilibrium, operation at temperatures up to 400°C became routine. Thus HB-40 became the reference coolant for WR-1 and CANDU-OCR power reactors, and the need for trace heating was eliminated.

It was recognized early that control of fouling – the formation of low thermal conductivity deposits on fuel heat transfer surfaces – would be one of the key feasibility questions to be answered if the CANDU-OCR was to become a viable concept. The ultimate result of severe fouling would be an increase in fuel cladding temperature and fuel failure. Research work identified two main types of fouling: mass transfer fouling which led to inorganic deposits, and particulate fouling which led to carbonaceous deposits. The mass transfer fouling was believed to be caused by impurities in the coolant that dissolved iron from the out-reactor circuit and then precipitated it as iron carbide or oxide on the hot fuel cladding. The mechanism for the particulate type of fouling was never fully explained, but it produced deposits that were mainly organic in composition. The rates of both types of fouling increased with increasing fuel cladding temperature, but that of mass transfer fouling increased with increasing coolant velocity, while that of particulate fouling decreased. If either type of fouling increased to a point where flow stagnation occurred, a third phenomenon took place, called coking. Coking resulted when high molecular weight compounds, produced by coolant polymerization, exceeded their solubility limit in the stagnation areas and precipitated as a hard carbonaceous solid. It was coking in the stagnant areas of the moderator that eventually led to the demise of the organic-cooled and -moderated concepts in the United States, but this would be a problem in the heavy-water moderated CANDU-OCR only if fouling could not be controlled on the fuel surfaces. Over time, a set of coolant specifications was developed that led to acceptable fouling performance. These specifications were: maintain the chlorine content below 50μg/kg, maintain the combined oxygen below 200μg/g, maintain the water content at 200-600μg/g, maintain the iron content below 500μg/kg, maintain the particulate content below 1,000μg/kg and maintain the coolant velocity above 5 m/s. To meet these specifications, all feed coolant had to be processed to remove dissolved oxygen, chlorine and particulate matter, a by-pass stream of coolant had to be continuously passed through attapulgus clay and glass spool filters for purification, chloride compounds had to be rigorously excluded from the reactor area, water had to be continuously added as it was removed in the degassing circuit, and coolant degassing and recycling systems had to be fabricated from stainless steel, to minimize the return of iron to the core. Because the mechanisms of fouling were so complex and the consequences of fouling so severe, it was deemed necessary to develop an on-line measurement technique to warn when the fouling potential of the coolant was increasing. The technique developed was the small-probe fouling test (SPFT). In this test a small flow from the reactor coolant system was passed over an electrically heated stainless-steel probe for a period of twenty-four hours. At the end of the period the probe was washed and the film was scraped from a fixed length of the hottest region. The weight loss on scraping was a measure of the fouling potential of the coolant. It was found that if the fouling potential was less than 5 mg/m^{2}h the films on the fuel surfaces would remain thin and would have no detrimental effects on operation. During the entire operation of WR-1 there was only one serious fouling incident. This occurred in 1968 when an experiment to determine the minimum amount of coolant chemistry control required coincided with the first attempt to raise the coolant temperature from 345°C to 400°C. Although the coolant chemistry was apparently adequate for successful operation at 345°C it was not adequate for operation at 400°C. By the time this was recognized, about 40 percent of the fuel from the coolant circuit involved[3] was rendered unusable, because of heavy fouling. When the circuit was subsequently taken to 400°C with better coolant chemistry control there was no problem. Eventually, one of the circuits was operated at a coolant temperature of 425°C, again with no fouling problems.

The heat transfer properties that are significant in discussing coolants for nuclear reactors are: the interface heat transfer coefficient, which determines the temperature difference between the bulk coolant and the fuel cladding; the heat capacity, which determines the flow of coolant required and the temperature rise across the core; the onset of nucleate boiling (ONB), which describes the conditions at which the coolant begins to boil; and the departure from nucleate boiling (DNB), which describes the

[3] The WR-1 reactor had three separate coolant circuits, and different experiments could be done in the different circuits simultaneously.

conditions at which the fuel surface temperature rises sharply, leading to rapid fuel failure. The heat transfer properties of organic coolants are very different from water and the significance of these differences are discussed below. Firstly, the heat transfer coefficients of organic coolants are much poorer than those for water, thus for equal surface heat fluxes[4] the difference between the fuel-cladding temperature and the bulk coolant temperature is much greater. In an organically cooled power reactor, where the outlet temperature will approach 400°C, the peak fuel cladding temperature will approach 500°C, as opposed to perhaps 315°C in a water-cooled power reactor with a coolant outlet temperature of 300°C. Secondly, the heat capacity of organic liquids is considerably less than that for water. Thus, to remove the same amount of heat, more organic coolant has to be moved through the core, or the temperature rise across the core has to be higher. Since there is a premium on neutron economy, this means, in practice, that in organically cooled systems both the coolant linear velocity and the temperature rise across the core are higher. Thirdly, in organically cooled systems, nucleate boiling results from the volatilization at the heated surface of low boiling fractions of the multicomponent coolant, while in water-cooled systems nucleate boiling results from the formation of steam bubbles at the heated surface. In both cases the bubble formation at the surface disrupts the surface film and improves heat transfer, but in the organic system the effect on sheath temperature is more pronounced, because of the poorer heat transfer coefficient. In the organic system the phenomenon is called simmering, and it can reduce sheath temperatures by 25-30°C, compared with a non-simmering system. Fourthly, in water-cooled systems departure from nucleate boiling is well defined, because the single-component coolant all boils at the same temperature. However, in organically cooled systems, each of the many components of the coolant mixture has its own boiling point and, after the low boiling components are boiled away, high boiling components remain to cool the fuel. Thus in organic coolants there is no precipitous increase in sheath temperatures at any given surface heat flux, but rather a continuously increasing temperature as the heat flux is increased and the more volatile components of the coolant are removed. This makes the organically cooled systems more tolerant to accidental overpower situations, and has a significant impact on safety analysis.

When the natural uranium CANDU-OCR power reactor was first conceived, the reference fuel was uranium carbide clad in sintered aluminum product (SAP), and the reference pressure tubes were made from SAP. SAP is a mixture of 15 percent aluminum oxide in aluminum metal, the aluminum oxide being added to give high-temperature strength. Zirconium alloys were not chosen as fuel cladding or pressure tube material, because it was felt that at the high temperature demanded, about 500°C for fuel cladding and 400-425°C for pressure tubes, excessive hydriding would make their use impractical. It was decided that neither uranium carbide fuel nor SAP pressure tubes could be developed in time for WR-1, so its fuel was to be enriched uranium dioxide clad in SAP and its pressure tubes were to be stainless steel. The plan was to change to uranium carbide fuel and SAP pressure tubes when they were developed. However, in irradiation testing, no matter what variables were tried, all SAP-clad UO_2 fuel failed, because of the low ductility of the SAP and, since time for fabricating the first fuel charge was growing short, the designers felt they would have to revert to stainless steel for cladding as well as pressure tubing. While the SAP-clad fuel was being tested, an AECL scientist was investigating the hydriding of zirconium alloys in organic coolant, and concluded that alloys of zirconium and niobium could be used as fuel cladding, provided that they were properly handled and the coolant chemistry was properly controlled. He prevailed upon the designers to try zirconium-2.5 percent niobium-clad UO_2 of the Douglas Point wire-wrap design (see chapter thirteen). From the first test these irradiations were successful. It turned out that, not only was the hydriding rate low enough in its own right to allow Zr-2.5Nb to be used as cladding, but, as an added bonus, practically all the hydrogen that was picked up

[4] Surface heat flux is directly related to fuel power; the higher the power the higher the surface heat flux.

migrated to the cooler wire wrap, where it did no structural damage. After only one test, a decision had to be made regarding the cladding for the first charge of WR-1, and Zr-2.5Nb was chosen. In WR-1 it proved highly successful and it soon became the reference cladding for the proposed CANDU-OCR power reactors. Further, this success refocused scientists' attention on the use of zirconium alloys rather than SAP as a material for pressure tubes, with the result that by 1971 all WR-1 stainless-steel pressure tubes had been replaced by tubes made from either Zr-2.5Nb or Ozhennite 0.5,[5] a zirconium alloy developed in the USSR that had a lower out-reactor hydriding rate than Zr-2.5Nb. It was later found that the in-reactor hydriding rates were virtually identical.

With the successful development of Zr-2.5Nb-clad UO_2 as fuel for WR-1, fuel effort now turned to the development of uranium carbide, which, it was expected, would give lower fueling costs in a natural-uranium-fueled CANDU-OCR power reactor. After an extended development program, a satisfactory fabrication technology, involving arc melting and casting, was developed at Eldorado Nuclear Limited, and irradiation testing started. It was determined that slightly hyperstoichiometric material (4.8-5.0 wt. percent C) performed best and could be irradiated to at least 400 MWh/kgU at linear heat ratings up to 100 kW/m without any problem. By 1971, the technology had reached the stage where Zr-2.5Nb-clad UC fuel could be chosen with confidence in CANDU-OCR power reactor studies, and by 1973 all necessary steps had been taken to allow UC fuel to replace UO_2 as the fuel for WR-1. This change took place in 1974, and from then until the reactor was shut down in 1985 (because it was no longer required in AECL's experimental program), the UC fuel performed very well.

One of the main differences between organic and water coolants is that the organic coolants are flammable. This feature threatened to eliminate the organic coolant program just as it was getting started. The initial organically cooled test facilities at Chalk River were plagued with spontaneous fires in the pipe lagging and it was felt that if this incidence of fires was the norm, the concept was not viable. However, it was discovered that the problem was caused by coolant leakage into the lagging, which provided enough insulation to keep the coolant hot and enough air ingress to allow it to ignite. The solution was good maintenance to eliminate leaks, good housekeeping to remove coolant-soaked lagging, and air-resistant paints on the lagging to restrict the access of air. With these precautions, the problem was almost eliminated from the test facilities and subsequently from WR-1. WR-1 was equipped with non-incendive[6] equipment and wiring, smoke detectors, temperature sensors and automatic sprinkler systems in potentially hazardous areas, and the occasional small fires that occurred were easily controlled. Even though the WR-1 operating experience with respect to fire was excellent, it was recognized that it would not be sufficient to obtain an operating licence for a CANDU-OCR power reactor. To obtain such a licence it would have to be proven that the consequences of even the most severe combustion event that could possibly happen would be acceptable. The two events of most concern were coolant autoignition and coolant vapour explosion. Work showed that the possibility of autoignition could be eliminated simply by keeping the ambient air temperature below 350°C, but that vapour explosion could occur if the coolant was dispersed in air to a concentration between 0.5 and 5.0 volume percent and an external source of ignition was present. Thus it was concluded that the power station would have to have a high-speed fire-suppression system if it was to be licensed.

By 1969, W.B. Lewis, who had started out only lukewarm to the CANDU-OCR concept, had become its strongest supporter. He was attracted by the good performance of WR-1, the demonstrated high coolant temperature and high power density capability, and the low radioactivity of the out-reactor circuits.

---

[5] Ozhennite 0.5 has a nominal composition by weight: zirconium-99.5 percent, tin-0.2 percent, iron-0.1 percent, nickel-0.1 percent, niobium-0.1 percent.

[6] Non-incendive equipment is designed to minimize the possibility of fires, but it is not as complex or expensive as the more familiar explosion-proof equipment. It was felt that explosion-proof equipment was not required in WR-1.

He felt that the CANDU-OCR was the ideal reactor for his high-burnup thorium/uranium-233 cycle (see chapter eighteen), and he developed a concept that would produce 1,500 MWe from a core the size of the 250 MWe Gentilly-1.

By the end of 1970, it was considered that all the feasibility aspects of a natural-uranium-fueled CANDU-OCR had been demonstrated and that a decision was required as to its future. Accordingly, in March 1971 the AECL board of directors authorized a detailed design and cost study of a 500 MWe CANDU-OCR generating station, to be done by the design team at AECL Power Projects, assisted by staff from the research laboratories. The AECL board of directors specified that the fuel was to be natural uranium, feeling that a design based on the thorium fuel cycle was premature. In the design, the fuel was Zr-2.5Nb-clad uranium carbide, the pressure tubes were Zr-2.5Nb and the coolant was HB-40. The coolant outlet temperature was 400°C, which led to a cost-optimized steam-cycle efficiency of 34 percent. The station had some attractive safety features relative to water-cooled reactors, including low radiation fields, low stored energy, low coolant pressure, no fuel ballooning and no dryout, but it had the unattractive feature of a flammable coolant. The study concluded that the CANDU-OCR would have a 10 percent advantage over the CANDU-PHW in both capital and unit energy costs, and the AECL board of directors had to decide whether that potential advantage was sufficient to continue the program. In January 1973, the board of directors decided not to continue the program, the main reasons being that the CANDU-PHW was working well, there was little utility interest in pursuing another concept, and there was little prospect for raising the $250 million required for a CANDU-OCR prototype. The CANDU-OCR technology was to continue to be demonstrated through the operation of the WR-1 reactor, but the development effort was to be switched to improving the CANDU-PHW.

It is perhaps unfortunate that the CANDU-OCR technology was not developed at the time that opportunities for prototype stations were available. In the mid-eighties an Advanced CANDU study team, headed by the AECL design organization, set out the properties of the ideal coolant for a CANDU reactor, and found that the organic coolant used in WR-1 approached the ideal much more closely than any other coolant they could find.

BIBLIOGRAPHY

Boulton, J. "Behaviour of Zirconium Alloys in Organic Coolants." *Canadian Metallurgical Quarterly* 11 (1972): 1-5. (Also AECL-4239.)

Hart, R.G., L.R. Haywood, and G.A. Pon. "The CANDU Nuclear Power System: Competitive for the Foreseeable Future." In *Peaceful Uses of Atomic Energy: Proceedings of the Fourth International Conference on the Peaceful Uses of Atomic Energy, Geneva, 6-16 September 1971,* vol. 5, 239-53. Paper A/CONF.49/P/151. New York: United Nations; Vienna: International Atomic Energy Agency, 1972. (Also AECL-3975.)

Pon, G.A. "CANDU-BLW-250." In *Heavy-Water Power Reactors: Proceedings of the Symposium on Heavy-Water Reactors, Vienna, 11-15 September 1967,* 201-15. Paper IAEA-SM-99/32. Vienna: International Atomic Energy Agency, 1968. (Also AECL-2942.)

Smee, J.L., V.R. Puttagunta, and R.F.S. Robertson. *Organic Coolant Summary Report.* AECL-4922. 1975.

Whiteshell Nuclear Research Establishment and Power Projects, Sheridan Park. *Summary Report on the Design of a Prototypical 500 MWe CANDU-OCR Power Station.* AECL-4441. 1974.

Chapter Seventeen

Heavy Water

H.K. RAE

THE NEED FOR R&D

Heavy water has been produced in nine countries using seven different processes in twenty-two plants. But heavy water has always been expensive, as the following examples illustrate:

LOCATION OF HEAVY WATER PLANT	YEAR	$ OF THE YEAR/KG	1985$/KG
Rjukan, Norway	1935	500	4,300
Trail, British Columbia, Canada	1945	130[1]	900
Savannah River, South Carolina, U.S.A.	1955	62	275
Port Hawkesbury, Nova Scotia, Canada	1970	48	150
Bruce, Ontario, Canada	1985	300	300[2]

The range in constant 1985 dollars is $4,300/kg to $150/kg. Contrast this range with the price of some Scotch whiskys – $35/kg in 1985 dollars, or of silver – $160/kg, and one can appreciate that this is a very costly form of water. Even a small heavy-water research reactor requires about 10 Mg of heavy water, while a CANDU 6 power reactor requires 450 Mg, worth about $135 million in 1985.

Thus, there has been, and continues to be, a large incentive to develop ever-cheaper processes for heavy-water production. This goal has been sought by many researchers in many countries over the last five decades, since the first heavy-water reactor operated in Chicago in 1944. This chapter will describe AECL's part in this worldwide search.

Deuterium occurs in nature with hydrogen at a concentration varying from about 80 to 160 parts per million atom ratio. All the world's heavy water has been produced from water or industrial hydrogen, although natural gas, petroleum and industrial ammonia are other large-scale potential sources. The high cost of heavy water stems from this low natural abundance of deuterium, requiring separating equipment with both a very large flow capacity and a high degree of enriching capability.

Having a compelling need for a reliable and economical supply of heavy water for its power reactor program, Canada produced more than 80 percent of the world's heavy water between 1970 and 1985. These heavy water plants were based on earlier American technology, and benefited from their operating experience.

---

[1] $60/kg operating cost plus 15 percent of the investment per annual kg of heavy water.

[2] The higher cost of heavy water in constant dollars at Bruce, relative to Port Hawkesbury, reflects the fact that construction costs, and especially energy costs, have risen from 1970 faster than the Canadian Consumer Price Index.

However, unlike reactor development, there has been no large technical effort elsewhere during the past two decades that could provide a basis for collaboration[3] in further development of heavy-water production technology. So Canada has been largely alone in this field.

Even more important than the search for a better heavy-water process was the major development program in the seventies, led by AECL, to assist the staff of Canada's four large heavy-water plants to achieve reliable and successful operation. This was a multidisciplinary team effort over ten years involving, at its peak in 1974, well over 100 scientists, engineers and technologists from AECL, Ontario Hydro, Canadian General Electric (CGE), and twenty other industrial organizations, government laboratories and university departments. It is a good example of the ability of a large mission-oriented research organization to develop the detailed scientific and technical understanding necessary to solve unexpected and complex operational problems.

HEAVY-WATER PRODUCTION IN CANADA

By the end of 1943, a small heavy-water plant (6 Mg/a) had begun production at Trail, British Columbia. It had been built by the Consolidated Mining and Smelting Company of Canada Limited (Cominco) to an American design under the direction of, and financed by, the United States War Department's Manhattan Project. It extracted deuterium from electrolytic hydrogen by a catalyzed vapour-phase exchange reaction between steam and hydrogen. By October 1945, Trail had produced 8 Mg of heavy water. The United States had built three other heavy-water plants, using the water distillation process, which had also begun production at the end of 1943; they provided a total of 21 Mg of heavy water by October 1945, when they were shut down because of their high operating costs. Of this total stock of 29 Mg, the United States transferred 20 Mg to Canada for use in ZEEP and NRX.

The Trail plant continued to be operated by Cominco for the United States Atomic Energy Commission (USAEC) until 1956, when Cominco's contract expired. They offered to produce heavy water for AECL at $76/kg, but the offer was not accepted and the Trail plant was closed. AECL had assurance of much larger quantities of heavy water from the USAEC at $62/kg. The small electrolytic cells used at Trail for final enrichment of heavy water (2-99.8 percent) were the basis for the design of a unit for reconcentrating (upgrading) recovered heavy water at Chalk River (see chapters two and fourteen).

In 1952, E.I. DuPont de Nemours and Co. Inc. began operating two large heavy-water plants at Dana, IN (410 Mg/a) and Savannah River, SC (450 Mg/a). The Girdler-Sulphide (GS) process (see Box A) was used. In 1955, an exchange agreement on atomic energy was signed by Canada and the United States, extending the collaboration that had begun in 1944. This agreement provided Canada with a heavy-water supply, at least for the initial phases of the CANDU program, and access to all the American heavy-water production technology.

This agreement turned the issue of securing a long-term, large-scale supply of heavy water at a reasonable cost for the CANDU program from process development to technology transfer. As a result, the small effort on heavy-water process development underway at Chalk River in the early fifties was terminated.

Once the large reactors at Savannah River were supplied with heavy water, the Dana plant and two-thirds of the Savannah River plant (SRP) were shut down over the 1957-59 period. This left Savannah River with an output of 180 Mg/a and it continued to operate until 1981.

Between 1955 and 1967, AECL purchased about 1,000 Mg of heavy water from the USAEC for ZEEP, NRU, ZED-2, WR-1, NPD, Douglas Point and the first Pickering unit.

A Canadian heavy-water production capability would be needed to sustain a major CANDU program, and government policy in the sixties required that this activity should be in the private sector. In 1963, AECL sought bids for the supply of 900 Mg over a five-year period at a price below $49/kg. The successful bidder would have a royalty-free license to the basic GS

[3] Collaboration with India, the only other significant heavy water producer in the eighties, was ruled out under Canada's nuclear non-proliferation policies.

BOX A THE GS PROCESS

When the USAEC selected the water-hydrogen sulfide exchange process to make heavy water for the Savannah River reactors, they asked the Girdler Corporation to operate a pilot plant and to design the first production plant, located at Dana, Indiana. The process became known as the Girdler-Sulfide, or GS.

The basis for this process is the exchange reaction between liquid water and gaseous hydrogen sulfide, $H_2O + HDS <====> HDO + H_2S$. The reaction occurs spontaneously in the liquid phase between the water molecules and dissolved hydrogen sulfide molecules, which are themselves in equilibrium with the gas phase. Deuterium transfers between the gas and the liquid until the two phases are in equilibrium with each other.

The process is carried out in large towers, analogous to distillation towers, containing many sieve trays (see later) on which the gas and water are intimately mixed so that their deuterium contents approach equilibrium. The two phases separate and the gas flows upward from tray to tray countercurrent to the water flowing down from tray to tray. This repeated countercurrent contacting leads to enrichment of the deuterium in one phase relative to that in the other. The tower (figure 17.A1) is divided into two temperature regions. In the upper, cold section, where feed water enters, deuterium is transferred from the gas to the water. At the bottom of this section some enriched water is withdrawn to feed a smaller, higher enrichment tower. The majority of the enriched water passes to the lower hot section where the direction of deuterium transfer is reversed. This hot section provides the enriched gas going to the cold section by stripping deuterium from the water. The gas-to-water flow ratio is carefully controlled (at about 2) so the gas leaving the top of the tower is close to equilibrium with the feedwater, and the waste water leaving the bottom of the hot section is close to equilibrium with this same gas that has been recycled to bottom of the tower. The deuterium concentration in the waste water, relative to that in the feed, thus approaches the ratio of the equilibrium constant in the hot section (130°C) to that in the cold (30°C), 1.8/2.3 or 0.8. Therefore, for these

FIGURE 17.A1
GS Process Tower

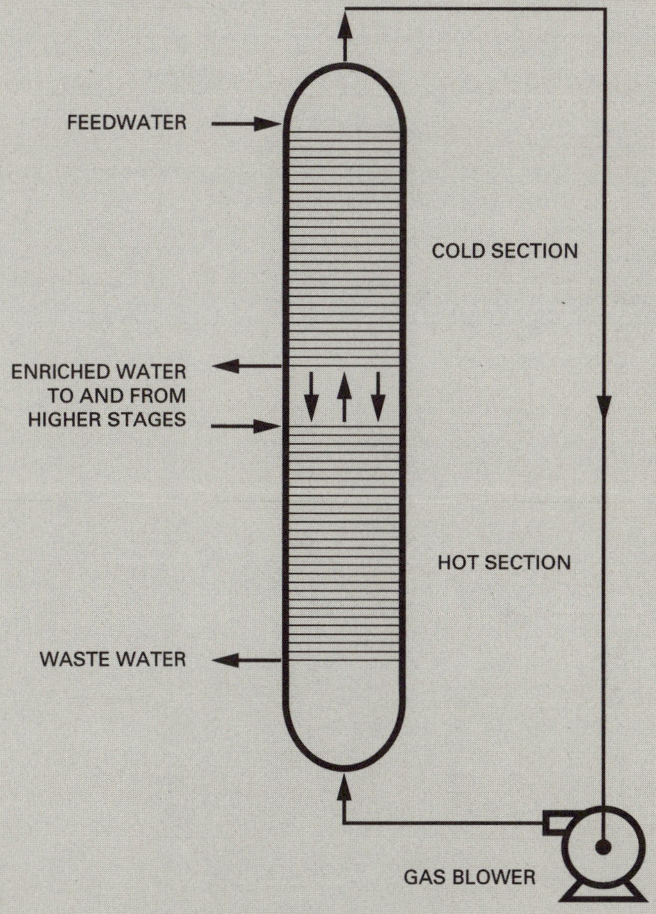

temperature conditions, deuterium recovery from the feed is limited to 20 percent. These temperatures and a pressure of 2000 kPa are close to the optimum process conditions which give a reasonable recovery without excessive energy requirements to maintain the hot section temperature; a lower cold section temperature is prevented by the formation of an ice-like hydrate of water and hydrogen sulfide.

A 400 Mg/a unit, such as the PHHWP shown in figure 17.A2 or one half of each BHWP, has three parallel first stage towers each 8.5 m in diameter and 75 m high to enrich the water by a factor of 4. A 7 m diameter second stage tower brings the enrichment to 24 times natural and two 3 m diameter third stage towers in series (one hot, one cold) produce 30 mole percent D_2O.

The criteria for the final enrichment process are simplicity, reliability and lower pressure. The holdup (inventory) of heavy water in this part of the plant can be worth millions of dollars so that the risk of any loss must be kept very small. For these reasons water distillation is the preferred process to provide reactor grade product (99.75 mole percent).

The key features of the GS process include a large flow of treated feedwater, large exchange towers with many sieve trays to provide intimate countercurrent contact of gas and water, substantial supplies of steam and cooling water, efficient heat exchange, materials suitable for the corrosive nature of aqueous hydrogen sulfide, and systems to handle toxic hydrogen sulfide safely and to minimize releases of it to the environment.

FIGURE 17.A2
Port Hawkesbury Heavy Water Plant

process patent, which had been assigned to AECL in Canada by the USAEC under the exchange agreement, and would have access to the American technology at Savannah River.

Inherent in AECL's status as a crown corporation, the federal government played an over-riding role in decisions on major expenditures, and these could be used to reinforce other government priorities. In Canada's first heavy-water plant, the government saw an opportunity to foster its new regional economic development policy. The result was that the contract was awarded to Deuterium of Canada Limited (DCL), once they had obtained strong financial support from the Nova Scotia government and revised their bid to propose a location at Glace Bay in Cape Breton, rather than in Alberta. The energy source for the plant would thus become subsidized coal, rather than cheap natural gas, but badly needed jobs would be created in an underdeveloped area. The president of DCL, J.S. Spevack, had invented the GS process during the Manhattan Project. As a result, the USAEC had the basic patent, but Spevack controlled several process improvement patents. Although his technical knowledge of the process was considerable, his company had little experience in managing the design, construction or operation of a large chemical plant. Despite strong reservations expressed both by AECL and the USAEC to granting the contract to DCL, AECL was instructed to do so. It was not an auspicious beginning, and it led ultimately to failure.

The Glace Bay project was plagued by problems – managerial, labour, financial, technical and organizational. It proceeded in isolation from AECL, and even contact with Savannah River was rather limited. The design was more complex than the American plants, in part because of the decision to use sea water both as feed and for cooling.

Meanwhile, in 1964 AECL invited another round of bids to establish a second heavy-water plant, since more capacity would clearly be needed. Once again, as in the first round of bids, the only company with chemical or petrochemical plant experience that bid was Imperial Oil Limited and, for a second time, their bid was rejected as being too high. All the other bids were found to be inadequate. In 1965, AECL obtained approval to negotiate with CGE for the supply of 4,500 Mg of heavy water over a twelve-year period; the contract was signed in 1966. CGE chose Port Hawkesbury in Cape Breton for their plant site, where steam from an adjacent oil-fired power plant was available, and a federal capital grant was provided for locating in a "designated area". The site had adequate fresh water both as a source of deuterium and for cooling.

Coincident with the CGE contract, AECL was instructed by the government early in 1966 to negotiate a new contract with DCL, to double the capacity of the Glace Bay plant to 360 Mg/a. The new contract called for DCL to deliver 4,500 Mg over twelve years at $41/kg. However, by the end of 1968, attempts to commission the plant at Glace Bay were abandoned. Spevack had left and Nova Scotia became the sole owner of DCL. An assessment of the plant by SRP personnel was arranged by AECL. Extensive corrosion by sea water, equipment failures and technical errors in design meant that substantial modifications were needed. After prolonged negotiations between the Nova Scotia and federal governments, in 1971 AECL was directed to oversee the rehabilitation of the plant and then operate it.

The Port Hawkesbury Heavy Water Plant (PHHWP) was designed and built by the Lummus Company of Canada Limited, the Canadian subsidiary of the designer of SRP. Where SRP had consisted of twenty-four parallel units, PHHWP had three, each with a six-fold larger output. This scale-up seemed justified in light of the progress in chemical plant experience between 1950 and 1966. Heavy-water production began in 1970. In the next few years a number of serious problems were encountered, which limited output to less than half the rated capacity until 1974.

With the decision to build the Bruce reactors, more heavy-water production capacity was required. In 1969, AECL obtained approval to build the Bruce Heavy Water Plant (BHWP), which Ontario Hydro would operate and have an option to purchase. AECL contracted with Lummus to design and build a plant essentially identical to the PHHWP. During the design phase the capacity was increased from 400 to 800 Mg/a, when it became clear that the Glace Bay plant would require extensive rehabilitation before it could operate. Steam was to be supplied from the Douglas Point reactor, backed up by oil-fired boilers, and later from the Bruce reactors. Production began in 1973, but was restricted during the first year of operation by some of the same problems encountered at the PHHWP. In June 1973, Ontario Hydro purchased the plant and declared it in-service.

Late in 1973, Ontario Hydro committed three additional 800 Mg/a heavy-water plants for the Bruce site, in anticipation of a continued high growth rate in the demand for electricity. By 1976 Bruce Heavy-Water Plant C (BHWP C) was cancelled and the schedule on BHWP D was extended; it was stopped in 1978 at 70 percent completion. BHWP B was completed and reliable production began in 1981; it incorporated many improvements based on the experience with the first plant, now called BHWP A.

In 1971, AECL awarded a contract to CANATOM MHG[4] for redesign and rehabilitation of the Glace Bay Heavy Water Plant (GBHWP). A new process arrangement was adopted that would optimize production from the existing towers. A reservoir for fresh feedwater was established, steam from a nearby power station was available, and heat rejection through cooling towers was adopted. The large towers, most heat exchanger shells, many pumps and some piping from the original plant were retained; the remainder was scrapped or sold. Production began in 1976, but satisfactory operation was not achieved until 1978.

AECL's Heavy-Water Projects Division was formed in 1971, with responsibility for the Glace Bay Heavy-Water Plant (GBHWP) and the completion of the BHWP A. AECL in 1973 shared Ontario Hydro's rosy expectations of future heavy-water demand. An 800 Mg/a plant was committed in 1974 for construction by CANATOM at LaPrade in Quebec. This plant, along with the PHHWP and the GBHWP, was to service anticipated sales of CANDU reactors in Canada outside Ontario and in the export market. Construction of LaPrade was slowed in 1976 and stopped in 1978 at 40 percent completion.

In 1975, CGE sold the PHHWP to AECL; they could no longer meet the heavy-water contract price as energy costs rose. Since AECL would soon be operating the GBHWP and was building LaPrade, the sale made better sense than renegotiating the contract. Thus, by 1975 the two organizations responsible for Canadian heavy-water production were Ontario Hydro and AECL.

The failure of the original Glace Bay plant and the production delays at the PHHWP meant that AECL had to scour the world for heavy water for Pickering units two, three and four. About 1,500 Mg were purchased or leased from other countries; e.g., Argentina, Sweden, the United Kingdom, the USSR and the United States. To maintain the Pickering schedule, it was also necessary to temporarily shut down other Canadian reactors, so that their heavy water could be loaned to Pickering; e.g., NRU, Douglas Point, Gentilly-1 (CANDU-BLW). The pressure to get the PHHWP and the BHWP A into full production was intense.

By the end of 1974, Canadian production began to exceed demand, and has done so ever since. Over the three decades from 1945 to 1974, Canada obtained close to 2,500 Mg of heavy water from abroad, about 80 percent of it from the United States. A total of 18,000 Mg had been produced in Canada to the end of 1990. With limited demand for heavy water outside Ontario in the eighties, AECL recommended that the PHHWP and the GBHWP be shut down in 1982; however, the government delayed their approval until 1985, when these plants were finally shut down and subsequently were dismantled. By 1985, AECL had a substantial stock of heavy water available to service future CANDU sales. Once sufficient heavy water for the four Bruce reactors was available, Ontario Hydro shut down BHWP A in 1984, and it has been dismantled. The BHWP B continued to operate, supplying heavy water for Darlington and as make-up to replace losses at all the Ontario Hydro nuclear stations.

THE AMERICAN CONNECTION

Canada had available the United States experience at Savannah River. To achieve as low a heavy-water cost as possible, the throughput per first stage tower in the Canadian plants was increased by a factor of eight over that at the SRP. Tower diameter was increased by a factor of 2.5 and gas velocity by 1.3. Stable operation at the higher gas velocity was demonstrated in 1966 in tests in one Savannah River tower, as part of a broad Canada-United

[4] A consulting engineering company located in Montreal, originally called CANATOM-MONMAX.

States cooperative development program on heavy-water reactor technology. The extraction of heavy water per first stage tower at the PHHWP and the BHWP was designed to be six times that at the SRP. That it was not eight times higher (directly proportional to throughput) was because the optimum hot tower temperature was found to be lower than used at the SRP, and the number of trays was reduced so that hot and cold contacting could be fitted into a single tower, which provided savings in piping, valving, hemispherical tower ends and tower supports.

Key staff members of the PHHWP and especially of the BHWP were attached to SRP to become familiar with the details of GS process operation and the various problems that had been encountered.

GS PROCESS DEVELOPMENT IN CANADA

Scope of the Development Program

In 1969, with the commitment of the BHWP, AECL began a small effort on GS process development at Chalk River. By 1971, CGE was experiencing hydraulic instability and foaming problems at the PHHWP which were severely limiting production. CGE began a development program to study the flow characteristics of sieve trays when foamy water was present, and to identify foaming agents present in the process; this work was done at the central laboratories of General Electric at Schenectady, NY. The AECL effort was expanded and was soon given a high priority when the extent of the difficulties being experienced became evident. By early 1973, with the BHWP also severely limited in its production, the AECL effort was expanded substantially, technical groups at each heavy-water plant were increased and other organizations were involved on a contract basis.

The total program was coordinated jointly by representatives of AECL, Ontario Hydro and CGE at the senior management level. Many collaborative working groups were formed to exchange information and plan experimental programs. There was good communication between the plants and the research sites, which proved to be an important element in the overall success of the heavy-water production program.

At Chalk River, a small pilot plant was built, with water treatment facilities and a 0.3-metre-diameter column that could operate at either hot or cold tower conditions. It was used for extensive studies of tray performance, of the effects of water quality and to screen a wide range of antifoaming agents. Experiments were done in the production plants to define detailed process behaviour. To obtain adequate information from these plant experiments, additional instruments and sample points were added to one of the towers at the BHWP, and later a tower at the GBHWP was similarly fitted. At the BHWP, AECL supplied and staffed a temporary special analytical laboratory.

The development program was wide ranging, including process analysis and control, process chemistry, analytical chemistry, materials behaviour, mechanical equipment, sieve tray hydraulics and efficiency, gamma scanning and water distillation.

All the Canadian heavy-water plants have experienced equipment, materials and process problems that, by and large, had already been encountered at Dana or the SRP; however, in the American plants these problems had not occurred to the same degree. The American plants had been designed very conservatively, because at that time the GS process was unproven and reliable production on a tight schedule was deemed to be far more important than reducing the capital costs. There was a large element of overdesign and the cost of the plants was high – the cost of the SRP ($130 million in 1951) was about twice that of the PHHWP or the BHWP in constant dollars per annual Mg of production capacity.

The very much larger equipment in the Canadian plants, the higher gas velocities in the towers, and different feedwater chemistry, all contributed to an enhanced effect of plant problems. While the Dana and SRP experience was invaluable in providing a sound foundation for the Canadian plant design and operation, it could not provide the depth of process understanding required to extrapolate reliably to the new conditions. However, this was not appreciated when the PHHWP and the BHWP were being designed, or even when the GBHWP was being redesigned. The large development program was necessary to gain this understanding, which helped to overcome the operating problems, and ultimately to fine-tune plant operation and develop process improvements.

BOX B SIEVE TRAY OPERATION

A sieve tray consists of thin metal sheets perforated by many small holes; the sheets are supported on a set of beams attached to the tower wall. The total area of the holes is about ten per cent of the tray area. Hole size has little effect on tray performance – diameters of 6, 10 and 13 mm have been used in the Canadian plants.

The kinetic energy of the gas passing through the holes, as it flows up the tower from tray to tray, maintains the liquid on the tray. The liquid enters one tray through two downcomers from the tray above located on chords at opposite sides of the tower, and flows inward across each half of the tray and over weirs into a downcomer located on the diameter perpendicular to the flow. This downcomer carries the liquid to the tray below where the horizontal flows across the tray are reversed. This arrangement is called a two-pass tray.

The velocity of the gas leaving the holes (4 m/s) creates a turbulent mixture of gas bubbles and liquid droplets having a large gas-liquid interfacial area. This mixture is called a froth. The amount of deuterium transferred between the phases depends on the area created and the length of time the gas and liquid are in contact – typically about a half a second for the 0.45 m tray spacing used. The tray efficiency, the ratio of the actual amount of deuterium transfer that occurs to the maximum possible if gas and liquid reach equilibrium, is a convenient measure of tray performance.

The froth height on the tray depends on the weir height and the head of froth flowing over the weir necessary to handle the total liquid flow rate in the tower. Obviously, froth density will also be a factor.

The disengagement space above the froth must be sufficient for most liquid droplets to settle out of the rising gas so that the amount carried to the tray above as entrainment is no more than a few percent. This entrainment represents backmixing of the downflow of liquid in the tower and reduces tray efficiency. The overall time for the gas to pass through the froth and then separate from it is about one second (see Figure 17B).

FIGURE 17.B
Schematic of an Operating Two-Pass Tray

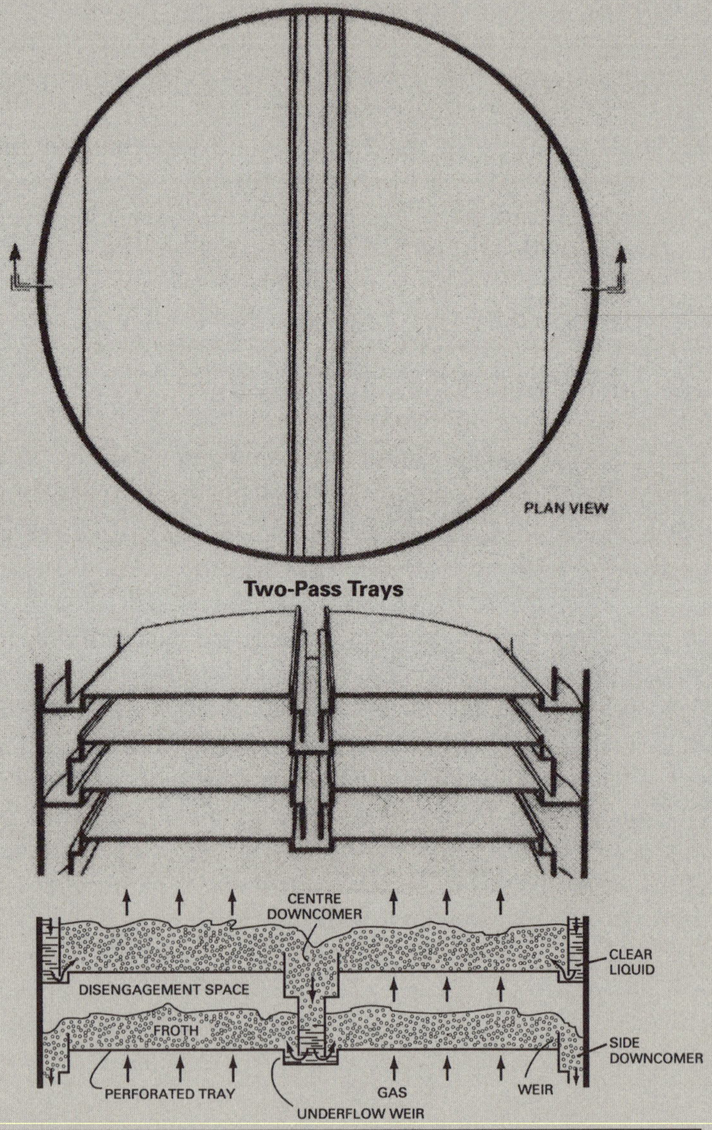

Froth entering the downcomer must also separate so that clear liquid enters the tray below, and backmixing of gas is avoided. Therefore, adequate downcomer area is needed and the flow resistance at the underflow weir must provide an appropriate depth of liquid and froth in the downcomer.

If the froth height on the tray is increased, the overall gas pressure drop will rise requiring a larger head in the downcomer. For a foamy system with a low froth density, this could lead to downcomer backup and the tower would flood. High froth height will also increase entrainment.

Stable tray operation can occur over a range of gas flow rates. The minimum is set by excessive weeping through holes ultimately dumping the liquid from the tray. The maximum occurs when the rise in pressure drop causes the tray to flood due to froth filling the disengagement space.

In general, mechanical equipment performance at the Canadian heavy-water plants has been very satisfactory. Valves, rotating equipment and instrumentation have all performed with a reliability better than or equivalent to that which is typical of the petrochemical industry. Heat exchangers ranked high as a cause of lost production at the PHHWP and the GBHWP, because of the time required to disassemble, repair and re-install.

Gas-Liquid Contacting and the Need to Control Foaming

The key to success in GS process performance lies in the design and operation of the sieve trays (see Box B and figure 17.1). These trays must satisfy conflicting requirements: intimate mixing of the gas and liquid on the tray as a froth with adequate residence time for the exchange reaction to occur, and essentially complete separation of the liquid and the gas as they flow to the tray below and the tray above, respectively.

These requirements can be met if the bubbles in the gas-liquid froth have a short lifetime, considerably less than a second. However, if materials are present that increase this lifetime, then the froth is said to be foamy, and the tray performance may suffer.

Foaming had been observed from time to time at the SRP and appeared to be associated with feedwater treatment problems or periods of high turbidity in the raw water. This foaming was adequately controlled by the addition of a silicone-based

FIGURE 17.1
Pre-installation Trial Assembly of a Bruce Heavy Water Plant 8.5 Metre Diameter Tray

antifoaming agent. In the design of the trays for the Canadian plants, it was assumed that the froth could be kept in a non-foamy condition with antifoamers or high-quality water treatment. This design was based on a large body of experience in the chemical and petroleum refining industries with sieve trays, as well as the data from SRP. A froth density equivalent to a non-foaming system was assumed which led to the choice of 100-mm-high weirs, to give sufficient gas-liquid contact time for deuterium exchange.

At both the PHHWP and the BHWP, early operation was plagued by hydraulic instability. This was manifested by a high and fluctuating pressure drop, which led to flooding and dumping of the trays. Stable operation was possible only at about 70 percent of the design flow rate, and deuterium extraction was less than half of the design rate. Foaminess was causing downcomer backup when flows were increased, and high froth height even at the reduced flow rate.

These tray problems were difficult to diagnose. Individual tray operation could not be observed or monitored by instruments. In the early operation of Port Hawkesbury, there were other problems – such as the plugging of the tray perforations by iron sulfide due to iron in the feedwater, necessitating a new water treatment system, and heat exchanger corrosion – which compounded the situation.

The R&D groups that were building up at this time soon found themselves in a crisis situation, responding to a variety of problems, and evaluating many possibilities in parallel to solve major problems, such as the hydraulic instability of the trays. When a small viewing port was installed on one tower at the PHHWP, foaming was observed. The first use of antifoam was successful only for a short time, and then foaming became even worse. It took some time to discover that emulsifiers present in the silicone antifoamer were hydrolyzed in the hot section of the tower to produce an accumulation of fatty acids that are foam promoters. The silicone antifoam specification was changed.

Adding this new silicone-based antifoamer to the feed water at a concentration of a few parts per million allowed stable operation at about 90 percent of the design flow, resulting in 70 percent of the design deuterium extraction rate. Using activated carbon at the PHHWP to remove trace organics from the feedwater was beneficial, but expensive. It was used for only a short trial period.

Before the second enriching unit at the BHWP A was put into operation in 1973, the weirs on the cold tower trays were reduced in height to 40 mm, so that froth height would be lowered. With antifoamer addition to this unit, stable operation at 110 percent of design flow and 93 percent of design extraction was soon demonstrated. The cold tower weirs in the first unit at the BHWP A and at the PHHWP were also reduced in height later in 1973. Once other equipment that was limiting water flow was modified in 1975, it was possible to operate the BHWP A towers routinely at 120 percent of design flow.

Surface chemistry studies demonstrated that the water-hydrogen sulfide system was inherently foamy at cold tower conditions, where the pressure is close to the liquefaction point. It was found that multilayer adsorption of the gas molecules onto the water surface could have the considerable effect of increasing bubble stability by several orders of magnitude relative to that for conditions far from liquefaction; e.g., in the hot tower or at low pressure. This inherent foaminess, apart from any added effects of foam-promoting impurities, meant that froth density was lower than assumed and froth height higher than expected. This did not impede stable hydraulic operation as long as antifoamers were used, but it had an important influence on tray efficiency, as discussed below.

Foaming has caused problems in other areas, such as the dehumidifier,[5] particularly at the GBHWP, and in the second or third stages. This was due to the accumulation of impurities that were carried down the cold tower, but had sufficient volatility in the hot tower that they did not leave in the effluent. They may originate in the feedwater, in the hydrogen sulfide gas supply, in oils that enter the process from compressor and pump shaft seals, or in the antifoamers added to the feedwater.

[5] The dehumidifier is the region of the tower where the gas is cooled before it enters the cold section.

A great deal of laboratory effort was devoted to identifying impurities that increased foaminess, and this led to improving the specifications for the purity of feedwater, gas supply, seal oil, etc. The Nova Scotia Research Foundation, in collaboration with AECL, played an important role in this work.

A large number of antifoamers were tested in the laboratory and the pilot plant. This work showed that several non-ionic surfactants were excellent antifoaming agents at process conditions. In the plants, these new antifoamers have further improved the stability of the operation and thereby increased production, and are also considerably cheaper than the silicone-based agent that had been used.

Sieve Tray Efficiency

Once the plants were operating stably at near design flow rates, it was clear that the tray efficiency was lower than design. Tray efficiency is the ratio of the actual exchange occurring to the maximum possible if the gas and liquid reached equilibrium before each left the tray. The design had been based on measurements of tray efficiency made at Dana. The technique used was to measure the change in the deuterium concentration profile in the towers that resulted from making small changes in the ratio of gas to liquid flows. It was not until the mid-seventies, when similar measurements were made in the Canadian plants and thoroughly analyzed, that the degree of uncertainty in the Dana results was fully appreciated. However, when the Canadian plants were being designed, the Dana values of 65 percent in the cold tower and 70 percent in the hot tower agreed with the sieve-tray efficiency correlations then available, and were accepted as reasonable. We now know that these correlations do not apply to the GS process cold tower trays, with their low froth densities.

The low efficiency was shown to be in the cold tower once pilot plant measurements were made, and this was confirmed by plant experiments which showed that the cold tower tray efficiency was only 50 percent. The effect at the PHHWP and the BHWP was that the recovery of deuterium from the feedwater was reduced to 90 percent of design. However, this was offset by operating at above the design flow-rate. The GBHWP design was more sensitive to reduced tray efficiency, and in this plant both cold and hot tower tray efficiencies were low. Thus, the design production rate could not be achieved with the existing trays.

A development program was required to improve tray performance. This program relied on the Chalk River pilot plant, where a wide range of tray operating parameters could be measured, and on plant experiments. The most important tool for obtaining plant data was the gamma-scanning equipment illustrated in figure 17.2. Originally used at the SRP to locate damaged trays, this technique was developed further at Chalk

FIGURE 17.2
Simplified Diagram of Gamma Scanner (installed on a third stage tower with single-pass trays)

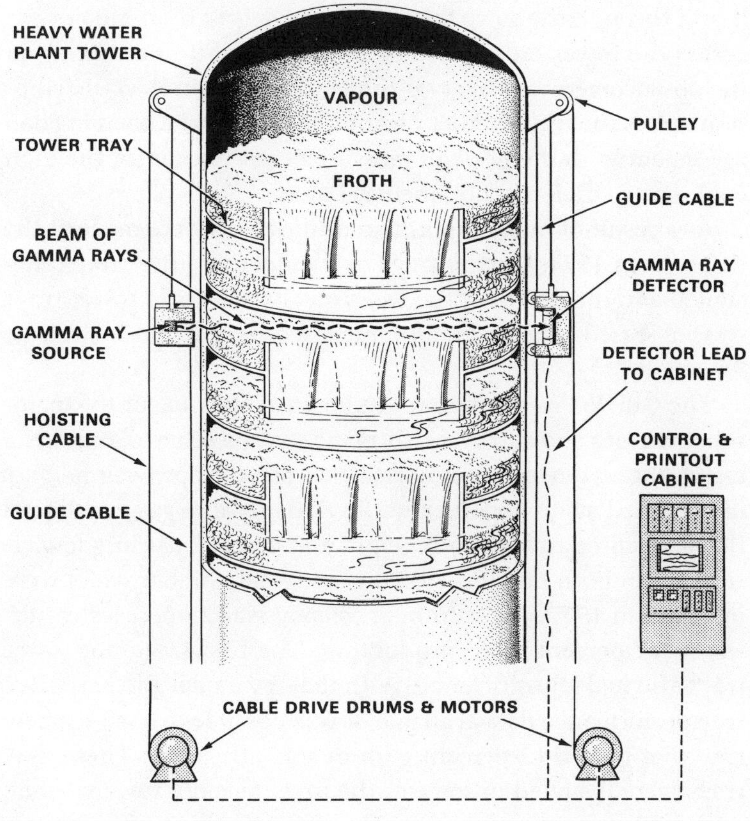

River to increase its sensitivity. The attenuation of gamma rays passing through the tower was measured as a function of height. Comparing attenuation through the froth with that through the disengagement space gave the froth density, and as the detector and source were raised could determine froth height on any tray during operation. The technique was also useful for locating regions prone to flooding, and had been useful in diagnosing some of the early problems at the PHHWP and the BHWP A.

To correlate all these data and apply them to making tray improvements, a fundamental mathematical model of the mass transfer and hydraulics of an operating tray was developed by AECL. This model predicted tray efficiency for a given set of conditions defined by pressure, temperature, gas and liquid flows and tray geometry. It took into account the effects of entrainment of liquid in gas rising to the tray above, weeping of liquid through the sieve tray holes, the degree of mixing of gas across the tower cross-section, the degree of liquid mixing in the downcomers and the two-dimensional liquid flow distribution across the tray. Predictions by the model have been in good agreement with direct plant measurements, except for the high cold tower efficiency obtained at Dana.

As a result of modeling, tray modifications were made at the PHHWP in 1977. Certain trays in the dehumidifier were redesigned and the height of the weirs on alternate cold tower trays was increased. These changes increased production by at least 5 percent.

The GBHWP was being re-designed at the time of the major uncertainties about tray performance in the other plants. As a result, the trays were designed conservatively. Low weir heights in the third stage, together with relatively low gas and liquid flows, resulted in low froth height and density, causing low efficiency in both hot and cold towers. When higher weirs were installed in 1977, the gain in efficiency was 23 percent resulting in 12 percent more production. The first-stage hot tower trays also had low efficiency. With the tray model and extensive tray performance data available, it was possible to design a new tray that gave a 12 percent gain in tray efficiency. These new trays were installed in some of the first-stage hot towers, along with higher cold tower weirs. However, by this time a surplus of heavy water was beginning to accumulate and further retraying could not be justified. If all the first- and second-stage towers had been modified it is likely that the GBHWP would have been able to reach its design output.

Process Analysis

A steady-state computer simulation model of an entire heavy-water plant was developed by AECL to assist in diagnosing problems and optimizing production. This was important, because a GS plant operates as an integrated whole, where all the towers are closely coupled and a number of interconnecting feedback loops exist; it was very difficult to analyze any one part in isolation. The models described the whole plant in detail, and were used to assess the effect of the variation of a local parameter at any point in the process on the overall operation, as well as to define local conditions at a point where no instruments existed. These simulation models were based on methods evolved at McMaster University.

Enthalpy, water, hydrogen sulfide and deuterium balances were calculated for approximately 1,000 trays by iteration. There were typically 2,200 streams to be defined, including a number of large, interacting recycle flows, which made the computation complex and the iteration lengthy.

Extensive plant data available from data loggers and numerous in-plant experiments enabled the simulation model to be tuned to the observed plant behaviour. Close collaboration between the process analysts and the plant operators was essential to ensure that the predictions were realistic.

The model showed how process control could be improved. Small perturbations from ideal operation have significant effects on production, particularly with the reduced tray efficiency. However, the effect of the perturbation of an individual process variable on plant output is very difficult to observe. The model was used to define the ideal settings of process variables and the precision with which they should be maintained. At the GBHWP, the role of the model in process control was more important, because of the complex interconnection of flows amongst towers with this design.

An early application of the simulation was the optimization of the temperature profile in the cold section of the towers. While they are nominally isothermal, there is a cooling effect due to the expansion of the gas from tray to tray. Simulation determined the correct gas and liquid inlet temperatures needed to locate the minimum temperature at its optimum position; experiments in the instrumented tower at the BHWP demonstrated that stable operation was possible within a degree of the hydrate temperature. Thereafter, all the cold tower sections at the BHWP were operated with this optimum temperature profile and the minimum temperature 2°C lower than design, which increased production by 3 percent (worth $12 million a year in 1985 dollars). Optimizing various temperatures and flows in the second and third stages, based on the simulation model, led to another 3 percent increase in output.

The simulation model has been used extensively to optimize steam consumption for a variety of situations, and to re-optimize process conditions to maintain an acceptable cost of production as steam costs have risen. The model played an essential part in assessing the value of the various tray modifications already discussed.

The detailed understanding of the process that has been gained through the simulation work has led to several potential improvements for future plants, and appropriate process patents have been obtained. The most important of these involves introducing the feedwater at the bottom of the hot section of the tower, and arranging for the heat transfer trays in the humidifier[6] to also provide deuterium transfer. With this design change, it should be possible to produce 10 percent more heavy water per unit tower volume and unit of process heat than in the best of the existing plants.

Materials

Carbon steel was used wherever possible in the GS plants. Its performance has been good, provided quality assurance specifications and welding procedures were rigorously followed. Blistering, due to hydrogen accumulation at voids or inclusions, has been rare. Following the instances of blistering found early in the operation of the BHWP B, specifications and inspection techniques were tightened, and would apply to any new plant.

Erosion-corrosion of carbon steel was generally avoided by using stainless steel in regions of high turbulence; however, a few unexpected failures did occur, requiring replacement or cladding by stainless steel. More troublesome was general corrosion of carbon steel piping in the BHWP A downstream of coolers, and the deposition of the iron so dissolved on hot tower trays and in process heaters. This iron transport was also a function of water chemistry and was not a problem at the Nova Scotia plants. A large laboratory effort by AECL to study the corrosion mechanism and the properties of the various sulfides of iron (see chapter seven) showed that the iron dissolution was due to the limited protection provided by the initial corrosion product on steel (mackinawite), and its very slow transformation in these cold regions at the BHWP to the protective sulfides (pyrrohotite and pyrite). A pre-operational conditioning treatment to passivate carbon steel surfaces was developed, and used successfully in the BHWP B; in this case, deposition on hot tower trays has been minimal.

Experience with stainless steel and Inconel in the GS process has been generally good. The use of low-carbon stainless steels minimized the risk of stress-corrosion cracking in sensitized regions, and type 316 was used to avoid pitting in regions where deposits may occur. Polythionic acids, formed by the interaction of sulfides, moisture and oxygen, were found by laboratory studies to be particularly aggressive. It was therefore recommended that all stainless-steel equipment should be flushed with a neutralizing solution before being opened to the air.

Development Program Costs and Benefits

The major benefits of the development program were increased heavy-water production, lower heavy-water cost and better process understanding. Three main contributions which R&D made to increased production were assessed in 1981: helping

[6] The humidifier is the region at the bottom of the tower where the incoming gas is heated and saturated with water vapour.

the plants achieve mature production rates sooner, defining optimum conditions that gave higher mature production rates, and determining process and equipment modifications that increased production, such as new antifoamers and tray changes.

The initial period of operation, when process instabilities and equipment shortcomings were overcome, was not included in the assessment. The increased production throughout this period was the result of the increasing experience of operators and designers.

Another important contribution from the R&D program, which was difficult to quantify, was the additional technical support available as the expertise of the R&D staff developed. The quick response that the R&D staff made in helping to solve urgent process and equipment problems at the plants contributed to higher capacity factors.

The total cost of the GS process development work to AECL and Ontario Hydro over the period from 1969 to 1980 was $64 million (1981 dollars), including overheads. This represented slightly less than 3 percent of the value of the 7,810 megagrams of heavy water that was produced during that period. A value of $300 (1981 dollars) per kilogram was used in this analysis.

The increased production up to 1980 due to the R&D was estimated to total 800 Mg, or about 10 percent additional heavy water. Of this amount, 320 Mg were ascribed to process simulation, 140 Mg to sieve tray improvements and 70 Mg to better antifoamers. The benefit-to-cost ratios for these specific activities were estimated to be 12, 6 and 20, respectively, while this ratio for the overall program was 4. These are conservative values, since no allowance was made to include ongoing benefits.

Although the simulation models cost several million dollars to develop, their benefits have repaid this investment handsomely. However, when the model development was committed, it was much more as an act of faith than a decision based on estimated payback.

The decision to commit a large R&D program to support the heavy-water plants was based on the need to ensure the security of a heavy-water supply. No payback analysis was made beforehand, although an adequate payback was expected. That judgement by AECL and Ontario Hydro management was based on many years of experience with mission-oriented R&D. It has been confirmed by the results described above and the success of the plants in meeting all heavy-water demands after 1973.

The severity of the difficulties in bringing the PHHWP, BHWP, and the redesigned GBHWP to mature operation was unexpected. The difficulties were certainly more severe than normally encountered in scaling up a chemical process in size and throughput. In retrospect, the effects of foaming and of low tray efficiency stand out clearly. At the time, it took a great deal of effort to disentangle these effects from a myriad of other possible causes of poor operation.

ALTERNATIVE PROCESSES

Early Work

In the late forties and early fifties, AECL had a small R&D effort to develop heavy-water production processes. Variants of the steam-hydrogen exchange process that could be operated independently of a hydrogen production facility were considered. While offering larger-scale production than was possible at Trail, such processes were found to be too costly.

AECL assessed a number of heavy-water processes in 1955, and concluded that the most attractive were likely to be water distillation, hydrogen distillation and several chemical exchange processes. Of the latter, exchange between water and hydrogen sulfide was mentioned, but not discussed at any length. This may seem surprising, considering that the two large American plants that had been started up in 1952 used this method, and that within a decade it was the technology to be adopted for Canada's new heavy-water plants. However, in 1955 the process was still classified as secret, and its details were not known by AECL.

When the scope of the American heavy-water production capacity became known in the mid-fifties, and with the signing of a Canada-United States exchange agreement on atomic energy in 1955, further heavy-water process development by AECL was halted in 1956.

One interesting project during the period of early process studies was the development of a parallel-plate, wetted-wall packing[7] for water distillation. A very low pressure drop was achieved, which could significantly reduce the energy required for a distillation system for heavy-water production, using a heat pump to recover the heat from the condenser at the top of the column and transfer it to the boiler at the bottom. However, this could not compete with the GS process. Although this packing was not pursued beyond the pilot scale by AECL, a similar wetted-wall principle was developed successfully into a more practical design by Sulzer Bros. in Switzerland. Their distillation units are used at many heavy-water plants and heavy-water nuclear power stations for concentrating heavy water from about 15 mole percent heavy water to reactor grade, 99.75 mole percent.

It might be concluded that AECL missed an opportunity to develop a product important to the nuclear program, and with applications to other chemical separations. However, in 1956 AECL could not divert scarce resources to such an undertaking and no other organization was interested in licensing the wetted-wall concept.

Process Evaluation

Over the years, AECL has screened a large number of possible methods for heavy water production. Most have been judged to be uneconomical. Several merited further study as alternatives to the GS process. Amine-hydrogen exchange and water-hydrogen exchange were both developed to the pilot-plant stage. During the early seventies, Polysar participated in these evaluations and related process development. Work continues on the water-hydrogen process, since it offers the greatest potential for cheaper heavy water. Laser isotope separation of deuterium was studied extensively by AECL and Ontario Hydro research division in the eighties, but does not appear promising.

Ammonia-Hydrogen and Amine-Hydrogen Exchange

Starting in 1966, a detailed process evaluation was made of the dual-temperature ammonia-hydrogen exchange process, by a group drawn from AECL and several chemical and engineering design companies. A large-scale plant based on water as feed was studied. High-temperature water-hydrogen exchange transferred deuterium from the feed to hydrogen, which then entered the ammonia-hydrogen dual-temperature enriching process. Exchange rates and other process parameters were based on laboratory results in the published literature. It was concluded that this process was unlikely to be cheaper than the GS process. However, the use of dual-temperature ammonia-hydrogen exchange to extract deuterium from ammonia synthesis gas was also studied, and appeared to be more attractive even though plant capacity would be limited to about 60 Mg/a by the size of a typical ammonia plant (900 Mg NH_3/d).

A small plant using a monothermal version of ammonia-hydrogen exchange[8] was being built in France at the time, and operated well for four years, starting in 1968. Since 1970, five ammonia-hydrogen plants have been built in India.

The use of amine in place of ammonia in a dual-temperature process appeared to offer advantages, and an experimental program to develop the dual-temperature amine-hydrogen process began at Chalk River in 1967. Methylamine was soon shown to be the best amine. The exchange reaction between the amino group of the methylamine (CH_3NH_2) and dissolved hydrogen is catalyzed homogeneously by potassium methyl amide dissolved in methylamine. However, at a process pressure of about 7 MPa (the pressure for ammonia synthesis), potassium hydride was found to precipitate, thus seriously reducing the concentration of the catalyst. This was also discovered in France, where amine-hydrogen exchange was also being studied. Scientists at AECL discovered that changing the catalyst to a mixture of lithium and potassium methyl amides avoided this problem, and gave AECL a key process patent. The catalyst is extremely reactive. Studies at Chalk River and on contract by Raylo Chemicals

7 A distillation column packing in which vapour flowed upward between many parallel, vertical plates, spaced 5-15 mm apart, on which thin films of liquid flowed downward.

8 In the monothermal process, the enriched gas that enters the bottom of the cold tower section (Box A) is produced by decomposing enriched ammonia, rather than by exchange in a hot tower section.

Limited, of Edmonton, defined the reactions with impurities such as water, oxides of carbon and oxygen. Raylo also measured rates of thermal degradation, which set the upper limit for hot tower temperature at 40°C.

Because of the low solubility of hydrogen in methylamine, the mass transfer rate at cold tower conditions is very much lower than that for the GS process. A gas-liquid contactor was developed by AECL, in which a gas-liquid froth flowed up through a deep (1 m) bed of knitted mesh packing on a sieve tray, and then the gas continued to the tray above and liquid flowed over a high weir to the tray below (figure 17.3). This arrangement gives a long gas-liquid contact time. Tests in a 25-cm-diameter contactor of this design demonstrated a tray efficiency of about 20 percent at -50°C. Sulzer had developed a contactor in which the gas and liquid were mixed by nozzles, to give a fine dispersion of liquid droplets in the gas having a very high interfacial area; the mixture was separated centrifugally and the gas went on to the next tray, while the liquid had to be pumped to the tray below, because of the high pressure drop in the nozzles. Some liquid recirculated to the nozzles (figure 17.3). This design, which had been used in ammonia-hydrogen plants in France and India, was tested by Sulzer with the amine system as part of a collaborative program with AECL. A tray efficiency of 70 percent at -50°C was obtained. Thus, the Sulzer contactor required a considerably smaller tower than the AECL contactor, and because of the high pressure and thick vessel walls, this saving more than offset the extra energy and complex tower internals required with the Sulzer contactor.

Sulzer prepared preliminary engineering designs for an amine plant to be attached to an ammonia plant in Alberta. Estimates of heavy-water production costs from this plant favoured the exploitation of the process. This was in 1976, just when the demand for heavy water was being revised downward, and concern about an excess of plant commitments was beginning to appear. As a result, work on the project and development of the process were halted and have not been revived.

The major advantages that the amine process offers over the ammonia-hydrogen process are a significantly higher rate of exchange and a lower cold tower temperature, so that the

FIGURE 17.3
AECL and Sulzer Contactors

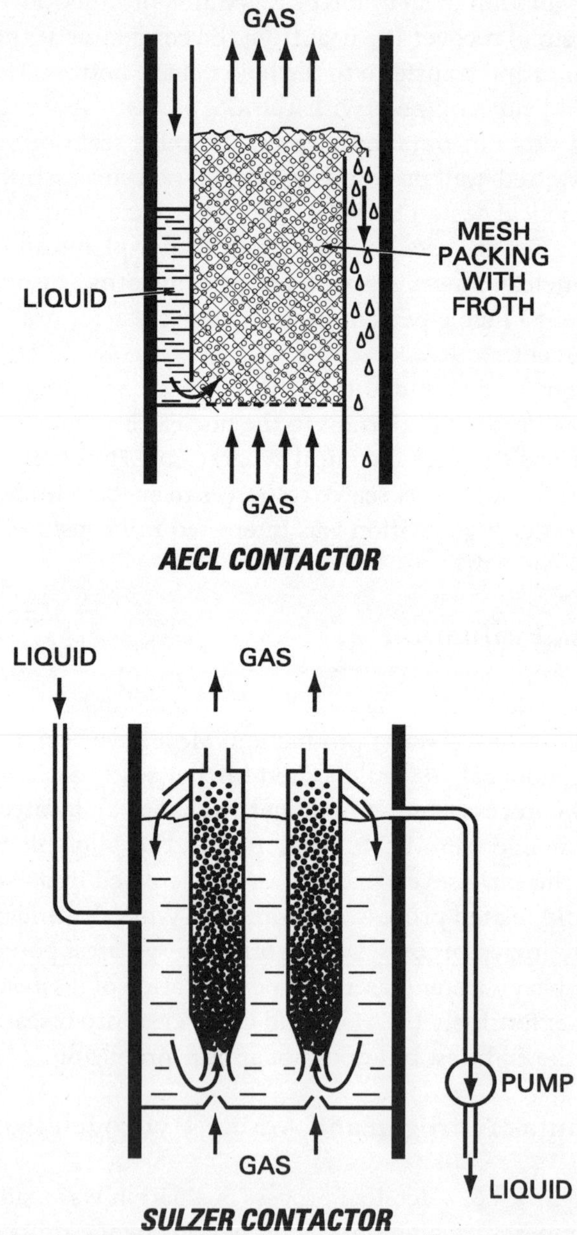

equilibrium constant is higher and fewer gas-liquid contacting steps are needed. The result is the potential for a smaller tower volume and lower capital cost. The disadvantage is a more complex catalyst chemistry with an increased operating cost and perhaps lower reliability. The amine process does not seem to have a clear advantage over the GS process.

Water-Hydrogen Exchange

The potential of the direct exchange between water and hydrogen has been recognized since early in the Manhattan Project. A bithermal arrangement similar to the GS process would have the advantages of larger equilibrium constants, less corrosive fluids and much lower safety concerns. However, until 1970, there was no suitable catalyst, and even by 1990 catalyst activity was not high enough for an economic bithermal process. Nonetheless, there are other, more attractive applications of water-hydrogen exchange.

Most catalysts for water-hydrogen exchange fail because they are wetted by water which prevents ready access of the gas to the catalytic sites. An AECL chemist invented a "wetproofed", or hydrophobic platinum catalyst in 1969.[9] The catalyst, platinum on carbon, is placed on the surface of a tower packing material, such as small alumina spheres. A coating of Teflon makes the catalytic surface hydrophobic. The platinum is thus accessible to the hydrogen and water vapour, so that deuterium can transfer from one to the other. Bare alumina spheres mixed in with the catalytic spheres provide a surface for water to spread on, where exchange between water vapour and liquid water can occur to complete the overall reaction. The catalyst performance has been steadily improved over the past two decades by extensive laboratory and pilot-scale programs. A number of different physical arrangements of the original concept are now available. All involve the countercurrent flow of hydrogen and water through a tower over the catalyst and an adjacent hydrophilic surface.

This exchange reaction can be coupled with electrolytic cells in a flowsheet similar to the Trail plant, to recover heavy water as a by-product of electrolytic hydrogen production. The exchange towers would be an order of magnitude smaller than those required at Trail, where the gas flowed alternately through dry catalyst beds and sieve trays to achieve transfer of deuterium from gas to water vapour to water. This application of water-hydrogen exchange, known as the CECE process (combined electrolysis and catalytic exchange), is undoubtedly the cheapest method to produce heavy water, provided that the cost of electrolysis is included in the value of the hydrogen produced. Thus, heavy water is a potential by-product of water electrolysis and a Canadian supplier of electrolytic units, Noranda-Electrolyser, has participated in the CECE process development. However, the potential production is small. Large electrolytic hydrogen plants (for fertilizer manufacture) are attractive only in a few locations, with cheap electricity, remote from markets. For example, the 180 MW Kima electrolytic plant at Aswan in Egypt could produce 20 Mg of heavy water per annum. In the early nineties joint studies of this possibility were being conducted by Canada and Egypt.

A somewhat similar combination of water-hydrogen exchange with a conventional hydrogen-producing steam-methane reformer[10] appears attractive. This is called the CIRCE process, for combined industrially reformed hydrogen and catalytic exchange. The process can best be applied to extracting deuterium from the large hydrogen streams produced for upgrading heavy oils in the tar sands, since here the hydrogen is not diluted with nitrogen. Process evaluation in collaboration with a

---

[9] He got the idea for the "wetproofed" catalyst while treating his basement walls with a silicone formulation, which, according to the advertisement, repelled liquid water but allowed water vapour to diffuse through it in either direction. He had been interested in heavy-water processes for many years and knew that platinum was a good hydrogen-water isotope exchange catalyst when the water was in the vapour phase, but that with liquid water the rate was low. The nub of his idea was to keep liquid water out but to let the hydrogen and water vapour reach the active sites on the surface of the catalyst.

[10] A reformer is a chemical plant unit in which steam and methane are converted to hydrogen and carbon dioxide, and the latter is removed to produce relatively pure hydrogen.

hydrogen producer has begun, and since 1992 CIRCE development has been given a higher priority so that the process can be demonstrated before the BHWP B reaches the end of its life.

The wetproofed catalyst can also be used for upgrading heavy water (a CECE process application,) and in processes to separate deuterium and tritium.

Laser Isotope Separation

When a compound of hydrogen is irradiated with photons of a precise energy that excites only the molecules containing deuterium, they can decompose to give an enriched product. An enrichment of 5,000 or more in a single step has been observed in the laboratory, and the production of a few micrograms of 30 mole percent heavy water from natural fluoroform (CF_3H) was demonstrated by AECL researchers. However, the development remains a long way from a practical process. Two key requirements for success are efficient use of the costly laser energy and finding a molecule that works well in the laser separation step and can be redeuterated by exchange with water or hydrogen. Unfortunately, common molecules like water, hydrogen or ammonia cannot be used directly in the process.

HEAVY WATER ANALYSIS

The two most sensitive methods for analyzing water for its deuterium content depend directly on the mass difference between hydrogen and deuterium atoms, rather than on the much smaller variation in a physical property with deuterium/hydrogen ratio. Obviously, one is mass spectrometry, where an ion beam is separated into components of different mass by a magnetic field; this method can be applied to any hydrogen compound. The other is infrared spectroscopy, based on the fact that the three molecular forms of water, H_2O, HDO and D_2O have different molecular vibration frequencies, and thus different absorption spectra, with the most useful absorption peaks occurring in the infrared region. Since deuterium is heavier, the molecules containing oxygen-deuterium bonds vibrate at lower frequencies than those containing oxygen-hydrogen bonds.

Deuterium concentration is usually expressed as atom fraction, D/(H+D), or atomic percent, 100 D/(H+D). In the case of water, the latter is often referred to as mole percent heavy water, even though the deuterium is present as both HDO and D_20 molecules, and at low concentration is almost all HDO.

Dr. H.G. Thode, at McMaster University, working in collaboration with the Montreal Laboratory in the mid-forties, developed a mass spectrometer for deuterium analysis and provided early heavy-water standards. Mass spectrometry was used routinely by AECL from the beginning to measure heavy water concentration. A small sample of water is injected into a heated volume, where it is totally evaporated, and the vapour is passed over uranium metal at 600°C to produce hydrogen, which then flows to the spectrometer where the mass 2 and 3 or mass 3 and 4 ratios are measured. In the sixties, simpler mass spectrometers became available that could be readily adapted for routine H_2-HD analysis. Chalk River developed several automated sample systems for these spectrometers, which circumvented the need for special sample preparation or purification, and provided rapid conversion of liquid to water vapour with negligible change in concentration. Modifying the new spectrometers for the requirements of heavy-water analysis in the 0 to 0.1 mole percent range, and combining them with an automated sampler, resulted in an overall analysis system that was reliable, precise and rapid. This overall system has been the standard for our heavy-water plants and research laboratories, and has given a routine precision of $\pm$ 0.00003 mole percent heavy water (0.3 parts per million) at natural concentrations.

In the early fifties, infrared spectroscopy methods were developed at Chalk River. Detailed spectra for H_2O and D_2O were determined and then, using samples of intermediate mole percent heavy water, the spectra for HDO was deduced by subtracting the contributions of H_2O and D_2O from the overall observed spectra. A double-beam differential technique capable of good precision was defined for the concentration ranges below 1 percent and above 99.5 percent. In this method, an HDO absorption peak of the sample is compared with that of a standard in a two-compartment cell. Later, a two-frequency technique was developed, in which absorption at an HDO peak was compared with absorption at a frequency where there is no discrimination amongst the three water molecules. The ratio of these two absorption measurements depends only on heavy-water concentration, and is independent of the effects of temperature drift, sample impurities or instrument instability.

The absorption peaks that enable good precision to be achieved at low concentration are very intense, and this limits their range of application to a narrow region at each end of the overall concentration range. However, a less precise infrared procedure was developed to cover the 1 to 99 mole percent range.

Several on-line instruments for heavy-water analysis in the range 99.0 to 99.9 mole percent were developed by AECL based on infrared absorption. Commercial models were produced by Barringer Research Limited and they have been installed in nuclear power stations and heavy-water plants. As with the infrared heavy-water leakage monitors discussed in chapter fourteen, it was found that the total system – the instrument, sample purification and conditioning equipment, and maintenance strategy – must be specifically engineered for each installation to achieve reliable operation.

The infrared spectroscopy methods were developed originally to be cheaper, faster and more precise than the early mass spectrometry. This was achieved for intermediate and high deuterium concentrations. However, the simple mass 2-3 spectrometer and its automated sampler displaced infrared spectroscopy for low deuterium concentrations.

In the late sixties, a systematic survey of the heavy-water concentration in Canadian waters was made. Ocean samples were found to contain 0.0155 mole percent heavy water. Values for fresh water varied from 0.013 to 0.015 percent. High values were found in coastal regions, or where enrichment by evaporation can occur, as in the shallower Great Lakes. The low values occurred in northern rivers and in the Rocky Mountain region, where rainfall derives from clouds already depleted in heavy water by previous precipitations. Natural gas in Alberta was found to vary from 0.011 to 0.0135 percent, while industrial hydrogen produced by steam reforming varied from 0.009 to 0.012 percent D/(H+D).

None of the analytical procedures measures absolute D/(H+D) ratios – all rely on calibration against standards. However, various methods have been evolved at AECL and elsewhere to determine the absolute concentration of some of the standards. For those in the concentration range near natural the absolute value has been determined at Chalk River to within $\pm$ 0.00005 percent. For example, Chalk River's value for Standard Mean Ocean Water (an international standard) is 0.01576 $\pm$ 0.00005, in agreement with many other laboratories.

The primary Canadian heavy-water standard is maintained at Chalk River. There is no recognized international standard. From 1963 to 1984, its concentration was defined as 99.95 $\pm$ 0.03 weight percent heavy water. However, in intercomparisons with other laboratories in the seventies, AECL values tended to be low. Then, in the early eighties, several AECL experimenters measuring various equilibria in near-pure heavy water found that they could achieve consistent results only if deuterium concentrations were adjusted upward. A new technique to measure the absolute concentration of HDO in heavy water was developed, and the Canadian standard was redefined as 99.9720 $\pm$ 0.0006 weight percent heavy water.

ACKNOWLEDGMENTS

I would like to thank a number of colleagues who have reviewed this chapter and given advice and assistance: K.T. Chuang, M.R. Galley, W.J. Holtslander, W.E. Lockerby, A.I. Miller, J.H. Rolston, D.A. Spagnolo, W.M. Thurston.

I also drew upon material from an earlier paper of mine published by the Canadian Society for Chemical Engineering in 1991. I appreciate the society's permission to include that material in this book.

BIBLIOGRAPHY

Benedict, Manson, Thomas H. Pigford, and Hans Wolfgang Levi. *Nuclear Chemical Engineering.* 2d ed. New York: McGraw-Hill, 1981.

Bayly, J.G., R.J. Booth, and W.H. Stevens. "An Improved Infrared Method of monitoring Heavy Water." *ISA Transactions* 4 (1962):315-9. (Also AECL-1632.)

Bayly, J.G., W.H. Stevens, and W.M. Thurston. *An Infrared Method for determining variations in the Deuterium Content of Natural Water.* AECL-2076. 1964.

Bothwell, Robert. "Pickering." Chap. 9 in *Nucleus: the History of Atomic Energy of Canada Limited.* Toronto: University of Toronto Press, 1988.

Brown, R.M., E. Robertson, and W.M. Thurston. *Deuterium Content of Canadian Waters. II.* AECL-3800. 1971.

Galley, M.R. "Future Trends in Heavy Water Production." In *Proceedings of the Fourth Pacific Basin Nuclear Conference, Vancouver, 11-15 September 1983,* 102-13. Toronto: Canadian Nuclear Association, [1983 or 4]. (Also AECL-7547.)

Galley, M.R., and A.R. Bancroft. "Canadian Heavy Water Production: 1970 to 1980." In *Proceedings of the 2nd World Congress of Chemical Engineering, Montreal, 4-9 October 1981*, vol. 2, 1-47. Ottawa: Canadian Society for Chemical Engineering, 1981. (Also AECL-7429.)

Ivanco, M., G.A. McRae, and R.D. McAlpine. "D_2O Production by Laser-Induced Selective Multiphoton Decomposition." In *Canadian Nuclear Society: Proceedings of the 13th Annual Conference, Saint John, N.B., 1-10 June 1992*, vol. 1, session 2. [Toronto: Canadian Nuclear Society] 1992.

Miller, A.I. "Deuterium Pilots New Hydrogen Technology." *International Journal of Hydrogen Energy* 9 (1984): 73-9. (Also AECL-8184.)

Miller, A.I. *Process Simulation of Heavy Water Plants – A Powerful Analytical Tool.* AECL-6178. 1978.

Neuberg, H.J., and K.T. Chuang. "Mass Transfer Modelling for GS Heavy Water Plants. Part II: Tray Efficiency on GS Sieve Trays." *Canadian Journal of Chemical Engineering* 60 (1982): 510-5. (Also AECL-7705.)

Rae, H.K. "Canada's Heavy Water Story." In *Chemical Engineering in Canada: An Historical Perspective,* 334-61. L.W. Shemilt, ed. Ottawa: Canadian Society for Chemical Engineering, 1991.

Rae, H.K., ed. *Separation of Hydrogen Isotopes: A Symposium Co-sponsored by the Physical Chemistry Division of the Chemical Institute of Canada and the Canadian Society for Chemical Engineering at the 2nd Joint Conference of the Chemical Institute of Canada and the American Chemical Society, Montreal, 30 May – 1 June 1977.* ACS Symposium series, 68. Washington, D.C.: American Chemical Society, 1978. (Also AECL-6053, 6054, 6055, 6056.)

Spagnolo, D.A. and K.T. Chuang. "Control of Foaming in Hydrogen Sulfide/Water Mixtures." *Canadian Journal of Chemical Engineering* 63 (1985): 572-7. (Also AECL-8724.)

Taylor, G.F. "The Corrosion of Steel in Heavy Water Production: Operating Experience and Laboratory Development." In *Proceedings of the 7th International Congress on Metallic Corrosion, Rio de Janeiro, 4-11 October 1978,* vol. 3, 1536-45. Rio de Janeiro: Associação brasileiro de corrosão, 1978.

Thurston, W.M. "Steam Film Sampling of Water for Mass Spectrometric Analysis of the Deuterium Content." *Review of Scientific Instruments,* n.s. 42 (1971): 700-3. (Also AECL-3818.)

Chapter Eighteen

Advanced Fuel Cycles

R.G. HART

INTRODUCTION

Nuclear power reactors produce electricity by harnessing the tremendous amount of energy released when unstable heavy atoms split to form two lighter atoms, the phenomenon known as fission. The only naturally occurring fissionable material is uranium-235, which exists at a concentration of 0.7 percent in uranium. However, nature also contains two materials, uranium-238 and thorium-232, which are converted to fissile material when they absorb neutrons. These are known as fertile materials. When uranium-238 absorbs a neutron it transforms to uranium-239 which decays first to neptunium-239 and then to the fissile material, plutonium-239. When thorium-232 absorbs a neutron it transforms to thorium-233 which decays first to protactinium-233 and then to the fissile material, uranium-233. The recycle of fissile, or fissile and fertile, material from used fuel is known as advanced fuel cycles. The fertile materials uranium-238 and thorium-232 are about 7,000 times more abundant in the earth's crust than is uranium-235, so it is possible to greatly increase the world's supply of nuclear fuel by using advanced fuel cycles. Further, recycling the fissile and fertile material can make the cost of fueling insensitive to the market price of uranium and thorium, allowing lower-grade ores to be used, thus further increasing the nuclear fuel supply.

In the early years of nuclear power it was universally accepted that the supply of economically recoverable uranium in the world was severely limited. Consequently, it was universally accepted that advanced fuel cycles would be a necessity. It was envisaged that the first generation of power reactors would be thermal neutron reactors fueled with either natural or slightly enriched uranium and that the plutonium produced in those reactors would be separated and recycled with fertile material to extend the available fuel supply. Accordingly, in most countries with nuclear energy programs early research programs included components on plutonium separation, the physics of reactors in which to use it, and the development of fuels in which it could be used. In many countries, early advanced-fuel-cycle work focused on fast breeder reactors, which it was thought could use all of the world's uranium. In Canada, early thinking on advanced fuel cycles was focused on thorium breeder fuel cycles, which it was thought could use all of the world's thorium. As early as 1950, W.B. Lewis was writing papers in which he envisaged low-cost electricity for thousands of years through the use of all the world's uranium in fast reactors and all the world's thorium in heavy-water reactors.

As the demand for uranium increased, more was found, and by the mid-sixties it was clear that enough economically recoverable uranium was available to fuel the world's reactors on once-through cycles for many decades. Advanced fuel cycle work now had to be justified more on economic grounds than resource grounds. In countries with adequate uranium reserves, work was phased down, commensurate with the longer-term nature of the need. In countries with poor uranium reserves, work continued with some priority, on arguments of security of fuel supply and possible foreign price gouging, but the emphasis shifted to recycle in thermal reactors. Fast-reactor programs continued in some countries where the perceived need was the greatest, most notably France and Japan, but with diminished priority. In Canada, there was an early flurry of activity on plutonium and uranium-233 separation and on recycling plutonium in the NRX reactor. Subsequently, the work was confined to the laboratory scale, with the objective of verifying the capability of the CANDU reactors to economically use advanced fuel cycles when that became necessary.

In 1968, a government/industry study on the Canadian incentive for fuel reprocessing and plutonium recycle concluded that plutonium recycle could be marginally attractive in the decade 1975-85. However, the attraction was not sufficient to change the overall thrust of the fuel-recycle program. Again in 1976, when the pressure for work on disposal of used nuclear fuel was mounting and the first oil shocks were reverberating through the economy, AECL asserted to the government that it would be irresponsible to dispose of the fuel without recovery of the tremendous energy value contained in the plutonium, and that the recycle program should be expanded. Reprocessing had, however, by that time, become publicly unpopular, so the only concession to the argument was that the waste management program could contain a component on the disposal of reprocessing waste.

By the mid-eighties the main interest in CANDU advanced fuel cycles was from foreign owners and potential foreign customers, and AECL was searching for a way to demonstrate the technology on a large scale. To do this, they turned to the slightly enriched uranium (SEU)[1] fuel cycle, which could provide a 20 percent savings in fueling costs over the natural-uranium fuel cycle in Canadian reactors. Although not an advanced fuel cycle, the SEU cycle could demonstrate at least the fuel behaviour and fuel management aspects of advanced fuel cycles on the scale required.

ADVANCED FUEL CYCLE PHYSICS

Lewis's early prediction of almost unlimited energy from the thorium-uranium-233 fuel cycle was based on relatively rudimentary hand calculations of neutron balances using the best cross-section data then available. However, by the mid-sixties these predictions were being made using the computer code LATREP (see chapter eleven), and LATREP had been combined with the power-reactor-cost estimating code, CANCAP, to allow various reactor designs and fuel cycles to be compared for cost and natural-uranium utilization. The predictions were, of course, more certain for the natural uranium cycle than for the others, as the physics could be confirmed by experiment (see chapter eleven) and the costs were better known. Nevertheless, the predictions were considered adequate to demonstrate the development potential of the CANDU system and its ability to respond to any future uranium shortage without any major change in reactor design. At the 1971 Geneva Conference, AECL presented papers emphasizing that natural uranium, plutonium-uranium and uranium-233-thorium fuel cycles could all be used in CANDU-PHW, CANDU-BLW and CANDU-OCR reactors with very little difference in fueling costs, but with substantial savings in uranium utilization.

In the plutonium-uranium cycles, irradiated natural-uranium fuels would be reprocessed to recover the plutonium and depleted uranium. The plutonium would be sent to a recycle fuel fabrication facility, where it would be mixed with either natural or depleted uranium to a fissile concentration of about 1.2 percent, converted to conventional oxide pellets, and fabricated into conventional CANDU bundles. In the reactor, the mixed oxide (MOX) fuel would produce about 480 MWh/kg of heavy elements (HE) before it would have to be removed for further recycle, as opposed to about 170 MWh/kg for natural uranium, and would give a saving of about a factor of two in uranium utilization. In the uranium-233-thorium cycle, the plutonium recovered from irradiated uranium would be mixed

[1] Slightly enriched uranium is the fuel of choice for conventional light-water-cooled and -moderated power reactors (LWRs).

with thorium to a fissile concentration of about 2 percent, converted to oxide pellets and made into CANDU bundles. After an irradiation of 240-900 MWh/kgHE it would be removed and reprocessed to recover the residual plutonium, the uranium-233 produced and the thorium. The uranium-233, plutonium and thorium would then be recycled in mixed-oxide form, with just enough extra plutonium to keep the cycle going. The amount of extra plutonium and thus the amount of natural uranium required to keep the cycle going was dependent on the frequency of reprocessing, with the high-burnup cycles giving about a factor of four saving in uranium utilization, and the low burnup cycles giving factors as high as 100. Lewis was particularly enthusiastic about the high-burnup thorium cycle in combination with the CANDU-OCR. He claimed that, because of the high allowable power rating of the fuel (100 kW/m), the high net conversion efficiency (37-39 percent) and the low heavy-water inventory (150 g/kWe), this combination would produce much lower power costs than any other combination possible.[2]

As better information became available throughout the seventies, it became clear that the costs of both reprocessing and recycle fuel fabrication had been grossly underestimated in the earlier predictions, and that fueling costs with advanced fuel cycles would be at least twice those of once-through cycles at then-current uranium prices. Thus Canadian utilities had no immediate interest in introducing them into Canadian reactors. Canadian utilities did, however, retain a long-term interest, and there was a more immediate interest in some foreign countries who were not blessed with indigenous natural uranium. Potential customers in those countries wanted to know that there were resource-conserving fuel cycles available if the cost of imported uranium became unacceptable. Thus the program continued. Further, as an aid to marketing CANDUs to utilities who already owned light-water reactors (LWRs), the scientists now conceived two fuel cycles that used the fuel discharged from LWRs as the feed fuel for CANDU reactors. The uranium discharged from LWRs contains about 0.9 percent uranium-235, and this can produce an additional 300 MWh/kgU in CANDU reactors. It can also be re-enriched and used again in LWRs, but this is complicated by the presence of uranium-236 and gamma activity[3] in the product, and recycle in CANDUs is simpler and gives better uranium utilization. The recycling in either reactor type involves reprocessing, and consequently higher costs than once-through fuel cycles, but it provides an insurance that has been attractive to some countries, most notably Korea. This cycle has been designated CANFURL, an acronym for CANDU Fueled with Uranium Recycled from LWR. An alternative to this cycle is to recycle both the plutonium and uranium separated from LWR fuel, in CANDU reactors. This material, which is about 1.5 percent fissile, would likely be diluted by about one-third with natural uranium, to be used in a CANDU reactor. This fuel, which would have a fissile content of about 1.2 percent, would produce an energy output of about 530 MWh/kgHE. This cycle has been designated the TANDEM fuel cycle, and again has been of interest to utilities owning both LWRs and CANDUs.

In the late seventies, in the United States, President Carter became very concerned that the spread of advanced fuel cycle technology – in particular, reprocessing technology – around the world would lead to nuclear weapons proliferation. He therefore persuaded the International Atomic Energy Agency (IAEA) to organize an International Nuclear Fuel Cycle Evaluation (INFCE), the objective of which was to develop resource-conserving fuel cycles that would avoid the need to separate fissile material in pure form. Preferred cycles were to avoid reprocessing completely. AECL put a lot of work into the study and, since CANDU is a good neutron-conserving reactor, developed several fuel cycles that gave improved uranium utilization relative to those currently in use. However, none produced by Canada or anyone else could approach the fuel utilization

---

2 For comparison, the natural-uranium-fueled CANDU-PHW power reactors have an established maximum fuel-element rating of 65 kW/m, a net conversion efficiency of just under 30 percent and a heavy-water inventory of 600 g/kWe.

3 Uranium-236 is produced by an n, gamma reaction with uranium-235. It is a strong neutron absorber and concentrates in the fuel during re-enrichment. The irradiated fuel also contains other uranium isotopes produced during irradiation that have gamma-active daughter products.

obtainable through complete fissionable material separation, and after a couple of years the study died a natural death.

By the eighties, AECL had conceived and developed the technology for several resource-conserving fuel cycles that were attractive to CANDU customers, but had been unable to demonstrate them on a large scale in Canadian reactors, because their costs were substantially higher than the once-through natural uranium cycle. However, one fuel cycle – SEU – had a lower cost than the natural-uranium fuel cycle and, although SEU was not an advanced fuel cycle, it would demonstrate the fuel performance characteristics of the advanced fuel cycles, and the reactor and fuel management technology for transition from the conventional to the advanced fuel cycle. Reactor analysis showed that a CANDU fuel enriched to 1.2 percent uranium-235 in uranium could produce 530 MWh/kgU in a CANDU-PHW reactor, with a 20 percent saving in fueling costs and a 30 percent saving in uranium utilization, and that the enriched fuel could be introduced to the reactor in a radial pattern, working from the outside in, without loss of reactor power. The analysis showed that fuel management, both during transition and at equilibrium, would be aided by the use of a fuel bundle that had smaller-diameter elements in the outer two rings and larger-diameter elements in the centre, which gave it more tolerance to neutron flux variations in the reactor.

Early in the advanced fuel-cycle program, the LATREP code used in the reactor physics calculations was verified against experiments with natural uranium and plutonium-depleted uranium fuel in ZED-2 (see chapter eleven), but there was a continuing desire to obtain experimental verification with all the other fuels proposed. This was one of the major rationales for the commitment in the late sixties to a recycle fuel fabrication laboratory (RFFL), which was to be used to produce multi-bundle-sized lots of plutonium-natural uranium, plutonium-thorium, uranium-235-thorium and uranium-233-thorium fuels in oxide form. However, because of the low priority of the advanced fuel cycle program, financing was limited and the RFFL did not produce its first fuel until 1978, and its first recycle fuel until 1979. Once fabrication started it went well, but the reactor physicists did not have all the fuel they required for all of their ZED-2 experiments until 1988. To attempt to ameliorate the delays, the physicists, around 1975, negotiated an agreement with CNEN (the Italian nuclear research organization) to fabricate (plutonium, uranium)O_2 fuel for ZED-2 experiments from plutonium separated from CANDU fuel in a cooperative reprocessing experiment. Fabrication of this fuel was also substantially delayed, and it was not available for experiments until 1985. Eventually, however, experiments were completed on (plutonium, uranium)O_2, (thorium, plutonium)O_2 and (thorium,uranium-233)O_2, and checked against code predictions. By this time, the type of calculation being done had changed from a cost comparison of different fuel cycles in different reactor types to a more detailed optimization of lattice parameters and fuel utilization with different fuel cycles in the CANDU-PHW. Accordingly, LATREP had been replaced by WIMS-AECL, a more sophisticated code originally developed at the Winfrith laboratory of the United Kingdom Atomic Energy Authority (Winfrith Improved Multigroup Scheme), which had been substantially modified by AECL to adapt it to CANDU lattices. In general, the agreement between WIMS-AECL predictions and the experiments was good, with the only apparent problem being the inability of the code to accurately predict the change in reactivity with temperature of fuels containing plutonium.

ADVANCED FUEL CYCLE FUELS

The reference fuels for advanced fuel cycles in CANDU reactors were plutonium-uranium, plutonium-thorium and uranium-233-thorium, all in oxide form. Until the mid-eighties, the reference bundle geometries were identical to those for natural uranium. Thus, all the development work for natural uranium fuels (see chapter thirteen) was directly relevant to fuels for advanced fuel cycles. The additional work required was to verify that the fuel could survive the higher burnups required in the advanced fuel cycles (500 vs 200 MWh/kgHE), to adapt the manufacturing techniques to accommodate the alpha activity of plutonium and the gamma activity of the uranium-233 feed,[4] to ensure that these

[4] The gamma activity comes from daughter products of uranium-232, which is produced by an n,2n reaction with uranium-233.

adaptations did not negatively impact on fuel performance, and to ensure that the behaviour of thorium-based fuels was at least as good as uranium-based fuels.

In 1965, work commenced on the fabrication of single-element (uranium-plutonium)O_2 and (thorium-plutonium)O_2 fuels for irradiation testing. It was expected that fuel pellets could be made by mixing the oxide powders, then pressing them into pellets and sintering in cracked ammonia, as was then done for uranium-dioxide fuel pellets. However, cracked ammonia is highly flammable, and there was concern about using it for large-scale production of pellets containing toxic plutonium. There was a desire, therefore, to determine whether pellets produced by sintering in a dilute hydrogen atmosphere (10 percent H_2 in argon) would give acceptable irradiation performance. Accordingly, pellets were made by both methods for comparison. The other variable tested was powder particle size. The fuel elements were irradiated in the X-2 loop in NRX in 1966. The results showed that there was no significant difference in performance between pellets made by the different sintering methods, or from the different particle sized powders, that there was no significant difference between the behaviour of (uranium-plutonium)O_2 and uranium dioxide irradiated at the same ratings, and that (thorium-plutonium)O_2 was somewhat more resistant to structural change than uranium dioxide irradiated at the same ratings. Subsequent tests showed that acceptable fuel could be made from powders arising from co-precipitation, and that oxygen/metal ratios should be kept close to stoichiometric.

The RFFL at Chalk River comprised three lines of glove boxes and fume hoods connected by appropriate conveyers for the product (see figure 18.1). Sintered oxide pellets were made, ground, sized and stacked (see chapter thirteen), and were loaded into commercially fabricated Zircaloy sheaths suitably equipped with brazed spacers and bearing pads, and with the first end-plug resistance welded in position. After loading and filling with helium, the second end-plug was inserted and welded in place using tungsten inert-gas welding. Bundles were then assembled by welding the elements to Zircaloy end plates using a commercially made resistance welding and assembly device. Product quality was assured by stringent control of

FIGURE 18.1
Glovebox and Fumehood, RFFL

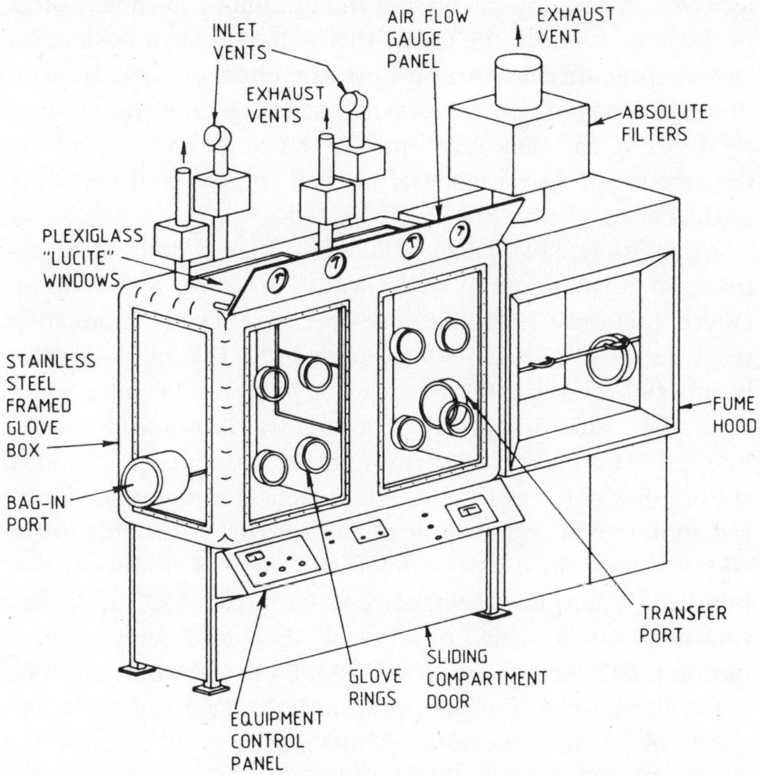

incoming powders for blending, along with careful control of process parameters to ensure consistency. The laboratory had its own dedicated analytical section, with the capability to do all the required quality assurance work. There was also a dedicated metallographic/ceramographic line for examination of fuel microstructure. Finished fuel elements were nondestructively assayed for fissile content using a commercial neutron interrogation device, and neutron radiography was available to provide additional information if other inspection techniques showed anomalies. Protection against the toxicological hazards of plutonium and uranium-233 was assured by handling them in dry-boxes until they were sealed in the Zircaloy sheaths, and by using personal air samplers to monitor the air in the vicinity of each worker. Protection against the gamma radiation emitted by the daughters of uranium-232 present in the uranium-233 feed was assured by processing the uranium-233 shortly after purification, and by the use of thermoluminescent dosimeters (see chapter three) worn on both the chest and the fingers. Protection against potential criticality incidents was assured by dividing the laboratory into fifteen zones and controlling the amount of fissile material in each zone to well below its critical mass.

As mentioned earlier, the RFFL was delayed in its construction and commissioning, and was not ready for operation on recycle fuel until 1978, but once operation started, fabrication went very well. From 1979 through 1980, fifteen 36-element bundles of (uranium-0.5 wt. percent plutonium)O_2 were completed for irradiation in NRU. Thorium was introduced into the line in 1981 and from 1981 through 1982 production included six bundles of (thorium-1.75 wt. percent, plutonium)O_2 for irradiation in NRU, one bundle of natural ThO_2 for irradiation in NRU (to study the distribution of uranium-233 produced), two bundles of (thorium-1.8 wt. percent uranium-235)O_2 for irradiation in WR-1, and two bundles of (thorium-2.3 wt. percent plutonium)O_2 for irradiation in WR-1. From 1982 through 1988, 1,332 elements of (thorium, plutonium)O_2 fuel and 1,350 elements of (thorium/uranium-233)O_2 fuel were fabricated for reactor physics experiments in ZED-2.

During the seventies some work was done on the fabrication of thorium/uranium-233 fuels by alternative methods that might be more compatible with remote fabrication. This would eliminate the need to fabricate shortly after purification. The methods studied were: 1) impregnation of porous pellets of thoria with a nitrate solution of uranium-233, 2) vibratory compaction of fully dense, mixed-oxide microspheres into the fuel cladding (the microspheres would be produced by an already proven process known as the sol-gel process), and 3) extrusion of a mixed-oxide clay, made by a process similar to the sol-gel process, into a mixed-oxide rod, which could then be cut into pellet-length slugs and sintered. After testing, the impregnation process was eliminated because it was incompatible with co-processing, vibratory compaction was eliminated because of poor irradiation performance, and extrusion was eliminated because experience showed it to be no easier to adapt to remote fabrication than the conventional pellet route.

Because of the delay in the RFFL, advanced fuel-cycle fuel bundles were not available for irradiation testing to the high-burnup required until the early eighties. However, relevant information could be obtained by taking SEU fuels to higher burnups than those required for natural uranium, by irradiating thorium mixed with uranium-235, and by following the fate of natural-uranium fuel that for various reasons was delayed in discharge from the power reactors. Since the large water loops at Chalk River were tied up with priority work on natural uranium fuel, irradiations were done in NPD and WR-1. In the seventies, the irradiation of four 19-element bundles of (thorium, uranium-235)O_2 was started in NPD and the irradiation of twenty-four 22-element bundles of (thorium,uranium-235)O_2 was started in WR-1. These bundles performed well and had reached burnups of 900 and 450 MWh/kgHE respectively, when the reactors were shut down in 1985 (WR-1) and 1987 (NPD). Unfortunately, because of the characteristics of these reactors, the fuel could be operated only at modest power ratings (36-48 kW/m). Ten 19-element bundles of SEU fuel were also irradiated in NPD, and they also worked well to 700 MWh/kgU, but again at low rating. A study of the history of the natural-uranium fuel in the power reactors showed that by 1985 over 3,000 bundles had been irradiated to burnups between 270 and 700 MWh/kgU, with many of them exceeding 500. All had been discharged and transferred to storage in the normal course of fuel handling with no problems. Unfortunately, the ratings of these are indeterminate, but most are likely low.

The plan for the irradiation of fuel fabricated in the RFFL was to put it into the NRU loops whenever there was appropriate space available, so that eventually it would achieve high burnup at high rating and, along the way, some bundles could be power ramped[5] as desired. Irradiation of (uranium, plutonium)O_2 and

[5] Moving quickly from low power to high power. This is an action that will occur in CANDU reactors and it stresses the fuel more than uniform power operation.

(thorium, plutonium)O_2 bundles started in 1981 and 1984, respectively. In addition, irradiation of three bundles of SEU and two bundles of (thorium, uranium-235)O_2 fabricated in industry started in 1980 and 1982, respectively, and in 1983 AECL was able to take over responsibility for an irradiation of five elements of (thorium, uranium-235)O_2 and nine elements of (thorium, uranium-233)O_2 started by Westinghouse Power Division (United States) in NRX in 1981. By 1985 SEU had been taken to over 700 MWh/kgU, (uranium,plutonium)O_2, (thorium, uranium-235)O_2 and (thorium, uranium-233)O_2 to over 500 MWh/kgHE, and (thorium, plutonium)O_2 to 115 MWh/kgHE, with only one failure, that in a SEU bundle that was power-ramped to a very high rating (70 kW/m). Most tests were to be continued to a target burnup of 1,000 MWh/kg.

These tests were augmented by special-effects tests in smaller loops, to study diametral changes, internal operating temperature, fission-gas release, defect behaviour and accident conditions, and all tests were compared against predictions made by computer codes developed for this purpose (see chapter thirteen). In most cases the predictions were accurate, with the only major discrepancy being in the predictions of the behaviour of (thorium, uranium-235)O_2 fuel made by the co-milling route. This fabrication route produced pellets with a coarser microstructure and a lower thermal conductivity than those made by the conventional method, so they operated at a higher central temperature. This fabrication route was discontinued. Subsequently, hot-cell examination of high-burnup fuel from a Bruce reactor identified a second area of discrepancy between prediction and performance, in that the fission-gas release was considerably higher than predicted. This was attributed to a burnup-dependent decrease in thermal conductivity, believed to result, at least in part, from the buildup of solid fission products in the uranium-dioxide matrix. The increase in fission-gas release was not enough to cause the fuel to fail, but it did introduce a factor that had not before been appreciated.

The excellent behaviour of natural uranium taken to high burnup in power reactors and the behaviour of SEU and (thorium, uranium-235)O_2 in NPD and WR-1, reinforced by the excellent behaviour of SEU, (uranium, plutonium)O_2, (thorium, uranium-235)O_2, (thorium, uranium-233)O_2 and (thorium, plutonium)O_2 in the loops, led, by 1985, to a high degree of confidence that fuel of the conventional CANDU design would give good performance in any of the advanced fuel cycles envisaged.

Although it was clear that fuel of conventional CANDU design could be used with confidence in advanced fuel cycles, it was also clear that this was not the optimum bundle design for these cycles. A better design would be one that tended to equalize the linear power rating of elements across the bundle. AECL therefore developed a new design that it called the CANFLEX bundle. The bundle was more subdivided than the conventional 37-element design, as it contained 43 elements, with the elements in the outer two rings being of smaller diameter than those in the centre. (see figure 18.2). The smaller outer elements allowed the bundle to operate at 1,250 kW without exceeding the 65 kW/m peak element linear rating that had been established as the maximum for CANDU fuel. Alternatively, if the CANFLEX bundle was used at the power limit of 1,035 kW, set for 37-element bundles, the peak element rating would be reduced by about 20 percent. This would provide increased flexibility to operate to

FIGURE 18.2
Canflex Fuel Bundle

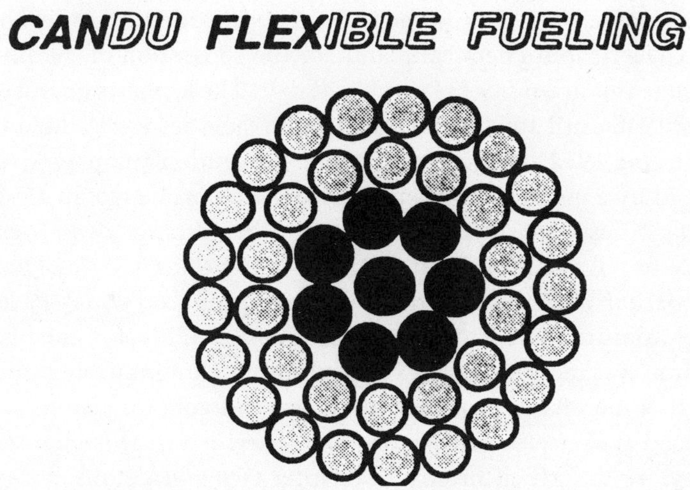

higher burnups, to withstand greater power ramping and to accommodate greater flux variations during transition from a natural-uranium fuel cycle to an advanced one. The advantages were such that the CANFLEX bundle was accepted as the reference design for all advanced fuel cycles.

Over the period, irradiation testing was also done on fuel types that, it was thought, could operate at the very high power ratings proposed by W.B. Lewis in some of his thorium reactor concepts. These fuels had either annular pellets or thin graphite discs between pellets, or a combination of both, to keep the temperature of uranium dioxide at acceptable levels. The fuels were irradiated in the NRX and WR-1 loops at powers up to 70 kW/m to burnups up to 800 MWh/kgU without any problems. In fact, two of the elements were inadvertently ramped to 90 kW/m for short periods without problem. The tests showed that CANDU fuels capable of operating at much higher power densities than that possible with the CANFLEX fuel were practicable and easy to fabricate, if required.

FUEL REPROCESSING

The wartime task given to the Canadian-United Kingdom team at the Montreal Laboratory was the pursuit of the heavy-water reactor route to plutonium and uranium-233 production. As a result, the early Chalk River laboratory included not only the NRX reactor but chemical plants for the separation of the plutonium and uranium-233 produced in it. These plants operated from 1948 until 1954. Final purification facilities were added in 1951 and 1952, and the plutonium separation plant was replaced by a much simpler one, based on ion exchange, in 1955. All facilities were shut down in 1957, when arrangements were made to sell used NRX and NRU fuel to the United States. Over the period 1948 to 1957, these facilities produced about 17 kg of plutonium and about 500 g of uranium-233. Most of the plutonium was recycled in NRX in plutonium-aluminum alloy fuel rods. None of the processes used was suitable for use in advanced fuel cycles so they are not described in this chapter. They are described briefly in chapter two and details can be found in the references given in the footnote.[6]

By 1950, W.B. Lewis was beginning to develop his first concept for a natural-uranium-fueled, heavy-water-moderated power reactor. In this concept, the fuel was to be uranium metal, which would have to be recycled with its contained plutonium after about 70 MWh/kgU, because of the dimensional instability of the uranium metal under irradiation. Lewis assigned the task of developing a process that could economically recover both uranium and plutonium to the chemical engineering branch at Chalk River.

At the time, laboratories in the United States had begun to work on a solvent extraction process that used tri-butyl phosphate (TBP) as the extractant, and this process was beginning to look very promising. Thus the TBP process was adopted by the chemical engineers as their reference process (see figure 18.3). In this process a 20 percent solution of TBP in a diluent such as kerosene was counter-currently mixed with the nitric acid dissolver solution from which it extracted both the uranium and the plutonium. The plutonium was then preferentially backwashed from the solvent using a dilute aqueous solution of a reducing agent such as ferrous sulphamate, following which the uranium was backwashed from the solvent with a very dilute aqueous solution of nitric acid. After a number of laboratory studies to optimize the process parameters, the chemical engineers designed and built a pilot plant to test the process on NRX fuel.

The heart of the plant was a series of mixer-settlers (see figure 18.4), in which the solvent and aqueous fluids were mixed and separated in counter-current flow. In theory the number of mixing and settling stages determined the extraction efficiency and product purity, but in practice these were determined by the presence of minute amounts of complexes that behaved differently than the bulk materials. Thus in practice it was necessary to have three extraction/backwash cycles, with conditioning between cycles, to obtain the product purity required. The pilot plant worked well and confirmed that uranium and

6 W.M. Campbell, Canadian Chemical Processing, Nucleonics, Vol. No. 9, September 1956. Also, O.J.C. Runnalls, Irradiation Experience with Rods of Plutonium/Aluminum Alloy, AECL-601, September 1958.

FIGURE 18.3
TBP Process Flowsheet – Single Cycle

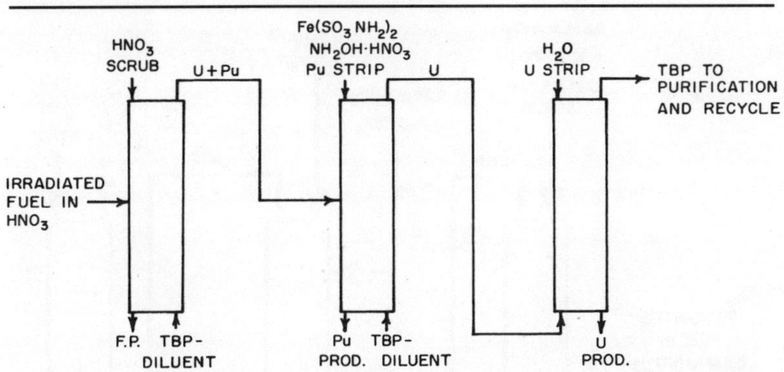

plutonium could be effectively and probably economically recovered for recycle, as required in the power reactor concepts being proposed. However, by this time, development work in other areas had indicated that uranium dioxide fuel could produce close to 240 MWh/kgU in a heavy-water-moderated reactor without recycling, and that this was sufficient to make a once-through fuel cycle economical. Thus the immediate need for recycle was removed and the TBP pilot plant was shut down.

The focus of AECL's reprocessing work now changed to that of putting on the shelf all the technology required to allow the rapid design and construction of a versatile commercial reprocessing plant when such a plant became necessary. The spectrum of fuels to be considered was broad, including uranium metal, uranium oxide, thorium metal, thorium oxide and plutonium-aluminum alloy, all clad in aluminum or zirconium. Since the TBP process offered the promise of being able to reprocess all of the above fuels in a single plant, it was adopted as the reference process. The work, which was done on a laboratory scale,

FIGURE 18.4
Schematic Flow Diagram for the Three Stage Mixer Settler

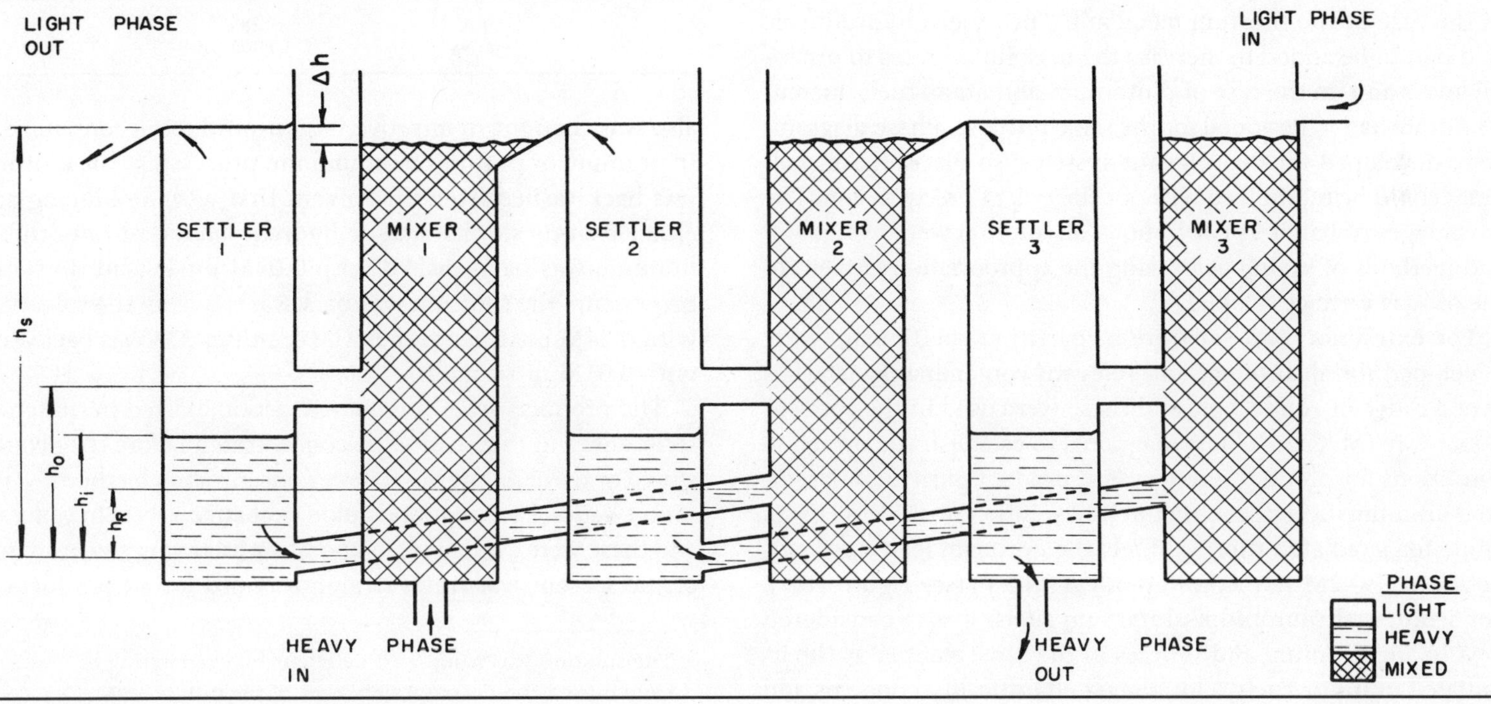

covered all aspects of the process, including decladding and dissolution of the fuels, optimization of the extraction and backwash conditions for each of the fuels, concentration of the final products, and recovery and reconditioning of the solvents for re-use.

Generally, the preferred methods for decladding were dissolution of aluminum in caustic soda and chopping the zirconium cladding into short lengths to expose the fuel. The alternative for aluminum was to dissolve the cladding in nitric acid, along with the fuel, and this was done with the addition of mercuric nitrate, to speed up the reaction. The alternatives for zirconium were dissolution in ammonium fluoride, dissolution in hydrofluoric acid, volatilization of the cladding as $ZrCl_4$ (hydrochlorination) and disintegration of the cladding by high-temperature oxidation. Reaction rates were obtained for all systems, and information was obtained on the problems with each method, and ways of controlling those problems.

For the TBP processes, all fuels had to be dissolved in nitric acid, and reaction rates were determined for all fuels of interest. In the case of the thorium metal and ThO_2 fuels, hydrofluoric acid had to be added to increase the dissolution rates to practical levels and, in the case of plutonium-aluminum fuels, mercuric nitrate had to be added for the same purpose. Phase diagrams were developed for the relevant systems, so that solution volumes could be minimized without the risk of precipitation. The advantages of batch vs continuous dissolution were evaluated, and methods of saving acid using the appropriate treatment of the off-gas were developed.

For extraction and separation, partition coefficients were developed for all product and relevant contaminant materials over a range of conditions, and these were used in the conventional way (McCabe-Thiele diagrams) to establish the optimum conditions for product recovery and product purity. For irradiated uranium fuels, the optimum feed solution was 2-3 M HNO_3, while for irradiated thorium fuels the optimum feed was acid-deficient[7] (-0.2M HNO_3, and 0.6M $Al(NO_3)_3$) (see figure 18.5). For irradiated plutonium-aluminum fuels, it was considered best to add uranium and process in the same manner as the irradiated uranium fuels. This alleviated criticality concerns and also was thought to improve fission-product decontamination. In uranium or plutonium-aluminum processing, the plutonium was backwashed from the solvent first with a reducing agent, either ferrous sulphamate or hydroxylamine nitrate, then the uranium was backwashed with 0.01M nitric acid. In thorium processing, the thorium was backwashed from the solvent first with 0.24M nitric acid, then the uranium-233 was backwashed with 0.01M nitric acid.

FIGURE 18.5
Thorex Flowsheet – Single Cycle

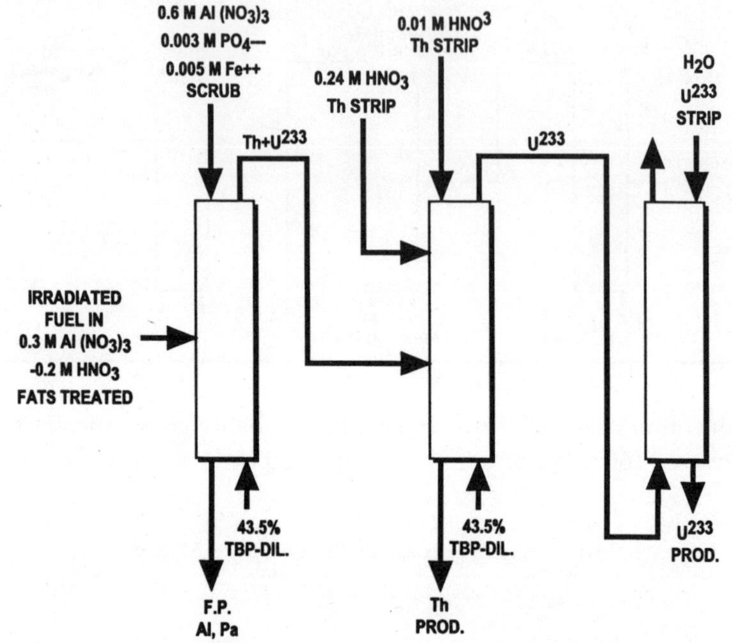

The products from solvent extraction existed in quite dilute solutions, and they had to be concentrated before they could be stored or used. At the time, three concentration methods – evaporation, cation exchange and anion exchange – were in general use and these were evaluated. Evaporation following steam stripping of the solvent was quite straightforward for all products, and

[7] The solution was made acid deficient by evaporating to molten thorium nitrate and decomposing some of the nitrate ions.

phase diagrams for product/nitric acid/water were developed for all products, to avoid overconcentration. The only significant problem was the formation of large amounts of hexavalant plutonium, which was undesirable in subsequent fuel fabrication steps. Some work was done on reducing agents that would put all the product in the desired tetravalent state, but the best solution turned out to be to hold the product in 7-8M nitric acid for a couple of months, during which time the product would reproportion to over 99 percent tetravalent. Both cation exchange and anion exchange would provide additional purification if used for concentration, with anion exchange giving the better purification and cation exchange the greater concentration. Auxiliary purifications, such as silica-gel removal of fission product zirconium and niobium, and thiocarbanilide-thioacetamide removal of ruthenium, were available to complement the ion exchange purifications if they were required.

Tributyl phosphate in use undergoes hydrolysis to form dibutyl phosphate, monobutyl phosphate and phosphoric acid, all of which can have a detrimental effect on process decontaminations, so it must be treated periodically to remove these impurities. It was found that a simple caustic wash was insufficient and that this would have to be followed by steam stripping. Work, both at Chalk River and elsewhere, showed also that the choice of diluent was quite important, as some were much more stable than others and their decomposition products affected process decontamination. Olefin-free normal dodecane and C12n-paraffin were specified as preferred diluents, and the specification called for them to be vacuum distilled, to remove impurities before use.

It was mentioned earlier that the microcomponents, not the macrocomponents, of the system governed the product decontamination obtained in each cycle, and that in practice three extraction/backwash cycles with intermediate feed conditioning were needed to give the product purity required. Some work was undertaken to identify these microcomponents, to determine whether they could be altered so that the number of extraction/backwash cycles, and thus the number of contactors in the reprocessing plant, could be reduced with resulting cost savings. Fission-product ruthenium and plutonium were the problem contaminants in the uranium product stream, while fission-product zirconium and its daughter niobium were the problem contaminants in the plutonium stream. It was discovered in other laboratories that ruthenium nitrosyl nitrate and colloidal zirconium were the problem fission-product species, and it was discovered at Chalk River that plutonium dibutylphosphate was the problem plutonium species. Work on the ruthenium and zirconium problems failed to produce any practical method of reducing the number of purification cycles required, but work on the plutonium problem at Chalk River produced a practical method of reducing the number of plutonium/uranium separation cycles from two to one, and the total number of contactors in the reprocessing plant from nine to seven.

By about 1960, the technology required to design and construct a multi-purpose reprocessing plant was reasonably in place and, since there was a need for all the effort that could be made available on problems related to the development of the once-through cycle CANDU, work on reprocessing was suspended.

In 1968, W.B. Lewis became a champion of the CANDU-OCR concept, which he felt was the best reactor for his Valubreeder fuel cycle, a fuel cycle based on the recycle of uranium-233 and thorium topped up with plutonium from uranium-fueled reactors. Since organic reactor technology was primarily a Whiteshell responsibility, Whiteshell also undertook to verify the technical aspects of the Valubreeder fuel cycle – in particular, the reprocessing aspects of that cycle. As mentioned above, technology had been developed for the recovery of plutonium and uranium from irradiated uranium fuels and uranium-233 and thorium from thorium fuels, but the Valubreeder fuel cycle required recovery of only plutonium from the uranium fuels, and plutonium, uranium-233 and thorium from the thorium fuels. Since there was concern about the hydrolysis and polymerization of plutonium in the acid-deficient thorium reprocessing system used at that time, it was felt that an acid flowsheet would have to be developed. Further, it was felt that the economics of the cycle would be improved if only plutonium was removed from the irradiated uranium fuel, and the uranium and fission products were denitrated and calcined to a form suitable for disposal.

Plutonium had already been effectively separated from uranium and fission products using anion exchange (see chapter two), and anion exchange resins were now available in liquid form. Thus it was felt that the old anion exchange process could be used, but now in a more convenient liquid/liquid extraction form. These "liquid ion exchange resins" were amines, and the process became known as the amine extraction process (see figure 18.6). The amine, diluted with diethylbenzene, was contacted in conventional counter-current solvent extraction equipment, with the dissolver solution conditioned to 2M nitric acid. The plutonium was extracted into the solvent, leaving behind the uranium and fission products. After scrubbing, to improve the decontamination, the plutonium was backwashed with dilute sulphuric and nitric acid, and then concentrated by anion exchange. The process was tested in a mini pilot plant constructed in the Whiteshell hot cells, and it worked well. It was found that two cycles with intermediate conditioning were sufficient to give the product purity required. In the early seventies, the opportunity arose to demonstrate the amine process on a larger scale in a versatile pilot facility in Italy. Fifty bundles of discharged Pickering fuel were acquired from Ontario Hydro and shipped to Italy, where the technical viability of the process was verified on the larger scale. Concurrent with the amine process tests, work was initiated on the denitration and calcination of the uranium/fission-product stream. This work showed that the uranium/fission- product waste could be conditioned for disposal using fluidized bed denitration and a rotary calciner. However, work was done only on inactive synthetic solutions of the appropriate composition, and the process could not be considered proven until demonstrated on the radioactive solution itself.

After the mini pilot-plant tests were conducted on the amine process, attention turned to reprocessing the thorium-uranium-plutonium fuels. Conventional McCabe-Thiele calculations using partition data developed in the laboratories indicated that thorium, uranium-233 and plutonium could be extracted with 30 percent TBP from a 2M nitric acid feed solution and backwashed with 0.01M nitric acid containing 0.3M hydroxylamine nitrate and 0.1M hydrazine, to reduce the plutonium to the trivalent state (see figure 18.7). The backwash solution was then to be adjusted to 1.5M nitric acid and contacted again with 30 percent TBP, which would extract the thorium and uranium, but leave the plutonium in the trivalent state behind. The thorium could then be backwashed from the solvent with 0.5M nitric acid leaving the uranium behind, then the uranium could be backwashed with 0.01M nitric acid. In tests in the mini pilot plant, it was found that the hydroxylamine nitrate could not hold the plutonium in the aqueous phase in the plutonium separation step, so it had to be replaced by 0.05M ferrous sulphamate. With this change, the process worked well, but the

FIGURE 18.6
Amine Process Flowsheet

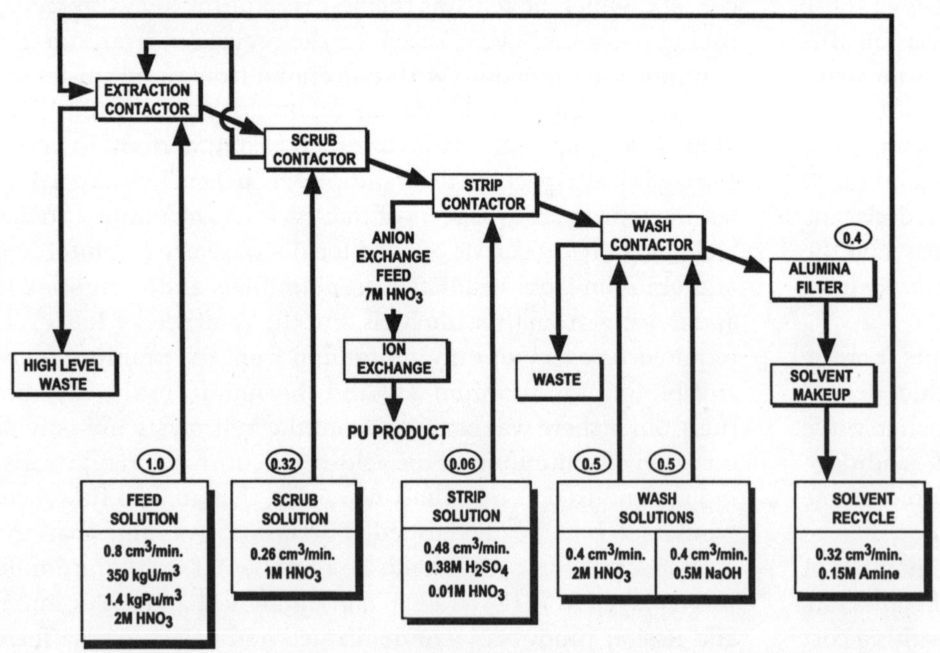

FIGURE 18.7
Modified Thorex Flowsheet for Th-U-Pu Fuel

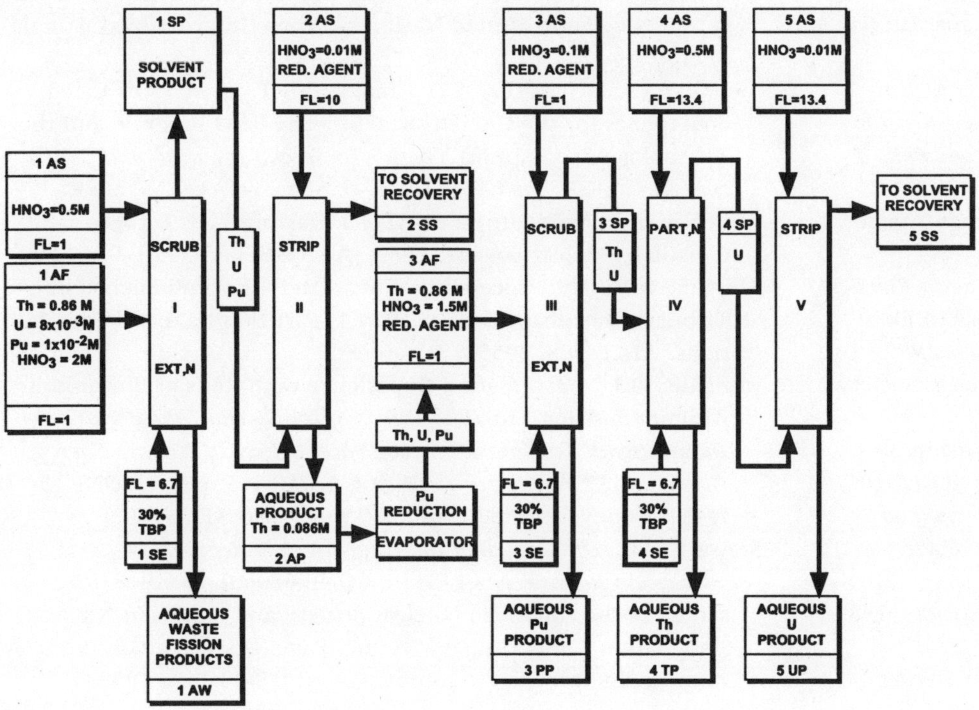

change added an additional step to the process to separate the iron from the plutonium, and it added iron to the solid waste. Another departure from previous practice was the diluent used for the TBP. Since a third phase formed when the conventional kerosene-type diluents were used with acid feed, a new diluent had to be found in which the thorium TBP complex was more soluble. It was found that either diethyl benzene or a commercial isoparafinic kerosene to which diethyl benzene was added were suitable diluents, which avoided the third-phase problem.

By the time these processes were demonstrated, in the late seventies, the immediate interest had again disappeared. Much more uranium had been found and Lewis had been unsuccessful in convincing the AECL board of directors that a demonstration CANDU-OCR power reactor should be built (see chapter sixteen). Accordingly, the technology was documented and added to the reprocessing technology already on the shelf, for use when it was needed.

BIBLIOGRAPHY

Campbell, W.M. "Canadian Fuel Reprocessing." In *Nucleonics* 14 (September 1956): 42-7. (Also AECL-357.)

Carter, T.J. "The Recycle Fuel Fabrication Laboratory at Chalk River." In *International Conference on CANDU Fuel, Chalk River Nuclear Laboratories, 6-8 October 1986: Conference Proceedings,* 168-76. I.J. Hastings, ed. Toronto: Canadian Nuclear Society, 1986.

Grant, G.R, W.W. Morgan, K.K. Mehta, and F.P. Sargent. "Heavy Element Separation for Thorium-Uranium-Plutonium Fuels." In *Actinide Separations: Based on a Symposium Sponsored by ACS Division of Industrial and Engineering Chemistry at the ACS/CSJ Chemical Congress (177th ACS National Meeting), Honolulu, Hawaii, 3-5 April 1979*, 351-69. James D. Navratil, Wallace W. Schulz, eds. ACS Symposium Series, 117. Washington, D.C.: American Chemical Society, 1980. (Also AECL-6712.)

Hart, R.G. *A Review of the Chemical Aspects of Aqueous Processing.* AECL-853. 1959.

Hart, R.G., L.R. Haywood, and G.A. Pon. "The CANDU Nuclear Power System: Competitive for the Foreseeable Future." In *Peaceful Uses of Atomic Energy: Proceedings of the Fourth International Conference on the Peaceful Uses of Atomic Energy, Geneva, 6-16 September 1971*, vol 5, 239-53. Paper A/CONF.49/ P/151. New York: United Nations; Vienna: International Atomic Energy Agency, 1972. (Also AECL-3975.)

Hastings, I.J., T.J. Carter, G.M. Macgillivray, R.D. MacDonald, and J. Judah. "Canadian High Burnup Fuel Experience." In *Proceedings of the Canadian Nuclear Society 6th Annual Conference*, 6.26-.32. P.M. French, G.J. Phillips, eds. Toronto: Canadian Nuclear Society, 1985. (Also AECL-8772.)

Jeffs, A.T. "Irradiation Behaviour of Experimental Fuel Assemblies of UO_2-PuO_2 and ThO_2-PuO_2." In *Plutonium as a Reactor Fuel: Proceedings of a Symposium on the Use of Plutonium as a Reactor Fuel, in Brussels, 13-17 March 1967*, 421-38. Paper IAEA-SM-88/29. Vienna: International Atomic Energy Agency, 1967. (Also AECL-2674.)

Jones, R.T., J. Griffiths, A. Okazaki, and A. Celli. "Reactor Physics Measurements in Support of Advanced Fuel Cycles." In *Canadian Nuclear Society 10th Annual Conference, 4-7 June 1989: Proceedings*, vol.2, 1.23-.32. Toronto: Canadian Nuclear Society, 1989. (Also AECL-10004.)

Lane, A.D., F.N. McDonnell, J. Griffiths. "The Relationship between Natural Uranium and Advanced Fuel Cycles in CANDU Reactors." In *Proceedings of the International Symposium on Uranium and Electricity: The Complete Nuclear Fuel Cycle, Saskatoon, Saskatchewan, 18-21 September 1988*. Ken H. Talbot, V.I. Lakshmanan, eds. Toronto: Canadian Nuclear Society, 1989. (Also AECL-9790.)

Lewis, W.B. *The Superconverter or Valubreeder: A New Breeder Uranium-Thorium Nuclear Fuel Cycle*. AECL-3081. 1968.

Lewis, W.B., M.F. Duret, D.S. Craig, J.I. Veeder, and A.S. Bain. "Large Scale Nuclear Energy from the Thorium Cycle." In *Peaceful Uses of Atomic Energy: Proceedings of the Fourth International Conference on the Peaceful Uses of Atomic Energy, Geneva, 6-16 September 1971*, vol. 9, 239-53. Paper A/CONF.49/P/157. New York: United Nations; Vienna: International Atomic Energy Agency, 1972. (Also AECL-4445.)

Moeck, E.O., J. Griffiths, A.D. Lane, and R.T. Jones. "Advanced Fuel Cycles for the CANDU Reactor: The R&D Program and the Technologies to Support Them." In *Second International Conference on CANDU Fuel, Pembroke, Ontario, 1-5 October 1989*: Conference Proceedings, 65-82. I.J. Hastings, ed. Toronto: Canadian Nuclear Society, 1989. (Also AECL-10115.)

Morgan, W.W. The Processing of Irradiated Thorium Fuels Using Tributyl Phosphate-Decalin. Part 1, Partition Data and Calculations. AECL-508. 1955.

Runnalls, O.J.C. "Irradiation Experience with Rods of Plutonium-Aluminum Alloy." In *Proceedings of the Second United Nations International Conference on the Peaceful Uses of Atomic Energy, Geneva, 1-13 September 1958*, vol. 6, 710-7. Paper A/CONF.15/P/191. Geneva: United Nations, 1958. (Also AECL-601.)

Stewart, D.D. *The Canadian Incentive for Fuel Reprocessing and Plutonium Recycle*. AECL-3136; CONF-680601-29. 1968. (Presented at the American Nuclear Society and Canadian Nuclear Association Joint Annual Meeting, Panel Discussion on Plutonium, a Nuclear Fuel Resource, Toronto, 9-13 June 1968.

WASTE MANAGEMENT

Chapter Nineteen

Waste Management

R.G. HART

LABORATORY WASTE HANDLING

By the time Canada's fledgling nuclear program moved to the Chalk River Laboratories in 1946, scientists were well aware of the hazards of radioactivity and knew that the wastes from the laboratory would have to be handled with special care if harmful effects to people and the environment were to be avoided. The wastes the scientists had to deal with were as varied as the operations of the laboratory itself, ranging from highly radioactive irradiated fuel and radioisotopes and highly radioactive solutions from fuel reprocessing and isotope separation, to slightly radioactive liquids from rod bay cooling, active-area process drains, and equipment and clothing cleaning, to radioactive paper, rags, glassware, rubber gloves and other solids arising from general laboratory operations and experimentation. The highly radioactive fuel reprocessing and isotope separation wastes could be stored for a time while processes for treating them were developed, but the lower activity wastes had to be handled immediately. The volume was simply too large to afford delay.

In specifying the methods of handling the wastes, the early scientists had to be aware of the radioactivity released to the environment from the remainder of the laboratory, in order that the total release would be well within the safety guidelines of the day. Forty years later it would be necessary to perform complex computer calculations to prove "beyond reasonable doubt" that the selected method for handling the wastes would not have deleterious effects on people and the environment. However, in 1946 only sound logic and good judgement was required. That the waste management facilities at Chalk River have withstood the test of over forty years of operation without even approaching any releases of concern is testimony to the soundness of judgement exercised by those early scientists.

The Chalk River property extends over a total area of approximately 37 km². The topography is typical of the Canadian Shield, consisting of gently rolling hills (mainly eroded bedrock outcrops) interspersed with several small lakes and marshes (see figure 19.1). In many areas between bedrock outcrops, pockets of glacial till and other sediments are covered by extensive sand deposits. The water table in those sandy areas is usually several metres below the surface. These well-drained sandy uplands were chosen as the best locations for the waste repositories. The good drainage would preclude the wastes sitting in water, thus minimizing leaching, while the sand would adsorb material that was leached, retarding its movement by groundwater flow. In the centre of the 37 km² property is a depression in the bedrock, from which all water drains into a small lake, called Perch Lake. A stream, known as Perch Creek, drains the lake to the Ottawa River. The early scientists decided to locate the waste management areas in the sandy uplands of the Perch Lake basin. Thus they had a situation where any radionuclides leached from the waste had to collect in Perch Lake, where the water could be monitored and if necessary processed. No such processing has ever been necessary.

FIGURE 19.1
Map of CRL Property Showing Waste Management Areas

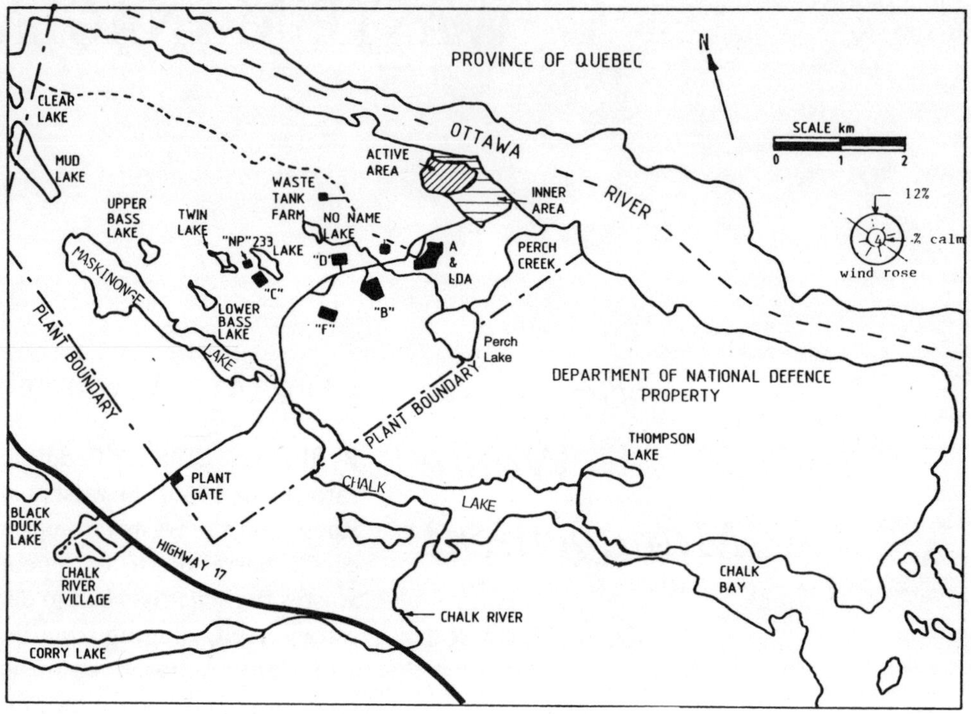

The first waste management area to be brought into service was a 1.2 hectare area located 750 metres from Perch Lake. It is now designated Waste Management Area A (WM Area A) (see figure 19.1). It was fenced-in in 1946 to restrict access by people and large animals, and was used from 1946 until 1952 for emplacement of low-level solid wastes, which were transported to the area in plastic or paper bags and buried directly in the ground in sand trenches. During this period, the amount of radionuclides going to the process sewers was limited by collecting them in glass bottles in each laboratory. The type and scale of laboratory activity at that time made this practical, and allowed most of the process sewer water to be discharged directly into the Ottawa River. These glass bottles were taken to WM Area A and buried in a large concrete tank, which contained sufficient limestone to neutralize any acid in the bottles if they broke. The area also contained a ceramic "tomb," which was used for the emplacement of cobalt-60 as metal. It was subsequently sealed all round with asphalt.

The 1952 accident to NRX prompted a drastic change in the use of WM Area A. After the accident, about 4,500 m³ of contaminated water collected in the NRX basement, and it had to be removed before cleanup and rehabilitation could commence. It was decided to pump the contaminated cooling water into WM Area A. A series of interconnecting ditches was dug over the existing sand trenches, and the contaminated water was pumped into these ditches via a temporary steel pipeline. Subsequently, the damaged NRX calandria and a miscellaneous collection of contaminated pipes, equipment, shielding and other material associated with the reactor dismantling and repair were also buried in WM Area A.

The use of the area for the NRX accident waste essentially terminated its use for routine emplacement of low-level solid waste, although bottles of laboratory liquid waste continued to be buried in the large concrete tank in the area until 1955. In 1954 and again in 1955, experimental quantities of fuel-reprocessing waste were released into the area to determine the hazards that could arise if a reprocessing-waste storage tank were to fail and allow its strong acid contents to penetrate the soil to the water table. It was found that the high-density acid solution sank rapidly through the soil to the water table, carrying most of the radionuclides with it. However, as soon as the acid was neutralized by the soil, the movement of radionuclides slowed to a small fraction of ground water velocity, and at that rate of movement would decay to an insignificant level by the time they reached Perch Lake. After 1955, all activities in WM Area A, with the exception of routine monitoring, ceased.

With the rebuilding of the NRX reactor in 1953, and the increased emphasis on nuclear power in the research program, the waste management requirements changed significantly. More experimental loops in NRX, more fuel examinations in reactor bays and hot cells and a general change from laboratory-scale to engineering-scale activities led to increased levels of radioactivity in both the process sewers and the solid wastes. As a result, it was necessary to include in the waste management area facilities for handling large volumes of slightly radioactive process sewer wastes and facilities for handling more highly radioactive solid wastes.

The process sewer waste was separated into two categories: (1) reactor process sewer waste, comprising water from the fuel rod bays, the reactor loops and other reactor systems, which contained few chemical contaminants, and (2) chemical process sewer waste, comprising liquid from the chemical plants and the chemical and metallurgical laboratories, which could contain acids, alkalis, complexing agents or salts in low concentration. Solutions containing large quantities of chemicals were not permitted in the process sewers.

The solid wastes were now also separated into two categories based on the amount of contained radioactivity. Solid waste that measured less than 0.01 mSv/h gamma 30 centimetres from the surface, or less than 10 mSv/h beta at the surface, and which contained no alpha contamination, was designated as low-level waste. All other solid waste was designated as high-level waste. Alpha-contaminated waste was included in the high-level category because it had to be assumed that the alpha irradiation came from toxic plutonium.

In 1953, a liquid dispersal area (LDA, see figure 19.1) was opened in the sandy uplands of the Perch Lake basin, close to WM Area A. The first pit in the area was a natural depression

FIGURE 19.2
Plan View of Reactor Pits and Liquid Dispersal Area

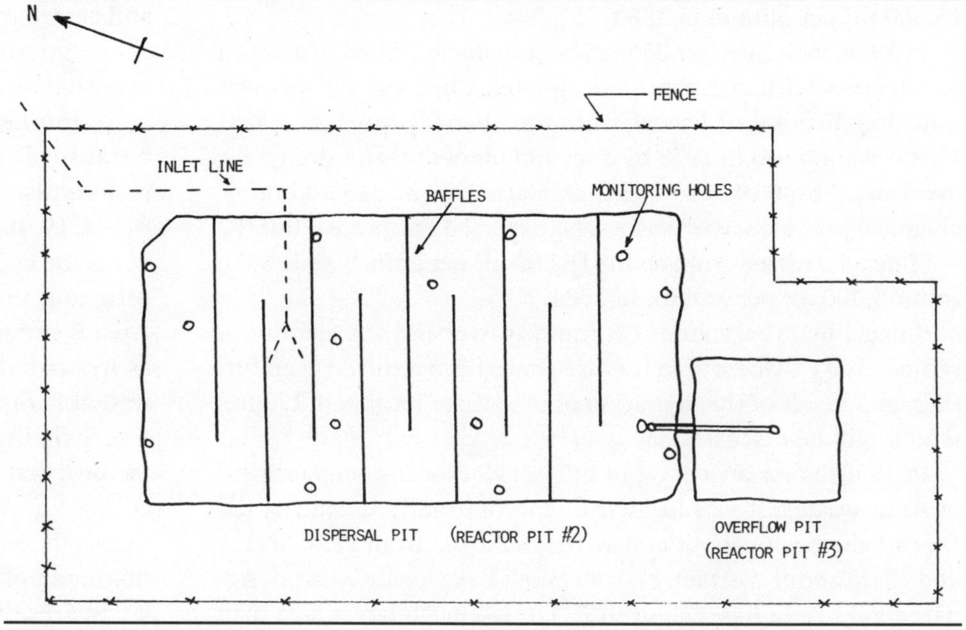

in the sand, approximately 45 metres in diameter and 3 metres deep. It was designated Reactor Pit #1 and was used between 1953 and 1956 to disperse 230,000 m^3 of low-level liquid waste from the reactor process sewer, mainly from the fuel-rod bays. When taken out of service in 1956 the pit was filled and covered with clean sand. It was replaced by a pit of improved design, designated as Reactor Pit #2. This was a man-made pit, baffled with asbestos board and filled with coarse aggregate (see figure 19.2). The liquid waste entered the pit about 1.2 metres below the surface and the waste seeped into the sandy soil. The aggregate provided means of preventing the sand walls from crumbling, acted as a gamma-radiation shield, controlled the spread of radionuclides by either wind or biota, and prevented freezing. In 1960 an overflow pit, Reactor Pit #3, was constructed to the same design, because it was felt that a buildup of sludge in reactor Pit #2 was going to limit its life. However, about that time, improvements in the systems feeding the reactor process sewer made Reactor Pit #3 redundant. From 1956 to the end of 1984, a total of 1,467,000 m^3 of reactor process sewer waste was dispersed in this area, but the annual volume decreased from about 210,000 m^3 per annum in 1960 to about 15,500 m^3 per annum in 1984.

In 1956, a circular pit 25 metres in diameter, filled to a depth of 4 metres with aggregate, designated Chemical Pit #1, was built for disposal of wastes from the chemical process sewer. This was followed in 1958 by a second pit of the same design for overflow. From 1956 to 1984, a total of about 315,000 m^3 of chemical process sewer waste was dispersed in the area, but the volumes decreased from about 18,700 m^3 per annum in 1960 to about 3,400 m^3 per annum in 1984.

Since 1984, the volume of liquid wastes and the amount of radioactivity dispersed in this area have been reduced even further, as a result of the operation of the Waste Treatment Centre, which will be discussed later.

In 1953, a new area was put into service for the emplacement of solid wastes. It was located in another sandy upland of the Perch Lake basin, about 600 metres removed from WM Area A and again about 750 metres from Perch Lake itself. It was designated WM Area B (see figure 19.1). It is a much larger area than Area A, comprising about 14 hectares, 8 hectares of which have to date been fenced in for use. Initially, all waste was buried in unlined sand trenches, as in Area A. However, as the operations of the laboratory changed and the waste was separated into high-level and low-level categories, new methods were developed for handling the high-level segments. The first method was to bury the high-level waste in asphalt-lined trenches in the sand. Once full, the trenches were covered with a layer of asphalt, and then covered with sand. This method was abandoned in 1959, when severe cracking of the asphalt covers was noted and water was found inside one of the trenches. A reinforced polythene sheet was installed above the asphalt to keep the water out, and the polythene was covered with sand to protect it. The asphalt trenches were followed in 1959 by reinforced concrete bunkers. Initially, a unit consisted of a rectangular bunker 60 metres long, 1.8 metres wide and 1.8 metres deep, divided into 12-metre long bays by cross walls. When in use, the structure was covered with a temporary roof. When full, the contents were covered with sand and a concrete roof. The structure was then covered with sand and graded (see figure 19.3). In 1963 the design was modified to a double bunker with cross and centre walls. The top 30 centimetres of the bunker were left above ground, and the cap was made removable to facilitate eventual recovery of the waste. In 1979 the design of the concrete bunkers was changed from rectangular to cylindrical because of concern regarding the strength of the unsupported rectangular bunker walls while the bunkers were empty (see figure 19.4). The cylindrical bunker is still in use.

As the experimental program, particularly the power reactor program, progressed, it became necessary to include in WM Area B storage facilities for intensely radioactive material such as irradiated fuel, hot-cell wastes, experimental fuel bundles, unusable radioisotopes, spent-ion exchange resin columns, active exhaust-system waste and fission-product waste from medical isotope production. These materials were stored in facilities known as tile holes which were concrete drain pipes set vertically on a poured concrete base at the bottom of a trench a minimum of 1.2 metres above the water table. The pipes were 4.9 metres deep and of varied diameters, up to 2.45 metres (see

FIGURE 19.3
Typical Single Concrete Bunker, Waste Management Area 'B'

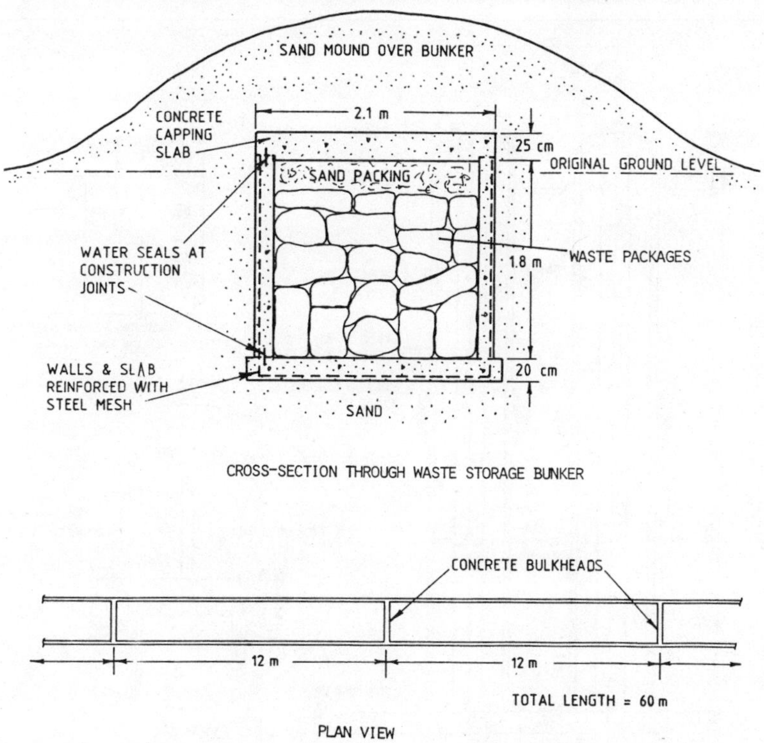

FIGURE 19.4
Typical Cylindrical Concrete Bunker – Waste Management Area 'B'

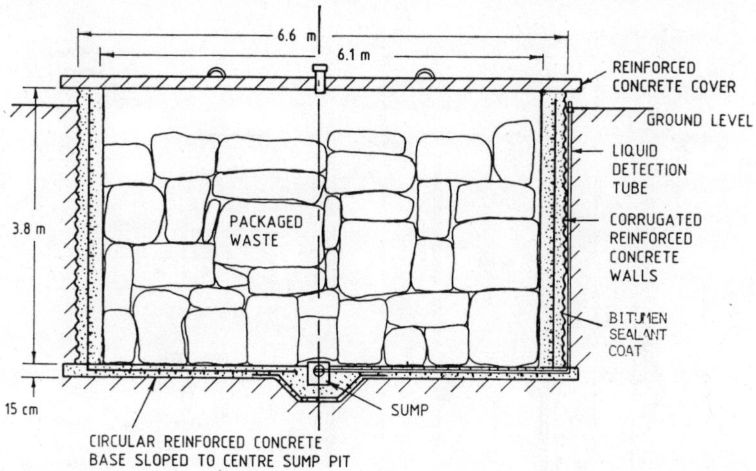

figure 19.5). The outer surface was waterproofed and the joint at the base was sealed. When the array of tile holes was assembled in the trench, the tops were closed with steel-and-concrete caps and shield plugs, and the trench was backfilled with sand to within a few centimetres of the top. The tile holes were then filled from transport flasks as required. For irradiated fuel, special tile holes were developed, with steel pipes within the tile and vented closures, to aid in the dissipation of heat and prevent a buildup of internal pressure (see figure 19.6).

Over the years, a number of special burials took place in WM Area B, including the second NRX reactor vessel in 1970, the first NRU reactor vessel in 1973, an evaporator used to reduce the volume of highly active reprocessing wastes in 1958, about 300 kg of natural-uranium scrap and turnings from fuel manufacture dating from 1962 and 1963, concrete made from mildly radioactive reprocessing wastes in 1955, and spent cobalt-60 sources from radiotherapy units returned between 1957 and 1960. WM Area B is still in service.

One of the earlier activities at the Chalk River laboratory was the experimental reprocessing of the irradiated fuel from the NRX reactor to recover the plutonium (see chapter two). Initially the plutonium was recovered by extraction with triethylene-glycol dichloride (the trigly process), which left as waste a highly radioactive solution of nitric acid, ammonium nitrate and uranium. In 1955 this process was replaced by an anion exchange process which left as waste a highly radioactive solution of nitric acid and uranium. Subsequently, the uranium was recovered from these solutions using tributyl phosphate, leaving two types of waste: one containing radionuclides, ammonium nitrate and nitric acid, and the other containing only radionuclides and nitric acid.

At the time, it was anticipated that reprocessing waste would be handled by evaporating off the liquid and immobilizing the remaining radionuclides in a leach-resistant glass, and processes to do this were being developed at Chalk River. The waste solutions

FIGURE 19.5
Typical Current Concrete Tile Hole in Waste Management Area 'B'

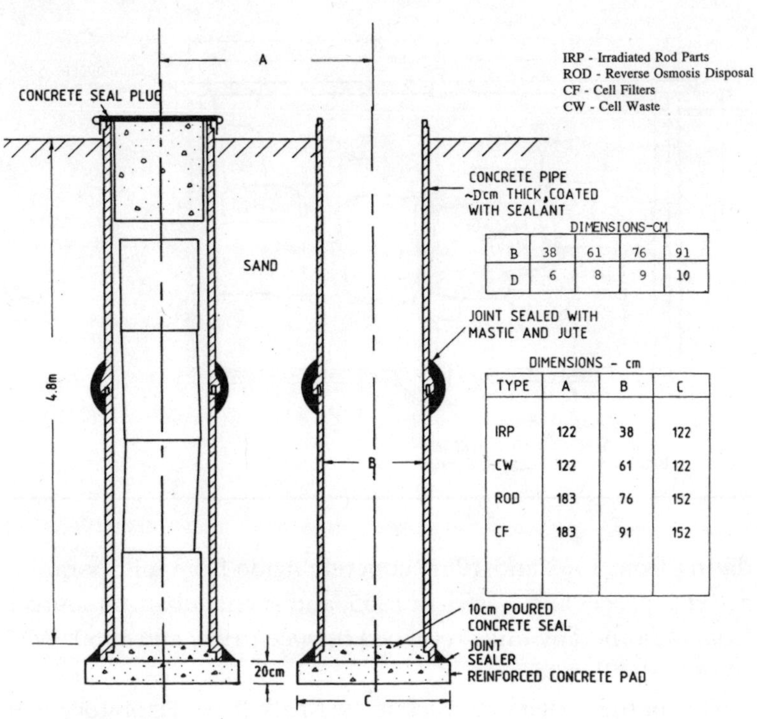

FIGURE 19.6
Typical Current Irradiated Material and Fuel Storage Tile Holes in Waste Management Area 'B'

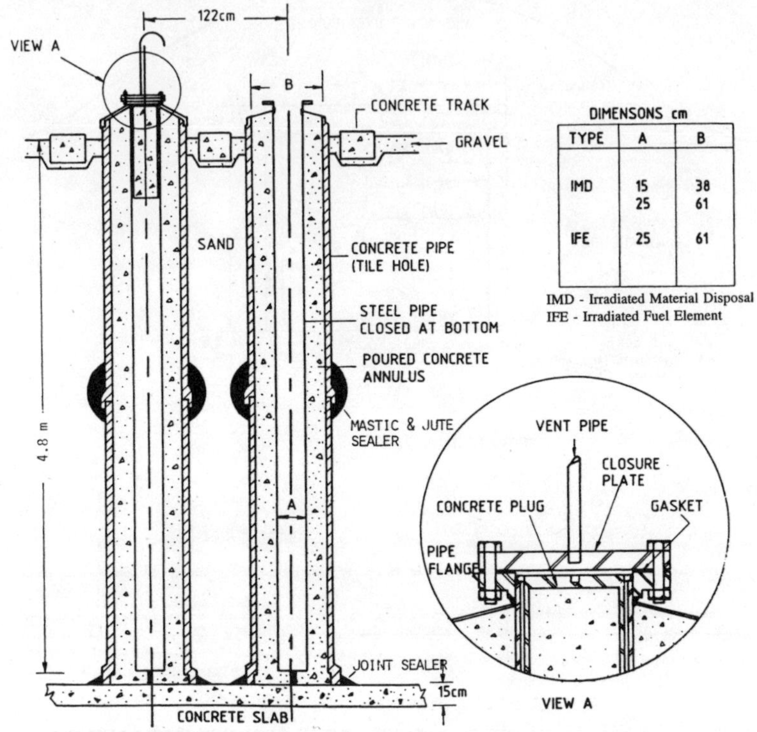

containing only radionuclides and nitric acid could be concentrated and glassified directly, but those containing ammonium nitrate could not. A process had to be developed for eliminating the ammonium nitrate before the trigly process wastes could be glassified. The development of this process led to the only work-related fatality to an AECL employee in the history of the laboratories. Ammonium nitrate will decompose in solution at the boiling point if chloride ion is added to catalyse the decomposition, but it cannot be allowed to collect in solid form anywhere in the reaction vessel without a risk of explosion. During the testing of the process on trigly wastes, an explosion occurred that killed a pipe-fitter who, at the time, was working in an unprotected area near the vessel. A plant was eventually constructed with appropriate blast protection in an isolated area of the laboratory property (see figure 19.1, N.P. 233), and the ammonium nitrate in the trigly process waste was successfully decomposed. The waste solutions from both processes were then concentrated.

However, only a small fraction of these wastes was converted to glass. In 1957 reprocessing of NRX fuel ceased as the result of an agreement to sell all irradiated NRX and NRU fuel to the United States, and it was felt that the small amount of reprocessing waste accumulated to that time could be stored until commercial conversion facilities became available. A test set of twenty-five nepheline syenite glass blocks was made in 1958, and these were buried in a 5x5 block grid below the water table in an unused area of the Perch Lake basin about 400 metres from the lake. When early results showed leaching rates to be too low to allow leaching patterns to be studied, a second set of twenty-five blocks was made of purposely poorer glass, and buried in a 5×5 block grid in a second unused area about 300 metres from the first site. After an initial high leaching rate, the leach rate decreased dramatically to an insignificant level. The initial higher leach rate was shown to be due to leaching from the near surface of the poorer glass. To date, no leaching has been detected from the first high-quality set of blocks. The remainder of the concentrated reprocessing waste was stored in double-lined tanks in a tank farm located in a sandy upland of the Perch Lake basin, about 2.5 kilometres from the lake (see figure 19.1), where it will remain until facilities are available to convert it to glass. Also in the tank farm are regeneration solutions from fuel-rod bay ion-exchange resins that have been concentrated ten-fold and then neutralized. These will remain in the tank farm until they are incorporated into bitumen.

In 1963, for the first time, waste was buried outside the Perch Lake basin. In order to preserve the area in WM Area B for high-level solid and special wastes, low-level solid waste burial grounds were located on a sandy upland in the Maskinonge Lake basin about three kilometres from the laboratories and about one kilometre from Maskinonge Lake. This area was designated Waste Management Area C (see figure 19.1). Maskinonge Lake is another lake on the Chalk River property, somewhat larger than Perch Lake. It has an outlet to the Ottawa River via Chalk Lake, Chalk River and Chalk Bay. As in Perch Lake, the release to the Ottawa River could be monitored in a stream connecting Maskinonge and Chalk lakes. The fenced-in area covers 4.2 hectares. As in Area B, the low-level solid waste was buried in unlined sand trenches. In addition to low-level solid waste, inactive acid, solvent and organic liquid waste was placed in specific sections of the trenches, or in special pits located along one edge of the area. In 1982, to conserve space, the method of burial was changed from parallel trenches separated by an intervening wedge of sand to a continuous trench similar to municipal landfills.

Over the years, surface and sub-surface monitoring has been done on a regular basis to study the migration of radionuclides in the waste management areas and assess the hazard they represent. It has been found that tritium travels at the rate of groundwater, that strontium-90 and cobalt-60 travel at about 3 percent of the rate of ground water, and that other radionuclides travel much more slowly. At this rate of travel, all will have decayed to insignificant levels by the time they reach Perch or Maskinonge lakes. The annual tritium, strontium-90 and cobalt-60 releases to Perch and Maskinonge lakes have never exceeded, respectively, 0.015, 0.001 and 0.001 percent of Chalk River's derived release limits for these species. The derived release limit is the calculated release of radionuclides from the plant boundary, which will ensure that the most exposed member of the general public will not receive a radiation exposure in excess of the Control Board regulatory limits.

When the Whiteshell Laboratories started up in 1961 in Manitoba, it was not blessed with sandy uplands well above the water table. Its property consisted of about 35 km^2 of flat grassland near the north-east boundary of the plains area of Manitoba, on the east bank of the Winnipeg River. The regional overburden is clay, deposited during the glacial lake period, overlying a compact layer of glacial till and bedrock. The water table is generally within one to two metres of the surface, so it was not possible to find an area where the waste could be buried above the water table. However, on the property, about four kilometres from both the laboratory buildings and the Winnipeg River, is a hydrogeologic discharge area; i.e., an area where the groundwater flow is to the surface. This area was chosen for the waste management area because if water flow was to the surface it could be channelled into a discharge ditch and monitored to ensure that excessive levels of radionuclides were not released to the Winnipeg River. An area of 4.6 hectares was fenced off for this purpose and no other area has subsequently been required.

The methods of handling the different categories of waste at Whiteshell were similar to those used at Chalk River, including unlined trenches for low-level wastes, concrete bunkers and concrete stand-pipes for medium-level wastes[1]

[1] Whiteshell nomenclature. This is waste that at Chalk River is designated as high-level waste, except that it cannot contain significant quantities of experimental irradiated fuel.

and stainless-steel tanks for high-level waste liquids, although there were minor differences in design to reflect the different environment. To minimize weather exposure and problems related to the high water table, low-level wastes were buried in the spring and in the fall. In the interim periods they were held in a steel Butler building with a storage volume of 300 m^3. The Whiteshell waste area also was equipped with an incinerator to burn organic liquids, primarily unusable fractions of the organic coolant used in the WR-1 reactor. The incinerator used a vortex burner, which allowed incineration without visible smoke or ash. Radioactive emissions were controlled by limiting the radioactivity in the feed to less than 15 Bq/mL. Organic liquids with radioactivity above that level were stored in stainless-steel drums on a concrete pad near the low-level waste building, until the radioactivity decayed to the required level. As at Chalk River, the waste handling facilities were modified with time to reflect experience. For example, concrete bunkers and stand-pipes are now surrounded by bentonite clay to keep water out and adsorb any radionuclides that might be released; concrete stand-pipes are now lined with galvanized steel to further protect against water ingress; and scrap and waste from irradiated fuel examinations are now stored in concrete canisters (to be described later) rather than concrete stand-pipes.

Surface, sub-surface and incinerator stack monitoring indicates that the only significant release of radionuclides from the waste handling facilities has been from the incinerator. That release has been less than 0.005 percent of the derived release limit for the laboratory, assuming that all the radioactivity was due to strontium-90, the most hazardous isotope of those potentially present.

All of these laboratory wastes at both Chalk River and Whiteshell are considered to be in safe storage; that is, they will present no hazard to the public as long as qualified people are present to look after them. However, safe storage cannot be equated with safe disposal. Plans for moving from storage to disposal will be discussed later in this chapter.

USED-FUEL STORAGE

From the earliest days of nuclear-reactor operations the accepted method of handling used fuel has been storage in water-filled storage bays. The water provides both cooling and shielding and its transparency permits visual examination of the fuel and the operations in the bay. The NRX storage bay consists of a long, narrow trench 37 × 1.35 × 3.75 m deep connected to a cut rod vertical storage bay, and a fuel-rod examination bay. The walls are thick concrete, painted white in some areas to improve visibility. Fuel is removed from the reactor in a lead shielded flask and is deposited in a storage block located over one end of the bay. After a short period of forced cooling in the storage block, the rod is discharged into the bay on a carriage mechanism, which orients it from vertical to horizontal and transports it to the cutting area of the trench. The inlet and outlet hardware is sawn off for reuse and a monorail transports the aluminum-clad fuel section to the storage area. Staff working with long tools from a bridge above the water surface can assist the operations and correct any malfunctions. Initially, the rods were held in the bay for a minimum six-month cooling period and were then transferred through a transfer chute to a dissolver, where they were dissolved for plutonium recovery. In the early years of operation the water in the bay could become quite radioactive, particularly if a lot of damaged fuel was being handled, and the water would have to be periodically replaced by feed and bleed. However, after side stream purification by ion exchange was introduced, this problem was eliminated. By 1957, reprocessing at Chalk River had ceased: arrangements had been made to sell the used fuel from NRX and NRU to the United States and ship it to the United States Atomic Energy Commission Savannah River facility for reprocessing. The rod bay was now used to load the shipping flasks. The flask was lowered into the bay with an overhead crane, its lid was removed, the rods were loaded, the lid was replaced, the flask was removed and was decontaminated before shipment.

The NRU bay is similar to the NRX bay, but it is more square and deeper. Some areas are painted white and others are surfaced with white tile. The fuel is removed from the reactor on-power and transferred to the storage block where it is force cooled by the heavy-water reactor coolant. After a suitable cooling period it is transported by flask to the bay where it is suspended in a vertical orientation. After the top is sawn off for reuse, the aluminum-clad fuel section is stored in a vertical orientation in the reactor bay. The NRU fuel was never reprocessed on site. It was sold to the United States. The fuel was loaded into the shipping flasks in the NRU bay.

When Canadian power reactors were built, they basically adopted NRX and NRU technology for used-fuel management. Typically, a CANDU 600 reactor fuel bay (see figure 19.7) has a storage area 21×12×8.4 metres deep and will hold ten to fifteen years' production of used fuel.[2] The walls of the bays are reinforced concrete, approximately 2 metres thick, and are lined with stainless steel or fibreglass-reinforced epoxy paint. The bay water is cooled through a heat exchanger and is kept clean by continuous circulation of a side stream through ion-exchange resins. The fuel in the bay is held in either large baskets or stacks of trays. When each basket is filled or each stack is completed, it is sealed with a zirconium alloy seal, the crystal pattern of which is unique. This seal was developed by AECL scientists to allow International Atomic Energy Agency inspectors to ensure that fuel from CANDU reactors, Canadian or foreign, was not being clandestinely diverted to weapons production. The inspector keeps a record of the signatures of the seals, and since no two crystal patterns are the same, the seals cannot be broken and replaced without the inspector's knowledge.

FIGURE 19.7
Typical CANDU Power Reactor Fuel Bay

The WR-1 reactor, built at Whiteshell in 1965, had a very small used-fuel bay, and by about 1972 the bay was filling up. Whiteshell had to either build a new bay or devise another method of storing fuel. They decided to experiment with dry storage of used fuel in reinforced concrete canisters, basically a concrete can with walls thick enough to provide the required shielding (see figure 19.8). The attraction of the canisters was that they could be built as needed and thus would avoid the large up-front expenditure that would be required for a new bay. The concern was that the thermal gradient across the concrete (hot on the inside, cold on the outside) would cause the inner concrete to expand and crack the outer concrete causing the exterior surface to weather rapidly or, in the extreme, lose shielding because of a crack through the wall. Fine exterior cracks were expected to relieve stresses, but it was the size and behaviour of the cracks that was important. The objective of the program was not only to demonstrate that used fuel could be stored in this way, but to develop and verify prediction methods that would allow future

[2] Storage bays in the multi-unit stations are larger, typically 32 × 17 × 9 m deep, and auxiliary bays have been built at Pickering and Bruce.

designs to be optimized. It was decided to build four experimental canisters: two cylindrical and two square, with bevelled corners (square prismatic). Although it was clear that cylinders would be most efficient in terms of heat transfer and utilization of material, all power-reactor fuel-handling equipment was based on a rectangular configuration, so it was deemed prudent to test a square design as well. One canister of each design was to hold fuel and the other was to be equipped with electric heaters to test the limits of applicability of the concept. The canisters were 5.26 metres high by 2.36 metres diameter or square, with interior mild steel liners 0.8 metre diameter or square. The fuel was to be loaded in a hot cell, into mild steel baskets 0.74 metre diameter or square, and a lid was to be welded on. The baskets were to be filled with inert gas, to prevent oxidation of the fuel should a fuel pin fail. Six baskets were to be placed in each canister. The design heat load was 2.0 kilowatts, or 4.4 tonnes of five-year cooled fuel.

Tests on the electrically heated canisters showed that both designs were practical and conservative. In fact, tests at 11 kilowatts, including cycling tests to represent summer and winter extremes, showed that cracks did not penetrate deeply and would close up as the heat decayed. Accordingly, the second canisters of each type were filled with fuel: the cylindrical one with WR-1 fuel and the square, prismatic one with Douglas Point fuel. These tests confirmed the viability of the concept and it was quickly adopted for the storage of excess WR-1 fuel. Optimized designs have since been used for the storage of Gentilly-1, Douglas Point and NPD fuel when those reactors were shut down. The first experimental concrete canisters at Whiteshell were left in an area of the property immediately adjacent to the parking lot, so that visitors could see for themselves that used fuel from nuclear reactors could be safely and economically stored for long periods.

USED-FUEL DISPOSAL

In the early years of nuclear power development it was universally accepted that the world's supply of economically available uranium was limited and that used fuel would have to be reprocessed to recover the contained uranium and plutonium for reuse. In fact, in the economics of early United States power stations, a credit of $10/gm was allowed for the plutonium contained in the used fuel. Accordingly, it was generally accepted that the waste that would have to be handled was a nitric acid solution of fission products from which the plutonium and uranium had been extracted, and that the handling method would

FIGURE 19.8
Cylindrical Canister #2 Containing WR-1 Fuel

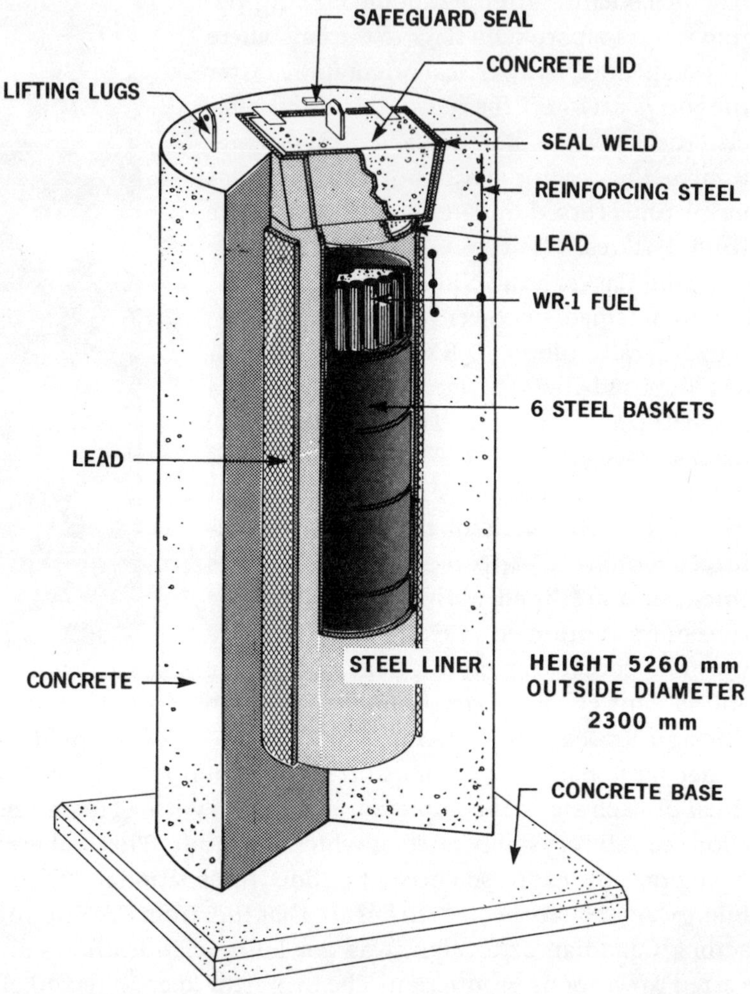

be to incorporate the fission products into leach-resistant glass blocks for burial in a suitable geologic formation. In the early to mid-fifties AECL scientists developed a process for incorporating the fission product waste into a glass comprising primarily nepheline syenite, a relatively abundant mineral widely used in the glass industry. The concentrated nitric acid solution containing the waste radionuclides was mixed with nepheline syenite powder in a ceramic crucible in proportions that would cause the mixture to form a gel. The crucible was then placed in a high-temperature furnace and the temperature was brought to about 1,350°C. At this temperature, the contained nitrates decomposed, leaving behind a relatively clear and highly leach-resistant block of glass. The off-gases passed through a liquid scrubber which removed entrained material and oxides of nitrogen, and an iron oxide bed which removed the volatile radionuclide ruthenium. As mentioned earlier, twenty-five such blocks were buried in sand below the water table and found to be so leach resistant that a second set of deliberately poorer quality blocks had to be made to allow leaching patterns to be studied. It may not be surprising that scientists at the time felt that the waste problem was solved in principle, and that there would be little difficulty when the time came to select a suitable site for final burial. History would show the naivete of this assumption but work on final disposal ceased and was not resurrected until the mid-seventies.

In the early seventies nuclear power stations were operating well in Canada but the social climate had changed. There was significant public opposition to nuclear power and one of the major focuses of this opposition was nuclear wastes. Critics claimed that there was no safe way of disposing of the wastes from the used fuel; the early glass block experiments were deemed insufficient and irrelevant because there had been essentially no accompanying geologic work and there were now no plans to reprocess the used fuel. Accordingly, in 1974, AECL, with strong government support, initiated a major program to prove "beyond reasonable doubt" that wastes from the used fuel could be managed safely. It would be necessary to shield the fuel for several hundred years, while the bulk of the gamma-emitting radionuclides decayed, and to isolate the fuel for a much longer period, to ensure that long-lived, toxic, beta- and alpha-emitting radionuclides did not escape in significant concentrations to drinking water. The challenge was to prove that long-term isolation was possible. The program was centred at Whiteshell and was to occupy a major portion of that laboratory's efforts for the next fifteen years.

Two objectives were established at the outset: one a safety objective and the other a responsibility objective. The first was that "the wastes should be managed in such a way that the potential hazards are minimized to the greatest extent practical."[3] The second was that "they should be managed in such a way that the trouble and concern to future generations in keeping them safely will be minimized."[4] A number of potential methods of meeting these objectives were reviewed, including transmutation to short-lived radionuclides, extra-terrestrial disposal, ice-sheet disposal, sub-seabed disposal and underground disposal. It was decided that underground disposal was the best option for Canada; it was deemed the method whose safety could most readily be proven "beyond reasonable doubt." AECL enlisted the help of the Geological Survey of Canada (GSC) to determine the type of geologic formation best suited for underground disposal in Canada. Around the world, a number of geologic media were being considered for disposal of nuclear waste, including clay, shale, tuff (volcanic ash), basalt (volcanic rock), salt, and hard rock. However, the scientists at GSC felt that a particular type of hard rock formation, a pluton, would be the most suitable formation for nuclear waste disposal in Canada. Plutons occur in abundance in the Canadian Shield, where they have been stable for hundreds of millions of years and are likely to remain stable for millions more. Their stability, uniformity, resistance to erosion, and abundance in areas of little commercial value and low population led to their selection by the GSC. The GSC recommendation was accepted by AECL and endorsed by an independent group of experts that had been

[3] The objective was to limit the risk to the most exposed individual to a small fraction of that from natural background.

[4] The objective was to preclude the need for any long-term care or monitoring.

commissioned by the Department of Energy Mines and Resources and headed by Dr. F.K. Hare, then Head of the Department of Environmental Studies at the University of Toronto.

AECL, with the assistance of Acres Consulting Services and Associates, next developed two reference systems for disposal of fuel wastes from CANDU power reactors: one assuming that the fuel would not be reprocessed, the other assuming that it would.

In the reference system for unreprocessed used fuel, the fuel bundles would be allowed to decay and cool for a minimum of ten years after removal from the reactor before transportation to the disposal facility. Transportation would be in casks designed to protect against radiation and the release of radionuclides, even if an accident resulted in severe stresses due to impact, fire or immersion in water (see figure 19.9). The casks could be transported by truck, train or barge. At the disposal facility, the bundles would be loaded on end into 1.15-metre-high x 0.91-metre-diameter titanium disposal containers in two layers of thirty-six bundles each. After welding on a lid with fill and vent holes, the free space in the containers would be filled with a molten lead-antimony alloy, which on solidification would immobilize the bundles and support the container walls. The vertical orientation of the bundles would facilitate complete filling of the free space. The fill and vent holes would then be welded shut and the filled container would be transported to the disposal vault. The vault would be built on a single level 500 to 1,000 metres below the surface. Exhaust shafts would be sunk at one end and a waste-lowering shaft, intake ventilation shaft and service shaft would be sunk at the other. Underground passageways, known as drifts, would be excavated to connect these shafts and to permit removal of excavated rock and the transfer of the waste and backfill materials. Further excavating and the emplacement of wastes would be carried out on a modular basis, starting at the end near the exhaust shafts and working toward the waste-bearing shaft (see figure 19.10). Excavation and waste emplacement would use different drifts, to separate construction and waste traffic. The finished vault would contain sixteen panels, each containing fifty-two rooms, each of which would hold 300 disposal containers in seventy-five rows of four. In each room the disposal containers would be placed on a one-metre layer of buffer material, and as the room filled,

FIGURE 19.9
Used Fuel Transportation System: (a) Transport Truck with Cask, (b) Currently Used Fuel Shipping Cask

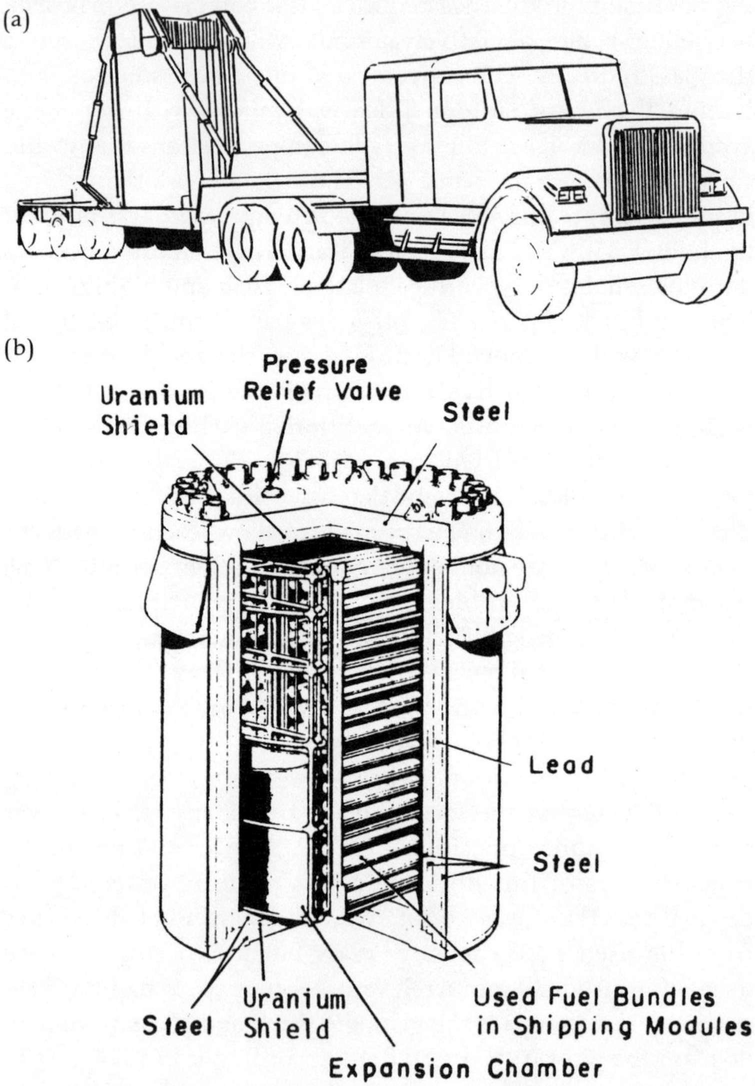

buffer material would be packed around and above the containers to a width and depth of at least one metre. The buffer material would consist of a sand-clay mixture designed to retard the movement of groundwater and radionuclides, and possibly to modify the chemical conditions in the vault to retard corrosion. Backfill material would fill the remainder of the rooms, the drifts and the shafts. The major component of the backfill would be crushed rock or sand, or a mixture of the two, with some clay added to achieve the desired low permeability. The reference unreprocessed fuel disposal vault would be 3,600 metres long and 1,700 metres wide, and it would hold 18 million fuel bundles. This was the then-projected total fuel arising in Canada up to 2055. After the vault was sealed, the surface facilities could be decommissioned and the site could be returned to close to its original state or used for another purpose.

In the reference system for the reprocessed fuel, the wastes would be incorporated into glass and poured into titanium containers 3.275 metres high × 0.457 metres diameter at the reprocessing facility. The containers would be sealed and then shipped to the disposal facility in transfer casks of the quality mentioned earlier. At the disposal facility, the containers would be transferred to a vault of similar design to the unreprocessed fuel vault, but the reprocessed waste containers would be placed in boreholes in the floor of the disposal rooms, rather than placed on the floor. Each room would have 480 boreholes, arranged in 120 rows of four, and one waste container would be placed in each borehole. The container in the borehole would be surrounded with buffer material to a thickness of at least 0.1 metre and then the room would be backfilled. The reference reprocessing waste vault would be 3,600 metres long and 900 metres wide, with fifteen panels, each having twenty-five rooms. This single vault could contain the waste from about 13 million fuel bundles, the projected total fuel arisings in Canada until 2050.[5]

FIGURE 19.10
Underground Layout of the Conceptual Disposal Facility

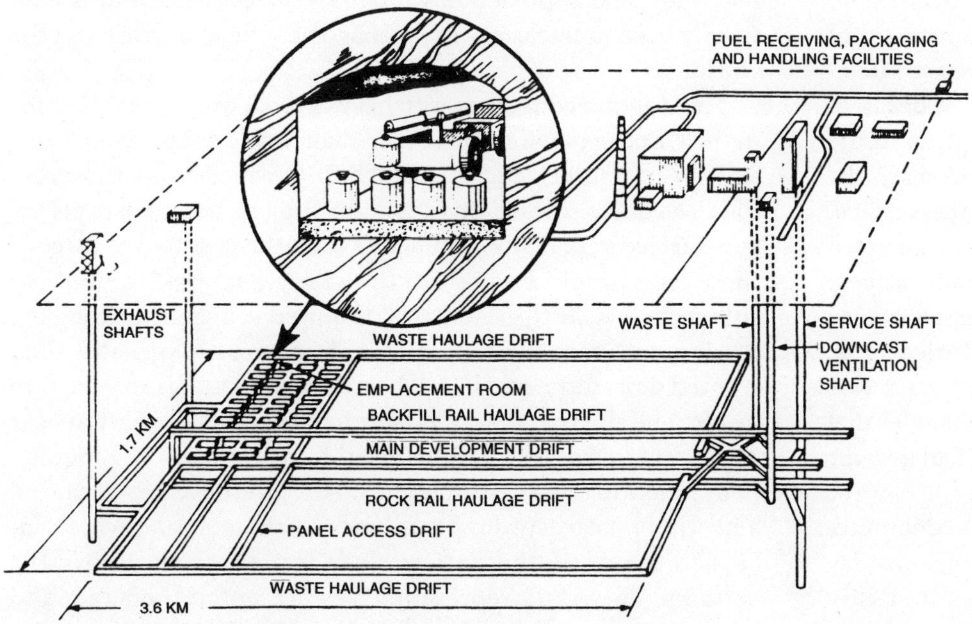

During vault construction, about 9 000 000 m^3 of rock would be excavated, half of which would be returned to the vault as part of the backfill. An area at the surface would be required for storage of rock and other materials. An additional area would be designated as an exclusion zone, so the total surface area required for the disposal centre would be about 27 km^2. The area of the vault itself would be about 7 km^2.

[5] An amusing incident occurred when the author presented the AECL program, at about this stage of development, to the Ontario Government's Royal Commission on Electric Power Planning (The Porter Commission). To ensure that the proceedings were seen to be scrupulously fair, Dr. Porter tended to let questioning of witnesses take its course without much restriction. Late one night, after what seemed like interminable and repetitive questioning from anti-nuclear intervenors, the author was asked: "How long do you think civilization will last?" Rather taken aback, the author paused for a moment and in that pause, the Commission Counsel was heard to mutter "Not as long as this commission."

Developing the reference system was relatively easy; the major challenge facing the scientists was to prove beyond reasonable doubt that the "radioactive material would be isolated such that the potential hazards to man and his environment would be negligible for all time." How could one prove in a few years that the system would present negligible risks for hundreds of thousands of years? To answer this challenge, the scientists devised a method they called a Systems Variability Analysis: it provides a probabilistic estimate of the radiation dose to people caused by radionuclide migration from the disposal vault. The exposure mechanism of concern is that in which groundwater penetrates the vault, corrodes the containers, dissolves the radionuclides in the waste and carries them to the surface environment, where they could disperse or concentrate and deliver via the food chain a dose of ionizing radiation to flora, fauna and humans. That radiation dose is dependent on a number of phenomena, including the rate of movement of groundwater, the rate of corrosion of the containers, the rate of leaching of radionuclides from the waste, the rate of decay of the radionuclides, the rates of sorption and desorption in the buffer material, the backfill and the surrounding rock mass, the rate of dispersion when the radionuclides reach the surface, and the rate of concentration in plants and animals consumed by people. Values of the parameters required to quantitatively describe these phenomena could be obtained from laboratory and field research, theoretical considerations, historical data, the opinions of experts, and simulation studies, and this information could then be fed into a mathematical model that represented the relevant physical processes and calculated the maximum radiation dose to people.

There is nothing unique about such an approach. Computer modeling to handle complex phenomena such as this is today the norm. What was unique about the Systems Variability Analysis was its method of handling the inevitable uncertainty in the values for the system parameters. To account for the natural variability within a pluton and probable errors in measurement, the values of the system parameters were expressed as probability distributions. To obtain parameter values from the probability distributions, the code randomly selected a single value from each distribution for each computer simulation. Many simulations were performed, to ensure a statistically significant use of the parameter distributions, and the results were expressed as a radiation dose versus probability plot. From such a plot one could determine, for example, the probability of the radiation dose to people ever exceeding 1 percent of natural background. From the analysis one could also determine the radionuclides causing the exposure and their rate of migration through each barrier, allowing modification of the system to possibly retard that particular radionuclide if it was considered a problem. To simplify the handling of the data and to optimize the value of the results, the system was divided into three physical regions: the vault, the geosphere and the biosphere. Sub-models were developed for each region, and then linked by an executive program called SYVAC (Systems Variability Analysis Code). The approach has proved very effective and is now widely used in nuclear waste management programs around the world.

The original concept was to have GSC identify 1,000 plutons in the Ontario portion of the Canadian Shield, select from maps and existing information the 100 best prospects from those and then select by surface investigation the ten best prospects for sub-surface study. The parameters from the best two or three of those sites would be used in the safety analysis. To do these sub-surface investigations, GSC required some rather complex equipment, some of which they had to develop, and they wanted a pluton near their laboratories in Ottawa in which to test it. The site they selected was the Mount Maria pluton near Madoc, Ontario. At a public meeting to explain the program, held in Madoc on 16 March 1977, AECL representatives faced a hostile, placard-waving audience of about 1,100, which included many of the anti-nuclear leaders from across the country. The AECL representatives presented their case and answered questions, and explained the 1,000, 100, 10 scenario, but if they convinced anyone that Mount Maria hadn't been already chosen as the Canadian nuclear waste vault site it wasn't apparent. The local MP and MPPs convinced the Minister of Energy, Mines and Resources to ban AECL and GSC from working in the area. Ironically, a group of local citizens, including some councillors, subsequently approached AECL and the minister to try to get the ban lifted, because they were interested in the jobs a disposal vault would create, and they felt they could turn public opinion around. However, too much damage had been done and the minister remained adamant.

The Madoc incident prompted a complete rethinking of the approach to site selection and the field research program. It was decided that it was both impractical and unfair to ask communities to consider hosting a waste disposal facility before they could be assured beyond reasonable doubt that it was safe. Accordingly, AECL decided that they would do the work necessary to generically prove beyond reasonable doubt the safety of the concept, and that they would submit that proof to the Control Board for review. Only after the Control Board agreed that the safety of the concept was proven beyond reasonable doubt would they ask communities to consider hosting a disposal site. It was recognized that this generic approach would increase the variability of the system parameters, because now many sites would have to be considered, but it was felt that the probabilistic treatment of the parameters could handle it. It was still necessary to gain access to appropriate plutons for field research, but it was to be made clear to neighbouring communities that these sites were for research purposes only, and that selection of a site for the disposal vault would not proceed until the concept had been accepted as safe by the Control Board. Further, it was deemed essential to ensure that local governments, regional politicians and the local public were fully aware of the research work that was to be done in the area, and were reasonably supportive before any work was done. At this time, AECL also decided that its work must have independent and formal peer review, beyond what it would get through publication, if its work on such a sensitive subject was to have public credibility. It therefore established a technical advisory committee, made up of independent experts in the appropriate key disciplines, recommended by the appropriate Canadian scientific societies. This group was to review the program and the work as it progressed, and publish an annual report on their findings. The committee is still active and their advice and criticism have been invaluable.

Atikokan, Ontario was the first site where the revised approach was used, and it met with enthusiastic response. The local government and population seemed to feel that if they supported the research work, they would have an advantage over other communities at the time of site selection, with resulting major economic benefits. The Atikokan site became one of AECL's main field research sites and contributed a wealth of information to the program.

The second site where the revised approach was used was Massey, Ontario, and the response was less than enthusiastic. The Atikokan pluton is made of granite and is representative of about 80 percent of the plutons in the Canadian Shield. The other 20 percent are made of gabbro, and it was felt that gabbroic plutons should be investigated to determine whether they had better properties for waste disposal than granite. A gabbroic pluton exists near Massey. Activists were able to convince many Massey citizens that AECL had given up on Atikokan because it was unsuitable and had selected the pluton near Massey as the Canadian waste vault site. However, with the help of strong support from regional politicians, the necessary work was done. The work showed that there was no significant difference between granite and gabbro for waste disposal; after about three years the Massey field research area was closed.

The third site of extensive field research was Whiteshell itself. The site is located on a granite pluton, the Lac du Bonnet batholith, which is about seventy kilometres long and twenty kilometres wide. Access to most areas of that pluton was obtained without major difficulty. An area about fifteen kilometres from the site and ten kilometres from Lac du Bonnet was selected for extensive study, and eventually became the site of an Underground Research Laboratory (to be discussed later). Although the Chalk River site was not on plutonic rock, some work was done there as well.

With the acquisition of research areas it was possible to proceed with the field research program that had been envisaged. The objectives of the program were twofold. The first was to develop non-intrusive methods of characterizing plutons; that is, to be able to predict from existing information, aerial surveys and surface observations the likely structure, fracture patterns, permeabilities and water flows beneath the surface. This was to aid in site selection. All such information would have to be verified by sub-surface investigation (drilling), but it was important from both a cost and a public relations viewpoint that approved sites not have to be subsequently abandoned for technical reasons. The second objective was to obtain accurate information on parameters such as water chemistry, fracture patterns and porosity, rock composition, water flow and rock uniformity, which were required in the analysis.

The non-intrusive investigation started with airborne measurements of magnetic and density variations to define the boundaries of the pluton. These were followed by careful examination of the surface within those boundaries, to identify any features that were likely to be the surface expression of faults and fracture zones in the underlying rock. Then at selected locations seismic and electromagnetic waves were transmitted from the surface into the underlying rock to determine the potential presence and orientation of fracture zones. The waves reflect differently from fractures than from a coherent rock mass. Core samples were then bored out of the rock mass at selected locations to confirm the picture of the underground structure predicted from the non-intrusive investigations. It was found that the non-intrusive methods gave a three-dimensional picture of the underground structure sufficiently accurate to screen potential sites, but not accurate enough for site selection without borehole confirmation.

The Systems Variability Analysis required methods for calculating the rate of failure of the containers, the rate of release of radionuclides from the waste forms, the rate of movement of radionuclides through the buffer, the rate of transport of radionuclides through the rock mass of the pluton (the geosphere), the buildup of radionuclide concentrations in water, food and air, and the consequent uptake of these radionuclides by people. It was the task of the researchers to provide the information required for these calculations. It was found that the variability of the values of all the parameters required could be adequately described using the six probability functions shown in figure 19.11.

To describe the rate of failure of the waste containers, a container failure function was used that gave the fraction of the containers failing as a function of time. The container failure function was based on the rates of uniform corrosion of the thin-walled titanium container, and had to take into account the space and time variations of temperature in the waste vault. It was found that the failure rate could be expressed as a truncated normal-probability distribution with a start time of about 16,500 years and a mean value of about 17,000 years. The corrosion rate as a function of temperature was determined in the laboratory using water of chemical compositions covering the range found in field research boreholes. The water from deep in the boreholes was reducing and strongly saline. This removed from consideration some of the lower-cost options for container materials, such as stainless steel and nickel alloys, which corroded relatively rapidly in such water.

To describe the rate of release of radionuclides from the unreprocessed used fuel, it was necessary to use a two-stage release model, to account for an instantaneous release and a long-term release. In the analysis it was assumed that container and fuel-sheath failure was simultaneous, and experiment showed that there was an "instant" release of iodine, cesium and selenium when the fuel sheath failed. This resulted from fractions of these elements that evaporated from the hot centre of the fuel pellet during operation and condensed in cracks in the fuel, gaps between the fuel pellets, and gaps between the fuel pellets and the sheath. The "instant release fraction" was defined by a lognormal probability distribution with a range of from 1.2 to 10 percent and a mean of 3.6 percent, and was based on results from a study on dissolution of CANDU fuel. Long-term radionuclide release was based on the assumption of congruent dissolution; that is, that the radionuclides are released as the uranium dioxide containing them is dissolved. It was found that the rate of dissolution of uranium dioxide was controlled by its solubility, which could be described by a lognormal probability distribution with a wide range and a low mean value. The wide range reflected the wide range of water chemistry that could possibly contact the fuel, and the low mean reflected the fact that the most probable water chemistry was the reducing saline composition found at depth.

To describe the rate of release of radionuclides from the reprocessing waste, researchers had to first specify the composition of the glass that would be used to immobilize the bulk of the waste and the other waste forms that would be produced in the reprocessing and glassification processes. Experiments were done to optimize glass stability and leach rate against ease of manufacture (low melting point), and it was decided that the optimum was achieved with a sodium calcium aluminosilicate glass. In the glass manufacturing process, radioactive iodine and carbon-14 are volatilized, and they have to be collected and immobilized and put in the waste vault as well. It was determined that bismuth oxyiodide and barium iodate could be used for iodine, and barium carbonate for carbon. Finally, in reprocessing, zirconium fuel sheaths, containing radioactive activation products, are left behind, and these have to be compacted and disposed of as well. The release of radionuclides from the sodium calcium aluminosilicate glass was described by a model conceptually similar to that for used fuel, with an "instant release" followed by a congruent release determined by the glass solubility. The instant release fraction was represented by a uniform distribution and the solubility was represented by a lognormal distribution with a low mean value. The release rates from the other waste forms – bismuth oxyiodide, barium iodate, barium carbonate and zirconium – were all controlled by their solubilities, which were represented by lognormal distributions.

Possible ways for the radionuclides released from either waste form to move through the buffer are convection and diffusion. Convection refers to the transport through the buffer by groundwater movement. Diffusion refers to the movement through ground water to regions of lower concentration, and is represented by the molecular diffusion coefficient for ions in water. Although both mechanisms were included in the model for transport through the buffer, experiment and analysis showed that migration was dominated by diffusion. This was because groundwater flow

FIGURE 19.11
Probability Density Functions Recognized by SYVAC

CLASS OF PDF	SAMPLE PLOT	REQUIRED DISTRIBUTION PARAMETERS
Constant	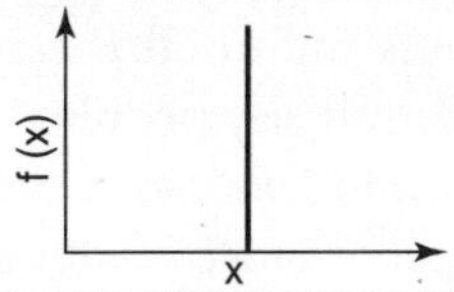	Constant value of x.
Uniform	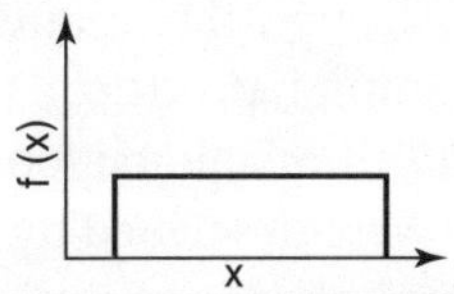	Lower limit of x. Upper limit of x.
Piecewise uniform	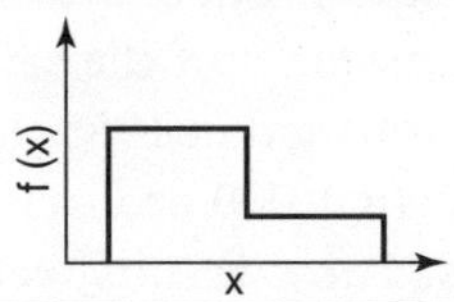	Number of ranges. for each range { Lower limit of x. Upper limit of x. Fraction of values of x that lie within the range.
Log uniform	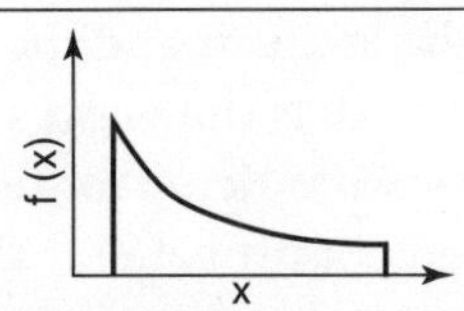	Lower limit of y, where $y = \log_{10}x$. Upper limit of y, where $y = \log_{10}x$.
Normal	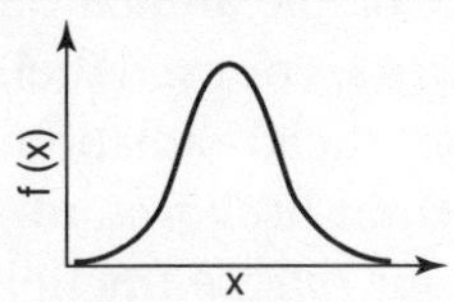	Mean, μ, of values of x. Standard deviation, σ, of values of x.
Lognormal	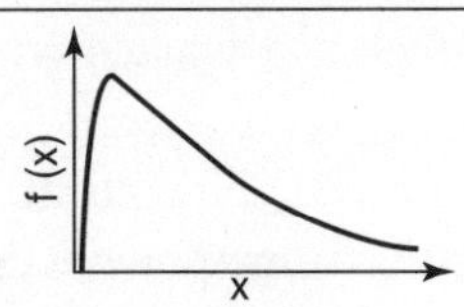	Mean, μ, of values of y, where $y = \log_{10}x$. Standard deviation, σ, of values of y, where $y = \log_{10}x$.

through the buffer was extremely slow. The molecular diffusion coefficient increased with temperature, and the temperature in the buffer would vary with time and with position over a range of 150-15°C. Values of the molecular diffusion coefficient were calculated over this range of temperature, and their probability was weighted by estimating the durations of temperature regimes after container failure. This resulted in their representation by a piecewise uniform distribution. Calculations showed that buffer layers as thin as only 25 centimetres would delay movement of dissolved or suspended radionuclides for thousands of years. The model also included a method for calculating retardation by sorption, which required the determination of partition coefficients for each radionuclide and their expression as appropriate lognormal distributions.

To describe the rate at which radionuclides leaving the buffer would reach the surface, it was necessary to determine the rate of movement of groundwater through the intervening rock (convection transport), the distance it would have to travel to reach the surface, the fraction that would be diverted from the main flow path by cross flow and diffusion (hydrodynamic dispersion), the fraction that would disappear by radioactive decay, the new radionuclides that would be produced as daughter products in the decay process, and the retardation that would occur by sorption and desorption on the walls of fractures in the rock. It was found that these phenomena could all be quantified using experimental observations made in the field research areas or the laboratories, and that each of the experimentally derived parameters required could be adequately described by one of the six probability distributions shown in figure 19.11.

The flux of radionuclides leaving the geosphere was used in a biosphere sub-model, to calculate the maximum annual radiation dose to an individual in the reference group. The reference group comprised the people who would live their entire lives in the area where radionuclides from the vault would reach the surface. It was assumed that a member of this group would eat locally grown food and that the source of household water would be either a well (well scenario) or a local lake (lake scenario). The biosphere sub-model consisted of transport and dose components. The transport component was used to calculate time-dependent concentrations of radionuclides in air, soil, lake and well compartments. The dose component was used to calculate the radiation dose from ingestion and inhalation of radionuclides and from external exposure. Variables affecting the radionuclide concentrations in the various compartments included size of the soil discharge area, soil depth, degree of soil saturation, net precipitation, soil type, soil density, soil infiltration function, soil distribution coefficient, lake area, lake water depth, ratio of lake drainage basin to lake depth, sediment depth, sediment density, sediment distribution coefficient, atmospheric dust load and deposition velocity from air. Variables affecting the dose calculations included the per capita water demand by both people and animals, peoples' total energy intake rate, the radionuclide holdup time in food products or drinking water, the energy provided by each food type in the diet, and transfer coefficients from soil to plants, from plants to animals and people, from animal products to people, from freshwater to aquatic food and from aquatic food to people. Values for all of these parameters were described by appropriate distribution functions based on experiment or available information.

The results of a SYVAC assessment of the safety of the reference disposal system were published in 1985. It was an interim assessment based on more than 1,000 scenarios each for used fuel and reprocessing wastes. These results were fed back into the system to improve both the analysis and the design, but the interim results gave confidence that the wastes could be safely handled. No consequences were predicted for tens of thousands of years and there was a 90 percent probability that no radiation dose would exceed 1 percent of the annual dose from natural background for ten million years. For used fuel, iodine-129 was the only radionuclide that produced radiation exposures even approaching 1 percent of natural background, and it was only the iodine-129 from the instant release fraction that produced the exposures. For fuel reprocessing wastes with iodine incorporated in bismuth oxyiodide, technetium-99 produced the highest doses, but when iodine was incorporated in barium iodate both iodine-129 and technetium-99 were contributors. No other radionuclides contributed as much as 0.1 percent of the dose from natural background. The main barriers limiting the dose were the waste forms, uranium dioxide, sodium calcium aluminosilicate glass and bismuth oxyiodide. In the biosphere the only significant contributor to dose was ingestion; the other sources were insignificant. For used fuel the only significant contributor was the household water from the well. For reprocessing waste, ingestion of technetium-99 in terrestrial plants also contributed.

The results of the analysis are consistent with analogues available in nature. For example, a few billion years ago uranium in the earth's crust was sufficiently enriched in the fissile isotope uranium-235 that natural fission reactors existed where conditions were optimum. One such location was Oklo, in Gabon, Africa, and scientists have studied that location to determine the fate of the wastes from those naturally occurring reactors. Their studies show that the plutonium and some of the most hazardous fission products had hardly moved before they decayed to insignificant amounts, even though the location is far from optimum for a disposal site. Closer to home, the stability of uranium oxide in groundwater is being studied at the Cigar Lake uranium deposit in northern Saskatchewan. The ore body is 2,000 metres across and 20 metres thick, and contains 150,000 tonnes of high-grade ore with an average uranium oxide concentration of 12 percent and peak concentrations as high as 60 percent. The uranium/oxygen ratio is close to that of the uranium dioxide in used fuel and the groundwater chemistry is reducing, similar to that found deep in plutons. Over the last 1.3 billion years the deposit has been in contact with groundwater and it is still intact. In fact, field measurements show that the uranium concentration in water only five metres away from the ore bodies is below the level specified for drinking water. These natural analogues give added confidence to the very low radionuclide release levels predicted by the SYVAC analyses.

It was estimated that the capital cost of a waste repository of the reference design would be $470 million (1979): $237 million for the surface facilities and $233 million for the underground facilities. The facility was projected to operate for thirty-five years to clear the backlog and the fuel arisings to 2050, and have an annual operating cost, including transportation, of $140-400 million (1979). The biggest variable was the cost of transportation, which was highly dependent on the location of the repository. Using these numbers, the cost of disposal to the electricity customer would be 0.01 to 0.03 cents/kWh (1979). The facility would employ a permanent labour force of about 480 and a peak force of about 800 during construction.

LOW- AND INTERMEDIATE-LEVEL WASTE DISPOSAL

Concurrent with its decision to initiate an extensive program on final disposal of used fuel and reprocessing waste, AECL decided to initiate a program on final disposal of low- and intermediate-level wastes. Until that time, all such wastes at AECL laboratories had been managed carefully, as discussed earlier, but they had not been, and were not being, managed in a way that would leave no burden of responsibility to future generations. Similarly, at power reactor sites across the country, such wastes were being compacted and stored in engineered facilities, but they were not being disposed of. The program AECL now initiated was to develop methods for volume reduction and final disposal that could be used not only for the wastes produced at or collected by its sites, but for similar wastes collecting at power stations. This program was centred at Chalk River.

Work initially concentrated on volume reduction of low-level wastes and subsequent immobilization of the concentrate for disposal, and this work focused on a new Waste Treatment Centre started at Chalk River in 1977. In this facility, combustible solid waste, which had previously been buried in sand trenches, was burned in a two-chamber incinerator to minimize radioactivity in the off-gases. In the first chamber, waste was pyrolyzed in a reduced supply of oxygen to minimize turbulence and the entrainment of radionuclides. The off-gases, mainly hydrogen and methane, were then burned in the second chamber, the afterburner, to produce carbon dioxide and water. Similar non-combustible solid waste was compacted in a fifty-tonne press to give a volume reduction of a factor of 7. The dilute liquid waste, which was previously dispersed in the reactor and chemical pits, was concentrated by ultrafiltration/reverse osmosis to 2-5 percent solids, and was then concentrated further in an evaporator. The ash from the combustion and the liquid concentrate were then combined with bitumen in the evaporator, to form a leach-resistant bitumen product for disposal. Experiments showed that volume reductions of about 20 and 2,500 could be achieved in the incinerator and the reverse osmosis units, respectively.

Once the Waste Treatment Centre was operational, attention turned to disposal. The plan was to develop methods for permanent disposal of the wastes being produced currently, and then to determine which of the wastes already in the waste management areas had to be recovered for further treatment, and which could be considered to be in permanent disposal in their existing state. The first thing required was a reclassification of the wastes. For waste management, the waste had been categorized primarily for handling, and primarily on the basis of external radiation fields. For waste disposal, it was necessary to categorize the wastes on the basis of the length of time the material would remain hazardous. In the new classifications, category 1 wastes would have radionuclide concentrations that would decay to innocuous levels within about 150 years, category 2 wastes would have radionuclide concentrations that would decay to innocuous levels within about 500 years, and category 3 wastes would have radionuclide concentrations that would remain hazardous for very long periods. It was decided that it was reasonable to consider disposing of category 1 and category 2 wastes at Chalk River, but that category 3 wastes would probably have to be sent to the fuel waste repository when one was built in Canada.

Three methods were considered for disposal at Chalk River: (1) an intrusion-resistant underground structure, (2) an improved sand trench, and (3) a shallow rock cavity. The intrusion-resistant underground structure was essentially a concrete box with an open bottom installed above the water table. In the open bottom was a layer of compacted sand/clay mixture that would pass water but adsorb many of the radionuclides. Thus if water gained access to the vault it would drain out but not carry radionuclides with it. The waste would be stacked in the structure and after it was full it would be capped with backfill, concrete, a polythene sheet, gravel, sand and topsoil. The cap and walls would be thick enough to discourage intrusion. The improved sand trench would be similar, except that the walls and cap would be concrete rip-rap or large boulders. It would be cheaper, but not as resistant to water ingress. The shallow rock cavity would be a vault excavated in the rock below the weathered zone (about 50 metres) but above about 200 metres. The vault would be below the water table, but the path for water would be tortuous enough that the water from it would not reach the surface for many centuries. It would be very intrusion resistant but much more expensive than the other options. Since it was considered that the improved sand trench would only provide adequate containment for category 1 wastes, and that the intrusion-resistant underground structure gave adequate security at a lower cost than the shallow rock cavity, it was decided to proceed with a prototype intrusion- resistant underground structure (see figure 19.12). The design was examined in more detail and a three-module arched-wall concept was found to be most cost-effective. The design was completed and a safety assessment was done in a similar manner to the SYVAC assessment for used fuel waste. It was submitted to the Control Board and at the time of writing was waiting for the Board's approval.

One of the major aspects of the research program associated with the disposal of laboratory and similar wastes is the development of methods of sorting them into categories 1, 2 and 3. To attempt to develop such methods, the wastes arriving at the Waste Treatment Centre are being analyzed to determine whether pre-sorting is necessary, and to develop economical methods of analysis to sort the wastes at the treatment centre, if that is necessary. Once such methods are available, they can be used to determine the concentrations of long-lived hazardous radionuclides in the various wastes already in the waste management areas. This information can then be used in a SYVAC type of analysis to determine what further treatment, if any, is required to accomplish disposal. Other elements of the research program are the characterization of possible components of the engineered containment system, such as immobilizing matrix, packaging and backfill materials, improving the knowledge of

FIGURE 19.12
IRUS (Intrusion Resistant Underground Structure) – Post Closure Phase

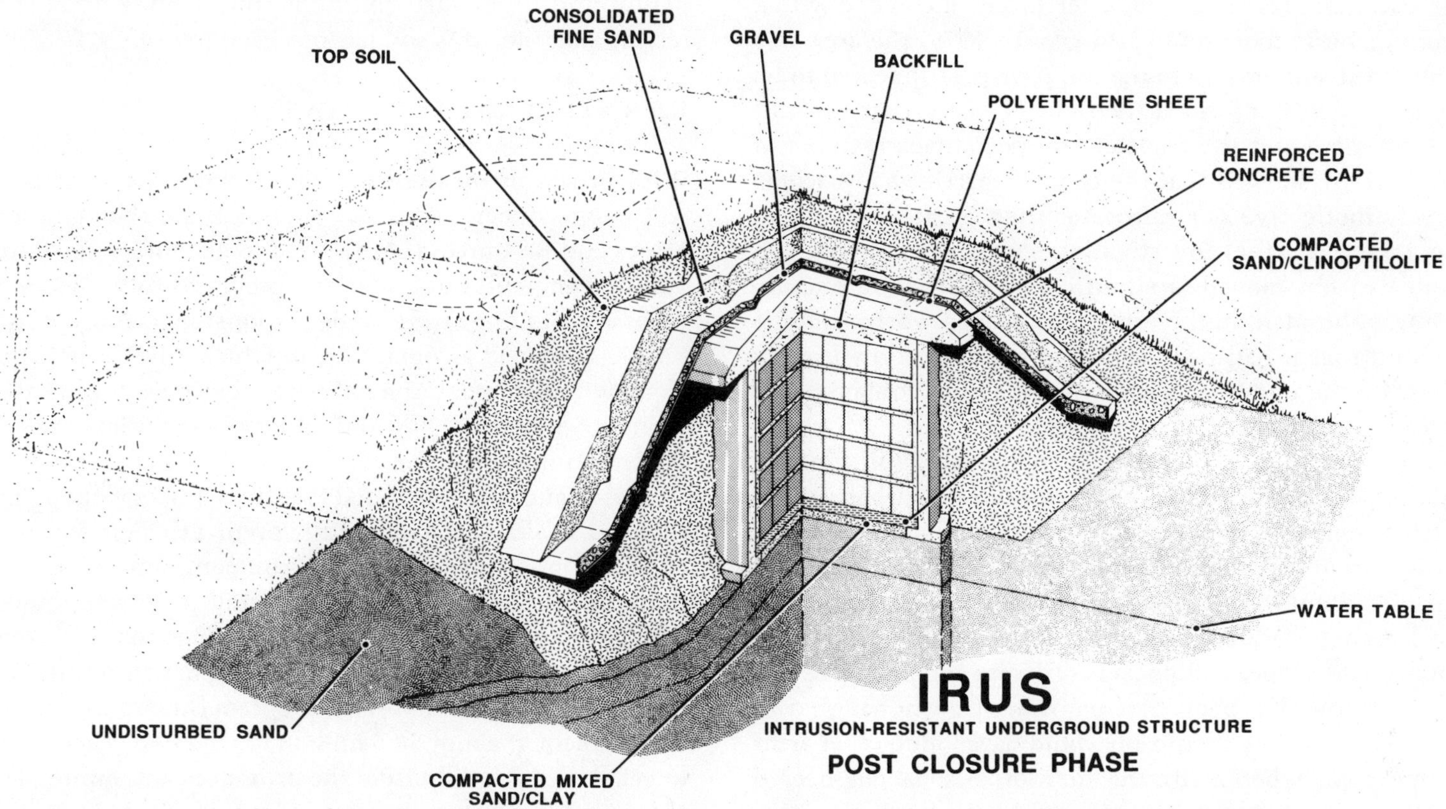

the characteristics of the radionuclide pathways from the disposal facility to people and establishing baseline data for the environs of the disposal area for comparison with data obtained after disposal has taken place.

In 1975, it was discovered that in the forties radioactively contaminated soil from the Eldorado Mining and Refining uranium refinery in Port Hope, Ontario, which was destined for the company's disposal sites, had been diverted for fill in the Port Hope area, and that significantly contaminated soil was widespread in that area. The Control Board, which in 1975 had responsibility for cleanup, asked AECL whether this soil could be disposed of on the Chalk River property, and AECL agreed. AECL subsequently also agreed to accept contaminated soil from niobium smelting operations in Ottawa, and radium dial-painting operations in Toronto. The site selected for the waste was a 1.5 hectare site located in a small valley between a sand ridge and a hard rock outcrop near the divide between the Maskinonge and Chalk Lake basins. It was designated WM Area F (see figure 19.1). The site was lumbered and cleared of slash, and bulldozers were used to spread and compact the contaminated soil in layers about 0.7 metres deep, to a total depth of about 12 metres. When the waste reached a level close to that of the adjacent sand ridge, it was covered with almost 30 centimetres of packed clayey silt to reduce penetration of water,

70 centimetres of sand to prevent erosion of the clayey silt, and 15 centimetres of sandy topsoil. It was then ditched to direct water into the sandy ridge, fertilized and seeded with a mixture of birds-foot trefoil and rye. In 1979, the area was becoming full and no end to the soil transport appeared to be in sight, so AECL informed the Control Board that it could accept no more and the transfer stopped. Investigations in 1980 and 1981 showed that the clay/silt barrier had become relatively ineffective as a rainwater shed because of drying and cracking. Further investigation showed that, while the radium had not moved, a significant quantity of arsenic, which was abundant in the waste, had moved. Fortunately, it was found that it had been collected by iron oxyhydroxide coatings on the underlying sand, and had not contaminated the local groundwater.

In total, over 118,700 tonnes of contaminated soil containing about 14 grams of radium were transferred to Chalk River; that total comprises 95,050 tonnes containing about 3 grams of radium from Port Hope, 22,180 tonnes containing about 10 grams of radium from Ottawa, and 1,530 tonnes containing about 1 gram of radium from Toronto. About 900,000 tonnes remain in Port Hope, and pockets of lesser contaminated soil have subsequently been discovered in Scarborough and Vancouver. Chalk River probably could have found other areas on its property where, with the addition of some engineered barriers, more contaminated soil could have been accepted. However, potential alternative solutions, such as transporting the soil to existing Eldorado waste sites in the Port Hope area – where it would have added only infinitesimally to the radioactive contaminants already present – seemed much more logical and cost-effective, and AECL's decision to accept no more Port Hope soil was at least partly motivated by a desire to force a more logical and cost-effective solution. This was probably a mistake. Shortly after the transfer of contaminated soil to Chalk River stopped, the government transferred responsibility for cleaning up these "historic wastes" from the Control Board to AECL, and AECL quickly found that the alternative "logical and cost-effective" solutions foundered on the emotional and highly political issue of one jurisdiction accepting another jurisdiction's wastes. Since that time, millions of dollars have been spent on study commissions, public participation and temporary solutions, but a publicly acceptable resting place for this soil has not been found.

URANIUM MINING WASTES

With programs on used fuel and low-level wastes in hand, AECL, in concert with the federal Department of Energy Mines and Resources (EMR), tried to encourage the Canadian mining companies to undertake programs to verify that uranium mine tailings were being responsibly managed. In 1981, EMR established a committee, in which AECL participated – along with representatives of EMR, the Ontario and Saskatchewan governments, and the mining companies – to outline action required on mine wastes.

This committee concluded that there was not enough scientific data available to confidently predict the long-term effects of uranium tailings on the environment, and recommended that a research program be conducted to fill the knowledge gap. The Government of Canada responded to this recommendation in 1982 by announcing a National Uranium Tailings Program. This was a five-year program funded by the Federal Government at about $8.5 million for the five years. Attempts to get shared funding from the provinces and mining companies failed, so the resources were limited to those provided by the Federal Government. A small group was set up within EMR to direct the work, which was all done on contract. Throughout the program, plans and progress were reviewed by two groups: a technical advisory committee (TAC), with representatives from five uranium mining concerns, three provincial governments and five federal institutions (AECL being one of the five), and a senior review board, with representatives from the Ontario Ministry of Environment, the Saskatchewan Ministry of the Environment, the Control Board and EMR.

The focus of the work was the development of a probabilistic predictive model, using the SYVAC approach, that would allow a comparison of the long-term implications of various tailings close-out strategies. The development of this model required a fundamental understanding of the scientific and engineering principles involved in contaminant migration, and some initial research work was done to provide this basic information. A preliminary model was then developed and tried for a hypothetical reference tailings site. Output from this was then used to further define information required, and the iterative process continued until the end of the program. Model development was the responsibility of SENES Consultants, to whom AECL acted as a sub-contractor. The model was designated UTAP, for Uranium Tailings Assessment Program. A particular concern of the TAC, particularly the industry members, was that the outputs from the predictive model would not be accepted by the regulatory authorities for licensing purposes. With the agreement of the provincial regulatory agencies, the Control Board was asked to represent the regulatory agencies on a modelling review group that reviewed and contributed input into UTAP, to ensure its understanding and acceptance by the regulatory agencies.

Following its development, UTAP was used to predict exposures to five different receptors for five different close-out alternatives at a reference site in the Elliot Lake area of Ontario, and used for two different receptors for four different close-out alternatives at a reference site in the Rabbit Lake area of Saskatchewan. The two reference sites were significantly different, in that the Elliot Lake tailings contain large quantities of pyrite, which over time oxidizes and produces sulphuric acid, while the Rabbit Lake tailings are primarily silica, gypsum and clays. The close-out alternatives considered for the Elliot Lake tailings were: (1) no remedial action, (2) waste-rock cover (0.6 m of crushed rock), (3) pyrite removal (cover by 2.0 m of depyritized tailings plus calcite), (4) glacial till cover (2.0 m), (5) water cover. The close-out alternatives for the Rabbit Lake tailings were: (1) no remedial action, (2) sandy till, top soil and vegetation covers, (3) soil, bentonite barrier, top soil and vegetation covers, and (4) waste rock cover. The results showed that the only treatment that reduced exposures relative to no treatment was water cover at the Elliot Lake site. However, in no case did the exposures exceed 1 percent of that from natural background.

At the termination of the program, the UTAP model was turned over to the Control Board for maintenance and general use by both the regulatory bodies and the industry.

SIGNIFICANCE OF THE WASTE MANAGEMENT TECHNOLOGY

Radioactive wastes must be safely managed. This imperative has resulted in the development of a technology and methodology which, if applied to other hazardous materials produced by industries, could provide significant financial return on the investment that has been made. The SYVAC type of analysis is not only applicable to radionuclides; it can be used to predict the environmental consequences of any disposal method for any contaminant from any industrial process. It forces one to quantify the biological hazard of the contaminant, consider all potential pathways for exposure, quantify the parameters affecting release, distribution and uptake and quantify the potential environmental effects. It allows the hazard to be quantified prior to process implementation, so that attention can be paid to contaminants that have a potential significant environmental effect.

BIBLIOGRAPHY

Beamer, N.V., W.T. Bourns, L.P. Buckley, W.H. Pettipas, K.A. Burrill, D.F. Dixon, and D.H. Charlesworth. *Studies of Reactor Waste Conditioning and Disposal at CRNL.* AECL-7551. 1982. (A compilation of papers presented at the Second Annual Meeting of the Canadian Nuclear Society, Ottawa, June 1981.)

Dixon, D.F., comp. *A Program for Evolution from Storage to Disposal of Radioactive Wastes at CRNL.* AECL-7083. 1985.

Dyne, P.J. *AECL's Responsibilities and Programs for Management of High Level Radioactive Wastes.* AECL-5111. 1975.

———. "Waste Management in Canadian Nuclear Programs." In *Canadian Nuclear Association Fifteenth Annual International Conference, Ottawa, 15-18 June 1975*, vol. 5, 3-43. Toronto: Canadian Nuclear Association, 1975. (Also AECL-5249.)

John, R.D. *National Uranium Tailings Program, 1982-1987: The Final Result.* Ottawa: CANMET, n.d.

McAuley, S.J. *A Safety and Hazards Analysis of the Chalk River Nuclear Laboratories Waste Management Areas.* AECL-Misc-306 (draft). 1985.

Ohta, M.M. *The Concrete Canister Program.* AECL-5965. 1978.

Wuschke, D.M., et al. *Second Interim Assessment of the Canadian Concept for Nuclear Fuel Waste Disposal.* 4 vols. AECL-8373. 1985.

ADAPTING TO A CHANGING ENVIRONMENT

Chapter Twenty

Business Development, Revenue Generation and the Impact of a Changing Environment

R.G. HART

OVERVIEW

At the time the fledgling nuclear research division of the National Research Council moved from Montreal to Chalk River in 1946 it did so in an atmosphere of enthusiastic support for science and technology in general and nuclear technology in particular. People had seen the atomic bomb bring an abrupt end to the second world war and, spurred by some rather science-fiction-like stories, were eager to see what applications might be developed for use in a peaceful world. The Chalk River laboratory was given no specific mandate to pursue any particular application but it was clear that C.D. Howe, the federal minister responsible for NRC, expected that practical peaceful applications would be developed and that he wanted such applications to be commercialized by private industry. In his view it was entirely appropriate that government should fund research and development and make any commercial opportunities which arose freely available to private industry for exploitation. To this end, when the crown corporation, AECL, was formed in 1952, he made sure that its board of directors was largely drawn from potential users; private business and the provincial utilities. Howe's views prevailed for the next two decades which saw the development of a flourishing radioisotope business, the development of the CANDU nuclear electric generating system and its adoption by Ontario Hydro, the development of a comprehensive nuclear power station design capability and the development of a self-regulating nuclear research reactor (SLOWPOKE) and its adoption by several Canadian universities. AECL made no attempt to negotiate royalties for these developments.

In the early seventies, the atmosphere began to change. Spurred by environmental and peace activism, a substantial fraction of the public became suspicious of science and technology in general and nuclear technology in particular. Their concerns focused on the safety of nuclear power stations, nuclear waste management and the potential for nuclear weapons proliferation through the export of power reactor and fuel cycle technology. With public disenchantment came political disenchantment and the expected expansion of the CANDU power system both within Canada and abroad was curtailed.[1] At the same time, the federal deficit was increasing and the user-pay principle was coming into favour. The Pickering generating station was working well and AECL began to get political feedback that the reactor development job was completed and that the bulk of government funding should be shifted to the areas of public concern – safety, waste management and non-proliferation. Further, the argument went, if more reactor development work was needed it should be paid for by the nuclear utilities whose customers were then benefiting from the low cost of nuclear electricity relative to that generated from coal. The utility executives on the AECL board, who in Howe's time were considered essential, were now considered to be in conflict of interest and were replaced by senior government bureaucrats. AECL knew that the reactor development job was not done and that technical problems could be expected down the line; hence, while it transferred a substantial amount of effort to the areas of public concern, it retained a substantial effort on reactor development and tried to supplement the government funds retained in this area with revenue from other sources. Since it shared the view that the beneficiaries of nuclear power should now be bearing more of the development costs, it pursued Ontario Hydro for some of this revenue. As a result, Ontario Hydro started to share, on a fifty-fifty basis, the cost of specific parts of the program that it thought were underfunded. The first programs to benefit from this arrangement were those on verification of reactor safety codes, long-term behaviour of zirconium alloy pressure tubes and behaviour of CANDU fuel.

In 1978 AECL was reorganized and the need for increased reliance on commercial revenue was reflected in the new structure. The various divisions of AECL were organized into five more or less independently operating companies, AECL Radiochemical (formerly Commercial Products), AECL Engineering (formerly Power Projects), AECL Chemical (formerly Heavy Water Projects), AECL International (formerly International Marketing) and AECL Research (formerly Chalk River Nuclear Laboratories and Whiteshell Nuclear Research Establishment), all reporting to an AECL Corporate Office. In AECL Research a business development office was established at its head office in Ottawa and commercial operations offices were established at each site. The mandate of these offices was to obtain commercial revenue that would assist in support of AECL's program objectives. A restriction was that they were not to accept work in direct competition with Canadian industry. In 1982 the AECL International, AECL Chemical and AECL Engineering companies were combined to form CANDU Operations. Over the decade following the reorganization AECL Radiochemical prospered with gross revenues increasing from $29 million in 1977/78 to $111 million in 1986/87. CANDU Operations fell on hard times with gross revenues falling from a peak of $448 million in 1980/81 to $88 million in 1986/87. AECL's two heavy water plants at Glace Bay and Port Hawkesbury had to be shut down and the staff level had to be reduced from about 3800 to 1400. This downsizing was done in such a way as to maintain the core capability to design CANDU stations when orders picked up. At the time of writing orders for three new CANDU units had been received from the Republic of Korea and CANDU Operations' (now AECL CANDU) fortunes were improving. In the Research company commercial revenue

[1] In the early seventies both BC Hydro and Manitoba Hydro seriously considered nuclear power; BC Hydro for Vancouver Island and Manitoba Hydro as an alternative to further development of the Nelson River. In both cases, public opinion played a significant role in its rejection. Also, AECL was restricted in some promising foreign markets because of Canadian government proliferation concerns.

increased from $8 million or 7 percent of total budget in 1977/78 to $41 million or 18 percent of total budget in 1986/87.

By 1979 Ontario Hydro's contribution to shared-cost research programs was approaching $2 million and the partners felt that some mechanism should be put in place to insure that the shared-cost funds were used to further the programs with the highest priority. Accordingly an agreement called CANDEV (for CANDU Development) was negotiated. A series of jointly staffed working parties, covering the spectrum of CANDU technologies, recommended programs to a steering committee composed of senior AECL and Ontario Hydro technical staff and chaired by a senior Ontario Hydro manager. The available funds were allocated by the steering committee and a full-time CANDEV manager coordinated the input, followed the progress of the programs and reported back to the steering committee. The arrangement was useful, not only in administering the shared funding, but also in ensuring user input into all of AECL's CANDU development programs and user access to the results. In 1983 AECL and Ontario Hydro succeeded in bringing the other Canadian nuclear utilities, Hydro-Quebec and New Brunswick Power, into the shared-cost program and the program was re-named COG (for CANDU Owners Group). Hydro-Quebec and New Brunswick Power contributed to programs in which they had an interest, in proportion to the number of CANDU reactors in their system relative to the total number in Canada. By 1986/87 the utilities were contributing about $8 million in shared-cost funding and about another $3 million in fully commercial contracts to the AECL CANDU development programs.

Canadian utility contracts were not the only, or even the major, source of AECL Research's commercial revenue. The growth of the various components of this revenue is shown in figure 20.1. In the early eighties much of the revenue came from CANDU Operations, primarily for computer rental and heavy water upgrading. However, by 1986/87, this had been largely replaced by revenue from research contracts with Canadian and foreign utilities, industries and institutions.

FIGURE 20.1
AECL Research Commercial Revenues; Constant 1986 $s

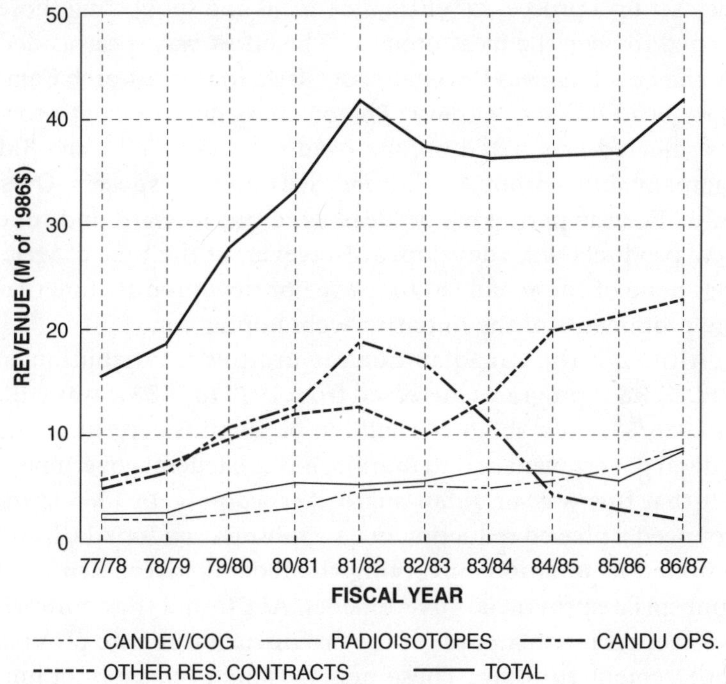

Thus by 1986 a significant fraction of AECL's capability to respond to problems in the Canadian nuclear industry was being maintained by contracts outside that industry.

When AECL was reorganized in 1978, senior managers felt that, to respond to the new environment in Ottawa, they should not only attempt to increase the company's reliance on commercial revenue but should also try to produce new industries from the company's stock of technology. They sensed that, although there was still a good deal of support resulting from the development of the radioisotope business and the CANDU reactor, there was beginning to be something of a feeling "but what have you done for me lately?" Management felt that new industries could be developed from the wealth of technology available within the company if they could only encourage the staff to look for such opportunities.

Accordingly they instituted a program of incentives for the identification of ideas which could lead to new industries and they set up a process for evaluating ideas and supporting those judged to have the most promise. The effort was spearheaded by the new Business Development Office in the Research Company head office in Ottawa. The business development process selected was based on one used successfully by the 3M company, but within AECL it met with limited success. Over an eight-year period many ideas were put forward and nine new products were developed. However, at the time of writing, none of these had led to the major new product line that the proponents of the initiative were hoping for.

Although the Canadian nuclear utilities' contribution to AECL's R&D programs increased from 1977 to 1986 it was still, in 1986/87, only about $11 million or about 5 percent of the federal government's contribution and the federal government felt that this was an unfair share. Accordingly, in 1986 it announced a phased reduction in its appropriation for AECL and initiated negotiations involving the federal government, the appropriate provincial governments, AECL, and the appropriate utilities to attempt to get the provincial utilities to provide replacement funding. These negotiations eventually culminated in agreements whereby AECL's base funding for R&D would remain at about $220 million (1990), escalated according to the consumer price index, through fiscal 1996/97 with the federal government providing 65 percent of the funding and the nuclear utilities 35 percent. The utilities contribution was to be made through COG and COG ground rules were to apply to its use. This meant that AECL had to match COG dollars fifty-fifty and that the COG steering committee effectively controlled 70 percent of AECL's program. Some AECL programs, including the new industries initiative, were casualties of this arrangement but adjustments were made and the arrangement worked reasonably well. AECL was of course free to increase its outside commercial revenue as it was able and, as a result, it had one of the largest commercial R&D operations in Canada.

The events outlined in this overview provide a wealth of interesting stories, a few of which are related in the remainder of this chapter.

BUILDING THE RADIOISOTOPE INDUSTRY

With the startup of NRX in 1947, scientists at Chalk River had the best vehicle in the world for large-scale production of artificial radioisotopes and they soon developed methods for their production and processing into forms that could be used for radiation therapy, radiation sterilization and a variety of tracer experiments, particularly in the medical research and life sciences fields. By 1951, the radioisotope business had grown to a size where transfer to private industry, according to Howe's dictates, could be considered. However, at the time the only Canadian industry with any experience in the radioisotope business was the crown corporation Eldorado Mining and Refining which had established a Commercial Products division to sell its by-product radium-226. Thus in 1951 an arrangement was made to have this division take over the marketing and distribution of the radioisotopes being produced at Chalk River (see chapter five). When AECL also became a crown corporation in 1952, AECL and Eldorado agreed that in the future the Commercial Products division would have much more in common with AECL than with Eldorado so the division transferred to AECL as the Commercial Products division of AECL. Over the period 1952 to 1955, the division established facilities in Ottawa for processing most of the radioisotopes to their final product form and the technology for this processing was transferred from Chalk River. Only the production, the more highly radioactive processing, and the waste handling remained at Chalk River. Chalk River was paid for its part of the work on a marginal cost basis,[2] while the profits accrued to the new division. This caused some considerable chagrin to some of the Chalk River scientists involved who were concerned that the R&D budget would end up subsidizing a major commercial activity. Later events would show that their concern was justified.

2 Marginal costing is a practice often used in industry whereby the main product, in this case R&D, bears the total overhead cost of the operation while the by-product, in this case radioisotope production, bears only its own direct cost.

In the early years of isotope production, when the demand was small, the production could be accommodated in the Chalk River reactors without difficulty. However, as isotope sales increased, isotope production increasingly dictated not only the use of research reactors but certain waste management activities as well. Accordingly the Research Company sought to change the basis of payment from marginal costing to full cost recovery and the urgency of this increased when initiatives were taken to privatize the Radiochemical Company (formerly the Commercial Products division). Subsequently, when the Radiochemical Company was privatized, an agreement based on full cost recovery for production and separation was achieved. However, the Research Company did not succeed in obtaining agreement on full cost recovery for waste handling and is facing substantial costs for waste treatment facilities, which it cannot recover.

BUILDING THE CANDU POWER STATION INDUSTRY

In the mid-fifties, when AECL and Ontario Hydro wanted to build a demonstration CANDU power station, the 20 MWe NPD station, the commercialization-by-industry philosophy was again attempted. The arrangement made was that Canadian General Electric Company (CGE) would supply the Nuclear Steam Supply System (NSSS) on a cost-plus, turnkey basis[3] and would make a $2 million contribution toward the cost; AECL would pay for and own the NSSS, and Ontario Hydro would pay for and own the balance of plant, operate the plant and purchase the steam from the NSSS at a price equal to the cost of steam at the J.L. Hearn coal-fired plant in Toronto. Any profits or losses on the NSSS part of the plant were to accrue to AECL. The idea was to have CGE become a turnkey supplier of CANDU generating stations, or at least the NSSS part of them, to the Canadian and foreign market. However, after NPD, Ontario Hydro made it clear that, for their stations, they did not want to be tied to a single supplier; hence, when the 200 MWe prototype station, Douglas Point, was built the arrangements were substantially different. AECL was persuaded to establish its own engineering division, Power Projects, to design the NSSS while the design of the balance of plant was contracted to Ontario Hydro. Further, AECL contracted construction management, construction, and subsequent operation to Ontario Hydro. AECL owned the station and Ontario Hydro agreed to purchase electricity from it at a price equal to the cost of generation at its Lakeview coal-fired plant in Toronto. Any profits or losses accrued to AECL. In subsequent CANDU stations built by the utilities, this design and construction pattern was generally followed except that, for stations outside Ontario, the balance of plant was generally subcontracted to CANATOM,[4] and the construction management of those stations was done by AECL. For a time CGE attempted to develop a business as a turnkey supplier of CANDU stations in the foreign market but without ongoing at-home business this proved impossible. Eventually they evolved into a role as a special systems designer and special components and fuel manufacturer and in these roles made a major contribution to the developing industry.

Making arrangements for design, construction and operation of prototype CANDU stations was only the start of the work involved in industrializing CANDU technology. The station is made up of a myriad of pumps, valves, pipes, heat exchangers, seals, fittings, instruments, all of which had to be made to a quality that was essentially foreign to Canadian manufacturers and critical components such as fuel, pressure tubes and calandria tubes had to be made from materials Canadian manufacturers had never even seen. To complicate matters the utilities wanted to have at least two qualified suppliers for each component so that they could have competitive bidding and the market for products of the quality required was, for the most part, not large. Industries were prepared to

---

[3] A turnkey supplier is one who takes a contract for supply of the complete system including design, construction and construction management. In the case of NPD, CGE did the design and construction management but contracted the construction to Canadian Bechtel.

[4] CANATOM is a private engineering company based in Montreal and jointly owned by several large Canadian engineering firms.

get involved but were not prepared to put out much of their own money to develop the new capabilities. Thus much of this capability had to be developed through development contracts. For example, a development contract would be let to a manufacturer to fabricate an experimental fuel assembly for testing in an NRX loop and laboratory scientists would work with the manufacturer to develop the technology required. These development contracts were often complemented by the attachment of manufacturer's staff to the laboratories to follow performance first-hand. In this way the capability to produce nuclear quality fuel, coolant tubes, calandria tubes, end fittings, valves, pumps, instruments, and a myriad of other components was developed in Canadian industries. That the method was successful is attested by the fact that over 90 percent of the cost of Canadian CANDU power stations is Canadian content.

THE SLOWPOKE RESEARCH REACTOR

Because of its large area and sparse population, Canada has a need for economical off-grid electricity and remote steam production. In 1964 AECL established a small group at Whiteshell to see if a small nuclear reactor could be developed to meet this need. The competition was packaged oil-fired boilers and the cost of the competition was determined by the cost of transporting the oil. Early in the study it became apparent that every practical innovative idea would have to be used if nuclear reactors were to meet the competition, with the most radical departure from conventional practise being completely automatic operation. At the size of reactor required to meet the bulk of the need, even one operator per shift would make the cost of the nuclear generated energy uncompetitive. After about eighteen months of study it was concluded that reactors could be developed to meet the competition in the more remote areas of the country but that the size of the market did not justify the development costs that would be required. Thus the group was disbanded.

However, some of the concepts lived on. A Chalk River physicist who had been attached to the group, returned to Chalk River with an idea for a research reactor which could operate unattended and be used by universities, hospitals and research institutes for teaching, activation analysis[5] and the production of small quantities of short-lived radioisotopes.

In early 1967, experiments at Los Alamos on critical masses had shown that when a uranium-235 polyethylene mixture was surrounded by a thick beryllium reflector the critical mass was surprisingly low. It was reasoned that if a beryllium reflector could be used to control the excess reactivity in a water-cooled core, a reactor could be designed in which any conceivable power transient would be self-limited by the negative temperature and void coefficients of the core. A reactor with such intrinsic safety features could be licensed for unattended operation even in large population centres where the universities, hospitals and research institutes were located. The design was developed through 1968 and 1969, and in 1970 the first prototype commenced operation at Chalk River. The reactor was named SLOWPOKE for Safe LOW POwer Critical(K) Experiment (see figure 20.2). At the maximum rated power of 20 kW, the reactor provided thermal neutron fluxes of 1×10^{12} and 4×10^{11} n/cm^2/sec to sample sites in, and just outside, the beryllium reflector respectively. During 1970, the prototype reactor was put through its paces at Chalk River and it performed as expected. Then in early 1971, the reactor was dismantled and transferred to the Environmental Science and Engineering Institute of the University of Toronto where it was reassembled in a renovated building adjacent to offices, laboratories and a library. On 4 June 1971 the University of Toronto received AECB operating licences which subsequently included approval to operate unattended for up to 18 hours. Remote surveillance was provided by an alarm in the building security console. Because of the intrinsic safety features of the system, i.e., no fuel failures even with complete loss of control, licensed operators with formal reactor training were not required. Instead "licensed

[5] The sample is irradiated with neutrons and the radioactive products are measured by beta- or gamma-ray counting. High sensitivity is possible for many elements (see chapter seven).

users" were allowed to start up and shut down the reactor, operate the water purification and gas purging systems and irradiate samples. However, only an authorized reactor engineer was allowed to adjust the beryllium reflector and one from AECL visited the University of Toronto every six to twelve months to perform this task.

Within AECL, the right to sell SLOWPOKE research reactors was granted to the Commercial Products group and in May 1971 a second SLOWPOKE reactor went into operation at their facilities in Ottawa. Commercial Products used their SLOWPOKE to provide a commercial irradiation service and to demonstrate the reactor to potential customers. A total of seven SLOWPOKE research reactors were eventually sold, five to Canadian universities, one to the University of the West Indies in Jamaica, and one to the Saskatchewan Research Council in Saskatoon.

FIGURE 20.2
A SLOWPOKE-2 Reactor Installation

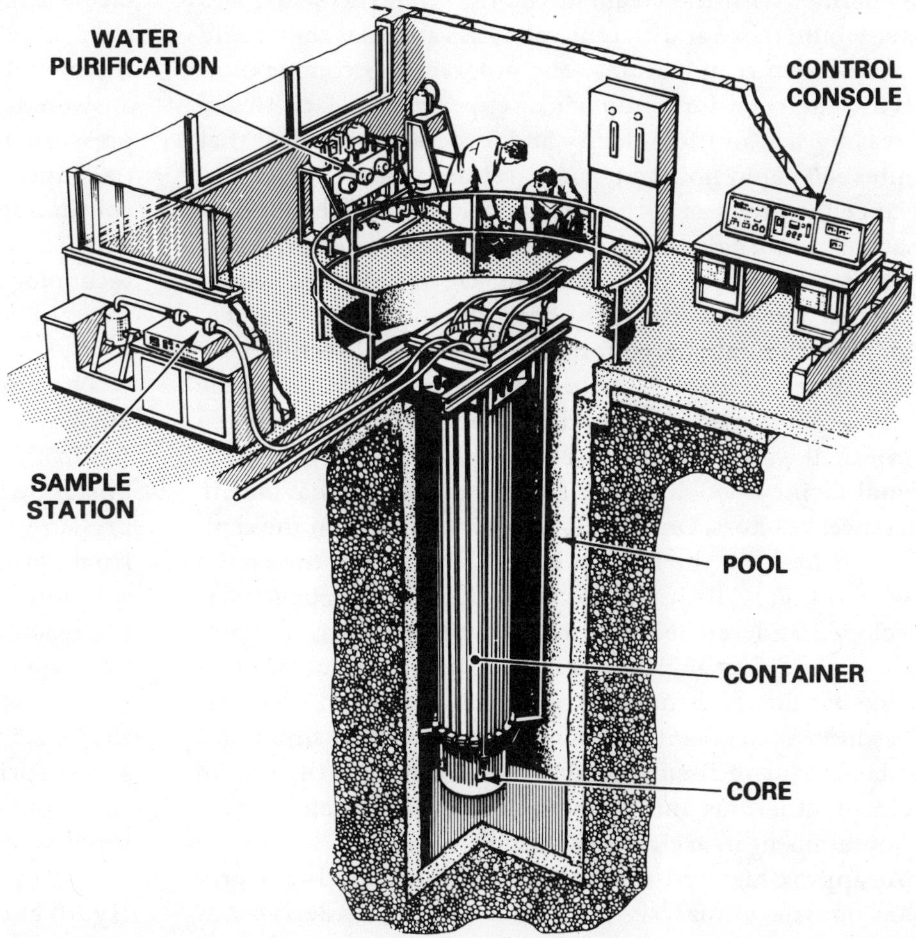

SOME INTERESTING SHARED-COST PROGRAMS

One of the first shared-cost programs initiated by AECL and Ontario Hydro was aimed at improving the computer codes used to predict the behaviour of the primary and emergency coolant in the event of a pipe break in the primary heat transport system of a CANDU power station (see chapter fifteen). In the early codes used to analyze such accidents the simplifying assumption was made that the steam and water produced in the accidents were uniformly distributed in the fuel channels and piping, moved at the same velocity and were of equal temperature, with the result that all fuel elements in the fuel channel were cooled at the same rate. To a number of scientists in both AECL and Ontario Hydro this assumption seemed to defy the laws of gravity, so when the CANDU-OCR program was terminated, Whiteshell was encouraged to establish a small group to investigate the problem. When experiments commissioned at Westinghouse Canada in 1973 verified that the steam and water would indeed separate in the channel causing uneven cooling of the fuel, the group at Whiteshell embarked on a program to develop a computer code that would take into account the fact that the water would settle to the bottom of the channel with the steam at the top, that the steam and water would move at different velocities and that they could be at different temperatures. The program was complex, involving the need for a number of experimental facilities of increasing size and complexity and it soon became clear that Whiteshell could not, on its own, release sufficient effort to advance the program at the rate Ontario Hydro felt was necessary. Accordingly an agreement was negotiated whereby the effort would be increased and Ontario Hydro would share the cost on a fifty-fifty basis. Included in the agreement was a rental charge for the experimental facilities required to verify that the mathematical representations of the physical phenomena observed were accurate. Indeed when Whiteshell was unable to raise the capital for a large experimental facility which would better represent behaviour in the power reactors, Ontario Hydro contributed half the capital to ensure that the facility would be built. In the end, a new code that realistically described the LOCA sequence was developed and verified. Subsequently the cooperative program in the safety analysis area was expanded to obtain better models for the behaviour of fuel channels and calandria tubes in loss of coolant/loss of emergency coolant situations, the behaviour of fission products released from the fuel in accident situations and the behaviour of hydrogen released to containment in accident situations.

At approximately the same time as the shared-cost program on safety analysis codes was getting under way at Whiteshell, Ontario Hydro experienced cracks near the end fittings in the Zr-2.5Nb pressure tubes in Pickering 3 and Pickering 4, which eventually led to the need to replace sixty-nine tubes in these reactors (see chapter twelve). This was a serious economic problem for Ontario Hydro so, in order to see the cause and solution determined as quickly as possible, they shared in the cost of the investigative program. Although the cause and a relatively easy solution were determined fairly quickly, the incident left a nascent worry in the minds of both AECL and Ontario Hydro people that perhaps our understanding of the long-term behaviour of this critical component was not as good as it should be. In fact at a meeting between senior AECL Research and utility staff in 1979, when asked what the three major concerns of the operating staff were, the Ontario Hydro representative answered, "#1, pressure tubes, #2, pressure tubes and #3, pressure tubes".[6] With this kind of feedback, AECL substantially increased its allocation to the fuel channel program and continuing financial support was provided by Ontario Hydro. The objective of the program was to put in place the technology that would allow the behaviour of the pressure tubes to be confidently predicted for the projected life of the reactor.

Then on 1 August 1983, a major rupture occurred in a Zircaloy-2 pressure tube in Pickering 2 (see chapter twelve). Again the cause was quickly traced, this time to hydride blisters which had formed at a line of contact between the pressure tube and the calandria tube, the contact resulting from out-of-place garter springs. In this case, however, the solution was not easy. It involved the complete retubing of the reactor, and nagging questions remained as to why the hydrogen concentration in the tube was higher than expected, whether the concentration would be high enough in the Zr-2.5Nb tubes to cause the same problem,[7] whether the garter springs in other reactors were out of place, if so how they could be relocated and how the tubes could be monitored to determine if they needed to be replaced. This resulted in an increase in shared-cost funding from Ontario Hydro and prompted the other Canadian nuclear utilities to get involved as well. After intensive investigation it was determined that many garter springs were out of place in all reactors and that the Zr-2.5Nb pressure tubes were susceptible to the same problem, although much later in life because

[6] L. Woodhead, Private communication.

[7] Only two of the commercial CANDU reactors built had Zircaloy-2 pressure tubes. All the rest had tubes made from Zr-2.5 percent Nb.

of the lower rate of hydriding. A method was developed for non-destructively monitoring the hydrogen concentration in the tubes so that it could be determined when they were approaching the danger point. Further, a joint CANDU Operations/Ontario Hydro Research program resulted in the development of equipment that could relocate the garter springs to their proper positions in existing reactors. For new reactors a design change was made to prevent the garter springs from moving.

SOME INTERESTING COMMERCIAL CONTRACTS

One of the critical parts of the CANDU program was the development of quality pump seals for CANDU reactors (see chapter fourteen). Because of the high cost of down-time for seal replacement the target was set that the seal should have an operating life of five years although existing seals were lucky to have a life of one year. The challenge was to maintain a constant, very small (0.25-2.5 micron) gap between a rotating face on the shaft and a fixed face on the housing for the five-year period. Using computer modeling to determine how the geometry could change with time as a function of temperature, pressure and material properties Chalk River engineers were able to develop a seal for CANDU pumps that met the target life. More importantly, since they had developed an understanding of why seals failed and why they did not, they were able to apply their technology to other pumps in other applications. By the late seventies many American utilities were having serious problems with pump seals and in response to the commercial initiatives of AECL Research the Chalk River team moved to sell their expertise in the United States. In the next few years, they obtained commercial contracts for over $18 million and resolved pump problems for Niagara Mohawk, General Public Utilities and Louisiana Power and Light. In 1986, the United States space shuttle Challenger blew up shortly after launch with the loss of seven lives. The problem was diagnosed as leakage of hydrogen fuel from an O-ring type seal at the base of the booster rocket. The development and testing of a new O-ring was contracted by NASA to the American company Morton Thiokol. As part of the pump seal development program Chalk River engineers had developed a considerable expertise in rubber O-ring performance and Morton Thiokol engaged Chalk River as a subcontractor to advise on the re-design and perform tests in their test rigs. In recognition of their contribution, two Chalk River scientists were invited to the first space shuttle launch in which the new O-ring was used. NASA has experienced no further difficulties from this source.

Over the period 1982 to 1985, the Chalk River biology division received grants totalling $1.4 million from the United States National Institute of Health to assist in the study of families having genetic disorders which made them abnormally sensitive to carcinogens (see chapter three). It was found that in people with a sunlight-sensitive inherited disorder (named xeroderma pigmentosum) the ability of cultured skin cells to repair sunlight damage to DNA was impaired and that the incidence of skin cancer in these people bore a direct relationship to the degree of impairment. In patients whose cultured skin cells were ten-fold more sensitive than normal to the lethal effects of the ultraviolet component of sunlight, the probability of contracting skin cancer at an early age is approximately 10,000 times normal. Subsequent testing showed that even in populations considered to be normal, the ability to repair DNA damage caused by ionizing radiation and certain other carcinogens varied by as much as a factor of five. This indicated that the radiation dose given to people in radiation therapy for cancer should be varied depending on their DNA repair capability and suggested that perhaps certain members of the population should take more care than others to avoid carcinogens. In 1987, a contract for $700 thousand was let by the AECB to the Chalk River biology division to assist in the development of a simple and economical test for DNA repair capability. The work was largely completed by 1989. The test involved taking a blood sample, separating and culturing the white blood cells, exposing a known number to a fixed dose of radiation, then periodically counting the number of cells to determine the rate of recovery. At the time of writing, it had not been decided how, or if, the test would be used in the general population, but considerable effort in other laboratories in the United Kingdom and the United States is being devoted to its applicability to radiation therapy for cancer.

In the early eighties the German particle physics community was attempting to obtain international participation in the construction and utilization of a large $0.5 billion, electron-proton colliding beam facility (HERA). The Canadian Institute for Particle Physics was interested in participating and, after some persuasion, the Canadian Natural Sciences and Engineering Council (NSERC) agreed to contribute funding. The arrangement made was that NSERC would pay for certain parts of the facility which could be built in Canada. After protracted negotiations, complicated not a little by a design change to HERA, AECL signed a contract to have its Accelerator Physics Branch at Chalk River provide some of these parts. HERA has three accelerating rings, each one taking the electrons and protons to higher energy, and AECL was to provide the 52 MHz radio frequency (RF) accelerating system for the second proton ring.[8] The value of the contract was $2.1 million. Subsequently AECL accepted a contract for the supply of the 52 MHz RF accelerating system for the third proton ring as well.[9] The value of this second contract was $2.6 million. The systems for the two rings were not identical and each had its unique problems. The basic system design was taken from an existing accelerator operating at similar frequency at the Fermi National Accelerator Laboratory (United States) but for the second ring at HERA it had to be adapted for a larger frequency swing and this took time. Further it was decided to incorporate a high quality ceramic pipe within the cavities to eliminate the need to evacuate the whole cavity. Unfortunately the quality of the first ceramic pipe used was not high enough and delays occurred while a replacement supply of higher quality was located. For the third ring the frequency swing was smaller so this was not a problem but the gap voltage had to be higher making the use of a ceramic pipe impractical and requiring evacuation of the entire cavity with attendant complications. Thus the work took longer than expected, each system taking over two years to complete. However, they did not hold up the commissioning of HERA and at time of writing both were performing to expectation. Because of the delays AECL did not make a profit on the HERA contracts. However, AECL developed technology and contacts that subsequently led to other substantial contracts in both accelerator hardware and other industrial RF applications.

In the mid-seventies, the United States Department of Energy wanted to obtain experimental confirmation of the computer codes used to predict the behaviour of American light-water reactor fuel under accident conditions and they contracted this work to Battelle Northwest Laboratories in Hanford, Washington. Battelle needed to do a number of in-reactor tests on instrumented fuel assemblies under increasingly damaging conditions for comparison against code predictions and NRU was one of the few reactors in the world where such tests could be safely done. Accordingly, a subcontract was let to Chalk River for the use of the U-2 loop in NRU for these tests. In total eleven tests were to be done over a period of about ten years. In the most severe tests, the fuel was starved of coolant and the reactor was shut down. The decay heat was then allowed to drive the temperature to the point where it peaked (around 2,400°C) and started to decrease. This indicated that the chemical reactions taking place had reached a point where they were limiting themselves without manual interference. Provisions were made to collect the hydrogen and fission products released in the test in order to check the code predictions of the mix of material released to power reactor containment after such an accident. After the

[8] The system includes accelerating cavities, RF supplies and controls.

[9] The third proton ring also includes a 208 MHz accelerating system provided by another contractor. The 52 MHz system in this ring is used to "bunch" the protons properly for injection into the 208 MHz system where they are accelerated to 820 GeV. The 52 MHz system can and, for a year before the 208 MHz system was ready, did accelerate the protons to their full 820 GeV but in this mode the bunch shape is not optimum.

tests, the severely damaged fuel was immobilized in an epoxy resin and examined in the Chalk River hot cells to determine the extent of damage and the extent and products of the reactions between coolant, cladding, fuel, control rod, and support grid materials. At time of writing, ten of the eleven tests had been completed without incident and the eleventh remained to be done. The contract value to AECL was about $14 million plus access to the test information, which was valuable to the Canadian program.

THE NEW BUSINESS DEVELOPMENT INITIATIVE

In 1978, a new business development process was initiated within AECL based on one used successfully by the 3M company (see figure 20.3). Ideas for new businesses were encouraged from anyone within the organization. These ideas were submitted to a New Applications Steering Committee (NASC) and, if the committee considered an idea had promise, it was sent back for business evaluation. This evaluation was done by the proponent and a business development officer and involved an estimate of the potential market and market share and, if necessary, some testing of the technology. After this evaluation, the idea was resubmitted to the Steering Committee where a decision was made as to whether or not the idea still had promise, and if it had promise, whether it could best be exploited as a service to be offered from the laboratory to increase commercial revenue, as a technology to be licensed to an existing firm or as a new business. If the decision was that it should be exploited as a new business, a business opportunity development (BOD) team was established to further develop the technology, confirm the market size, share and entry requirements and prepare a business plan for furthering the opportunity. The business plan could suggest operation as an internal project centre within the laboratory, immediate launch as a spin-off business, or incubation within a business unit. If the suggestion was business incubation and that suggestion was accepted, an AECL-funded business unit would be established which would have to

FIGURE 20.3
The Business Development Process

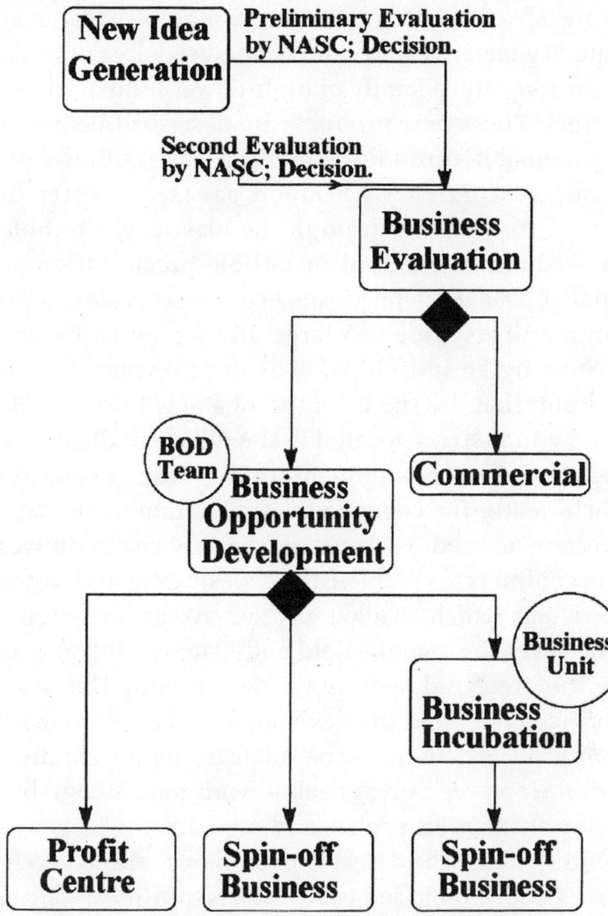

confirm the technology by prototype construction, obtain initial orders from clients and establish business operations. Experience at 3M indicated that only about one idea in 500 would result in a successful new product line but this rate of return had been enough to keep 3M well ahead of its competitors in new product development.

In the eight-year period the initiative was in place several hundred ideas were put forward and nine new products were developed. The nine new products were a personal neutron dosimeter, a line of eddy current probes for boiler and heat exchanger inspection, a new type of research reactor, a self-regulating space heating reactor, an engine wear meter, a steam quality meter, a wetproofed catalyst, a line of internally cooled clothing and a family of high-power industrial electron accelerators. These new products are described below.

The personal neutron dosimeter is a small cylinder of clear plastic supersaturated with Freon gas (see chapter three). When a neutron passes through the plastic, gas bubbles are precipitated and the extent of bubble precipitation is proportional to the neutron dose. Prior to this development, neutron monitors were too large and heavy to be conveniently worn by an individual. The development resulted in the incorporation, by the inventor, of a new business, Bubble Technology Industries, located in the village of Chalk River; it employs about fifteen people. In hindsight the dosimeter was not nearly ready for commercial application at the time the company was formed. Fortunately the new company was able to obtain contracts to complete the development and to produce other products which enabled it to survive its early years.

By the seventies remote-field eddy current inspection had become the preferred method for determining the condition of in-service boiler and heat exchanger tubes, with major applications, in particular, in the nuclear and air-conditioning industries. However a great deal of work remained to be done to develop inspection probes and signal processing systems that could separate the defect signals from extraneous signals. Defect conditions included weld defects, pitting, axial and circumferential cracks and tube thinning caused by wear against the support plates. Extraneous signals could come from tube sheets, support plates, corrosion products or simply probe "lift-off" from the surface. Complications were also introduced by the tube materials themselves, with ferromagnetic materials causing particularly difficult problems. By the mid-eighties the Chalk River group had developed a series of probes (including one for ferromagnetic materials) that could separate the defect signals from the extraneous signals and a business unit within the laboratory was marketing them. The quality of the probes attracted the attention of the United States Westinghouse non-destructive examination unit, ECHORAM, who were looking for a line of probes to complement their signal processing equipment. An agreement was reached whereby AECL would license its probe technology to Westinghouse and Westinghouse would establish a probe manufacturing facility in the local area. This facility was established in the town of Deep River as the Nuclear Services Division of Westinghouse Canada; it employs about fifteen people.

The new type of Research Reactor developed is called MAPLE (for Multi-purpose Applied Physics Lattice Experiment). It can be built in sizes from 1 to about 30 MW (thermal) to meet the program requirements of the customer. A 1 MW reactor could meet the major artificial radioisotope needs of many countries while a 20 MW reactor could perform most of the functions of NRX and be built and operated at a much lower cost than an NRX replacement. One was sold to the Korean Advanced Energy Research Institute but a second, planned for construction at Chalk River, has been cancelled.[10]

The self-regulating space-heating reactor resulted from the upgrading of the 20 kW (thermal) SLOWPOKE research reactor to 10 MW. Although the reactors operate on essentially the same principles, the excess reactivity required to over-ride xenon poisoning in the higher powered heating reactors resulted in a demand for safety systems that were more complex than had been anticipated. This made the reactors uncompetitive with oil and gas at today's prices. The technology has been documented so that it can be resurrected when it is needed (see figure 20.4). A 2 MWt prototype was built at Whiteshell and operated at low power for some time, to confirm the design parameters.

The engine-wear meter monitors iron in oil on-line and can be used to determine when most kinds of rotary machinery require maintenance. This eliminates the need to base maintenance on mileage or hours of usage. The potential market for such an instrument is large, and AECL set up an internal

[10] Subsequently the construction of two MAPLE reactors at Chalk River was commissioned by Nordion International for radioisotope production.

FIGURE 20.4
Proposed 2 MWt SLOWPOKE Heating Reactor

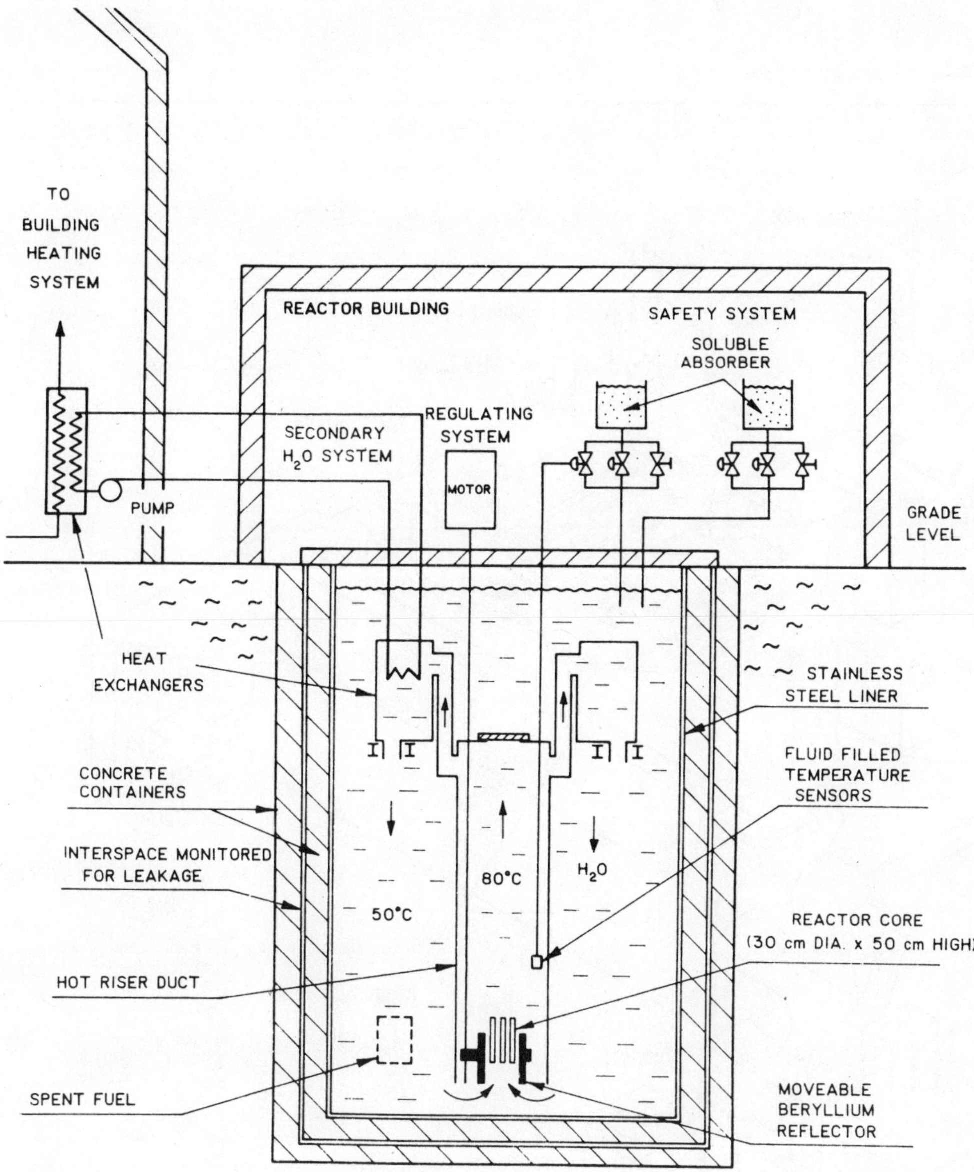

business unit, SENSYS, to try to tap that market. SENSYS succeeded in getting prototype wear meters installed in several applications notably aircraft engines, helicopter rotors, and natural gas pipeline pumps but the operation of the prototypes was not trouble-free and the problems took time to resolve. Thus, in the three years that SENSYS operated, it never obtained the major sale it required to make it profitable. Its assets were sold and the technology was licensed to an Ottawa engineering company, Gastops Inc.

The steam-quality meter measures the amount of entrained water in steam by neutron attenuation. The meters were developed to measure steam quality in reactor safety experiments, but also have application in enhanced recovery operations in the petroleum industry. AECL set up sales offices in both Calgary, Alberta, and Bakersfield, California to market this product but field tests showed that the electronics developed for the reactor-safety application were not suitable for the petroleum application. The sales offices were therefore closed while new electronics were developed.

The wetproofed catalyst was developed to promote hydrogen/deuterium exchange in the liquid phase for heavy-water production but it can also be used to catalyze both oxidation and reduction reactions at low temperatures. Several applications of this property were studied including the recombination of hydrogen and oxygen to prevent the buildup of explosive mixtures in oxygen generators and aluminum batteries, the reduction of nitric oxide in fertilizer manufacturing-plant off-gases and the oxidation of carbon monoxide in a carbon monoxide-monitoring instrument. Batches of catalyst are sold on a continuing basis for these applications and the carbon monoxide-monitoring instrument has been licensed to the Asahi Co. for marketing.

The internally cooled clothing product arose from a search for applications of the wetproofed catalyst. The recombination of hydrogen and oxygen on the catalyst produces heat

FIGURE 20.5
Impela Electron Beam Facility

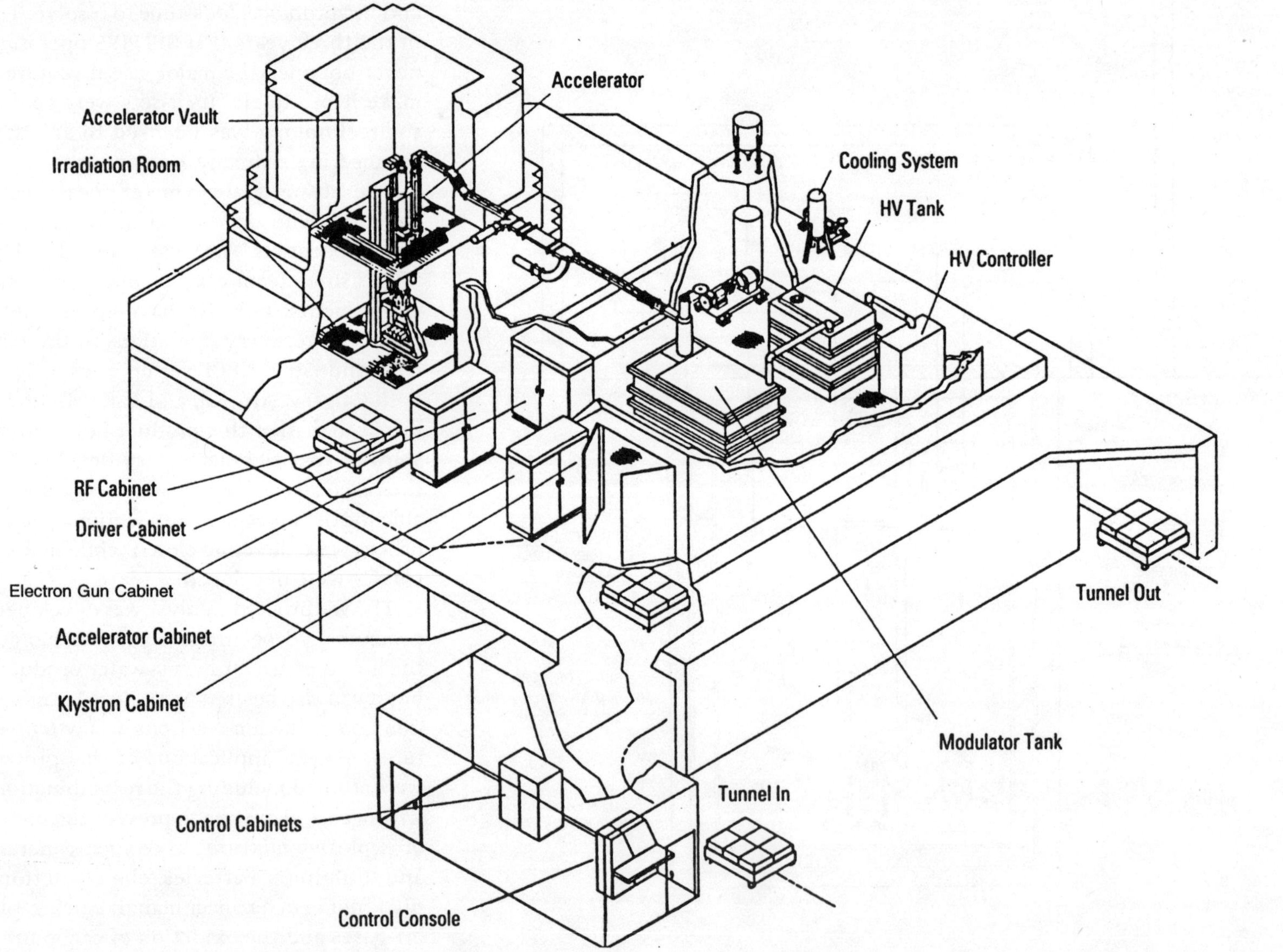

and the original concept was to produce a line of internally heated clothing for underwater and arctic applications. Methods were developed to incorporate small-diameter flexible tubing into clothing to carry the heat transport fluid, and small pumps were developed to circulate it. However, after a time, the two people working on the program recognized that there was likely a larger, and certainly a more immediate, market for internally cooled clothing and that in many applications the cooling could be conveniently provided by an external source such as dry ice or an ice bath. Accordingly they incorporated a new business, EXOTEMP Ltd., in the city of Pembroke to produce internally cooled clothing. Originally they thought the main users might be people in motor sports and people who had medical disorders that prevented them from perspiring, but with the advent of the Gulf war their main market became the military, for personnel who had to wear chemical-warfare-resistant clothing in a hot climate. To meet the demand their staff level increased to about 35 but it has since decreased to less than half that number.

The family of high power industrial electron accelerators is an out-growth of the post-ING transition (see chapter nine) and has been given the name IMPELA (for Industrial Materials Processing Electron Accelerator) (see figure 20.5). The family includes machines with beam powers from 25 to 250 kW and electron energies from 5 to 18 MeV. The power determines the amount of material that can be processed in a given time and the energy determines the penetration of the electrons. The accelerators have advantages over cobalt-60 for industrial irradiation in that they can be turned off and become non-radioactive when they are not in use and they can process much more material per unit time. Applications include existing ones such as sterilization of medical supplies and polymerization of plastics; possible future applications include food irradiation, the breakdown of lignin in paper manufacture, manufacture of composite materials, and sterilization of hospital and airport wastes. The accelerators are being marketed by an AECL business unit located in Kanata, Ontario. The business unit employs about twenty-five people who do the marketing, design and assembly and oversee the installation. The components are all fabricated on contract, mostly to local suppliers.

The 3M process applied within AECL did not produce the major impact product or products its proponents were looking for. Thus when programs had to be rearranged to reflect the new shared-funding agreement with the provincial utilities the initiative was terminated.

BIBLIOGRAPHY

Aikins, E., AECL, and P. Brown, EXOTEMP. Private communication on internally cooled clothing.

Atomic Energy of Canada Limited. *Annual Report*.

Bothwell, Robert. *Nucleus: The History of Atomic Energy of Canada Limited.*

Toronto: University of Toronto Press, 1988.

Campbell, P., and D. Faulkner, AECL. Private communication on the engine wear meter.

Gentner, N., AECL, and D.K. Myers, AECL retired. Private communication on carcinogen sensitivity.

Hilborn, J.W., and G.A. Burbidge. *SLOWPOKE: The First Decade and Beyond.* AECL-8252. 1983.

Hilborn, J.W., AECL retired, and W.T. Hancox, AECL. Private communication on SLOWPOKE reactors.

Ing, H., Bubble Technology Industries. Private communication on Bubble Technology Industries.

Lane, A.D., AECL. Private communication on the U.S. Department of Energy tests.

Lennox, C.G., AECL retired, G.F. Lynch, D. Desjardins, and E. Morrison, AECL. Private communication on commercial revenues.

Lennox, C.G., and J.A.L. Robertson, AECL retired. Private communication on the business development process.

Lidstone, R.F., AECL. Private communication on the MAPLE research reactor, and MAPLE brochures.
Ormrod, J., and G.C. Hanna, AECL retired. Private communication on HERA.
Pothier, N., AECL. Private communication on pump seals.
Sharpe, L.,and D. Angus, Westinghouse Canada. Private communication on eddy current probes, and Westinghouse Canada brochures.
Stirling, A., and J. McKeown, AECL Accelerators. Private communication on IMPELA accelerators, and AECL Accelerators brochures.
Ullyett, Bernard. "Radioisotopes." (Chapter five of this book.)
Woodhead, L., Ontario Hydro. Private communication.

A RETROSPECTIVE

Chapter Twenty-one

Learning from History

J.A.L. ROBERTSON

INTRODUCTION

Canada's emergence as a major industrialized country coincided with the early period of this history. In 1950, North America was beginning to enjoy the fruits of massive wartime industrialization converted to peacetime purposes. For the next two decades Canada's booming economy allowed impressive expansion of the social programs, notably health-care and education. However, by 1970 the manufacturing sector was in relative decline, with exports of raw materials increasingly paying for imports of manufactured goods. Now, in the nineties, Canada is facing economic competition not just from the United States and Europe but even more from the Pacific Rim countries: virtually all consumer electronics and many automobiles are imported, while the aviation industry is in serious trouble.

Canada's future will be crucially dependent on developing and nurturing technically sophisticated industry, but there is little agreement on how this can be achieved. Many people believe that science and technology (S&T) is an essential ingredient, but again there is little agreement on what should actually be done. There are interminable declarations and conferences on S&T policy, but no effective S&T policy, and no evidence that relevant lessons are being learned from history.

Perhaps the most important lesson to be learned is that Canada is capable of fostering technically sophisticated industry, from hydroelectricity generation to telecommunications technology. At the Canadian Engineering Centennial in 1987, the top ten achievements of Canadian engineering during the previous hundred years were honoured. One of those achievements was the CANDU reactor. AECL's experiences in directing and managing its activities, especially the

R&D leading to the CANDU reactor, offer many useful lessons for S&T policy in Canada.

The CANDU reactor is one of only two internationally competitive nuclear power systems. It was developed at a fraction of the cost spent by other industrialized countries to develop or adapt the other system (see Table 21.1). How could a country of only 14 million people (in 1950), with a resource-based economy, pioneer a high-technology industry?

A retrospective of the Canadian program, comparing it with other national programs, identifies several factors contributing to this achievement, some external and some internal.

EXTERNAL FACTORS

The first factor that contributed to the unique Canadian program for nuclear energy was the wartime decision that Canada would investigate the potential of heavy-water-moderated reactors. This alone did not determine the future design of CANDU reactors, but it did give Canada a head start in the technology and, in NRX, a facility that allowed Canada to become pre-eminent in some areas of nuclear research and in the supply of radioisotopes.

The next important factor was the Canadian government's post-war decision not to develop nuclear weapons, and the derivative policy not to develop nuclear propulsion units for military purposes. In the United States, the nuclear-weapons program continued at high priority, while in the United Kingdom and France programs to develop their own independent nuclear weapons were initiated. In those two countries, graphite-moderated, gas-cooled reactors were used to produce the plutonium for the nuclear weapons; it was the design needing least development.

World War II left Europe desperately short of all forms of energy. The United Kingdom, before the discovery of North Sea oil and natural gas, was critically dependent on dwindling coal resources, not only as a fuel in its own right, but also for electricity and coal-gas. Coal shortages, exacerbated by strikes at the mines and in the distribution systems, were causing cuts in all three fuels, and three-day weeks for industry. Under these circumstances, nuclear energy was seen as a boon that had arrived at just the right time to be exploited as quickly as possible. To do so, the plutonium-producing design was modified to allow the heat produced to generate electricity. France followed a similar path to develop its own design of gas-graphite power reactors.

TABLE 21.1
Total Nuclear Energy R&D Expenditures to 1985 (Billion $US)

U.S.	31.6
France	12.2
West Germany	11.2
Japan	9.6
U.K.	6.0
Italy	3.4
Canada	2.5

The United States suffered no such energy shortages. Instead, the imperative there was to exploit nuclear energy for submarines with greatly extended range and quiet engines. For this application, a small reactor core is desirable and the gas-graphite design is too large. Thus the light-water-reactor (LWR) was developed, with its enriched-uranium fuel cooled by ordinary water in a compact pressure vessel. This design was feasible for the United States with its wartime uranium-enriching technology and an industry capable of manufacturing the pressure vessels

In the fifties and sixties, the United States Atomic Energy Commission investigated the potential of a wide variety of power-reactor designs, carrying several to the prototype stage. However, when it was time to offer reactors in the marketplace, both domestically and for export, it was the multinational corporations such as Westinghouse and General Electric that decided the reactor type. Westinghouse, which had developed the LWR for the United States navy, chose to stick with that type for civilian applications: General Electric, which had developed an unsuccessful, liquid-metal-cooled type for the navy, marketed boiling water reactors, a variant on the LWR in which the coolant boils in the core. Both the United Kingdom and France developed and built gas-graphite reactors for civilian applications, but subsequently turned to the LWRs that were dominating the international market.

The way in which reactor types were developed in the three countries was an example of "industrial inertia." Once a technology is established in industry, it is easier to expand and extend that technology than to introduce a new one: there is a huge financial investment, facilities exist, expertise and experience have been gained, and there is a natural tendency to stick with the familiar and not to strike out into the unknown. This is not necessarily bad. The industry's customers – electrical utilities – are conservative, and rightly so, since they have an obligation to provide a reliable service.

INTERNAL FACTORS

In Canada there was no urgency to develop nuclear energy for either military or civilian applications. Ontario was running out of its traditional source of electricity, hydroelectric sites, but it could import plentiful coal from the United States. Consequently, nuclear energy was viewed as desirable, to reduce dependence on a foreign energy source, but only if it were economically competitive. The Canadian pioneers faced a tougher challenge than their colleagues elsewhere.

It was at this stage that individuals became important, and one in particular, W. Bennett Lewis, known as "WB" throughout AECL, was pre-eminent in defining and developing what became the CANDU reactor. In 1951, he put together three pieces of information to argue that a heavy-water-moderated reactor (HWR) could conceivably produce economically competitive power (chapter one). One of the pieces of information, the practicability of Zircaloy-clad fuel elements (containing uranium then, uranium-dioxide later), was largely due to two individuals in the United States, Admiral Hyman Rickover, who directed their navy's nuclear program and Ben Lustman, of Westinghouse's Bettis laboratory. However, the availability of this information to Canada was largely the result of the excellence of the NRX reactor in which Bettis was testing its fuel.

Lewis argued from basic principles that the capital cost of a nuclear-electric plant could never be low, so that the fueling cost would have to be kept very low if the electricity produced was to be competitive. He identified neutron economy as the key criterion to achieving this target, and this provided the direction and the discipline for the Canadian program. Emphasis on neutron economy led to the selection of heavy water as moderator: a recognized merit rating for moderators is some thirty times better for heavy water than for light water.

It was not that Canada was alone in appreciating the intrinsic advantage of HWRs. The United States, United Kingdom, France and other countries all developed HWRs at one time or another, but subsequently dropped them, largely due to industrial inertia. The subsequent adoption of LWRs by the United Kingdom and France exemplified international industrial inertia supplanting the domestic variety. Canada, coming later, had the luxury of being able to choose what appeared to be the best design on technical and economic grounds, without the influence of extraneous political factors.

Other individuals vital to the success of the Canadian program besides Lewis were Harold Smith and John Foster, who led the engineering team that converted the scientific concept into safe and reliable CANDU reactors; and Lorne Gray, who managed the engineering support as well as providing the administrative and political expertise discussed by Robert Bothwell in "Nucleus".

Thousands more, scientists, engineers and others, worked in a highly cooperative environment at the Chalk River site.[1] This spirit of cooperation can be partly attributed to the site's geography. Although it occupies a total area of 40 square kilometres, most of the buildings are crowded into a relatively small space between the river and surrounding rocky slopes. Employees' cars had to be parked outside this inner area. The result was that buildings were close together, making visits to other divisions easy. Also, when expansion of the laboratories was agreed upon in 1960, a deliberate decision was taken to

1 When I moved to Chalk River from the United Kingdom's equivalent, Harwell, in 1957, I was struck by the ease with which one could work with those in other disciplines, and how the reactor operators went out of their way to help experimenters.

establish a separate site in Manitoba, rather than make the Chalk River Laboratories larger than was thought to be ideal.

Individual laboratories in the United States and the United Kingdom had complements more than twice the size of Chalk River's, which was about 2,500. At the Argonne National Laboratory in the United States, one would normally drive between buildings, each of which had its own parking lot; at the Hanford site many of the buildings were out of sight from each other. The Harwell laboratory was less dispersed than those in the United States, but still less compact than the one at Chalk River: the divisions were more independent than at Chalk River, and individual divisions within Harwell tended to promote different reactor types.

There is another way in which the relatively small size of the Canadian program was beneficial. With only one company designing power reactors for non-competing electric utilities, and all of these federal or provincial crown corporations, there was not the commercial competition seen in the United States. There, if something went wrong, the first reaction by both vendor and utility would be to shut down communications between them, to protect their positions in case of legal action. In Canada, on the other hand, not only AECL and the utilities but also the component suppliers would cooperate openly in solving problems that arose. A good example was the rapid response by AECL, Ontario Hydro and the fuel fabricators to the detectable increase in fuel failures at Douglas Point in 1970, leading to the introduction of CANLUB fuel (chapter thirteen).

The same non-confrontational atmosphere existed between the regulator, the Atomic Energy Control Board, and its licensees. Most disputes were resolved at the technical level between the appropriate staffs of the regulator and the licensees, assisted by AECL as a consultant: when agreement was not reached, the five-person board acted as the final arbiter. The legal court system is not involved in licensing in Canada, as it is in the United States. For most of the period covered by this history, the Control Board employed only one lawyer. In contrast, the United States, with ten times the population, employed about one hundred lawyers in its nuclear regulating agency.

An important benefit of the nonconfrontational approach is an openness that encourages reporting of failings, both equipment failures and human errors, within the licensee's organization and by the licensee to the regulator. An open admission of failings, followed by their analysis and corrective measures, results in a continuous improvement in true safety. This mechanism for effective feedback of operating experience, combined with the defence-in-depth principle decreases the probability of a serious reactor accident.

Concentrating on a single reactor type, the HWR, was obviously one reason why Canada could achieve so much with so little. Also, because CANDU reactors have the potential to extend nuclear fuel resources by operating on advanced fuel cycles (chapter eighteen), Canada did not have to develop fast-breeder reactors with their challenging technology, considered necessary by countries with LWRs.

Yet another reason was to be found in the nature of the development programs. In several disciplines, e.g., reactor physics, thermalhydraulics, fuel development, and even Lewis's original concept assessment, AECL's design process was apparently less sophisticated than equivalents elsewhere. This was because AECL's approach was more empirical, less theoretical. With the advantage of excellent facilities, particularly the test reactors, AECL developers were able to derive simple empirical correlations from experimental results, drawing on an understanding of the physical phenomena to determine which variables would be important. The outcome was simpler and better rooted in reality.

Over the decades covered here, the design models have become much more sophisticated, so that they now differ little from those for other reactor types. The principal benefit is the ability of the designers to reduce the margins that had to be assumed to allow for uncertainties in various physical limits, e.g., critical heat flux and maximum power per unit length of element (chapter thirteen). However, much of the increased sophistication has been required to convince the regulator that various assumptions and engineering judgements were justified, largely in the area of predicting the consequences of postulated accidents (chapter fifteen). This has resulted in a few instances where the design should be modified, but generally the increased sophistication has served only to confirm that even a very severe reactor accident would not be expected to cause serious harm to the surrounding population. Unfortunately, none of these technical assessments seems to have reassured the public.

The excellence of the NRX and NRU reactors provided another benefit, by attracting researchers from many other countries to perform experiments at Chalk River. As a result, AECL had access to detailed information on other national programs that would not otherwise have been available. The most obvious example was AECL's access to Bettis' development of uranium dioxide and zirconium alloys just in time for the redesign of the Nuclear Power Demonstration reactor.

Most of the lessons from AECL's experience are based on successful developments but it is just as important to learn from one's failures. Despite AECL's relatively small size and generally good cooperation, its most expensive failures can be at least partly attributed to poor communications and coordination. It is arguable that the pressure-tube failures due to delayed hydride embrittlement (chapter twelve) would never have occurred had the designers, the metallurgists and the commissioning engineers been fully conversant with what the others were doing. Similarly, the recent fuel failures at Ontario Hydro's Darlington Station (chapter thirteen) might have been avoided had researchers familiar with hydraulic vibrations been more involved in the design of the primary coolant circuit. In each case, no single individual had put together the various pieces of relevant information, and the design audit had failed to involve effectively all those disciplines that could potentially contribute to a complex technology.

Drawing a more constructive lesson from the same experience, one appreciates the importance of a competent champion to any project. The champion must not only be an enthusiastic leader capable of convincing others of the rightness of the project, but must also be right in the scientific direction of the project. CANDU and Lewis are the example par excellence. Had Lewis not been scientifically sound in his analyses, no amount of enthusiasm on his part would have made CANDU successful: but it is unlikely that any design, however excellent technically, would have succeeded without his omnipresent and omniscient leadership. A disadvantage in having such an outstanding individual directing a project is that his lieutenants may rely on him too much for overall coordination, and not institute the necessary procedural checks, such as adequate design audits.

An example from AECL experience of the lack of an effective champion is the failure to develop commercially successful medical accelerators (chapter nine). Several desirable criteria for success were present, including a strong R&D team thoroughly experienced in accelerator technology, an identified market need and a well-defined mission. Competent champions for different approaches existed at both Commercial Products and Chalk River, but neither had the authority within AECL or the outstanding abilities of Lewis to bridge the division between the two sites. There was never the close collaboration between Commercial Products and Chalk River at the management level that there was between Power Projects and Chalk River, possibly because Commercial Products management personnel, unlike those of Power Projects, had never been located at Chalk River.

AECL's attempts from 1978 to produce new businesses from its technology (chapter twenty), as distinct from obtaining commercial revenues for its nuclear work and other R&D contracts, have to be judged unsuccessful in that very few new businesses were generated. Seen from the perspective of S&T policy, however, it can be regarded as a valuable experiment on a very large scale. Despite the best efforts of a successful R&D organization, and with enthusiastic support of senior management, no new means of generating industry from research were discovered. AECL's experience here deserves more analysis. Some believe that greater involvement by people with business experience would have made the difference, while others believe that trying to push out research into the marketplace is ineffective so one must look to market-pull to define the R&D needed.

This same experience illustrates another lesson important to S&T policy: failures are an integral and inevitable part of research. If the results of a research program could be predicted, there would be no point in doing the research. In this respect, research can be compared to geological exploration, where hundreds of holes have to be drilled before one profitable mine or oil-well can be developed.

Some of AECL's programs cannot yet be assigned to either the success or failure categories. Even the CANDU program can be compared to a glass that is half-full or half-empty. CANDU reactors represent about a 5-percent share of the world's power reactors, and 10 percent of those under construction in 1993, but this is substantial compared with Canada's share for other high-technology industries. The major, decades-long program on waste management has resulted in a concept for safe and responsible disposal of fuel wastes (chapter nineteen) that is currently being assessed by a Federal Environmental Assessment Review Process Panel. Advanced fuel cycles (chapter eighteen) and small reactors (chapter twenty) have been developed to the point that they could be made commercially available quickly when the need arises: their development costs to date represent premiums paid on an insurance policy against uranium shortages and high oil prices. As for industrial irradiation, public acceptance of food irradiation could result in a rapid expansion of the market, for which AECL is well positioned with its accelerator development (chapter twenty).

Nuclear energy has acted as a pathfinder through some of the social problems of our technological world, with AECL playing a key role. Nuclear energy defined the health risks and proposed means of protection before the industry was established. The ICRP (chapter three) pioneered the simple principle, that these health risks should be kept as low as reasonably achievable (ALARA), social and economic factors being taken into account, and recommended that ionizing radiation should be assumed to be potentially harmful at any level, however low, as a prudent measure in framing regulations. Even today, guidelines for some other environmental contaminants assume that there is a threshold level below which they are harmless. Nuclear energy introduced the concept that its hazardous wastes should be contained from the start, and held in supervised storage until means for safe disposal can be agreed upon: the industry proposed means on its own initiative. Other industries could benefit from two innovative approaches to the disposal of nuclear wastes (chapter nineteen):

- a deliberate separation of approval of the disposal concept from the selection of a disposal site, to reduce the socially paralyzing effects of the "NIMBY (Not In My Back-Yard) Syndrome."
- a siting process that does not impose a waste-disposal facility on a community, but allows potential host communities to volunteer, stating the conditions under which they would accept the facility.

In what follows, specific lessons derived from AECL's experience and relevant to S&T policy are itemized and explained. Those concerned with funding may have restricted relevance in the private sector, since AECL's R&D was very largely funded directly by the federal government. However, in Canada and elsewhere, much R&D is indirectly funded by the government, by tax provisions, military contracts, etc.

GENERAL LESSONS

Lesson #1. R&D that is intended to provide economic, industrial or social benefits (i.e., mission-oriented R&D as opposed to basic research) should be determined by the mission's requirements; it should be planned, directed and rigorously selective, with the selection based on sound principles.

In its early years, Chalk River was fully occupied operating the NRX reactor and the chemical extraction plants – activities arising from its wartime mission to demonstrate the feasibility of a heavy-water-moderated reactor – and conducting basic research. From the start, there had been an interest in the possible generation of economical nuclear electricity. However, it was not until the early fifties that NRX experience, both the behaviour of the reactor itself and of tests being performed in it, pointed the way to a practicable approach. Without disrupting the basic research, an R&D program was established with the clearly defined mission of developing an economical nuclear-power plant. The R&D program, small relative to others, was narrowly channeled towards this mission, occasionally being redirected and modified to maintain its overall aim.

Throughout AECL's history the provision of radioisotopes for medicine, industry and research – for both the domestic and export markets – has been another example of a well-defined mission leading to a commercial success.

A more laissez-faire model for R&D management (such as that employed by the federal government's Natural Sciences and Engineering Research Council, and the Panel on Energy R&D of the seventies) makes funds available to researchers proposing peer-endorsed projects with only general relevance to stated objectives (e.g., economical energy conservation and supply). This approach may produce useful results, but lacks the drive and coherence needed to ensure that all the necessary R&D and associated activities are performed and completed to meet the overall schedule. AECL's unsuccessful attempts in the eighties to seek applications for R&D outside its specific mission demonstrated the difficulties of this approach.

Lesson #2. Potential users of the R&D should participate in formulating the broad policy for the R&D program.

In the formative days of the CANDU R&D program, the involvement of representatives of electrical utilities on the board of directors of AECL, hitherto essentially a research organization, helped to focus the program and keep it down to earth. More recently, the CANDU Owners Group (COG) has played a similar role. Greater involvement by potential users might have made more productive the search for spin-off businesses (chapter twenty).

Lesson #3. Managers of R&D programs must be vigilant in assessing alternatives and terminating those that prove disappointing.

The Canadian program explored alternatives, but on a smaller scale than other countries, and always within the program's general direction. Examples that were deliberately terminated include the pressure-vessel design for the NPD reactor, uranium-metal fuel, aluminum-alloy sheathing, and the boiling-light-water and organic-cooled reactors. The fusion alternative was repeatedly visited, but assessments based on good knowledge of international progress concluded that a major program could not be justified.

Lesson #4. The development of modern technologies requires large interdisciplinary programs.

AECL's early strength in science continued to serve it well, particularly in providing sound guidance for the mission-oriented programs. With commitment of the first power reactor (NPD) came recognition of the need for more engineering skills: the CANDU concept would stand or fall on the performance of the fueling machines, sophisticated engineering devices in advance of anything elsewhere; excellence in engineering is the first line of defence in nuclear safety; metallurgical engineering was vital for the fuel-channel components, as was chemical engineering for heavy-water production.

AECL's program, unlike many other research programs, was constantly reviewed for economic feasibility; technical feasibility alone was not sufficient. As part of the concern for economics, technology transfer was integrated into the program from the start, with the result that researchers had to include in their jobs the training of staff attached from industry, the supervising of industrial contracts and the importation of industrial know-how, when appropriate. All these skills were integrated into a coherent program, making the AECL of the sixties unlike the popular idea of a scientific laboratory.

Most modern technologies require large, interdisciplinary programs with expensive, efficient and helpful support services, combining underlying and applied science; engineering development and design; safety, economic and market analysis; and the operation of test facilities as well as pilot and prototype plants. In Canada, universities and industry can rarely fill this role, and some excellent government laboratories have filled it. A major factor in CANDU's success has been the integration in a single program of all necessary components from basic science through marketing. An incidental benefit of such integrated programs is that they provide the sort of "one-stop shopping" for Canadian technologies that is attractive to newly industrialized countries interested in importing new technologies.

Lesson #5. Even in a mission-oriented program there is need for research at a fundamental level.

AECL found its underlying-research component invaluable. Initially, it helped attract suitable researchers. Later, it provided a sound understanding of relevant phenomena, a "mobile reserve" of people capable of tackling from first principles the inevitable problems that arise in a major endeavour, intelligence about discoveries elsewhere through the international science subculture, a means of evaluating potential programs, and the basis for future advances.

Lesson #6. Applied research is an essential step in going from the international pool of basic research to the development of technologies for society's benefit.

Despite AECL's strength in basic research, most of the basic discoveries leading to the CANDU reactor were made elsewhere. They include nuclear fission, neutron moderation, heavy water, zirconium alloys and uranium dioxide. AECL was weak in basic patents but strong in adopting and adapting others' discoveries. This was achieved through an informed awareness of international developments and by application of the knowledge to Canadian objectives through applied research.

Lesson #7. A substantial R&D effort is required to absorb and maintain imported technology.

At first, the decision to produce heavy water in Canada seemed to involve no more than importing an already developed technology from the United States. In practice, however, this proved to be a very expensive oversimplification. The design of the plants, and the conditions under which they were to operate, differed in significant respects from those in the United States. The problem is exacerbated when the technology has been established empirically, without a broad understanding of the underlying phenomena. Reliable heavy-water production was eventually achieved, largely through competent management of the plants, supported by an R&D strength in Canada (chapter seventeen).

Uranium-dioxide fuel in zirconium alloys is a more positive example, for which the original technology was imported at an earlier stage and into a strong R&D capability. This led to the development of CANDU fuel, with its excellent performance. AECL and its associated industrial companies (chapter thirteen) eventually contributed much more to both these technologies than had been imported.

Lesson #8. There is an ongoing need for R&D far beyond the innovation stage.

The unforeseen in-service rupture of pressure tubes is the single biggest failure of the CANDU program. While every effort must be made to avoid or mitigate such failures – and several detailed lessons have been learned from the incidents – the fact is that in any multi-billion dollar endeavour, such as a major energy system, some such failures must be expected. Their likelihood can be minimized by sound engineering, the application of familiar materials wherever possible, well-established, reputable manufacturers, quality assurance programs, etc., but provision must be made for their occasional occurrence.

In the response to the pressure-tube failures, AECL's multidisciplinary laboratories proved invaluable: had they not been able to diagnose the fault rapidly, it might have been necessary to shut down all CANDU reactors as a precautionary measure.

Any technology that is stagnant will soon become obsolete. To remain competitive, a technology must continue to develop and improve itself, thus requiring ongoing R&D. When Ontario Hydro's Pickering nuclear station proved so successful, it was difficult to convince the funding authorities and the public that much R&D remained to be done, not only in waste management but also in continuing development of the CANDU design, to reduce costs and accommodate changes in the market and in regulatory requirements. CANDU's foreign competitors were continuing to devote large sums to the development of their products.

Lesson #9. An R&D program has to plan from the start for the transfer of technology to industry.

Laboratories may produce the ideas for new products, but most do not produce the products themselves. AECL developed and designed the CANDU reactor, but Canadian industry manufactured the hardware. The transition from laboratory to industry is a constant subject for debate in discussions of S&T policy. AECL established several techniques to ensure an efficient transfer of the CANDU technology to industry, including development contracts, attachment of industry's staff to the laboratories and cooperative work on standards. CANDU fuel provides an excellent example of efficient technology transfer.

However, technology once transferred cannot be recovered, so financial and other arrangements must be agreed upon before the transfer. The development of CANDU fuel took place in the sixties when government policy was that government laboratories, such as AECL's, should simply assist in the development of Canadian industry, without obtaining any return of public funds. Since then, the policy has changed to one of user-pay. Although some hundreds of millions of dollars worth of fuel based on AECL developments is used each year, AECL derives no licensing fees to reinvest into further development.

Lesson #10. Some R&D is needed for the development of S&T policy.

While a major purpose of S&T policy is to determine what R&D should be done, R&D has an important role to play in determining S&T policy. Should the government devote substantial resources to developing a new technology? Usually, an informed decision requires the performance of at least exploratory R&D. Often, the open literature will give an over-optimistic impression of the new technology's potential, not revealing the important problems or the probable costs. A large R&D program by AECL in the seventies on nuclear-waste disposal provided the basis for a decision by the federal and Ontario governments to pursue deep geological burial for the disposal of nuclear-fuel wastes. Outside AECL's sphere, renewable energy sources are a good example: much exploratory R&D is needed to establish technical feasibility and economic promise before making these technologies part of an S&T policy.

FUNDING LESSONS

Lesson #11. Even when all foreseen components of an R&D program have been allowed for in the budget, a substantial contingency fund is essential.

The nature and magnitude of the R&D needed for something like CANDU is generally not well understood; even the researchers most involved have habitually underestimated requirements. To provide the necessary heat-transfer data for CANDU required several multi-million dollar engineering rigs not foreseen in the early days. If public trust is to be deserved, R&D managers must provide more realistic estimates. The public sees gross cost overruns as evidence of incompetence or even dishonesty.

Uncertainties in cost estimates are inherent in the research process and must be recognized. However, past experience over many industries provides some guidance in what R&D expenditures have to be expected, and therefore what contingencies are reasonable. Low-technology industries, with which Canada is most familiar, devote only a few percent of product cost to R&D, while for high-technology industries, such as electronics and aerospace, the R&D expenditures may be around 20 percent.

Lesson #12. Funding for mission-oriented R&D should be determined by the mission's objectives: that for basic research should be a percentage of Gross Domestic Product (GDP) concordant with that devoted to basic research by comparable countries.

The S&T debate has been oversimplified by a demand that the government increase spending on R&D to some arbitrary percentage of GDP, as if that alone would ensure that S&T provided all the benefits expected of them. Government-funded applied R&D is wasted unless it is performed as an integral part of the government's industrial and social policies that are prepared to exploit successful R&D. A defensible S&T policy presupposes clearly formulated industrial and social policies. AECL experience suggests that finding an efficacious S&T policy will require changes in attitude at all levels.

Part of the trouble is the habit of talking of "S&T" and "R&D" as if these were homogeneous topics. It is true that one can trace a continuous path from pure scientific research to the final product, but the nature of the activities and their needs vary greatly along the way. The greatest distinction along the path is determined by the motivation for the work, between pure, or curiosity-driven, research on the one hand, and mission-oriented programs on the other. For the latter the mission should be defined by industrial and social policies, so that objectives and beneficiaries are clearly identified: the R&D activities are a necessary part of the program, so the mission manager should determine the funding out of the program budget which should contain an appropriate allowance. R&D is one means, not the end.

Pure research, while vital for the advancement of mankind in the long term, rarely benefits directly those who fund it. European research of the nineteenth century provided the basis for United States industrialization in the twentieth century, and Japanese industry is now benefiting from the science the United States performed earlier in this century. Canada, as an affluent country, has an obligation to support curiosity-driven pure research in proportion to its relative wealth, as an altruistic activity.

Most of AECL's R&D was mission-oriented, to apply nuclear energy to benefit Canadians. Some fundamental research was performed to support these missions, to provide a better understanding of the underlying phenomena (e.g., radiation-induced creep), and for other reasons explained under Lesson #5. AECL, acting within the mandate under which it was established, deliberately devoted a fraction of its budget to curiosity-driven pure research in areas where it had unique resources; e.g., the NRX and NRU reactors for neutron diffraction.

Lesson #13. Accountability is necessary, but over-accountability is counterproductive.

An increasing recognition over recent decades that researchers must be accountable to those who provide their resources, ultimately the Canadian public, is a healthy trend. However, AECL experience in the eighties was that requirements for planning and accountability reached a level where a sizeable fraction of the total effort was devoted to the activities involved. The climate tended to discourage unplanned initiatives and cooperation. A judicious three-way balance must be struck between accountability, responsibility and trust.

Lesson #14. Program commitments should match the resources available: when funding is cut, programs may have to be eliminated, not schedules simply extended.

R&D programs have to be stopped as well as started. How this is done is important to staff morale. AECL found that displacing a low-priority program by a high-priority one was more effective than simple cancellation. This approach helped to maintain staff enthusiasm and support, something that is vital to a healthy R&D program.

Not all AECL experience was so satisfactory, however: when funding was reduced, cuts were sometimes imposed across the board; more selective suspension of whole programs would have resulted in the remaining programs being more productive. Also, during some cuts, a reluctance to reduce personnel levels resulted in too small a fraction of the budget being devoted to new equipment and capital facilities.

Lesson #15. The source of R&D funding needs careful and explicit consideration.

Some R&D activities may be appropriately provided from a levy on the beneficiary, while others may be a legitimate charge against the public treasury. These latter can include activities where the potential beneficiaries cannot be identified, and where work is needed to determine government policies.

LESSONS ON PUBLIC PERCEPTION AND ACCEPTANCE

Lesson #16. A research institute and its staff must be prepared to explain and defend its work to the public.

AECL, in common with the rest of the science and engineering community, was slow to recognize that a new technology, however sound technically, is not acceptable until the public accepts it. Even before the CANDU program started, AECL was doing research to protect people and the environment, simply regarding this as an obvious responsibility, but failing to publicize the fact. At that time, the public generally regarded technology as beneficial, and was content to leave its approval and control to anonymous government officials and elected representatives.

Since then, the public has become suspicious of technology and wants a say in important decisions. Thus, to be accepted, any new technology must not only be fully defensible, it must be defended and even promoted. Furthermore, the advocacy should not be left solely to public relations people who usually lack an exhaustive understanding of the subject. Their work must be complemented by articulate scientists and engineers who are willing to listen as well as talk. Public relations must be a two-way process.

Lesson #17. Researchers must identify those who will benefit from their research, then sell themselves, either direct or through others.

R&D practitioners, both institutions and individuals, have to accept that the world does not owe them a living, paying them for what they enjoy doing. They have a responsibility to explain themselves to their patrons, to show why they should be supported, and to answer concerns. This attitude was foreign to most R&D institutions in the fifties. It is now an integral part of AECL's corporate culture.

Lesson #18. An R&D program for a new technology should include the necessary research on health and environmental effects.

The scope and character of many technologies are such that, without proper safeguards, they can produce serious effects on health, or major disruptions of the environment. Because of the known effects of radiation, from the earliest days of nuclear research, care was taken to protect the public and workers from undue exposure to radiation. The lack of a similar appreciation of the hazards of other technologies has resulted in serious impacts on the global environment, so that technology in general has got a bad reputation.

ACKNOWLEDGMENTS

While the opinions in this chapter are those of its author, they owe much to what is in other chapters, and to comments from their authors.

Index